## 🐾 Zoology

F. Pough, J. Heiser, W. McFarland
*Vertebrate Life, Fourth Edition (1996)*

P. Moyle and J. Cech
*Fishes: An Introduction to Ichthyology,*
*Third Edition (1996)*

R. Wallace, W. Taylor, J. Litton
*Invertebrate Zoology Laboratory Manual,*
*Fourth Edition (1989)*

## 🐾 Anatomy and Physiology

F. Martini
*Fundamentals of Anatomy and Physiology,*
*Third Edition (1995)*

F. Martini and M. Timmons
*Human Anatomy (1995)*

M. Timmons
*Anatomy Text and Laboratory System (1995)*

G. Tortora
*Anatomy and Physiology Laboratory Manual,*
  *Fourth Edition (1994)*
*Laboratory Excercises in Anatomy and Physiology*
  *With Cat Dissection, Fifth Edition (1996)*
*Laboratory Exercises in Human Anatomy With*
  *Cat Dissections, Third Edition (1993)*

G. Tharp
*Experiments in Physiology, Sixth Edition (1993)*

M. Hadley
*Endocrinology, Fourth Edition (1996)*

## 🐾 Microbiology

T. Brock, M. Madigan, J. Martinko, J. Parker
*Biology of Microorganisms,*
*Seventh Edition (1994)*

J. Black
*Microbiology: Principles and Applications,*
*Third Edition (1996)*

M. Jensen and D. Wright
*Introduction To Microbiology For The Health*
*Science, Third Edition (1993)*

R. Stanier, J. Ingraham, M. Wheelis
*The Microbial World, Fifth Edition (1986)*

G. Wistreich
*Microbiology, Fifth Edition (1988)*
*Laboratory Exercises in Microbiology (1996)*

J. Levy, H. Fraenkel-Conrat, A. Owens
*Virology, Third Edition (1994)*

G. Stine
*Acquired Immune Deficiency Syndrome:*
  *Biological, Medical, Social and Legal Issues,*
  *Second Edition (1996)*
*Aids Update 1996, Third Edition*

## 🐾 Plant Biology

D. Woodland
*Contemporary Plant Systematics (1991)*

C. Reiss
*Experiments in Plant Physiology (1994)*

E. Moore-Landecker
*Fundamentals of the Fungi, Fourth Edition (1996)*

D1279141

# Ecology: Theories and Applications

Second Edition

## Peter D. Stiling
*University of South Florida*

Prentice Hall
Upper Saddle River, New Jersey 07458

**Library of Congress Cataloging-in-Publication Data**

Stiling, Peter
  Ecology : theories and applications / Peter Stiling.—2nd ed.
    p. cm.
  Rev. ed. of: Introductory ecology, 1992.
  Includes bibliographical references and index.
  ISBN 0-13-221939-5
  1. Ecology.    I. Stiling, Peter    Introductory ecology.    II. Title.
QH541.S675   1996
574.5—dc20                    95-17406
                                     CIP

Senior Editor: Sheri L. Snavely
Editorial-production service: Electronic Publishing Services Inc.
Manufacturing buyer: Trudy Pisciotti
Cover design: Design W Inc./Wendy Helft
Cover art: Maxine Fumagalli and Brendan Japantardi for CHURINGA Designs; © RipTide, Inc.
      Represented by RIMPACIFIC

© 1996 by Prentice-Hall, Inc.
Simon & Schuster/A Viacom Company
Upper Saddle River, New Jersey 07458

Previous edition © 1992 under the title *Introductory Ecology*

Printed in the United States of America

10 9 8 7 6 5 4 3 2

**ISBN 0-13-221939-5**

Prentice-Hall International [UK] Limited, *London*
Prentice-Hall of Australia Pty. Limited, *Sydney*
Prentice-Hall Canada Inc, *Toronto*
Prentice-Hall Hispanoamericana, S.A., *Mexico*
Prentice-Hall of India Private Limited, *New Delhi*
Prentice-Hall of Japan, Inc., *Tokyo*
Simon & Schuster Asia Pte. Ltd., *Singapore*
Editora Prentice-Hall do Brasil Ltda., *Rio de Janeiro*

# About the Author

*The photograph shows the author with daughters Zoe and Leah at the coast, near his field sites.*

Peter Stiling is an associate professor of biology at the University of South Florida, Tampa. He has taught classes in Ecology, Environmental Science, and Community Ecology, and in 1995 received a teaching award in recognition of classroom excellence in these areas. Dr. Stiling obtained his Ph.D. from University College, Cardiff, Wales, and completed postdoctoral research at Florida State University. It was while teaching Ecology at the University of the West Indies, Trinidad, that the idea for this book was conceived. Dr. Stiling's research interests include plant–insect relationships, parasite–host interactions, and biological control. He has published many scientific papers in journals such as *Ecology, Oikos,* and *Oecologia.* His current field research, supported by the National Science Foundation, focuses on insect communities on salt marsh plants.

*To Don and Dan*

# Contents

# *Preface to the First Edition*

## WHO STARTED ECOLOGY?

Most authors, including the Oxford English Dictionary and the Encyclopedia Britannica, ascribe the term Ecology to Ernst Haeckel who, in 1866, defined it as "the comprehensive science of the relationship of the organism to the environment"—a useful definition even today. However, the first recorded usage was in a letter by Henry David Thoreau in 1858. These early uses of the term ecology seem oblique and Goodland (1875) argues convincingly that the Dane, Eugen Warming (1841–1924), should be regarded as the founder of the science of ecology because of his pioneering research in Brazil and because he wrote the first book on the subject in 1895 (Warming 1895).

## WHAT IS ECOLOGY?

To most people today, ecology is associated with the broad problems of the human environment, especially pollution. To others, ecology is synonymous with conservation and saving the whales or the forests. Yet, ecology does not simply deal with pollution or conservation, it is related to the environment much as physics is to engineering. Ecology provides the scientific framework upon which conservation programs or pollution monitoring schemes can be set up. The scientific framework is often erected, however, not from studies of rare, exotic animals or studies of oil spills but from studies of invertebrates such as insects, and small, relatively unappealing plants because these best can be manipulated experimentally to test ecological theories. For example, it is rather difficult to experiment with blue whales or Florida panthers, of which there are relatively few. This book is, thus, peppered with studies of the teeming hordes of common life.

The biosphere is composed of some 1.8 metric tons of living tissue and covers the surface of the globe in a layer proportionately thinner than the skin of an apple. If all the material were evenly distributed, it would be less than 1 cm thick and weigh about 3.6 kg (Anderson 1981), but the Earth's biomass is not evenly distributed. It shows great variation, from about 45 kg in tropical forests to 0.003 kg in the oceans. Ecology is concerned with explaining this variation; it asks why plants and animals are found in certain areas and what controls their numbers. The study of ecology can be divided into two areas. Functional ecology asks how populations are maintained at a particular size and touches on behavior, competition, predation, and other contemporaneous factors. Historical ecology asks how populations have come to be this way and places great emphasis on biogeography (the study of global distribution patterns and shared characteristics), which presupposes a knowledge of continental drift and of evolution.

## TEXT DEVELOPMENT

In 1983, I began teaching ecology to students in the Department of Zoology at the University of the West Indies, Trinidad. Most had little prior ecological background, so I started from scratch. I wanted to give the students a full overview of ecology and not specialize in one particular area; I needed to present an introductory ecology course. As previously noted, a full understanding of ecology requires some knowledge of historical ecology and, thus, of evolution, so, to begin, I introduced a little evolutionary ecology. Also, I wanted to include some ecology of animal behavior, a fascinating area and one enjoyed by most students. An ecology course should also encompass population biology, the regulation of plant and animal numbers, which is the core material of most texts and the area where most ecologists publish (Stiling 1994). Coverage of communities was also necessary as it is this area which is noted as of most interest to ecologists in surveys (Travis 1989). And finally, I thought there should be some reference to applied ecology. For most of these students, a knowledge of applied ecology would be essential in the real world to which they would soon belong. Having set this outline for my introductory course, I found that there was no textbook incorporating evolutionary ecology, behavioral ecology, population ecology, community ecology, and applied ecology. This book is an attempt to remedy that situation. It assumes knowledge of college-level biology, perhaps that offered by an introductory biology course in which some knowledge of genetics and heredity, physiological adaptation of plants and animals, and the diversity of life have been touched upon.

## THEMES

This book has several main themes. First, it is broad in coverage. Few other ecology texts attempt as diverse a treatment in explaining the distribution and abundance of plants and animals. Take for example the distribution patterns of the polar bear. Why is it not found in the Antarctic? Clearly, climate and topography are not sufficient answers. For this species, a knowledge of evolutionary ecology is vital. For many other species, particularly those with good dispersal abilities, such as plants, physiological factors are the main constraints on observed distribution patterns. Within these distribution patterns, the abundance of species can be affected by food supply, competitors, predators, and parasites and, for animals, territorial behavior. Thus, a sound knowledge of population ecology is essential. Ultimately population fluctuations may be influenced by higher-order community interactions—hence, the value of examining ecological processes at these large spatial scales. Finally, so many ecological processes are impacted by people, especially in these times of exploding population growth, that it is worthwhile to examine ecological phenomena as they are affected by humans—applied ecology. Breadth, of course, has its drawbacks. By combining many elements into an introductory course, the text could become overly long. I have resisted this temptation by presenting ideas as succinctly as possible. The result is, I hope, a streamlined introductory ecology text.

A second main theme of this book is to present both sides of an argument. Many texts are content to tell a "just so" story. This is often true when ecological processes are modeled. There is a great temptation either to present a model and then the evidence to back it up or to spend a long time concocting a model to fit available data. It is very easy to present an ecological principle with a "happy ending," that is, a nice tidy explanation with no loose ends, that appeals to the student and is intellectually satisfying. Unfortunately, as most field ecologists know, the actual data are rarely neat and often fit no clearcut hypothesis. Even if the data do fit a hypothesis, the next set of data taken in another system is just as likely to show the opposite trend. To combat this, I have tried, wherever possible, to present a variety of evidence both for and against proposed theories.

## CHAPTER FORMAT

It is difficult, in such a wide-ranging book as this, to stick to one particular format for each chapter. There are, however, some generalities of approach that are worth bringing out. In each chapter, the ideas and theory are discussed first, then some examples follow, both for and against the proposed hypotheses. I have tried to keep these examples to a minimum. Often, modern examples are used but sometimes older work is referred to. There are many older pieces of work that will forever remain standards. In the field of competition theory, for example, the displacement of one species of parasitic *Aphytis* wasp by another in Southern California will probably forever remain a classic example of competitive displacement in the field. Nevertheless, references are kept as updated as possible; over 12% are from the 1990s. This is, therefore, an unusually current treatment of ecology. Following the examples, I discuss any recent overviews of the subject. Ultimately, one is swayed either for or against a theory by the weight of the evidence. I cannot present all the pertinent data for and against one theory in this book, but I can refer the reader to the research papers that do. For example, in discussing the frequency of density dependent parasitism (Chapter 11), I point out that two independent reviews (Stiling 1987; Walde and Murdoch 1988) have shown that in nature density dependence does not occur very frequently despite the fact that many ecologists have based much of their work on modeling density dependence. Similar reviews are referred to in the section on competition (Connell 1983; Schoener 1983; Denno et al. 1995) and predation (Sih et al. 1985). It is my hope that in the not-too-distant future all subject areas of ecology will be methodically reviewed in this way so that ecologists will no longer be content to pull contrasting examples out of a hat and that there will be a review that will show a preponderance of the evidence going one way or the other.

## CHAPTER ELEMENTS

In determining the actual layout of each chapter, I have relied heavily on the use of tables, figures, and photographs to present the evidence. As Connor and Simberloff (1979a) noted, "You can't falsify ecological hypotheses without data." I personally chose each table, figure, and photograph that is used in the book (and had to write all the letters of permission to use them!). Each piece of illustrative material is directly related to a point raised in the text. To maintain the brevity of style, however, the text does not usually go over the same ground. Each table, figure, or photograph is intended to be self-explanatory. Other more superficial learning tools, such as boxed essays and asides, are absent. Material that could be presented in this way is simply integrated into the text. I have tried to lessen the effects of ecological jargon by providing a glossary. Terms that are boldface, plus some that are not, are defined at the end of the book. Though the book tries to be brief and not burden the student with unnecessary baggage, which they would have to sift through with a highlighter, it is well referenced so that the interested reader can always find where to read further on a subject. Despite its size, there are more references in this text than in many of its competitors.

## KEY CONTENT

There is a necessary tradeoff between classical material and recent developments. In ecology, subjects such as biodiversity, conservation, restoration, bioengineering, acid rain, and global warming get much of the press. These are all important issues. But it is worth remembering that predicting the effects of such things is often based on more traditional disciplines. Thus, conservation ecology relies heavily on population ecology. To predict the effects of genetically engineered organisms, many experts have used as analogs the results of releasing exotic species into novel environments. Physiological ecology can best give us the likely answers to questions of how the distribution patterns of species will change in the event of climatic alterations. This book attempts to integrate new concepts with new and older theory.

It has often been said that biology only makes sense in the context of evolution. I begin this book with a treatment of evolutionary ecology, the end point being to discuss the reasons for extinction of plants and animals today. This is followed by Section Three, Population Biology, in which I discuss the multitude of effects of the environment, competition, predation, herbivory, parasitism, and other factors on the abundance of plants and animals, and tie these together by comparing their effects in the final chapter entitled "The Causes of Population Change." Section Four, Community Ecology, discusses the integration of populations into communities, the units in which they occur in nature. Things such as species diversity, diversity gradients, stability, succession, and biogeography are discussed here. In Section Five, Ecosystems Ecology, I explain the flow of energy and nutrients through communities and the assemblage of trophic links in ecosystems. With this background established, we are in a position to discuss the impacts of humans. The last section, Applied Ecology, documents the main effects of people; habitat destruction, exploitation of wildlife for its own sake, pollution, and the introduction of exotic organisms.

# ACKNOWLEDGMENTS FOR FIRST EDITION

This book could not have been completed without the help of many friends and associates. I would like to thank Dana Bryan, T. S. Carter, K. R. McKaye, A. Murie, R. H. Reeves, P. M. Room, D. Simberloff, D. A. Sutton, and J. O. Wolff for kindly lending me their own photographs. The following people provided invaluable help in locating photographs: Lavonda Walton, Mae Goff, Robert Hailstock, and Nancy Chedester of the United States Department of Agriculture; Chuck Frazier of the Florida Game and Fresh Water Fish Commission; Joan Morris of the Florida Archives; Barbara Mathe of the American Museum of Natural History; Raymond Rye of the Smithsonian Institution; R. W. Paugh of the United States Coast Guard; Giuditta Dolci-Favi of the Food and Agricultural Organization of the United Nations; Tracy Hornbein of the Florida Department of Natural Resources; and Dale Connelly of the United States National Archives. I am also grateful to Caroline Reynolds and Elizabeth Fairley for tracking down some hard-to-find references. All the authors whose tables or figures are reproduced here freely gave permission for their use.

I am indebted to the following reviewers for their useful comments and suggestions: Stanley H. Faeth, Arizona State University; Nicholas J. Gotelli, University of Oklahoma; Robert P. McIntosh, University of Notre Dame; David M. Gordon, University of Massachusetts—Amherst; Richard Tracy, Colorado State University; Mark A. Hixon, Oregon State University; Thomas H. Kunz, Boston University; William Rowland, Indiana University.

Finally, I should like to offer my sincere appreciation to Anne Thistle for a meticulous job of typing and editing this book and for making me look much better grammatically. Heartfelt thanks to Sharon Strauss, University of Illinois, who had the fortitude to read the entire thing. My editors, Betty O'Bryant at Technical Texts and David Brake at Prentice Hall, provided many helpful suggestions. Students at the University of West Indies, Trinidad, and at the University of South Florida have been quick to point out any inconsistencies. However, I would be pleased to hear about errors in fact or interpretation, omission of material, further examples, or other relevant ideas.

# *Preface to the Second Edition*

The changes for this second edition have been quite extensive—there are 62 new figures and tables and over 300 new references. New ideas, new concepts and new examples have been included in the text to keep the book current. However, this revision is not just an "in with the new, out with the old" exercise. Many existing sections have been thoroughly reworked for clarity. The annoying little errors that are often present in first editions, such as typographical errors or mistakes in equations, have been fixed. But again, this edition is not simply about fixing mistakes. It incorporates major conceptual changes. Although the text still has an "applied ecology" section, I have made a strong effort to integrate much "applied" material into the entire text. Thus, there is coverage of global warming in Chapter 5, conservation biology in Chapters 2 and 3, biodiversity in Chapter 14, and restoration ecology in Chapter 19.

As regards specific sections, among the first and most important changes are to be found in Section Two, Evolutionary and Behavioral Ecology. Both evolution and behavior are important in the study of ecology. What I have done in the second edition is to make this material more like evolutionary and behavioral ecology and less like straight evolution and behavior. For example, the seemingly simple species concept is shown to be far from simple and the relevance of this to conservation biology underscored.

Many chapters in Section Three, Population Biology, have been updated. The chapter on population growth, Chapter 6, has been thoroughly stripped down and rewritten and now includes material on time-lags and chaos. The biggest change of all in Section Three is in the last chapter, Chapter 12, which gives a broader synthesis of the causes of population change and population regulation. Factorial experiments are introduced as a good way to compare the strength of mortality factors. Population equilibrium and the idea of metapopulations are introduced to students. Density vagueness and the $CV^2$ rule are contrasted with ideas about density dependence. The HSS, OF and MS models are compared, top-down and bottom-up effects are contrasted and indirect effects are introduced in their own subsection.

The fourth part of the book, Community Ecology, has been expanded to include material on rarefaction, cardinal versus ordinal indices and rank-abundance diagrams. The material on ecosystems has been lengthened and given more emphasis in its own two-chapter section. Biogeochemical cycles are given increased coverage with new subsections on the phosphorus and carbon cycles.

Most of the environmental problems that exist in today's world can be traced to a still increasing human population. It is for this reason that, in Section Six, Applied Ecology, the topic of human population growth has been dealt with in more detail. Like Section Two, this section has been thoroughly stripped down and rewritten. The discussion on design of nature preserves now incorporates the concept of Minimum Viable Population (MVP) and prevention of poaching, as well as more traditional areas like the SLOSS debate. There is now a new subsection entitled Restoration Ecology. Students enjoy talking about restoration ecology because many of them envisage careers in this area.

The following users of the first edition provided critiques which have proven invaluable in the preparation of the second edition: Gary K. Clambey, North Dakota State University; Daniel F. Doak, University of California-Santa Cruz; Kristina A. Ernest, Central Washington University; Margaret H. Fusari, University of California-Santa Cruz; Nicholas J. Gotelli, University of Vermont; Jeffrey R. Lucas, Purdue University; Peter Smallwood, Bryn Mawr College, Pennsylvania; Karen Olmstead, University of South Dakota; and Roy A. Stein and Alison Snow, Ohio State University. My new editor, Sheri Snavely, was instrumental in upgrading the production values for the book, including a two-color format and color photographic inserts. The whole feel of the book benefited from her efforts. I particularly appreciate the efforts of Electronic Publishing Services Inc. in the copyediting, proofreading, photo research, and production of the book. And finally, Jacqui Stiling deserves special mention—without her constant efforts and computer expertise this second edition would not have been possible.

# Introduction

Life is not distributed evenly on Earth, but rather in patches. Ecology seeks to explain this phenomenon. Here, in the Okavango delta, Botswana, papyrus reedbeds and other vegetation is distributed throughout the region. Islands may begin as termite mounds. Abiotic variables, such as water levels, may affect the distribution of plants as may herbivores. (Dimijian, Photo Researchers, Inc. 7L5327)

There is a widespread belief that people of preindustrial civilizations did far less damage to their environment than do their modern industrial counterparts. This belief supposes that hunter-gatherers lived in harmony with nature, practicing a conservation ethic and somehow avoiding short-sighted, destructive exploitation. They did not. For example, on every oceanic island for which we have adequate knowledge, the first arrival of humans was quickly followed by extermination of all or most large animals (Diamond 1986a). Easter Island, home of the famous monolithic statues, was once covered with palms, trees, and shrubs. Polynesians reached the island around A.D. 400. By 1500, 7,000 people lived there; they had deforested the island so completely that its tree species are now extinct (see photo on following page). The deforestation had serious implications: No logs were available to be made into canoes, so offshore fishing was curtailed, and the huge statues could not be erected without log levers. Once the population exceeded the carrying capacity of the island, warfare was rampant, as were chronic cannibalism and slavery. Spear points were manufactured in enormous quantities, and people reverted to living in caves for defense.

The scale on which destruction occurred was not limited to small islands. In the deserts of the U.S. Southwest stand huge, empty communal houses or pueblos, relics of the Anasazi, one of the most advanced pre-Columbian civilizations in North America. When construction began, the cliffs were covered with pinyon-juniper woodland. Collection for firewood and construction denuded the area completely to a radius of 40 to 70 km by the time the site was abandoned.

Now the affects of humans have begun to change the entire globe. Acid rain is carried from one country to another. Carbon dioxide pumped out by the industrial centers of developed nations has increased the atmospheric $CO_2$ levels worldwide from the poles to the equator. Pesticides, powerful human-made poisons, have been detected in human breast milk and in tissues of penguins—both of whom were completely unintended targets. Now, more than ever, there is a strong impetus to understand how natural systems work, how humans change those systems, and how in the future we can reverse these changes.

Ecologists are among the best-equipped scientists to study natural systems. Before 1960 ecologists were few in number and their activities were dominated by taxonomy, natural history, and speculation about observed patterns. Their equipment included sweep nets, quadrats, and specimen jars. Since that time, ecologists have become active in investigating environmental change on regional and global scales. They have embraced reductionist analyses and experimentation and have adapted concepts and methods derived from agriculture, physiology, biochemistry, genetics, physics, chemistry, and mathematics (Grime 1993). Their equipment includes portable computers, satellite-generated images, and

*Easter Island, off the coast of Chile, is one of the most isolated islands on Earth. At one time, this island was covered with palms, trees, and shrubs. Deforested by Polynesian colonists by* A.D. *1500, the lack of trees on this island is testament to the destructive capabilities of humans.* (Zuckerman, Tom Stack 1690-4-4.)

sophisticated chemical autoanalyzers. The challenge is for ecologists to come together and agree on solutions to the world's ills. The alternative, as Stanford University ecologist Hal Mooney was quoted as saying (Baskin 1994), is "Frank Sinatra" science where ecologists always try to do things "my way."

# Why and How to Study Ecology

Aside from the concerns expressed over general development which result in general phenomena, like an increase of atmospheric $CO_2$, specific projects in themselves can have dire consequences. For example, in 1970, after 11 years and an expenditure of $1 billion, construction of the Aswan High Dam, depicted in Photo 1.1, was completed. It is the largest dam of its kind in the world, located in southern Egypt on the world's longest river, the Nile (Fig 1.1). It contains more than four times the capacity of Lake Mead, the lake behind Hoover Dam, the largest dam in the United States. The Aswan High Dam was projected to provide several years of irrigation reserve, to add 1.3 million acres to the arable lands of Egypt, to produce 10 billion kilowatts of electric power annually, and to protect the country from catastrophic floods.

Today the dam does produce more than 50 percent of Egypt's electrical power. The reservoir of water saved the rice and cotton crops during the droughts of 1972 and 1973. It has facilitated the cultivation of two or three crops annually rather than one (Azim Abul-Atta 1978), thus increasing productivity by 20 percent for some crops and by 50 percent for others and also increasing governmental and national annual income from agriculture by 200 percent. One million hectares (ha) of additional farm land can be irrigated year round, and 380,000 ha of desert are being irrigated for the first time (Moore 1985).

In many ways, however, the dam has been argued to stand as a monument to ecological ignorance. First, loss of water to seepage through bedrock meant that the dam was still not full by 1988, though it had been predicted to fill by 1970. This problem may have been compounded by 20 years of below-average rainfall in the area (Wright 1988). Until unusually heavy rains fell in August 1988, the volume of water in the impoundment had fallen so low that serious conservation measures were about to be implemented, and the electric turbines

4

Photo 1.1   *The Aswan High Dam, Egypt—economic boon or ecological nightmare? The dam produces much of Egypt's electrical power and counteracts the crippling effects drought has on agriculture, but it is also thought to reduce silt deposition (and thus fertility renewal in the valley) and fish catches in the Mediterranean Sea, and to increase the frequency of schistosomiasis, a severe parasitic disease of humans.*  (WFP photo by J. Van Acker, courtesy of FAO.)

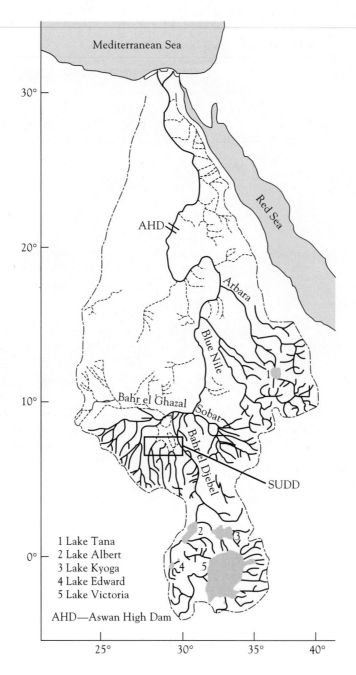

**Figure 1.1**   *The River Nile and its tributaries and the location of the Aswan High Dam.*

were about to stop, which would have caused widespread power cuts. The volume of the reservoir had dropped to around 3 billion m³, out of a capacity of 110 billion m³.

Second, the incidence of the tropical flatworm causing schistosomiasis (a debilitating parasitic disease) in the area was argued to have increased from 47 to 80 percent because the parasite's secondary hosts, snails, reproduce year-round in the reservoir and thus are no longer reduced by drought (van der Schalie 1972). This problem may since have abated (Moore 1985) and, according to some sources, may never have been as severe as was thought (Walton 1981). The variety of schistosomiasis now prevalent, however, is a much more severe one.

Third, reduced flow into the Mediterranean reduced phytoplankton blooms and fish harvest in the discharge area. Sardine catches alone dropped from 15,000 tons annually to 500 tons (George 1972), and yields from the new fishing areas behind the dam have been low.

Fourth, reduced silt deposition along the floodplain (Petts 1985) has increased the need for commercial fertilizers to the tune of $100 million annually. The new fertilizer plants use much of the hydroelectric power from the dam.

Fifth, most farmers overwater their land and as the water evaporates the salt is left behind. In 1986 almost half the irrigated area in Egypt was affected by salt (Kishk 1986).

It is interesting to note, however, that many of the adverse opinions of the dam are based on Western standards; the Egyptians themselves have chosen to look more positively upon the dam (Fahim 1981). Nevertheless, it is clear that a deeper ecological knowledge would have helped predict what would happen in nature.

Theoretical ecology has, more and more, been winning its spurs in the real world (Moffat 1994). Roy Anderson and Robert May of Oxford University in England modeled the spread of disease in nonhuman hosts. Following the explosion of the AIDS epidemic, these two scientists tried to model its spread. They were among the first to show that a relatively few promiscuous individuals could be disproportionately responsible for the rapid spread of AIDS.

Population biologist Russell Lande of the University of Oregon in Eugene modeled how big the size of the home ranges of the spotted owl ought to be in the old growth forests of Pacific Northwest to prevent extinction. The models showed that the proposed set asides of the Forest Service and Bureau of Land Management were too small. After these agencies were forced by the federal courts to consider these findings, new, larger patches of forest were set aside.

## 1.1  ECOLOGY AND DIVERSITY

If theoretical ecology is winning its spurs, one of the toughest rides will be to determine how much biodiversity matters. During the latter half of the 1980s, the reduction of the Earth's biological diversity emerged as a critical issue and was perceived as a matter of public policy (U.S. Congress 1987). A major concern was that the loss of plant and animal resources would impair future development of important products and processes in agriculture, medicine, and industry. For example, *Zea diploperennis*, an ancient wild relative of corn, could be worth billions of dollars to corn growers around the world because of its resistance to seven major diseases that plague domesticated corn. Two species of wild green tomatoes discovered in an isolated area of the Peruvian highlands in the early 1960s have contributed genes for a marked increase in fruit pigmentation and soluble-solids content currently worth $5 million per year to the tomato-processing industry (U.S. Congress 1987). Loss of tropical forests could mean loss of billions of dollars in potential plant-derived pharmaceutical products. About 25 percent of the prescription drugs in the United States are derived from plants, and as long ago as 1980 their total market value was $8 billion per year. On a smaller scale, individual species often thought worthless can actually be very valuable for research purposes. Armadillos, for example, are the only known species, other than humans, that can be used in research on leprosy (Photo 1.2). Desert pupfishes, found in the U.S. Southwest, tolerate salinity twice that of seawater and are valuable models for research on human kidney diseases. The technology does not exist to recreate ecosystems or even individual species. Once a species or a system is gone, it is lost forever.

More than this, humans use not just individual species but whole ecosystems too. Forests soak up carbon dioxide, preserve soil fertility, and retain water, preventing floods. The loss of biodiversity could disrupt an ecosystem's ability to carry out these functions. Having convinced governments of the value of biodiversity, we now have to determine just how far ecosystems can be altered before they cease to function in an acceptable way. There are two contradictory theories about this. The "rivet hypothesis" of Stanford ecologists Paul and Anne Ehrlich likens species to rivets on an airplane. The loss of each rivet weakens the plane a small amount until it eventually crashes. The "passenger" hypothesis of Australia's Brian Walker asserts that species are like people on the plane, not rivets. Most are superfluous to requirements, and only a few key species (the pilot and crew) keep the plane in the air. Only recently have data become available to indicate that increased diversity increases ecosystem performance. Naeem et al. (1994) built multitrophic-level communities of 9, 15, and 31 species, with the species-rich communities representing the more diverse communities. The reduction in diversity was cut across each of four trophic levels so that all communities had similar numbers of trophic levels. The experiment ran for six months and all plants grew from seedlings to flowering adults. A variety of ecosystem attributes such as retention rates for nitrogen, phosphorous, and other elements were measured, along with plant productivity. Plant productivity increased two to threefold with biodiversity—perhaps because increased diversity increased the number of plant-leaf canopy levels and hence light interception. However, this experiment was done in an indoor laboratory setting. It still remains a challenge for ecologists to show that diversity increases productivity in the field. The temperate forests in the Northern Hemisphere show great differences in species richness

Photo 1.2    *The nine-banded armadillo. This is the only animal other than humans to contract leprosy and is, therefore, valuable to medical researchers. (Photo © copyright by Florida Game and Fresh Water Fish Commission, A.V. Department, negative no. 4365.)*

(East Asian forests include 876 tree and shrub species, North American forests 158, and European forests 106), but they are virtually identical in productivity. Species diversity may increase productivity up to a certain point, but beyond that not much happens because a lot of the essential "niches" have been filled.

Finally, good arguments can be made against ecological mismanagement and loss of biotic diversity on moral and ethical grounds. We simply have no right to destroy species and the environment around us. Philosophers such as Tom Regan argue that animals are to be treated with respect because they have a life of their own and therefore have value apart from anyone else's interests (Gunn 1990). Other philosophers such as Christopher Stone, a law professor at the University of Southern California, have argued that entities such as trees, or even natural features such as lakes, could be given legal standing just as corporations, by a legal fiction, are treated as persons for certain purposes. Corporations can sue and be sued; the "interests" of corporations are represented legally by actual persons.

Can animals (or plants) "count" in their own right? Consider a tragic accident in 1974 in which a schooner sank off the eastern coast of the United States (Johnson 1990). The captain, his wife, their 80-pound Labrador retriever, and an injured crewman occupied a lifeboat to which two youths, aged 19 and 20, as well as a 47-year-old Navy veteran, tied themselves with ropes and floated in the freezing waters. The captain refused repeated requests by the swimmers that he throw the dog overboard to make room for (some of) those in the water. He later explained that he could not bear to do it. After nine hours, the youths perished from exposure. All the occupants of the lifeboat, dog included, were subsequently rescued. After an initial investigation, the Coast Guard recommended that no criminal action be brought. However, in May 1975, the captain was indicted in a federal court for manslaughter for refusing to eject his dog to make room for some of the swimmers who died. Can a dog count morally, count directly, for its own sake, rather than because of some human's interest in it? Can wildlife count for more than humans in some cases?

A more utilitarian argument is that about 95 million U.S. citizens participate every year in nonconsumptive recreational uses of wildlife (observation, photography, and so on). In addition, 54 million people fish and 19 million hunt. These figures may have more impact when it is considered that in 1991 $24 billion was spent on lodging, transportation, licences, stamps, tags, permits, and equipment for fishing, $18 billion for nonconsumptive uses, and $12 billion for hunting. Recreational hunters in North America pursue some 90 species (Prescott-Allen and Prescott-Allen 1986). In other countries, money from the observation of wildlife can be extremely important to the economy. In 1985, Kenya netted about $300 million from almost 300,000 visitors, making wildlife tourism the country's biggest earner of foreign exchange (Achiron and Wilkinson 1986).

Some techniques for maintaining existing ecological habitats and plant animal species are listed in Tables 1.1 and 1.2. Prominent among these are the preservation of natural areas. The role of the U.S. government in passing legislation and providing funds to help maintain diversity is potentially substantial. As of 1987, 29 federal laws impinged on the maintenance of biological diversity (U.S. Congress 1987), but gaps still existed in the laws, the national policies, and the data needed to address these issues. A specific edict regarding diversity and public policy was lacking.

Perhaps one of the soundest arguments for conservation of biodiversity is that, at the least, it keeps our options open for the future.

**TABLE 1.1** *Management techniques for maintenance of biological diversity. (After U.S. Congress 1987.)*

| Onsite | | Offsite | |
|---|---|---|---|
| *Ecosystems maintenance* | *Species management* | *Living collections* | *Germplasm storage* |
| National Parks | Agroecosystems | Zoological parks | Seed and pollen banks |
| Research natural areas | Wildlife refuges | Botanic gardens | Semen, ova, and embryo banks |
| Marine sanctuaries | In-situ genebanks | Field collections | Microbial-culture collections |
| Resource development planning | Game parks and reserves | Captive-breeding programs | Tissue-culture collections |

Increasing human intervention ⟶

⟵ Increasing emphasis on natural processes

**TABLE 1.2** *Management goals and conservation objectives. (After U.S. Congress 1987.) Items to be maintained.*

| Onsite | | Offsite | |
|---|---|---|---|
| *Ecosystems maintenance* | *Species management* | *Living collections* | *Germplasm storage* |
| –a reservoir or "library" of genetic resources | –genetic interaction between semidomesticated species and wild relatives | –breeding material that cannot be stored in genebanks | –convenient source of germplasm for breeding programs |
| –evolutionary potential | –wild populations for sustainable exploitation | –field research and development on new varieties and breeds | –collections of germplasm from uncertain or threatened sources |
| –functioning of various ecological processes | –viable populations of threatened species | –offsite cultivation and propagation | –reference or type collections as standard for research and patenting purposes |
| –vast majority of known and unknown species | –species that provide important indirect benefits (for pollination or pest control) | –captive-breeding stock of populations threatened in the wild | –access to germplasm from wide geographic areas |
| –representatives of unique natural ecosystems | –"keystone" species with important ecosystem support or regulating function | –ready access to wild species for research, education, and display | –genetic materials from critically endangered species |

# 1.2  SCIENTIFIC METHODS

Can science in general and the study of ecology in particular help overcome the loss of biotic diversity and all the problems it entails? Yes, it can. Although acid rain, toxic wastes, air and water pollution, and nuclear radiation are often seen by the public as direct results of scientific "progress"—and science is seen not as the "hero" but as the "goat"—one must bear in mind that these ecological problems are the results of the misuse of scientific knowledge. Solutions to ecological problems can also be found through scientific methods.

The scientific method is at the heart of the acquisition of scientific knowledge. Generally, the first step in the scientific method is sound observation, repeated to determine the frequency of an event and verified and confirmed by independent observers. After observations, the next step is construction of a hypothesis to explain the observed events. Good hypotheses can be tested by further observations or experiments. If a hypothesis stands up to repeated testing, it may reach the status of a theory or law. Popper (1972a) has emphasized the important point that science progresses not by trying to confirm theories but by attempting to falsify them. It is usually possible to find at least some confirmatory evidence for any hypothesis; one piece of negative data, on the other hand, refutes the hypothesis absolutely. Unfortunately, manuscripts that report the nonoccurrence of something are often distrusted—reviewers apparently regard it as easier to overlook a real phenomenon than to find a spurious one. As a result, it is often easier to get confirmatory papers accepted than to publish "negative" results. Mahoney (1977) submitted two sets of contrived research papers differing only in the results they reported. He found that papers reporting "negative" results were less likely to be accepted for publication and less likely to be rated methodologically sound than were those reporting confirmatory results.

The scientific method, of course, seems rather formal and tedious to the layperson and sometimes even to the scientist. Roughgarden (1983) and Quinn and Dunham (1983) argue that sometimes hypotheses can be developed less formally, by the type of commonsense logic that we use in everyday life, which usually involves the research for confirmatory evidence. Simberloff (1983) has replied that, although this type of approach is most seductive, it is likely to be wrong. "Commonsense" leads millions to conclude the existence of a deity and millions more to deny such existence. As Strong (1983) noted, commonsense led many to believe the world was flat. Nowadays, we have books with such titles as *The Common Sense of Drinking* (Peabody 1931) and even *Commonsense Suicide* (Portwood 1978). At the very least, formalization should help us become more efficient in using our research time. Besides, Simberloff points out, even perception and the search for confirmatory evidence itself imply the tacit mental construction of some part of a hypothesis.

In practice, scientific advances probably occur by several routes. Isaac Newton claimed that scientists work from the particular to the general, first observing phenomena and only later deriving generalizations from them. Popper argued that imagination comes first; then hypotheses are tested by experimentation. Both avenues have undoubtedly proved valuable in research.

Popperian science has its drawbacks, of course. Popper (1972b) has denied that Darwin's theory of evolution is a valid scientific theory; it seems untestable as a whole and is best formulated from many separate lines of evidence (see Section Two). The same arguments apply to the science of astronomy. Creationists have exploited this view in an attempt to refute the entire theory of evolution. For this reason, some philosophers (for example, Suppe 1977) and ecologists (Dunbar 1980) have abandoned Popper's type of approach. Fagerström (1987) has even made the controversial suggestion that it would be better to erect a theory and throw out the data if they are not in agreement. He argues that this approach is not as absurd as it at first sounds. For example, ecological data are not always "hard facts"; they are produced, digested, and interpreted with the aid of theories and apparatus and, by people who are constrained by prejudices and previous experiences. Data are, in actuality, theory laden. Ecological theories are judged more often by their simplicity, beauty, and intuitive appeal than by strict agreement with data. Even the simplest ecological statement assumes more about nature than can be concluded from observations alone. There can never be a complete match between theories and data no matter how much empirical evidence is gathered. Murray (1986) suggests that, in fact, most ecologists interpret their data by formulating the question, "How can I explain my data in terms of theory X?" a procedure that often leads to a rash of "me too" papers in the wake of publication of a new theory and that can slow scientific progress by leading to the uncritical acceptance of any plausible hypothesis. It would be preferable simply to ask, "How can I explain my data?"

Dyson (1988) has suggested that among scientists in all fields there are "diversifiers," who are content to explore the details of phenomena, and "unifiers," who strive to unearth general principles. He believes that biology is the natural realm of diversifiers, much as physics is the realm of unifiers. As a result, biology lacks general themes. Dyson suggests that physicists "have a driving passion to find general principles," but biologists "are happy if they leave the universe a little more complicated than they found it."

It has been further argued that biologists think inductively and that physicists and chemists think deductively. For example, from a series of observations, a physicist might predict the next element in the periodic table. Biologists group facts together, draw conclusions from them, but are loath to make subsequent predictions. Einstein and Infeld (1938) illustrated this by discussing the motion of a cart. If we give the cart a push, it will travel a certain distance. If we reduce friction by oiling the wheels and smoothing the road, it will travel further. If we remove all sources of friction, the cart will never stop. We can never do this experiment; we can only think it. Murray (1992) has argued that biologists would spend their time performing experiments to determine the roles of slope of the road, the number of wheels on the cart, and so on, on the motion of the cart, while a physicist would be the one to make the ultimate jump to a law of inertia, much as Newton did in 1729: "Every body continues in its state of rest, or of uniform motion in a straight line, unless it is compelled to change that state by forces impressed upon it."

Strong (1980) has suggested that the set numbers of particles in matter makes explanations simpler for physicists than for biologists, in whose science the number of particles, or species, is very much higher. However, I believe that many biologists, especially ecologists, *have* been making predictions and testing them, especially over the past 10 to 20 years. In applied ecology especially, politicians and policy makers are constantly forcing uncomfortable ecologists to put numbers and estimates to rates of environmental degradation. Very often, general hypotheses are tested in ecology.

## Null Hypotheses

Often, the working hypothesis that is set up in ecological research is that of "no effect"—the hypothesis that observed ecological patterns arise by chance and not through any effect of the forces of nature. This hypothesis of "no effect" is called the null hypothesis (Connor and Simberloff 1979; Strong, Szyska, and Simberloff 1979), and there exists a wide range of statistical tests designed to challenge it (Conover 1980; Fleiss 1981; Zar 1984). Of course, this type of hypothesis is not the only one that is useful in ecology. For example, one might have reason to erect the hypothesis that the disjunct distribution of two species in a particular habitat is due to competition and that they therefore live in mutually exclusive zones. However, the null hypothesis is often the most logical place to start (Simberloff 1983).

With the right type of observation, experimentation, and attempts at falsification, hypotheses can be solidified into ecological theories. J.B.S. Haldane (1963) has observed, rather cynically, that this process is normally reflected in four stages through which the hypothesis passes in the regard of the scientific community:

1.  This is worthless nonsense.
2.  This is an interesting, but perverse, point of view.
3.  This is true but quite unimportant.
4.  I have always said so.

## Ecological Questions

What types of ecological questions should we be asking? Slobodkin (1986a) has suggested that the "big questions" in ecology and evolution (What is life? How do higher taxa evolve? What determines the number of species in one locality?) may be too large to be amenable to theoretical formulation. He maintains that the most useful approach is to confine ourselves to more specific questions, such as, "How does meat dissolve in the stomach of a kite?" He suggests that in such minimal systems, it may be possible to see meaning and derivations more clearly and to examine the criteria for theoretical quality. Colwell (1984), Slobodkin (1986b), Bartholomew (1986), and others have all urged a return to focusing on organisms and real problems rather than on the external criteria of philosophers and mathematicians. Those "pesky biological details" matter a great deal; they often violate the assumptions of mathematical theory, and theorists who ignore them risk wasting their efforts.

There have been several recent surveys of the memberships of ecological societies to try to determine what ecologists think are important issues (Table 1.3). It is hard to be precise in comparing these surveys because each survey contained a somewhat different list of concepts that members were asked to rank. Nevertheless, it is clear that most ecologists seem to believe subjects like succession and ecosystems (community-oriented concepts) to be more important than competition and plant-animal interactions (population-oriented concepts). They also believe both theoretical ecology and life-history theory (ecological theory) and conservation biology and ecosystem fragility (applied ecology) were important. So much for what people think; what do they actually do? Stiling (1994) answered this question by surveying published papers in three mainstream ecological journals *Ecology, Oikos,* and *Oecologia* (Table 1.4). There is some disparity between what ecologists think and what they do because population-oriented concepts and autecology are actually studied more than community ecology and applied ecology. Perhaps this is because the former are more easy to study, despite the interest of the latter, or perhaps the contributors to the journals *Ecology, Oikos,* and *Oecologia* do not represent mainstream views. Stiling (unpublished) surveyed ecological articles from 26 other journals over roughly the same time period (Table 1.4). The concordance of rank of concept between the two sets of journals was suprisingly high, suggesting that most

**TABLE 1.3** *Recent surveys of members of ecological societies, Cherrett (1989) British Ecological Society, Travis (1989) Ecological Society of America, and Sugden (1994), readers of the magazine* Trends in Ecology and Evolution. *Respondents checked off what they thought were important topics from lists of various concepts (n values). Lists contained somewhat different items in each survey.*

| Rank | Cherrett 1989 (n=37) | Travis 1989 (n=40) | Sugden 1994 (n=33) |
|---|---|---|---|
| Top | | | |
| 1 | The ecosystem | Community ecology | Conservation biology |
| 2 | Succession | Ecosystems studies | Animal ecology |
| 3 | Energy flow | Animal population ecology | Life history theory |
| 4 | Conservation of resources | Plant population ecology | Behavioral ecology |
| 5 | Competition | Plant-animal interactions | Microevolution |
| 6 | Niche | Theoretical ecology | Biogeography |
| 7 | Materials cycling | Plant physiological ecology | Macroevolution |
| 8 | The community | Phytosociology | ? |
| 9 | Life history strategies | Conservation biology | ? |
| 10 | Ecosystem fragility | Ecosystem theory | ? |
| | | | |
| Bottom | | | |
| 1 | The Diversity/Stability hypothesis | Landscape architecture | Microbial ecology |
| 2 | Socioecology | Plant systematics | Oceanography |
| 3 | Optimal foraging | Regional planning | ? |

ecologists do study population-oriented phenomena, regardless of taxonomic discipline or area of interest. However, bear in mind that most ecologists publishing in these journals were still scholars or researchers, not "practical" or "applied ecologists." The authors of the 26 ecological journals surveyed were primarily in education (78.4 percent) or research (13.6 percent), and only 8 percent were practicing conservation biologists, agriculturalists, foresters, or fisheries people.

There is no doubt that for just about any ecological question more data are needed to answer it. Simon (1986) and others have noted this lack of information, and they use it to say, for instance, that the data are insufficient to show that deforestation is causing a great reduction in the numbers of plant and animal species on Earth. To take another example, Parker and Douglas-Hamilton have independently used the same data to discuss elephant conservation and the ivory trade. Parker concludes that a substantial portion of harvested tusks come from natural mortality and that hunting for profit is a serious threat in only a few areas. He further claims that increases in hunting deaths are due to the elephants' competition with Africa's rapidly expanding human population. Douglas-Hamilton disagrees with these contentions on every point. He claims that Parker considerably exaggerates the number of deaths due to natural mortality, that elephants are overhunted in all but the most inaccessible areas, and that the high price of ivory is to blame. When such disparate positions

TABLE 1.4 *Top ten, and bottom three, concepts studied by ecologists in 3,108 papers published between 1987 and 1991 in the journals* Ecology, Oikos *and* Oecologia *(after Stiling 1994). Same data from 2289 ecological papers from one year over the same period in the journals* American Naturalist, American Midland Naturalist, Trends in Ecology and Evolution, Journal of Animal Ecology, Journal of Ecology, Biological Conservation, Conservation Biology, Journal of Wildlife Management, Wildlife Society Bulletin, Journal of Applied Ecology, Journal of Biogeography, Biotropica, Marine Biology, Ecological Entomology, Journal of Fish Biology, Annals of the American Entomological Society, Auk, Condor, Herpetologica, Copeia, Journal of Mammalogy, American Journal of Botany, Animal Behavior, Behavioral Ecology and Sociobiology, Evolutionary Ecology *and* Evolution. *The total number of different concepts scored was 29 in each case.*

| Concept Rank | From *Ecology, Oikos, Oecologia* | From 26 "subject-oriented" journals |
|---|---|---|
| **Top** | | |
| 1 | Ecological adaptation, physiological ecology | Life history strategies |
| 2 | Life history strategies | Ecological adaptation |
| 3 | Plant-herbivore | Habitat selection, spatial variation |
| 4 | Competition and coexistence | Optimal foraging |
| 5 | Habitat selection, spatial variation | Predator prey |
| 6 | Predator-prey | Mating behavior |
| 7 | Nutrient cycling | Competition and coexistence |
| 8 | Population regulation | Dispersal, migration |
| 9 | Optimal foraging | Plant-herbivore |
| 10 | Stability, disturbance | Stability, disturbance |
| **Bottom** | | |
| 1 | Restoration | Restoration |
| 2 | Pest control | Pest control |
| 3 | Pollution | Pollution |

can be reached from identical information, more data are needed. Parker and Douglas-Hamilton's work is discussed in a paper by Pilgram and Western (1986).

## 1.3 EXPERIMENTS

As outlined in discussions of the scientific method, ecological hypotheses are best tested by experiments. Hairston (1989) has noted that the percentage of field studies involving experiments has risen from less than 5 percent in the 1950s to around 10 percent in the 1970s and to more than 30 percent in 1987. Experiments can be classified in several ways. Diamond (1986b) distinguishes three main types: laboratory experiments, field experiments, and natural experiments. In practice, these types form a continuum. Diamond further divides the natural experiments into two categories: Natural trajectory experiments are comparisons of an ecosystem or species before and after some dramatic perturbation such as a storm, a volcanic eruption, or the introduction of another species. Natural snapshot experiments compare natural areas that differ from one another in only one or two characteristics, for example, presence or absence of certain predators. Such differences have often been maintained throughout recent history. Strengths and weaknesses of these different types of experiments are outlined in Table 1.5. For example, the spatial scale of laboratory experiments is likely to be limited to the size of a constant-temperature laboratory room, around 0.01 ha, and that of field experiments to usually less than 1 ha. Natural experiments, however, may be virtually unlimited in scale and often use large islands or continents.

Laboratory experiments can regulate exactly all abiotic factors from light, temperature, and moisture to available nutrients. They are valuable in investigations of these factors. The biotic community represented in a laboratory experiment, however, is likely to be limited at best. Laboratory experiments are best used to study the physiological responses of individual animals rather than the population dynamics of reproducing populations.

Field experiments are conducted outdoors and have the advantage of operating on natural rather than synthetic communities. The most commonly used manipulations include local elimination of a species, local introduction of a species, and erection of a fence or cage. Darwin used a field experiment to demonstrate that either mowing or the introduction of grazing animals increases plant species diversity on a lawn (by preventing some species from outcompeting others). Field experiments commonly manipulate systems through use of phenomena (such as cages or fences) that are unlikely to be generated by nature itself.

Natural experiments are usually the sole technique for following the trajectory of a perturbation beyond a few decades. Simberloff (1976) was able to examine defaunation and recolonization on mangrove islands for several years, but only on the island of Krakatau has the process been followed in the long term, for more than 100 years (see Chapter 8, Part 8.2).

**TABLE 1.5  *The strengths and weaknesses of different types of experiments in ecology. (After Diamond 1986a.)***

|  | Laboratory experiment | Field experiment | Natural trajectory experiment | Natural snapshot experiment |
|---|---|---|---|---|
| 1. Regulation of independent variables | Highest | Medium/low | None | None |
| 2. Site matching | Highest | Medium | Medium/low | Lowest |
| 3. Ability to follow trajectory | Yes | Yes | Yes | No |
| 4. Maximum temporal scale | Lowest | Lowest | Highest | Highest |
| 5. Maximum spatial scale | Lowest | Low | Highest | Highest |
| 6. Scope (range of manipulations) | Lowest | Medium/low | Medium/high | Highest |
| 7. Realism | None/low | High | Highest | Highest |
| 8. Generality | None | Low | High | High |

One of the few exceptions is the experimental fertilization of selected experimental plots at Rothamstead Experimental Station, England, for more than 100 years, beginning in fact in 1843 (Williams 1978). The weather is frequently shown to be vital in influencing the population densities of many species, but we cannot manipulate the weather. Natural experiments involving drought situations or floods commonly provide the best types of data on this subject. Furthermore, natural experiments often have general implications for the ecological system because they sample from a wider range of natural variation among sites than do field experiments. In summary, it is apparent that there is no best type of experiment; the choice depends on what one is investigating. This point has ramifications relating to the preceding section on the scientific method—no one "right" type of experiment has inherent superiority over others.

Krebs (1988) has suggested that, of the three types of experiments, field experiments are preferable. He has detailed how laboratory experiments in the 1940s and 1950s failed to provide any useful insights into the population dynamics of rodent populations, but that field experiments begun in the 1950s have been successfully used to test single- and multi-factor hypotheses of population regulation. That experiments must be correctly set up should be obvious. A simple example illustrates the point. Otto Korner (cited by Sparks 1982) performed experiments in the early 1900s to see whether fish could hear. He engaged a well-known opera singer to perform before his aquaria. He watched for the signs of enthusiasm from the fish that would surely result as their piscine hearts were uplifted. None was forthcoming. Korner concluded that fish could not hear. It was left to Von Frisch to show that if fish were "given a reason" to respond to sounds—perhaps by association of sound with food—they would respond.

Experiments should be planned for greatest usefulness. For example, simple removal of one species to examine possible competitive effects on another might document a phenomenon, say the elevation of density of one species, and a hypothesis might then be developed that species $a$ and $b$ do compete, but no idea of the mechanism involved is provided. In such cases, the ability to predict the outcome of other pairwise or multispecies interactions is limited because no idea of mechanisms has been gleaned. Experiments that take the phenomenological approach and merely document a phenomenon (such as competition) are less useful than those taking mechanistic approaches, which attempt also to explain why the experimental results are obtained (in this case, say, resource competition, allelopathic effects, or effects of shared predators), because the level of prediction obtained in the latter is greater (Tilman 1987).

## 1.4  THE EFFECTS OF SCALE

Just as laboratory experiments, field experiments, and natural experiments are performed on different scales, it has become obvious that the effects of scale on ecological research and conclusions are staggering (Wiens et al. 1986, Levin 1992). What is the "right" spatial scale over which to look for a phenomenon? Ecologists must address this question in planning their experiments.

The size of the study area and the duration of an investigation can limit what one can see of an ecological system (Dayton and Tegner 1984). Wiens et al. (1986) have made the analogy that studying ecology is comparable to what it would be like to study chemistry if the chemist were only a few angstroms long and lived for only a few microseconds. If the chemist were no larger or longer-lived than the molecules and processes under study, the overall course of chemical reactions would be difficult to distinguish from the random collisions of molecules. Wiens et al. emphasize that scale is, of course, a continuum but that five major points on that continuum can be recognized:

A space occupied by a single individual sessile organism or a space in which a mobile organism spends its entire life.

A local patch occupied by many individuals.

A region large enough to include many patches or populations linked by dispersal.

A space large enough to contain a closed ecosystem (one receiving no migration)—an unlikely scenario in practice.

A biogeographic scale, large enough to encompass different habitats and climates.

Investigations on these different scales yield answers to different types of questions, and it is important to realize this point (see Table 1.6). In studying the distribution and abundance of zooplankton, different phenomena cause different patterns at different scales (Fig. 1.2). For example, large-scale patterns may be caused by ocean currents or climatic events. Small-scale patterns may be caused by local upwelling or diel vertical migration.

Addicott et al. (1987) suggest that the correct scale depends exactly on the question being asked, so there will never be a single ecological neighborhood for a given organism but rather a number of neighborhoods, each appropriate to a different process. The correct scale might depend on the ambit of the organisms involved. For example, imagine ten trees each 50 meters apart (Fig. 1.3). Each tree contains caterpillars and wasps that parasitize the caterpillars. In a study of how the wasps utilize the caterpillars (perhaps in a study of density dependence, Chapter 15), one might examine all ten trees to see if the wasps aggregate on the tree with the most caterpillars. However, if the wasps can only fly a few meters, they would not be able to fly from tree to tree. They would only be able to response to density differences of caterpillars within a tree. The correct scale of ecological investigation might then be within trees not between them.

Just as with spatial scale, the appropriate choice of temporal scale depends on the phenomenon and the species to be studied. One would choose a relatively short time over which

**TABLE 1.6  *The effects of scale on ecological investigations.***

| Scale | |
| --- | --- |
| 1. Individual space | Physiological ecology, sociobiology, foraging ecology, reproductive biology (Section 3) (special problems arise for migratory species, for which important behavior may occur at another location entirely). |
| 2. Local patch/ecological neighborhood | Predation, herbivory, parasitism, and pollination (Section 4). |
| 3. Regional scale | Immigration, emigration, outbreaks, habitat preference (Section 4). |
| 4. Closed system | Nutrient cycling, ecological energetics (Section 5). |
| 5. Biogeographical scale | Climatic limits, evolutionary ecology (Section 2). |

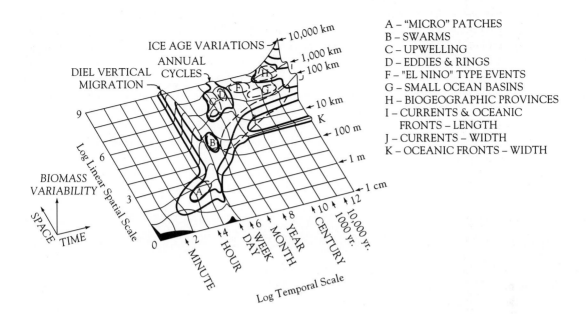

Figure 1.2    *Stommel diagram of spatial and temporal scales of zooplankton biomass variability.* (From Haury et al. 1978.)

to study behavioral responses, a longer one for population dynamics, and a still longer scale for studies of genetic change and evolution (Wiens et al. 1986).

Finally, the issue of scale can relate to scale of effort involved in research. If an investigation involves relatively few observations or experiments, we have less confidence in the results than if the experiment were well replicated. If the investigation has few experimental replicates, it has low statistical power—even if treatment had a relatively dramatic effect, we might not be able to detect it. Conversely, if a treatment had a negligible effect but we performed scores

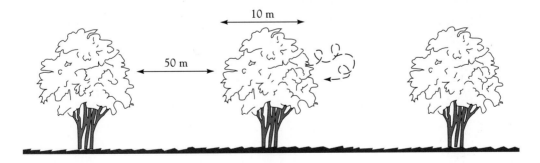

Figure 1.3    *The affect of neighborhood distance on spatial scale of investigation. If the ambit of a parasitic wasp is only 10 m, then it will not be able to respond to differences in densities of caterpillars, its prey, between trees spaced 50 m apart.*

of replicates, even a relatively unimportant effect would show up as significant—as having a meaningful effect. For example, Stiling et al. (1991) detected parasitism rates of 8.33 percent and 8.24 percent on two species of planthoppers inhabiting salt marshes. Because the sample sizes were 16,044 and 7,637 respectively, this difference in parasitism rate was statistically different. But is it biologically meaningful? Sometimes we cannot perform many replicates of some experiments—for example, defaunations of whole islands, nutrient additions to large lakes—and even though effects are large, they are not statistically significant. This is why it is sometimes difficult to use strict statistical procedures to attach meaning to results of experiments. The results may depend on the scale of the investigation in terms of how much effort was involved in replication.

## *Summary*

1. Ecology is to environmental science as physics is to engineering. A knowledge of ecology is therefore important in understanding such phenomena as the conservation of species, preserving biodiversity, maintaining ecosystem functions, preserving soil fertility, studying the effects of pollutants, preventing global warming, and calculating sustainable yields in fisheries, forestry, and hunting.

2. Are all species integral to maintaining life on Earth? There is still much debate as to whether species act as rivets on a plane, where the loss of each rivet weakens the aircraft, or whether species are like the people on board—only a few (the captain and other crew members) are necessary and the rest are, like passengers, superfluous. The truth is obviously somewhere in between, but to which end of this spectrum does the truth lie?

3. Progress in understanding ecological systems can be gained through observations of natural systems, through field or laboratory experimentation, through mathematical models, or from reviews and syntheses of existing studies. The effect of the scale of these investigations can have a large influence on the conclusions.

*Discussion Question:* How can we determine when loss of biodiversity will affect ecosystem function?

# Evolutionary and Behavioral Ecology

Conservation of many endangered species, such as this Siberian tiger, may depend on a knowledge of genetics and how to overcome loss of genetic diversity, as well as how to preserve suitable habitats. A knowledge of evolutionary and behavioral ecology is therefore vital to modern ecologists. (Leeson, Photo Researchers, Inc. 2S6960)

I f ecology is concerned with explaining the distribution and abundance of plants and animals and with control of their numbers, then evolutionary ecology is an important part of the discipline. For example, one may argue about what controls penguin numbers in the Southern Hemisphere, but a nagging question remains—Why are there no penguins in the Northern Hemisphere? The answer is not insufficient food or too many predators. Penguins simply evolved in the Southern Hemisphere and have never been able to cross the tropics to colonize northern waters. Bromeliads (members of the Bromeliaceae, the pineapple family) are virtually all neotropical, and the Myrtaceae (**Eucalyptus** and its relatives) are restricted to Australia. The Columbidae (the doves) are virtually cosmopolitan. Flight is not the only reason—the Todidae (small kingfisherlike birds) can fly but are restricted to the Caribbean islands, whereas skinks (Scincidae) are entirely terrestrial lizards but are cosmopolitan in most tropical and temperate areas. South America, Africa, and Australia all have similar climates, ranging from tropical to temperate; yet each is clearly characterized by its inhabitants. South America is inhabited by sloths, anteaters, armadillos, and monkeys with prehensile tails. Africa possesses a wide variety of antelopes, zebras, giraffes, lions, and baboons, the okapi, and the aardvark. Australia, which has no placental mammals except bats and the introduced mouse and dingo, is home to a variety of marsupials such as kangaroos and the peculiar egg-laying mammals or monotremes, the duck-billed platypus and the echidna. A plausible explanation is that each region supports the fauna best adapted to it, but introductions have proved this explanation to be incorrect; rabbits introduced into Australia proliferated rapidly. The best explanation is provided by macroevolution and **continental drift.** Other, more recent geological phenomena can explain the distributions of some plants and animals. For example, during the glacial periods of the Pleistocene, lakes covered much of what is now desert in the U.S. Southwest. With the retreat of the ice, the lakes disappeared. Desert pupfishes *(Cyprinodon)* and other aquatic organisms were once widely dispersed throughout the Death Valley region but now occur only in isolated springs. Similarly, tapir fossils are found widely over the globe, and their present spotty distribution in South America and Malaya merely represents the relict populations of once-widespread groups. A knowledge of evolution and historical ecology is clearly of paramount importance to contemporary ecology. A study of evolution also necessitates a knowledge of genetics.

A knowledge of genetics can also be important in conservation biology. First, the rate of evolutionary change in a population is proportional to the amount of genetic diversity available. Reduction of genetic diversity reduces future evolutionary options.

Second, heterozygosity, or high genetic variation is often positively related to fitness (Allendorf and Leary 1986; Koehn et al. 1988; but see also Hedrick and Miller 1992). However,

*A mother cheetah and her cubs resting at Nairobi National Park, Kenya. Is low genetic diversity to blame for low population densities or is the problem heavy predation on juveniles?* (Edwards, Animals Animals 678314-M.)

many species have naturally low levels of heterozygosity and there is no set "norm"; loss of heterozygosity, however, may indicate cause for concern. An often, cited example of a correlation between low heterozygosity and low fitness is the African cheetah. Fifty-five cheetahs from several populations examined had no detectable genetic diversity at 47 allozyme loci and very low heterozygosity of 150 soluble proteins (O'Brien et al. 1983, 1985). These animals were so genetically uniform that skin-tissue transplants were routinely accepted among all individuals—their immune systems could obviously not distinguish between themselves and other individuals. It was claimed that the animals had difficulty breeding in captivity, that the males had sperm counts ten times lower than related cat species, and that there were higher rates of infant mortality in the wild than in captive populations. Low heterozygosity due to population bottlenecks was blamed.

However, as is often the case in ecology, there is a flip side to the story. Other scientists (Caro and Laurenson 1994) presented convincing evidence that low population densities and poor recruitment in the wild are due to heavy predation on juveniles rather than lack of genetic diversity. They have suggested that poor performance in captivity is merely due to poor husbandry and that male cheetahs in zoos, despite abnormal sperm, can fertilize females with high efficiency. The bottleneck responsible for the low heterozygosity in cheetahs may be due to events following the retreat of the last ice sheet thousands of years ago. Human assaults on the cheetahs' range and numbers have certainly not aided its recovery from the effects of this bottleneck. But it is interesting that similar effects of inbreeding resulting from recent population bottlenecks are seen in relic populations of lions in the Gir forest of western India and in the natural Ngorongoro crater in the Serengeti of Tanzania.

Thirdly, the global pool of genetic diversity represents all of the information for all of the biological processes on Earth. Wilson (1985) calculated that all the bits of information encoded in this genetic library would, if translated into English text, fill all the editions of the *Encyclopaedia Britannica* published since 1768. Conservation programs of rhinoceroses, grizzly bears, Siberian tigers, and many other organisms are benefiting from the input of geneticists.

# The Diversity of Life

As Grant (1977, p. 3), said, "The world of living organisms exhibits several general features that have always aroused feelings of wonder in mankind." First, there is a tremendous diversity of life forms—more than 4,000 species of mammals, 9,040 species of birds, and about 19,000 species of fish, which together with other organisms give a total of 43,853 known vertebrates. Diversity of other groups, especially the arthropods and mollusks, is even more staggering (Table 2.1). It has been tentatively estimated that at one point or another during the history of life there have been in the neighborhood of 1 billion species.

Second, among these organisms there is a great structural complexity and an apparently purposeful adaptation of many characteristics to the environment. Darwin's own example was that of the woodpecker, with its chisel bill, strengthened head bones and neck muscles, extensile tongue with barbed tip, and stout tail for balance while chiseling. Another classic example is provided by the variety of feet in birds, from the perching feet of warblers, to the grasping feet of eagles, the walking feet of quail, the wading feet of herons, the swimming feet of ducks, and of course the specialized feet of woodpeckers (see Fig. 2.1). How can we explain this diversity?

## 2.1 HISTORICAL BACKGROUND

### Lamarck

A number of naturalists and philosophers, beginning with the ancient Greeks, had supposed that many forms of life evolved from each other. The first actually to formalize and publish a theory of **evolution**—"transformism," as he called it—was Jean-Baptiste Lamarck (1744–1829), though similar ideas had been thought of by others, including the Comte de Buffon (1707–1788) (see Greene 1959). (See Photo 2.1.)

## TABLE 2.1 *Numbers of described species of living organisms.*

| Kingdom and Major Subdivision | Common Name | No. of Described Species | Total |
|---|---|---|---|
| *Virus* | | | |
| | Viruses | 1,000 (order of magnitude only) | 1,000 |
| *Monera* | | | |
| Bacteria | Bacteria | 3,000 | |
| Myxoplasma | Bacteria | 60 | |
| Cyanophycota | *Blue-green algae* | *1,700* | 4,760 |
| *Fungi* | | | |
| Zygomycota | Zygomycete fungi | 665 | |
| Ascomycota (including 18,000 lichen fungi) | Cup fungi | 28,650 | |
| Basidiomycota | Basidomycete fungi | 16,000 | |
| Oomycota | Water molds | 580 | |
| Chytridiomycota | Chytrids | 575 | |
| Acrasiomycota | Cellular slime molds | 13 | |
| Myxomycota | Plasmodial slime molds | 500 | 46,983 |
| *Algai* | | | |
| Chlorophyta | Green algae | 7,000 | |
| Phaeophyta | Brown algae | 1,500 | |
| Rhodophyta | Red algae | 4,000 | |
| Chrysophyta | Chrysophyte algae | 12,500 | |
| Pyrrophyta | Dinoflagellates *DIATOMS* | 1,100 | |
| Euglenophyta | Euglenoids | 800 | 26,900 |
| *Plantae* | | | |
| Byrophyta | Mosses, liverworts, hornworts | 16,600 | |
| Psilophyta | Psilopsids | 9 | |
| Lycopodiophyta | Lycophytes | 1,275 | |
| Equisetophyta | Horsetails | 15 | |
| Filicophyta | Ferns — *Pterophyta* | 10,000 | |
| Gymnosperma | Gymnosperms | 529 | |
| Dicotolydonae | Dicots | 170,000 | |
| Monocotolydonae | Monocots *angiosperms* | 50,000 | 248,428 |
| *Protozoa* | | | |
| | Protozoans: Sarcomastigophorans, ciliates, and smaller groups | 30,800 | 30,800 |
| *Animalia* | | | |
| Porifera | Sponges | 5,000 | |
| Cnidaria, Ctenophora | Jellyfish, corals, comb jellies | 9,000 | |
| Platyheiminthes | Flatworms | 12,200 | |
| Nematoda | Nematodes (roundworms) | 12,000 | |

*Amoeba perimecium*  *malaria?*  *sleeping sickness*

| TABLE 2.1 (continued) *Numbers of described species of living organisms.* | | | |
|---|---|---|---|
| Kingdom and Major Subdivision | Common Name | No. of Described Species | Total |
| Annelida | Annelids (earthworms and relatives) | 12,000 | |
| Mollusca | Mollusks | 50,000 | |
| *Echinodermata* | Echinoderms (starfish and relatives) | 6,100 | |
| Arthropoda | Arthropods | 751,000 | |
| Insecta | Insects | | |
| Other arthropods | | 123,161 | |
| Minor inverebrate phyla | | 9,300 | 989,761 |
| *Chordata* | | | |
| Tunicata | Tunicates | 1,250 | |
| Cephalochordata | Acorn worms | 23 | |
| Vertebrata | Vetebrates | | |
| Agnatha | Lampreys and other jawless fishes | 63 | |
| Chrondrichthyes | Sharks and other cartilaginous fishes | 843 | |
| Osteichthyes | Bony fishes | 18,150 | |
| Amphibia | Amphibians | 4,184 | |
| Reptilia | Reptiles | 6,300 | |
| Aves | Birds | 9,040 | |
| Mammalia | Mammals | 4,000 | 43,853 |
| TOTAL, all organisms | | | 1,392,485 |

Whereas Buffon conceived variation as a process of degeneration or random deviation from innumerable ancestral forms, Lamarck viewed it as evolution from simple beginnings. Lamarck's (1809) work, however, was largely ignored because the mechanism he proposed to explain how evolution works was based on the inheritance of **acquired characteristics**. For example, Lamarck supposed that giraffes, in their continual struggle to reach the highest leaves on trees, stretched their necks by a few millimeters in the course of their lifetimes. This increase in neck length was passed on to their offspring, who continued the process until the necks of giraffes reached their current proportions. In a similar way, Lamarck explained racial variation in skin color by assuming that the suntan developed by ancestral races was transmitted to their descendants, who were in turn a little darker than their parents.

In 1988 Lamarchian evolution received a boost when John Cairns, a geneticist at Harvard University, and his colleagues, published a study of spontaneous genetic mutations in the bacterium *Escherichia coli* (Cairns et al. 1988). They put bacteria that was unable to digest lactose into a petri dish with only lactose for nourishment—and found that the bacteria preferentially acquired the crucial mutations they needed to become lactose eaters. Later Barry Hall of Rochester University, New York, found a similar occurrence in other bacteria (Hall 1991), and scientists uncovered similar work by other geneticists in the 1940s and

**Figure 2.1** *Different adaptive types of bird feet.* (**a**) *Perching foot (Audubon warbler, Dendroica audubon).* (**b**) *Grasping foot with strong talons (horned owl, Bubo virginiarius).* (**c**) *Climbing foot (acorn woodpecker, Melanerpes formicivorus).* (**d**) *Walking and scratching foot (California quail, Lophortyx californicus).* (**e**)*Wading foot (green heron, Butorides virescens).* (**f**) *Swimming foot (pintail duck, Anas acuta).* *Drawings not to the same scale.* (Redrawn from Grant 1963.)

1950s (Luria and Delbruck 1943; Ryan 1955). However, as Lenski and Mittler (1993) have pointed out, *directed* mutation has still not been demonstrated. Rates of mutation in bacteria and yeast certainly do increase during starvation, but mutations may still be random. There are now just more chances of a mutation to be advantageous. Nevertheless, the very fact that mutation rates increase under stress is fascinating. Is this adaptive, in the sense of having evolved by natural selection for alleles that specifically promote increased genetic variation under stress, or is the process merely an unavoidable consequence of physiological deterioration under starvation or of the induction of mechanisms to repair damage to DNA? What is needed is an assay to detect mutations—at present, it's hard to prove the absence of other mutations because none of them survive.

## Darwin

Lamarck's theories were replaced by the ideas of Charles Robert Darwin (1809–1882), the founder of modern evolutionary theory (Photo 2.2). Darwin was born into a prosperous background. (His birthday, February 12, 1809, was the same as that of Abraham Lincoln.) His

Photo 2.1   *Jean Baptiste Lamarck, 1744–1829.*   (Photo negative no. 124768, courtesy Department of Library Services, American Museum of Natural History.)

father was a successful doctor, and his mother and his wife were both Wedgwoods, of china fame. His family's fortune enabled Darwin, educated at Edinburgh and Cambridge universities, to accept an unpaid job as scientific observer on board HMS *Beagle*, which sailed on a five-year world survey from 1831 to 1836, concentrating on South America. In some ways, Darwin was "primed" to accept the theory of gradual biological change and evolution because he had with him a copy of *Principles of Geology* (1830), newly published by Charles Lyell, who had taken the unprecedented step of describing the physical world as one that changed gradually through physical processes, not through a few catastrophic events (the view held firmly up until that time by the clergy and the masses).

During the voyage of the *Beagle* up and down both coasts of South America, Darwin was able to view diverse tropical communities, some of the richest fossil beds in the world in Patagonia, and the Galápagos Islands, 600 miles west of Ecuador. The Galápagos contain a fauna different from that of mainland South America, exhibiting tortoises and other animals different in form on virtually every island. By the time he had finished the expedition, Darwin had amassed a wealth of data, described an astonishing array of animals, and built up a vast collection of specimens.

A year after his return, Darwin read a revolutionary book on human population growth by the English clergyman Thomas Malthus, written (anonymously) in 1798, which argued that populations had the capacity for geometric growth (that is, they could grow as rapidly as the series 1, 2, 4, 8, 16); yet food supply increased only arithmetically (that

Photo 2.2    *Charles Robert Darwin, age 51, 1860.*  (Photo negative no. 326668, courtesy Department of Library Services, American Museum of Natural History.)

A.R.W.—31 yrs 1860

is, like the series 1, 2, 3, 4). Malthus proposed that—because the Earth was not overrun by humans as it should be—food shortage, disease, war, or conscious control must limit population growth. Darwin quickly established that the **Malthusian theory of population** would apply to animal populations. He made the logical deduction that these factors would act to the detriment of weaker, less well-adapted individuals and that only the strongest would survive.

Darwin had formulated his theory of **natural selection,** survival of the fittest: a better-adapted plant or animal would leave more offspring. In this way, giraffes born, by genetic chance, with longer necks would be better fed and able to reproduce more successfully. This trait, because it was genetically determined, would be passed on to their offspring, and long necks would become common. Only rarely would such long-necked mutants evolve independently; much more commonly, distinct traits, such as neck length, are inherited unchanged.

Incredibly, Darwin did not immediately publish his theory. He waited for nearly 20 years, collecting data on a wide range of organisms. Eventually, he was pushed into publication (Darwin 1859) by the arrival of a manuscript by Alfred Russel Wallace, who had independently and years later arrived at the same conclusions.

## Wallace

Alfred Russel Wallace (1823–1913) was born into poverty and farmed out to an older brother in London at age 14, after only six years of schooling (Photo 2.3). He held down a succession of jobs until his brother died and left him some money. Wallace set sail immediately for the Amazon. A year later, a second brother joined him. Both men contracted yellow fever, and the brother died. After four years in the jungle, Wallace sailed for home with his precious collections. En route the ship caught fire and sank. After ten days in the open sea in a small boat, Wallace was saved, but four years of labor went down with the ship. Back in England, Wallace began to prepare for a second voyage, this time to the Malay Archipelago as a professional collector and naturalist in the company of W.H. Bates (for whom Batesian mimicry is named). It was there, during another bout of fever, that Wallace conceived the idea of natural selection. Wallace's one major advantage over Darwin was that he was persuaded before he left on his voyages that species evolve; Darwin did not abandon his creationist beliefs until after his return. Thus, Wallace was able to gather data with an eye to his evolutionary hypothesis.

Unfortunately, Wallace's earlier papers had been ignored by the scientific community, and he was faced with the problem of lack of recognition. His solution was to send his manuscripts to Darwin, with whom he had previously corresponded. Darwin's higher standing in the scientific community made it more likely that he would be taken more seriously and that he would garner most of the credit—a phenomenon that still probably

Photo 2.3   *Alfred Russel Wallace, age 46, 1869.*
(Photo negative no. 326812, courtesy Department of Library Sciences, American Museum of Natural History.)

exists today. Darwin immediately sought the advice of friends (geologist Lyell and botanist Sir William Jackson Hooker) and, as a result of their suggestions, Darwin and Wallace presented their theories jointly at a historic meeting of the Linnean Society of London on July 1, 1858. One year later, Darwin at last published his *On the Origin of Species by Means of Natural Selection*, an abbreviated version of the manuscript based on his 20 years of work.

Although Wallace deserves full credit as a codiscoverer of the chief mechanism of evolution, Darwin's subsequent work continued to explore the same ideas and principles inherent in the original work. Furthermore, Darwin has been credited as a pioneer of the hypothetico-deductive method, that is, the testing of hypotheses through use of observations to confirm deductions from them. In Darwin's day, most science was done by induction— conclusions were drawn from an accumulation of individual observations.

Darwin's main conclusions about the origin of species were two. First, all organisms are descended with modification from common ancestors. All the prominent scientists of the day were convinced of this point within 20 years, although the religious community, of course, was skeptical and still is. (For those wishing to read about the fascinating and ongoing creationism-evolution debate, an excellent entry into the literature and the rhetoric is provided by Numbers [1982], Godfrey [1983], Wilson [1983], and the National Academy of Sciences [1984].)

Second, the mechanism for evolution was natural selection. This conclusion convinced few people at the time and was not fully accepted until the late 1920s, partly because of a widespread belief in blending inheritance, in which the traits of the parents were thought to be blended in the offspring like the colors of two paints blending to produce an intermediate color. Natural selection would not work in such a system. For example, if a long-necked giraffe mated with a short-necked giraffe, the offspring would have a neck of medium length, and the advantage of a long neck would be lost. Furthermore, the belief that environmentally induced variation could be inherited was widespread and provided an alternative to natural selection. Darwin's theory of natural selection had one serious flaw: He knew nothing of the causes of hereditary variation and could not well answer questions on that subject. The evidence of genetics and Mendel's laws of heredity were available but had passed into obscurity and were only resurrected in the early part of the twentieth century.

## Mendel

Gregor Johann Mendel (1822–1884) was an Austrian who entered monastic service at Brno in 1843. As the monastery was a center of learning, Mendel was sent to Vienna University and later taught at Brno Technical School during 1856–1864, where he performed his revolutionary work on peas. He became abbot in 1868 but still carried on his hybridization work on plants, although he never published the results.

Mendel crossed tall and dwarf strains of peas and found in the second generation a 3:1 ratio of tall to dwarf plants. He could therefore conclude that the parents differed with respect to a single gene (known today to be a single unit of DNA) that controlled size. The gene for height in pea plants existed in two different forms (or alleles), tall and short. We know today that these alleles can always be found at the same point on a chromosome, the locus of that gene. During meiosis, DNA is replicated (doubled in quantity), and the resulting two sets of chromosomes recognize each other and become precisely aligned on the meiotic spindle. During the first meiotic division, homologous chromosomes are drawn to opposite poles. Mendel's experiments formed the basis of a genetic understanding of the acquisition of inherited traits, although much controversy was still to exist for many years because his simple experiments did not work for all species. For example, the equivalent cross of tall and short humans does not yield a neat 3:1 ratio in the second generation because in more complex species, features like height are often controlled by many genes, each of which alone has only a slight effect on size. Each of the individual genes controlling these so-called polygenic traits follows the principles Mendel outlined, however. Genes that contribute to a single character or to characters that are functionally related often show strong linkage; that is, they are located close to one another on the same chromosome and are often inherited together. Such clusters of genes are often called **supergenes.** There are only a few cases in which parental experience can influence offspring characteristics. These are usually due to the maternal origin of the cytoplasm of the egg (the sperm contributes almost nothing except DNA) and involve self-replicating bodies such as mitochondria and chloroplasts.

Mendel's work showed that inheritance is generally particulate, with the exception of genes on sex-linked chromosomes. He showed that inherited factors did not contaminate one another and could be passed down from ancient ancestors in the same form. Furthermore, genes appeared exceptionally stable. As Weismann (1893) was to show, germ plasm is often entirely separate from, and immune to any influences from, the soma (the rest of the body), so environment has no influence on heredity.

Yet some variation must occur in populations if selection is to work at all, and it is readily obvious that not all members of the same species are identical. Before addressing this point, it is worth reconsidering the many apparently purposeful adaptations discussed at the beginning of this section.

It is important to realize that the development of adaptations does not always proceed via natural selection for those adaptations. Feathers, for example, may have evolved as insulation for warm-blooded birds and only secondarily become useful in flight. Sometimes, chance alone may influence evolution. Perhaps more commonly, the developmental biology and other morphological constraints may impose selection by coupling one trait with another. Because genes exert their effects on the **phenotype** through biochemical reactions, other reactants must be incorporated into the equation. A phenotypic effect at one locus often depends on the genotype at one or more other loci: this is the phenomenon of **epistasis.** Finally, some features may actually be neutral with respect to "fitness." For example, blue eyes and brown eyes confer equal visual ability on the majority of humans; they are merely different morphological solutions to the same problem.

## 2.2  HOW VARIATION ORIGINATES

It is genetic variation that can produce either an increase or a decrease in the variability of a population. In all organisms (except RNA viruses), the genetic material is DNA, deoxyribonucleic acid. In **prokaryotes** the DNA exists in the form of a single circular chromosome; in **eukaryotes** it is arranged into a set of linear chromosomes that reside in the cell nucleus. (The exact structure of the famous double-stranded DNA molecule is given in most introductory biology texts and should be familiar to biology students.) When the DNA code is copied for delivery into the gametes, mistakes are possible; these mistakes are the source of much genetic variation.

Increases in the amount of genetic variability present in the gene pool arise chiefly from **mutations** during copying. There are two kinds: gene or point mutations (the most important in enriching the gene pool) and chromosome mutation (the most important in rearranging it). Most mutations result in a loss of fitness; the effects of mutations are at random with respect to adaptiveness (but note again Cairns et al. 1988), and the chances are small that a random change to an already well-adapted system would result in an improvement.

## Point Mutations

A point mutation results from a misprint in DNA copying (Fig. 2.2). Most point mutations are thought to involve changes in the nucleotide bases that make up the DNA base pairs (adenine, thymine, guanine, and cytosine) at single locations only (cistrons), for example, from adenine to guanine or from adenine to thymine. Two such changes, which result in the change of the sequence GAA to CUA, combine to substitute the amino-acid valine for glutamic acid, the change that causes the abnormal beta chains in sickle-cell hemoglobin. Because of the redundancy of the genetic code, about 24 percent of these codon (base triplet) substitutions do not change amino-acid sequences and thus do not alter phenotype. The genetic code is universal among prokaryotes and eukaryotes (Jukes 1983); the same triplets code for the same amino acids in all organisms.

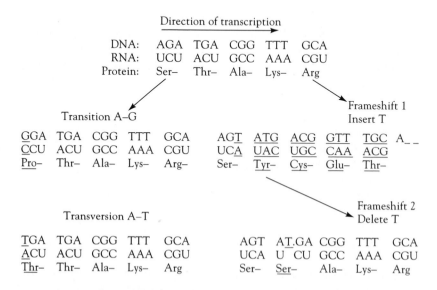

Figure 2.2  *Types of point mutation.*

More drastic changes in amino-acid sequences are caused by frameshift mutations. An addition to (or deletion from) the amino-acid triplet sequence alters the whole reading frame, leading to drastic and often fatal mutations. A second such change at another site may rectify the original reading frame so that only a short nucleotide sequence is then read in altered triplets. Genes themselves consist of many long sequences of base pairs. A gene for collagen in chickens is 40,000 base pairs (bp) (40 kilobases, kb) long. The size of the complete genome (all the genes) varies greatly among organisms from less than 400 bp in some viruslike particles to more than $10^{11}$ bp in some vascular plants. It has been estimated that the average mammalian genome is sufficient to code for more than 300,000 genes, that of *Drosophila*, 100,000. In *Drosophila*, however, only about 10,000 genes are recognized. Much DNA therefore codes for repetitive sequences or codes for nothing; it is essentially junk DNA. In prokaryotes, there are many single copies of each DNA sequence.

Although mutations may be accelerated by human-made radiation, UV light, or substances such as colchicine, in nature such mutagens (usually in the form of weak cosmic rays) are too rare to produce many mutations. Nevertheless, it has been conservatively estimated that mutations occur in nature at the rate of one per gene locus in every 100,000 sex cells (Dobzhansky 1970; Neel 1983). Higher organisms contain at least 10,000 gene loci, so one out of ten individuals carries a newly mutated gene at one of its loci. Of course, most mutations are deleterious—they arise by chance, and the genes cannot know how and when it is good for them to mutate. Only one out of 1,000 mutations may be beneficial, and thus one in 10,000 individuals carries a useful mutation per generation. In actuality, every individual *Drosophila* fly, corn plant, and human being seems to carry at least one abnormal, mutant allele (Spencer 1957; Morton, Crow, and Muller 1956), some originating in that organism and some inherited from ancestors.

If we estimate 100 million individuals per generation and 50,000 generations for the evolutionary life of a species, then 500 million useful mutations would be expected to occur during this span. It has been estimated that only 500 mutations may be necessary to transform one species into another, so only one in one million of the useful mutations needs to be established in a population in order to provide the genetic basis of observed rates of evolution. The chief factor limiting the supply of variability is therefore not the rate of new mutations. In fact, variability is limited mainly by gene recombination and the structural patterns of chromosomes. Even more astonishing is that some scientists have argued, with good evidence, that pure genetic drift is probably sufficient to cause directional genetic change in species with effective population sizes of under 5 million (Charlesworth 1984), which is probably greater than the population size of many mammals. Lande (1976) showed mathematically how even very weak selection on horses' teeth, two selective deaths out of one million individuals per generation, would be sufficient to explain the dramatic evolution of horse teeth through the ages.

## Chromosomal Mutations

Chromosomal mutations do not actually add to or subtract from gene-pool variability; they merely rearrange it. Followed by natural selection, chromosomal mutations make certain adaptive gene combinations easier for the population to maintain.

Chromosomes can undergo four types of changes (deletions, duplications, inversions, and translocations) in which the order of base pairs within the gene is unaffected but the order of genes on the chromosome is altered (Fig. 2.3).

A deletion is the simple loss of part of a chromosome. A deletion is usually lethal unless, as in some higher organisms, there are many duplicated genes.

When two chromosomes are not perfectly aligned during crossing over, the result is one chromosome with a deficiency and one with a duplication of genes. The change may be advantageous in that greater amounts of enzymes may be coded for. In yeasts, for example, an increase of acid monophosphatase enables cells to exploit more efficiently low concentrations of phosphate in the medium.

An inversion occurs when a chromosome breaks in two places and then refuses with the segment between the breaks turned around so that the order of its genes is reversed with respect to that on the unbroken chromosome. Such breaks probably occur at prophase, during which the chromosomes are long and slender and often bent into loops. In translocations, two nonhomologous chromosomes break simultaneously and exchange segments.

Together, gene and chromosomal mutation provide most of the genetic variability in a population. However, some other mechanisms for promoting genetic diversity do exist and will be mentioned briefly. If the first meiotic division fails to occur, unreduced **diploid** (2n) gametes are formed. Normally the union of a diploid gamete with a normal **haploid** (n) gamete forms a triploid (3n) zygote. Such individuals, if they can exist at all, are usually sterile; they cannot produce gametes with effectively balanced complements of chromosomes. However, if two unreduced (2n) gametes unite, the resulting tetraploid (4n) zygote may be fertile. Such a phenomenon can occur in some plants where, because of a doubled gene dosage, these polyploids are often bigger and more robust than diploids (Levin 1983). Tetraploidy is also found, rarely, in animals that are **parthenogenetic** (those in which females can produce daughters without fertilization by male gametes), like brine shrimp, *Artemia salina*, in which tetraploids, pentaploids, octoploids, and even decaploids have been found.

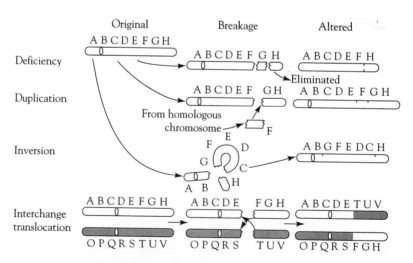

Figure 2.3   *Chromosome breakage and reunion can give rise to four principal structural changes.*

The existence of all this variation was something of a shock in the 1920s and 1930s, before which genetic uniformity had been taken for granted, and it led to a new ground swell of opinion, led by Theodosius Dobzhansky, that populations were in fact huge collections of diverse genotypes (Futuyma 1986). How much change in the amount of DNA, in the genome size, or in chromosome structure is necessary to cause speciation? There is no simple correlation. Two species of deer in the same genus, *Muntiacus reevesi* and M. *muntjac*, even have vastly different complements of chromosomes, $2n = 46$ and $2n = 6$, respectively (M.J.D. White 1978)!

## 2.3   HOW VARIATION IS MAINTAINED

### How Variation Is Maintained Without Selection

Although variation is created by mutation and chromosomal rearrangements, given simple Mendelian genetics and dominance, how is it that a dominant allele, responsible for a 3:1 numerical ratio of phenotypes, does not gradually supplant all other types of alleles, given that all alleles confer equal fitness on the organism? That is, if a gene pool in one generation consists of 70 percent A alleles and 30 percent a alleles, what stops the proportion of A alleles from increasing dramatically? What will the proportion of alleles be in the next generation? This very question was posed and independently answered by G.H. Hardy and W. Weinberg in 1908. They assumed three things: (1) populations are large, which thus negates sampling errors; (2) individuals contribute equal numbers of gametes; and (3) mating is random. In short, gametes carrying different alleles combine in pairs in proportion to their respective frequencies in the gamete pool.

### The Hardy–Weinberg Theorem

Assume that a diploid (normal) population polymorphic for A contains the following proportions of genotypes: 60 percent AA, 20 percent Aa, and 20 percent aa. The genotype frequencies are 0.60 AA + 0.20 Aa + 0.20 aa = 1. The allele frequencies ($q$) must then be

$$\frac{0.60 + 0.60 + 0.20}{2} = 0.70$$

$$\frac{0.20 + 0.20 + 0.20}{2} = 0.30$$

The paired combinations of matings are then

| Female Gametes | | Male Gametes | | Sygotic Frequencies | |
|---|---|---|---|---|---|
| 0.70 A | × | 0.70 A | × | 0.49 AA | |
| 0.70 A | × | 0.30 a | × | 0.21 Aa | |
| 0.30 a | × | 0.70 A | × | 0.21 Aa | } = 0.42 Aa |
| 0.30 a | × | 0.30 a | × | 0.09 aa | |

The gene pool of the second generation now contains the two alleles in the following frequencies:

$$\frac{0.49 + 0.49 + 0.42}{2} = 0.70$$

$$\frac{0.42 + 0.09 + 0.09}{2} = 0.30$$

The allele frequencies are the same as they were in the first generation, which is why the proportion of alleles is unchanging.

In general, if $p$ is the frequency of allele A and $q$ the frequency of allele a $(q = 1 - p)$, then the combinations of A and a gametes will produce zygotes in proportions given by the expansion of the binomial square $(p + q^2 = p^2 + 2pq + q^2$, that is, $p^2$ AA, $2pq$ Aa, and $q^2$ aa. This generalization is known as the Hardy-Weinberg theorem. For three alleles whose frequencies are $p$, $q$, and $r$, the equilibrium frequency of genotypes is given by the trinomial square $(p + q + r)^2$. In general, for an $n$-ploid organism, the genotype frequencies are given by $(p + q)^n$.

## The Effects of Nonrandom Mating

Of course, genotype frequencies are not always constant over time. Species change and evolve as a result of deviations from the Hardy–Weinberg assumptions. For example, not all mating is random or **assortative.** Bateson, Lotwick, and Scott (1980) have shown that individual Bewick's swans (*Cygnus columbianus*) can distinguish other individuals by their face markings. Swan families tend to have characteristic family markings, and young birds prefer to mate with partners whose faces are clearly different from their own. This preference reduces inbreeding, which can dramatically increase the proportions of genotypes in favor of homozygotes (Grant 1977). Furthermore, sexual selection in many organisms ensures that individuals do not always contribute equal numbers of gametes. In territorial species some males contribute more genes than others; in those with harem-holding males, just one individual will fertilize the majority of the females. New genes can also often enter the population via the migration of new individuals from different areas. The result is that there are very few truly **panmictic** species, in which all the individuals from one species form a single randomly mating population. One of the few examples is the common eel, *Anguilla rostrata*, in which individuals from U.S. eastern-seaboard drainages and Europe as well migrate to one area of the ocean near Bermuda, the Sargasso Sea, to breed.

## Environmental Variance

In addition to disruption of the Hardy–Weinberg assumptions by nonrandom mating, each genotype may be phenotypically variable to some extent because its development is directly affected by the environment. Fly weight depends often on the amount of medium available to the larvae and whether or not they have been competing for it. Certain crop plants and domesticated animals perform better in particular climates and agricultural regimes. This dependence is termed environmental variance, $V_E$, and affects variance of phenotypes in nature, the Hardy–Weinberg theorem notwithstanding. It is in contrast to the genetic variance $V_G$. Total phenotypic variance $V_P$ would thus be

$$V_P = V_G + V_E$$

Which type of variance is most important? This is the old "nature versus nurture" or genetics versus environment debate. The answer usually is "both."

Sometimes, the reaction to a difference in environment differs among genotypes, so there is a genotype-by-environment interaction $(V_{G + E})$, which also contributes to the phenotypic variance. In other words, certain genotypes do better in some habitats, and other genotypes do better in others. Thus:

$$V_P = V_G + V_E + V_{G + E}$$

For example, resistance in different varieties of wheat to the wheat-stem sawfly was correlated to the solidness of the straw; larvae could not move to obtain sufficient food if the pith was firm and compact (Painter 1951). However, expression of stem solidness was influenced strongly by the amount of light that the stem received when it was elongating (Platt 1941). Maddox and Cappuccino (1986) showed that differences in aphid numbers on clones of goldenrod were apparent only when the plants were well watered and the effect was not found in drier environments. However, Kennedy and Barbour (1992) stated that relatively few studies clearly document the occurrence of genotype-by-environment interactions as they affect resistance in natural systems. In a study of the insect community on coastal plants in Florida, Stiling and Rossi (1995a) found only one out of 15 possible genotype × environment interactions to be significant.

In most populations, it is hard to know whether a gene exists for a particular trait, and it is hard to study its properties. The reason is that alternative alleles at that locus are hard to identify; studies of variation have therefore focused on distinct polymorphisms such as red eye versus white eye in *Drosophila*. In nature, however, species are seldom found with two or more discrete phenotypes that lend themselves well to Mendelian crosses. The reason and studies of them are the subject of Chapter 3. In many of these cases, the two phenotypes or morphs were originally described as distinct species. The differential survival of these **morphs** in nature and how sources of mortality act differentially upon them fall into the realm of natural selection.

## How Variation Is Maintained with Selection

When selection acts at a locus and a homozygous genotype is most fit, the less advantageous alleles should be eliminated. This is the concept of natural selection (which will be developed further in Chapter 3). This situation contrasts with the Hardy–Weinberg theorem, which showed how two different alleles are maintained in the same population, generation after generation, if one is not favored over the other. Natural selection is obviously a common and potent force in the real world. How then does variation continue to exist in nature? Much of the effort of population geneticists has been devoted to this question. There are three common answers:

1. Selection acts on the locus so as to maintain a stable polymorphism, in which different genotypes are most fit under different situations; the "wild type" allele is not always most fit (see Chapter 3).
2. Fixation by selection is counteracted by mutation (for example, mutation may explain the reoccurrence of albinos in populations from which they are eliminated by natural selection).
3. Fixation by natural selection is counteracted by **gene flow.**

## 2.4  HOW MUCH VARIATION EXISTS IN NATURE?

Given the existence of so many mechanisms by which variation is produced, it is interesting to speculate on how much genetic variation exists in nature. For a long time, such speculation was difficult because it was impossible to tell how many loci exhibited no genetic variation—no clues were given by the phenotype. In the 1960s it was realized that most loci actually code for proteins, especially enzymes. Different forms of, say, alcohol dehydrogenase are coded for by different alleles. Two individuals with the same form of enzyme are presumably genetically identical at that locus. Therefore, invariant gene loci can be found through a search for invariant enzymes. The most common technique for distinguishing different genetic forms of enzymes (allozymes) is gel electrophoresis. In this technique, specially prepared samples of specimen tissue are placed in a porous gel, and an electrical potential is applied, causing the electrically charged enzymes to migrate through the gel along the lines of electrical force. Because slightly different forms of the same amino acid differ in charge, they migrate at different rates and separate into bands at different distances from the original specimen. These bands become visible when the gel is first flooded with a substrate on which the enzyme acts and then stained with a substance that colors the reaction products. Such enzyme differences have proved to be of genetic origin, and experienced workers have identified much genetic variation among individuals in populations whose phenotypes looked identical. For example, as Photo 2.4 shows, morphologically identical strains of bacteria have proved to be distinctly different when examined electrophoretically.

The first assays of genetic variation by protein electrophoresis were published in 1966 by Harris, who reported on variation in ten enzyme loci in humans, and by Lewontin and Hubby (1966), working on *Drosophila*. The results were something of a surprise; no one had suspected just how much variation existed in nature. In *Drosophila* the average population was polymorphic at about 30 percent of its loci, with about two to six alleles per polymorphic locus. An average fly was likely to be heterozygous at 12 percent of its loci, so any two flies would, on average, differ at about 25 percent of their loci. In humans, Harris's data indicated that 30 percent of the loci were polymorphic and about 10 percent were heterozygous. Moreover, it has since been shown that the enzyme products of some alleles have similar electrophoretic mobility, and variation would not be distinguished by the gel technique (Coyne 1976; Coyne, Felton, and Lewontin 1978). Therefore, existing estimates of variation in nature could err on the low side.

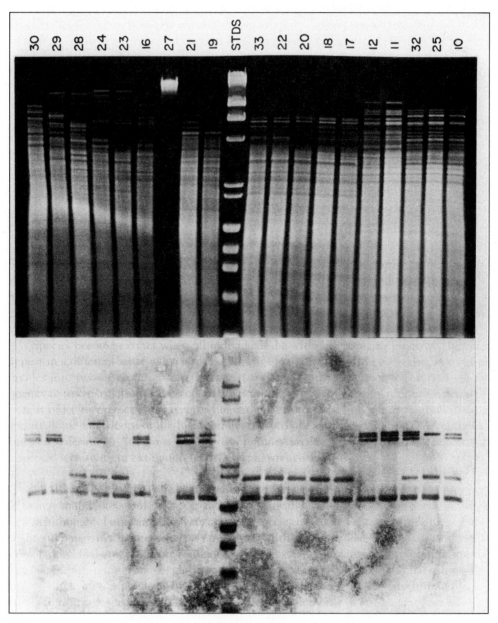

Photo 2.4  *Genetic variation within a species. The upper photo is DNA from 19 different cultures of* Xanthomonas maltophilia *taken from a Tallahassee, Florida, hospital in 1989. The lower photo is the same DNA, subjected to an additional treatment. All the cultures in each photo were treated in the same way (according to a method that produces a DNA "fingerprint"). Despite the superficial similarity of the cultures, their DNAs are clearly different. The additional techniques used in the lower photo show that even some of those that appear identical in the upper photo (for example, 10, 25, and 32) are unique.* (Photos courtesy of R.H. Reeves, Florida State University, Tallahassee.)

Genetic variation for a wide variety of taxa is summarized in Table 2.2. In general, vertebrates are less polymorphic than invertebrates; species that form small local populations or are inbred have reduced heterozygosity. It must always be noted, however, that some enzymes are not soluble and cannot be examined electrophoretically; whether the same degree of polymorphism exists in these is not known.

Very recently, development of DNA technology has provided us with a means to sample genetic variation directly at the DNA level. At present, however, these methods are more expensive and generally more difficult than allozyme techniques. Therefore, there are not yet the large amounts of data on species variation available from allozyme studies. This situation will probably change as the technology becomes less expensive and more widely available.

Because different parts of DNA evolve at different rates we can choose to study particular segments to answer particular questions. Some genes change very slowly and can be used to study relationships among groups of organisms that diverged from one another hundreds or even thousands of millions of years ago. Other regions of DNA change at such a rapid rate that most individuals in a population are distinct. Still other regions of DNA show intermediate levels of variability that are useful in studies of variation

**TABLE 2.2  *Genetic variation at allozyme loci in animals and plants. (After Selander 1976.)***

| | Number of species examined | Average number of loci per species | Average proportion of loci | |
|---|---|---|---|---|
| | | | Polymorphic per population | Heterozygous per individual |
| Insects | | | | |
| Drosophila | 28 | 24 | 0.529 | 0.150 |
| Others | 4 | 18 | 0.531 | 0.151 |
| Haplodiploid wasps[a] | 6 | 15 | 0.243 | 0.062 |
| Marine invertebrates | 9 | 26 | 0.587 | 0.147 |
| Marine snails | 5 | 17 | 0.175 | 0.083 |
| Land snails | 5 | 18 | 0.437 | 0.150 |
| Fish | 14 | 21 | 0.306 | 0.078 |
| Amphibians | 11 | 22 | 0.336 | 0.082 |
| Reptiles | 9 | 21 | 0.231 | 0.047 |
| Birds | 4 | 19 | 0.145 | 0.042 |
| Rodents | 26 | 26 | 0.202 | 0.054 |
| Large mammals[b] | 4 | 40 | 0.233 | 0.037 |
| Plants[c] | 8 | 8 | 0.464 | 0.170 |

[a] Females are diploid, males haploid.
[b] Human, chimpanzee, pigtailed macaque, and southern elephant seal.
[c] Predominantly outcrossing species.

within and between populations of a species or of variation between closely related species.

## Restriction Fragment Polymorphisms (RFPLs)

The DNA-based techniques most widely used for studies of within-and-between population variation make use of the properties of enzymes derived from various species of bacteria that use them to protect themselves from infection by viruses. The viruses are almost pure DNA, and they produce by inserting their own DNA into the bacteria DNA having the host replicate it, essentially producing more viruses. The cutting (restricting) enzymes of the bacteria are very specific in the DNA sequence. They recognize and cut foreign DNA as they attempt to cut out the viral portions. These enzymes form the background of the technology of DNA manipulation. If DNA from an individual is extracted and cut with a restriction enzyme and the resulting fragments are separated by length in an electrophoretic gel, a pattern is obtained. Other individuals might have slight changes in their DNA that produce additional or fewer sites that are recognized by the enzymes, and a different pattern of restriction fragments is seen on the gel. Thus, by repeating this process with many individuals and different restriction enzymes, patterns of variation can be seen and analyzed to estimate the amount of variation in the DNA sequences among the individuals.

## DNA Sequencing and PCR

A yet more powerful and more expensive method of assessing genetic variation is to sequence a portion of DNA itself. This has been made possible by the advent of the polymerase chain reaction technique (PCR), which can be used to make millions of copies of a particular region of DNA. Because of this, it is now possible to obtain DNA sequence data from a wide variety of organisms much more quickly than was possible previously. Because PCR techniques can amplify even minute amounts of DNA, we can obtain data from very small organisms that contain too little tissue to use with RFLPs. We can also use tiny samples of tissue from larger organisms without having to kill or otherwise injure them. For example, a drop of blood, a hair root, or a feather are now adequate material for DNA sequence-based work. This has obvious importance in dealing in rare and endangered species and is much in vogue in conservation ecology. The finest scale of analysis of genetic material is known as genetic fingerprinting and was discovered by Dr. Michael Jeffreys and colleagues in the 1980s at the University of Leicester in the United Kingdom (Jeffreys et al. 1985). Genetic fingerprinting makes use of a common, but peculiar DNA circumstance known as minisatellites. These are disposed throughout the genome and consist of randomly repeated copies of short sequence units. High levels of variation in the numbers of these repeated units are exploited in fingerprinting to identify close relatives. Since this discovery, the "genetic fingerprinting" of individuals has found a number of applications in human genetics, including forensic medicine and paternity analysis. Ecologists have also found it useful in determining kinship and relationships of family members in behavioral ecology studies.

Very often, allele frequencies differ at loci more from one population than they do between populations. Genetic diversity in a species thus consists of within-population diversity and among-population diversity (Fig. 2.4). A simple genetic model of this diversity is

$$H_t = H_p + D_{pt}$$

where $H_t$ = total genetic variation in the species, $H_p$ = average diversity within populations and $D_{pt}$ = average diversity among populations across the total species range (Nei 1975).

The point is that a species of little genetic variation may be partitioned into component parts: within- versus among-population diversity. Within that approach, one can determine how variation is spatially distributed and thus define areas of particular conservation interest. For example, Stangel et al. (1992) found that the endangered red-cockaded woodpecker (*Picoides borealis*) in the Southeastern United States had an overall mean allozyme heterozygosity level of 0.078 ( or 7.8 percent), which is about typical for most bird species that have been sampled (Photo 2.5). Of the total genetic variation measured, 14 percent consisted of among-population differentiation, and 86 percent was mean genetic diversity within populations. This is certainly true, but the 14 percent is relatively high for birds: most birds have widespread dispersal that results in high genetic exchange and little local genetic differentiation. The woodpeckers are more site-specific than many birds, and consequently local populations tend to diverge genetically. Conservation programs for this species should therefore protect both components of genetic diversity—among and between populations—in order to retain the maximum amount of variation and to maintain a natural population genetic structure. Studies of other organisms such as plants bear out the generationzation that within-population diversity is much greater than among-population diversity. A review of the available alloyzme literature from 449 species of plants showed that 78 percent of the diversity was found within populations (Hamrick and Godt 1990).

Variation among populations exists on much larger geographic scales as well. Commonly, the farther apart populations are, the more different they are in allele frequencies and in phenotype characteristics. A gradual change along a geographic transect is called a **cline.**

**Figure 2.4**   *Partitioning of total genetic diversity, $H_t$, into within- and among- population variation. This figure represents a species with three populations, each with some level of within-population heterozygosity ($H_1$, $H_2$, and $H_3$); mean heterozygosity is $H_p$. Among-population divergence ($D_{12}$, $D_{23}$, and $D_{13}$) is represented by the arrows between populations; mean divergence is $D_{pt}$.*

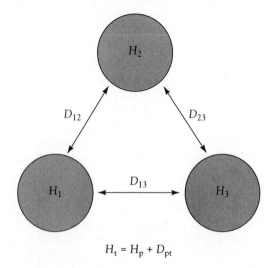

$$H_t = H_p + D_{pt}$$

Photo 2.5    *The endangered red-cockaded woodpecker about to enter its nest on a slash pine in Florida. A relatively high proportion of genetic variation is found between populations of this species, which means that in order to preserve genetic diversity, it will be necessary to conserve many areas and many populations.* (Lepore, Photo Researchers, Inc. 7X5584.)

Clines may extend over the whole geographic range of a species. For example, body size in white-tailed deer (*Odocoileus virginianus*) increases gradually with increasing latitude over most of North America—this phenomenon is common in mammals and birds and has been termed **Bergmann's rule.** In the clover *Trifolium repens*, (Photo 2.6) the proportion of plants

Photo 2.6    *White clover,* Trifolium repens, *can produce cyanide in its leaves. Freezing temperatures rupture cell membranes allowing the release of this toxin, killing the plant. For this reason, the production of cyanide increases in warmer climates where the likelihood of freezing decreases.* (Degginger, Animals Animals/ Earth Scenes.)

that produce cyanide increases in warmer locations in Europe (Fig. 2.5). This difference in frequency is caused by a balance between the advantage cyanogenic plants derive from being protected from herbivory and the disadvantage of being killed when frost disrupts the cell membrane, releasing toxins into the plant's tissues (Foulds and Grime 1972 a,b, Jones 1973, Dirzo and Harper 1982). Cyanogenesis occurs in more than 800 plant species spanning 100 plant families, including the rubber tree (Lieberei et al. 1989), so this is not a unique phenomenon. However, in some other species the distribution patterns are not influenced by environmental gradients (Compton et al. 1983), so a general explanation of the distribution of these plants is, as yet, lacking.

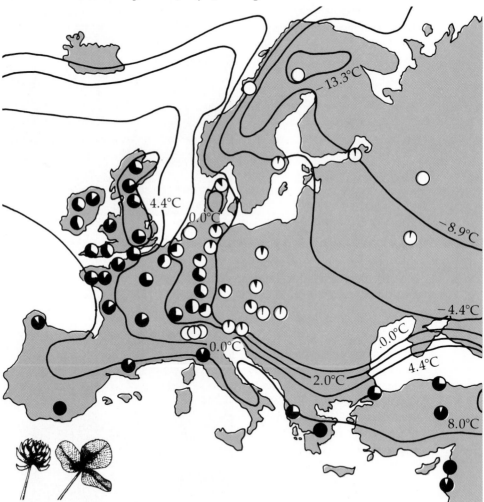

Figure 2.5   *Frequency of the cyanide-producing form in populations of white clover (Trifolium repens), represented by the black section of each circle. The cyanogenic form is more common in warmer regions. Lines are January isotherms.* (Redrawn from Daday 1954.)

## 2.5 REDUCTION IN VARIATION

Reduced genetic diversity can arise from four factors that are all a function of population size: inbreeding, genetic drift, neighborhoods, and bottlenecks.

### Inbreeding

The phenomenon of inbreeding depression has long been known. **Inbreeding** (mating among close relatives) is certainly a reality in many social animals (Chapter 7). One severe form is self-fertilization or "selfing" in plants. The effects are usually deleterious, and, as is dramatically illustrated in Figure 2.6, crossbreeding can reverse those effects. Viability and especially **fecundity** (or yield of crops) decline as populations become more inbred in the laboratory (Table 2.3). Generally, the more inbred the mating system, the quicker the proportion of heterozygosity in the population drops (Fig. 2.7). In human populations, the consequences include higher mortality, mental retardation, albinism, and other physical abnormalities (Stern 1973). The reason that inbreeding is so disadvantageous is that the frequency of homozygotes for recessive alleles is thought to increase. Recessive alleles are more likely to be deleterious for the simple reason that dominant deleterious traits are more quickly selected out of the population, leaving behind their recessive counterparts. Ralls and Ballou (1983) showed the effects of inbreeding on juvenile mortality in captive populations of mammals (Fig. 2.8). In ungulates, primates, and small mammals, there was higher mortality from inbred matings than from noninbred matings. Data from domesticated animals indicate that an

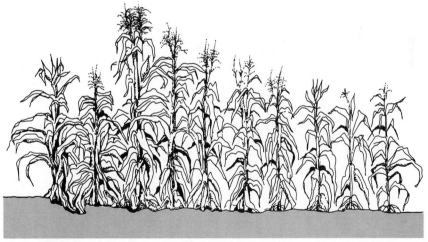

**Figure 2.6**    *Inbreeding depression and hybrid vigor. The two corn plants at the left are of two inbred strains; the larger plant to their right is a hybrid of the two. All the plants to the right are successive self-fertilized generations from the hybrid.* (Drawn after a photo by Jones 1924.)

**TABLE 2.3** *Inbreeding depression in rats, after Lerner (1954). The years 1887–1892 span about 30 generations of parent x offspring and sib matings.*

| Year | Nonproductive matings (percent) | Average litter size | Mortality from birth to four weeks (percent) |
|---|---|---|---|
| 1887 | 0 | 7.50 | 3.9 |
| 1888 | 2.6 | 7.14 | 4.4 |
| 1889 | 5.6 | 7.71 | 5.0 |
| 1890 | 17.4 | 6.58 | 8.7 |
| 1891 | 50.0 | 4.58 | 36.4 |
| 1892 | 41.2 | 3.20 | 45.5 |

expected increase of inbreeding per generation of 10 percent will result in a 5 to 10 percent decline in individual reproductive rates such as clutch size or survival rates; in aggregate, total reproductive attributes may decline by 25 percent (Frankel and Soule 1981).

Not all inbreeding is cause for alarm. Many natural populations have apparently experienced low levels of inbreeding for many generations with no ill effects. Inbreeding depression is probably more important in a species or population with historically large population sizes than now occurs in small populations. In historically small populations, such as those on islands, species have had to deal with relatively high rates of inbreeding for thousands of years.

The opposite, outbreeding, which increases the number of heterozygotes and conceals more of the recessive alleles, is termed **heterosis.** As well as inbreeding depression, we can get outbreeding depression. Often, local populations adapt to their regional

**Figure 2.7** *Decrease in heterozygosity due to inbreeding. Systems of mating are exclusive self-fertilization (curve A), sib mating (curve B), and double-first cousin mating (curve C).* (Redrawn from Crow and Kimura 1970.)

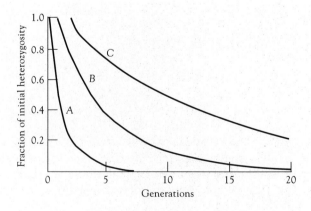

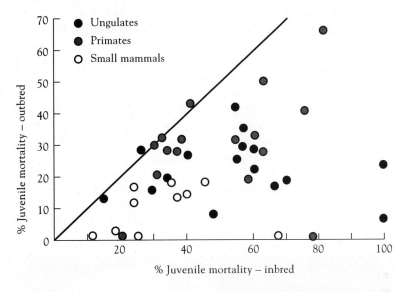

**Figure 2.8** *The effects of inbreeding on juvenile mortality in captive populations of mammals. Each point compares the percentage of juvenile mortality for offspring of inbred and noninbred matings. The line indicates equal levels of mortality under the two breeding schemes. Points above the line represent higher mortality from noninbred matings; points below the line, higher mortality from inbred matings. The distance of a point from the line indicates the strength of the effect of level of inbreeding.* (From Ralls and Ballou 1983)

environment, particularly if dispersal among populations is limited. Local adaptation, or the formation of demes, has been shown to occur for species of insect on long-lived plants (Karban 1989a). Species such as insects can undergo hundreds of generations on hosts such as trees or clonal bushes that live for hundreds of years. The insects can therefore remain one step ahead of the plants in the evolutionary arms race. Deme formation may also be more likely where limited gene flow occurs between demes—where the insects are brachypterous or without wings. However, in a review of ten studies testing the deme formation hypothesis (Karban 1992) only one study showed support for the existence of demes. Since that review, three additional studies have been performed with one showing lack of support for the deme hypothesis (Cobb and Whitham 1993) but two showing support for it (Mopper et al. 1995, Stiling and Rossi 1996). It is especially interesting that these last two studies involved mobile, winged insects, which underlies the fact that demes may form despite the presence of some gene flow. Hybridization between individuals from different local populations can result in individuals that are not adapted to any, or are to the wrong local environment. One of the best examples here is the ibex (*Capra ibex*) that became extinct locally in the Tatra Mountains of Czechoslovakia through overhunting (Photo 2.7). Ibex were then successfully transplanted from nearby Austria, which had a similar environment. However, years later some additional animals were imported from Turkey and the Sinai Peninsula, areas with a much warmer, drier climate. The introduced animals readily interbred with the Tatra Mountains herd, but the resulting hybrids rutted in the early fall instead of the winter (as the native Tatra and Austrian ibex did), and the resulting kids of the hybrids were born in February, the coldest month of they year. As a result, the entire population went extinct (Greig, 1979).

The effects of inbreeding are more extreme in small populations. The smaller the population (N), the faster heterozygosity declines (Fig. 2.9). This result has important consequences in the real world where animal populations are constantly declining because of shrinking habitats, and conservation science has become particularly concerned with the genetics of small populations. A rule of thumb has been that a population of at least 50 individuals is necessary to prevent inbreeding for the immediate future. One species of concern is the California condor, shown in Photo 2.8. It remains to be seen whether the offspring of 26 captive individuals—the sole remnants of the species in 1986—can overcome the effects of inbreeding and form a fit and

Photo 2.7    *The ibex,* Capra ibex, *exists as many races, each adapted to a different area in Europe and Africa. Shown here is the Nubian ibex. Threatened with extinction, some populations have been bolstered by the addition of individuals from another race. However, the resultant hybrids may not always survive well in their given climate, illustrating some of the problems inherent in hybridization.* (Dick, Animals Animals 455123-M.)

**Figure 2.9** *Decrease in heterozygosity due to finite population size. Variation is lost randomly through genetic drift. N equals population size.* (Modified from Strickberger 1986.)

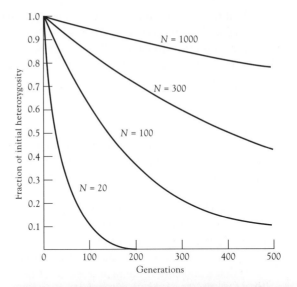

**Photo 2.8** *California condor. Because the total population of condors was only 26 birds in 1986, many conservationists warned that insufficient genetic diversity existed to ensure the continued success of the species.* (Ferrara, Photo Researchers, Inc. 2R7134.)

healthy population when released into the wild. As of 1993, 49 California condors had been bred in the San Diego Wild Animal Park and Los Angeles Zoo, and eight had been released back into the wild (Toone and Wallace 1994). One bird was killed by ethylene glycol, a toxin found in car antifreeze, but the other seven were doing well, feeding on provided carcasses.

## Genetic Drift

In small populations there is a good chance that some individuals will not mate and be represented in the next generation. Isolated populations will lose some percentage of their original diversity over time, approximately at the rate of 1\2N per generation. A population of 1,000 will retain 99.95 percent of its genetic diversity in a generation, while a population of 50 will retain only 99.0 percent. Such losses seem insignificant but are magnified over many generations. Thus, after 20 generations a population of 1,000 will still retain more than 99 percent of its original variation, but the population of 50 will retain less than 82 percent. For animals that breed annually, this could mean a substantial loss in variation over a very few years. Once again, this effect becomes more severe as the population size decreases. Again, a rule of thumb is that a population size of at least 500 is necessary to lessen the effects of genetic drift. Thus the "50/500" rule has entered the conservation literature as a "magic" number in conservation theory (Simberloff 1988) (50 being the critical size to prevent excess inbreeding and 500 the critical size to prevent genetic drift [Franklin 1980]). Lacy (1987) showed that the effects of genetic drift could be countered by immigration of individuals into a population. Even relatively low rates of immigration of one immigrant every generation would be sufficient to counter genetic drift in a population of 120 individuals (Fig. 2.10). One of the best tests of the

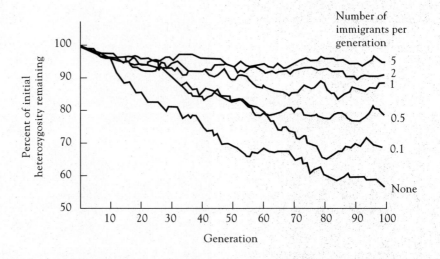

Figure 2.10   *The effect of immigration on genetic variability in 25 simulated populations of 120 individuals each. Even low rates of immigration (1 immigrant per generation) can prevent the loss of heterozygosity from genetic drift.* (After Lacy 1987.)

idea that 50 individuals may be a minimum number for conservation purposes was provided by Berger's (1990) study of 120 bighorn sheep (*Ovis canadensis*) populations in the U.S. Southwest. The striking observation was that 100 percent of the populations with fewer than 50 individuals went extinct within 50 years, while virtually all of the populations with more than 100 individuals persisted for this time period (Fig. 2.11). The exact causes of extinction were difficult to pinpoint and undoubtedly included disease, predation and starvation as well as inbreeding and genetic drift.

Genetic drift has often been argued to apply more to neutral alleles (those that do not differ in their effect on survival or reproduction) than on other alleles. Whether or not many alleles conform to this assumption is a subject of considerable controversy. Kimura (1983a,b) and Nei (1983) hold that much molecular variation in allozymes, DNA, and proteins is neutral and that any divergence detected among species is likely to be the result of genetic drift.

An interesting aside is that population theory predicts that loss of genetic variation by genetic drift is not as important as loss of rare alleles from the population. Rare alleles may contribute relatively little to overall genetic variation, but they may be important to a population during infrequent or periodic events, such as temperature extremes or exposure to new parasites or new pathogens, and can offer unique responses to future evolutionary changes. The analogy here is that global climate change may not affect populations very much, but the frequency of extreme events such as hurricanes, extreme drought in the summer, or freezing temperatures in

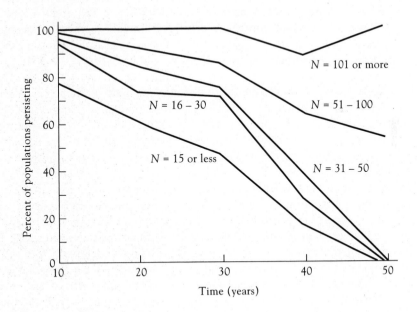

Figure 2.11   *The relationship between the size of a population of bighorn sheep and the percentage of populations that persist over time. The numbers on the graph indicate population size (N); populations with more than 100 sheep almost all persisted beyond 50 years, while populations with fewer than 50 individuals died out within 50 years.*  (After Berger 1990)

the winter could well be the straw that breaks the camel's back for many species. So it is with lack of diversity. The lack of diversity may not be as big a problem as loss of rare alleles. Rare alleles are lost much more rapidly from small populations, even though much of the overall genetic diversity is retained (Denniston, 1978).

## Neighborhoods

Even in large populations, the effective population size may actually be quite small because individuals only mate within a neighborhood. The number of individuals in a neighborhood is given by $4\pi s^2 D$, where $D$ is the **density** of the population and s is the standard deviation of the distances between the birth sites of individuals and the birth sites of their offspring. By marking individual deer mice (*Peromyscus maniculatus*), Howard (1949) showed that at least 70 percent of males and 85 percent of females breed within 150 meters of their birthplaces. Even migratory species, such as birds, usually return to the vicinity of their birthplaces to breed. Barrowclough (1980) has also shown that, for noncolonial birds such as wrens and finches, local effective population sizes range from 175 to 7,700 individuals, and Levin (1981) has argued that, despite seed and pollen dispersal, effective size of local populations for plants is often in the dozens or hundreds.

Furthermore, even within a neighborhood, some individuals do not reproduce. If only half the individuals in a population breed, half the heterozygosity is lost per generation. In a population of 50, if only half the members breed, then the population has an effective size $N_E$ of 25. In territorial species and those with harems, $N_E$ is even lower because the few males that reproduce contribute disproportionately to the subsequent generations and genetic drift is inflated. If a population consists of $N_m$ breeding males and $N_f$ breeding females, the effective population size is $N_E = (4\,N_m N_f) / (N_m + N_f)$. In a population of 500, a 50:50 sex ratio and all individuals breeding, $N_E = (4 \times 250 \times 250) / (250 + 250) = 500$. However, if 450 females bred with 50 males $N_E = 180$ or 36 percent of the actual population size. Many species have mating structures that greatly skew $N_E$. A knowledge of effective population size is vital to ensure the success of many conservation projects. Notable among these are the reserves designed to protect grizzly bear populations in the contiguous 48 states of the United States. The grizzly bear (*Ursus arctos horribilis*) has declined in numbers from an estimated 100,000 in 1800 to less than 1,000 at present (Photo 2.9). The range of the species is now less than 1 percent of its historical range and is restricted to six separate populations in four states. Some computer simulations of population size suggested a target of less than 100 individuals in a population. However, research by Fred Allendorf of the University of Montana (Allendorf 1994) indicated that populations of 100 would likely lose heterozygosity because effective population sizes are only about 25 percent of actual population sizes. Thus, even fairly large, isolated populations such as the 200 bears in Yellowstone National Park are vulnerable to harmful effects of loss of genetic variation. Allendorf and his colleagues proposed that an exchange of bears between populations or even zoo collections would help tremendously in promoting genetic variation. Even an exchange of two bears between populations would greatly reduce the loss of genetic variation. Allendorf was left to conclude that viable populations could only be maintained by maintaining large reserves, by promoting artificial gene flow among reserves, and by promoting gene flow between zoos and reserves. A combination of various techniques would thus become necessary for the grizzly bear and probably for the preservation of many large animals in the United States.

Photo 2.9   *Because not all individuals in a population of grizzly bears breed, effective population sizes may be much less than real population sizes. An exchange of bears between populations, or even the addition of bears from zoos, might do much to alleviate the problems of inbreeding.* (Leeson, DRK Photos 221761.)

## Bottlenecks

Effective population sizes can be further lowered if populations vary in size from generation to generation. The effective size $N_E$ in this instance is estimated by the harmonic mean-population size

$$\frac{1}{N} = \left(\frac{1}{t}\right) \sum_{i=1}^{t} \left(\frac{1}{N_i}\right)$$

Thus, if a population goes through five generations of size 100, 150, 25, 150, and 125 individuals, $N_E$ is about 70, as opposed to the arithmetic mean of 110. $N_E$ is more strongly affected by lower than by higher population sizes. The actual magnitude of the loss of genetic diversity in a bottleneck is about 50 percent when $N_E$ is reduced to 1 but is virtually zero percent with a large $N_E$ of 1,000. This assumes the bottleneck is only maintained for one generation. If the bottleneck is maintained for longer and the population growth of the population afterwards is low, some additional genetic variation is lost. However, a bottleneck rarely has a severe genetic or fitness consequence if population size quickly recovers in a generation or two. It is a surprising fact that the gametes of just one individual can carry, on average, 50 percent of the genetic diversity of the population.

The northern elephant seal (*Mirounga angustirostris*) went through a severe bottleneck in the 1890s when its numbers were reduced to only about 20 by hunting. Remarkably, it recovered, and more than 30,000 are alive today. Because of the harem mating system, the effective population size must have been less than 20. Bonnell and Selander (1974) were able to demonstrate by electrophoresis that there is no genetic variation among today's northern elephant seals in a sample of 24 loci, whereas in the southern elephant seal (*M. leonina*) normal genetic variation exists. Southern elephant seals were never drastically reduced in numbers.

In many colonizing species, a bottleneck occurs when a new habitat is colonized for the first time. This type of bottleneck is known as the **founder effect.** In dispersing species, for example, it is not unusual for only one or two members of a population to make it successfully to an island. A colony founded by a pair of diploid individuals can have at most four alleles at a locus, although there may be many more in the population from which they came. Bottlenecks and founder effects are, therefore, common occurrences for weeds and pests and should not always be regarded as universally deleterious.

The main conclusion is that, despite limitations by bottlenecks and neighborhoods, much variation exists in nature. The question of whether much of it is evolutionarily meaningful or is maintained by selection is open to debate (Lewontin 1974; Koehn, Zera, and Hall 1983). Probably the answer is that variation is neutral on prevailing genetic backgrounds in many environments but affects fitness on different genetic backgrounds in some environments. The cases in which it does are examples of natural selection operating in nature.

## *Summary*

 1. The distribution patterns of plants and animals on Earth depend not only on contemporary factors such as climate, habitat quality, and the presence or absence of competitors, predators, and diseases, but also on the evolutionary past. It is valuable to know where species originally evolved and whether or not they have been able to disperse to areas outside of this center of origin.

2. Variation in nature results from gene and chromosome mutations. A knowledge of how much of this variation is maintained within and between populations is vital to conservation programs. If little variation occurs within a population and large variation exists between populations, then many populations will need to be conserved to preserve this variation.

3. How big do populations have to be to prevent inbreeding or genetic drift? The 50/500 rule suggests that 50 is the critical size to prevent inbreeding, and 500 is the critical size to prevent genetic drift.

4. Is it feasible for humans to move individuals from one wild population to another to simulate natural dispersal? Some models suggest that moving even one or two individuals per generation would counter genetic drift.

*Discussion Question:* Why are there so many different species of plants and animals? Why aren't there fewer?

# Natural Selection and Speciation

Natural selection, the backbone of Darwin's theories on evolution and the mechanism that can explain how evolution occurs, is in essence the differential survival and reproduction of some individuals in a population and the death without issue (or with fewer issue) of others. Important though random changes and genetic drift are, a more active process, **natural selection,** is best invoked to explain many natural features of evolution in the real world. The feet of different types of birds are features that fit them to their ways of life and are best explained by natural selection. Never, however, is there a predestined form to which plants or animals are shaped as they evolve (Dawkins 1986); as the environment changes so usually do the organisms that live in it—no moral or ethical forces impinge on natural selection.

## 3.1 NATURAL SELECTION

Theoretically, natural selection can operate whenever different kinds of self-reproducing entities differ in survival or reproduction. Sometimes, the genes themselves have been regarded as these entities (*selfish gene theory*, Dawkins 1989), but more commonly individual organisms are regarded as the units (see Sober and Lewontin [1982] for a critique of the selfish gene theory). Natural selection, then, occurs when genotypes differ in fitness.

## Industrial Melanism: The Case of the Peppered Moth

One of the best-analyzed examples of natural selection of genotypes in operation is the change in color that has taken place in certain populations of the peppered moth, *Biston betularia,* in the industrial regions of Europe during the past 100 years. Originally, moths were uniformly pale grey or whitish in color; dark-colored or melanic individuals were rare and made up less than 2 percent of the total individuals (Kettlewell 1973; Bishop and Cook 1980). Gradually, the dark-colored forms came to dominate the populations of certain areas—especially those of extreme industrialization such as the Ruhr Valley of Germany and the Midlands of England. Genetic tests showed that the dark allele was dominant and that crosses of dark with pale individuals produced typical Mendelian segregation. Pollution did not directly affect mutation rates. For example, caterpillars who feed on soot-covered leaves did not give rise to dark-colored adults. Rather, it promoted the survival of dark morphs on soot-covered trees. Melanics were normally quickly eliminated in nonindustrial areas by adverse selection; birds found them conspicuous. This phenomenon, an increase in the frequency of dark-colored mutants (carbonaria forms) in polluted areas, is known as **industrial melanism.** The North American equivalent of this story is the *swettaria* form of *Biston cognataria* which showed up in industrialized areas such as Philadelphia, New Jersey, Chicago, and New York in the early 1900s. By 1961, it constituted more than 90 percent of the population in parts of Michigan (Owen 1961, 1962).

The operation of natural selection on the peppered moth was illustrated by Professor H.B.D. Kettlewell of Oxford University. He argued that normal pale forms are cryptic when resting on lichen-covered trees, whereas dark forms are conspicuous. In industrialized areas, lichens are killed off, tree barks become darker, and the dark moths are the cryptic ones. Figure 3.1 illustrates the two forms of *Biston betularia.*

Kettlewell suspected that birds were the selecting force, and he set out to prove it by releasing thousands of moths marked with a small spot of paint into urban and industrialized areas (Kettlewell 1955). In the nonindustrial area of Dorset, he recaptured 14.6 percent of the pale morphs released but only 4.7 percent of the dark moths. In the industrial area of Birmingham, the situation was reversed: 13 percent of pale morphs but 27.5 percent of dark morphs were recaptured. Birds were clearly implicated in differential predation, eating more pale morphs in industrial habitats and more dark morphs in nonindustrial areas. As a test of his field observations, Kettlewell and companions set up blinds and watched birds vora-

(a)                                                              (b)

Figure 3.1   *Industrial melanism. (a) Typical (light-colored) form of* Biston betularia. *(b)* Carbonaria *(dark-colored) form of* Biston betularia. (Drawn from photo by Kettlewell 1955.)

ciously gobble up moths placed on tree trunks. The action of natural selection in producing a small but highly significant step of evolution was seemingly demonstrated.

However, the black form has not become fixed even in the most industrial of locales, so other factors may play roles in the maintenance of melanic frequencies (Lees 1981). Furthermore, many authors have pointed out inconsistencies in the peppered-moth story (summarized in Berry 1990). Kettlewell was a general medical practitioner who took up a fellowship at Oxford University at the age of 45 years. He was "the best naturalist and almost the worst professional scientist," making rapid diagnoses and refusing to be diverted by what he regarded as irrelevant evidence (Berry 1988).

Various authors have, at one time or another argued for:

a. a physiological advantage of *carbonaria* even in rural areas (Lees and Creed 1975).
b. a higher dispersal rate of *carbonaria* which influences correlations of pollution levels with melanic frequencies (Brakefield, 1990).
c. a failure of adult moths to select an appropriately camouflaged background (Grant and Howlett 1988; Liebet and Brakefield 1987).

There is clearly more to melanism than meets the eye.

Interestingly enough, the white form of the peppered moth is making a strong comeback. In Britain, a Clean Air Act was passed in 1965. Sir Cyril Clarke has been trapping moths at his home on Merseyside, Liverpool, since 1959. Before about 1975, 90 percent of the moths were dark, but since then there has been a steep decline in *carbonaria* forms, and in 1989 only 29.6 percent of the moths caught were melanic (Fig. 3.2a) (Clarke, Clarke, and Dawkins 1990). The mean concentration of sulphur dioxide pollution fell from about 300 $\mu g$ m$^{-3}$ in 1970 to less than 50 $\mu g$ m$^{-3}$ in 1975 and has remained fairly constant since then (Fig. 3.2b). If the spread of the *typica* (light-colored) form of the moth continues at the same speed as *carbonaria* spread in the last century, then the melanic form will again be only an occasional mutant in the Liverpool area by the year 2010.

Although it appears that peppered moths may be a reasonable indicator species of high or low levels of environmental pollution, it is disconcerting that the numbers of dark morphs only decreased (1978) when pollutant levels had been at a low value for a number of years.

The case of the peppered moth notwithstanding, much evidence still shows that many alleles or genotypes vary in fitness according to their environment. There is no automatic "better" genotype (remember the genotype x environment interactions discussed in Chapter 2). Other cases of industrial melanism are known (Bishop and Cook 1981), and other examples of rapid evolutionary change have become apparent as, for example, more and more pests become resistant to insecticides and more and more bacteria become resistant to antibiotics. Antibiotics are chemicals produced by organisms, generally fungi or bacteria, that are inhibitory to the growth of other microorganisms. Although antibiotics have been known since 1928 when Sir Alexander Fleming in England isolated penicillin from the mold *Penicillium*, the industry did not undergo rapid expansion until after the Second World War. Today, several thousand types of antibiotics and their chemical derivatives are known, and use of antibiotics has revolutionized the treatment of infectious disease. Indeed, antibiotics have been misused in medicine by their administration to patients with nonspecific ailments or virus-caused conditions such as the common cold or influenza. Misuse has been particularly brazen in animal husbandry, where antibiotics have been incorporated into livestock feed as "preventatives" against pathogens. As might be expected, the upshot of the continuous presence of antibiotics in the

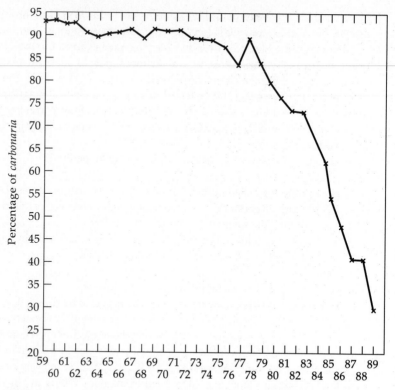

**Figure 3.2a** *Decline in the proportion of form carbonaria in West Kirby. The total number of B. betularia trapped in the 31-year period was 14,882.* (After Clarke et al. 1990.)

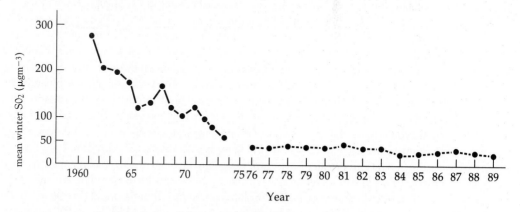

**Figure 3.2b** *Sulphur dioxide concentrations in West Kirby (1976-1989) and near West Kirby (1961-1974).* (After Clarke et al 1990.)

bacterial environment has been the evolution of antibiotic-resistant strains. This same thing has happened in the pest control industry. Growers were encouraged by chemical companies to use sprays on their crops as a preventative measure. Eventually, the insect pests became resistant, and more sprays of higher toxicity had to be used against them. This in turn led to the development of still more resistant types of pests spawning the term *pesticide treadmill*.

Even such a normally harmful allele as sickle-cell hemoglobin (which causes severe anemia) can be advantageous in areas where malaria is prevalent because heterozygous individuals, who carry the allele but are not severely affected by the anemia, are more resistant to malaria than are normal individuals. This mechanism for the maintenance of a polymorphism is known as *heterozygous advantage*. In the absence of malaria, the sickle-cell trait is quickly lost. In Norway rats, the allele for resistance to the pesticide warfarin lowers the animals' ability to synthesize vitamin K. Resistant varieties are thought to be at a 54-percent disadvantage to wild types in nature (Bishop 1981), but the allele is maintained in the population by the advantage it gives individuals that encounter warfarin. Finally, trappers in Canada preferred the pelts of silver foxes over red foxes (Elton 1942). Because of this, the proportion of silver foxes in the catch declined from 15 percent in 1834 to only 5 percent in 1933.

## Balanced Polymorphism

The unstable existence of two or more morphs in a population is called a transitional or directional **polymorphism;** one allele is replacing another in the population. In many cases, however, one allele does not completely replace the other, and the stable intermediate frequency is called a **balanced polymorphism.** An example can be seen in the land snail *Cepaea nemoralis* in which six alleles affect the color of the shell, which can be several shades of brown, pink, and yellow. (See Fig. 3.3.)

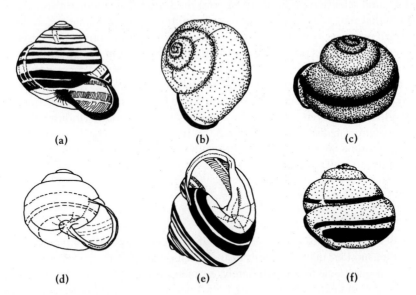

(a)                     (b)                     (c)

(d)                     (e)                     (f)

Figure 3.3    *The highly polymorphic land snail* Cepaea nemoralis. *(a) A yellow shell with five bands and a dark lip at the mouth of the shell. (b) A pink shell with no bands and a dark lip. (c) A brown shell with only the central band present. (d) A yellow shell with the bands present but unpigmented, making them translucent (the lip is also unpigmented). (e) A yellow shell with pigmented bands but an unpigmented lip. (f) A pink shell with the first two bands missing so that it has only the central and two lower bands present.*

The relative abundances of the various forms differ, even between localities less than a mile apart, and fossils show that this polymorphism has persisted at least since the Pleiostocene epoch (Diver 1929). This European snail is common in a variety of habitats, woods, meadows, and hedgerows, and the maintenance of the polymorphism has again been shown to be due to bird-predation pressure, this time by the song thrush, *Turdus ericetorum* (Cain and Sheppard 1954a). Thrushes hunt by sight, and shell color plays the central role in the concealing coloration of these snails. The birds break open shells on suitable stones, or anvils, providing the experimenter with an ideal opportunity to compare the proportion of varieties in a colony with those taken from it by song thrushes. In beech woods, the leaf litter is red-brown, and the snails are brown and pink. In grassland habitats, red and brown forms are rare, and yellow forms predominate. The story is somewhat complicated by the fact that genotypes differ in their susceptibility to extreme temperatures (Jones, Leith, and Rawlings 1977), and yellow shells may also be more common in colder areas (Cameron 1992). In addition, the snails have a complicated pattern of thin black bands. This characteristic is again linked to habitat type, as unbanded snails are cryptic in uniform habitats and banded ones are cryptic in "rough" habitats (as are zebras) (Cain and Sheppard 1954a). Interestingly, an African land snail exhibits a similar variation in shell color and banding patterns (Owen 1966), so the phenomenon is likely to be of wide ecological occurrence. Similar predator-driven polymorphisms may exist in insects too (Stiling 1980).

The peppered-moth and land-snail examples show, fairly conclusively, the operation of natural selection on populations and the survival of the fittest individuals. They also show that natural selection can operate in more than one way; it can drive populations toward one type of morph or act to maintain more than one. Often it is difficult to tell whether variations on a theme in nature represent truly different species or merely different morphs of the same thing. Many different morphs of species are so different that they have been categorized as different species. In general, three types of natural selection are recognized: stabilizing, in which one morph is favored; directional, in which the population is driven to exhibit a different type of morph (industrial melanisms); and diversifying or disruptive, in which, as in the case of the land snail, more than one morph can be favored (Fig. 3.4). Presumably, when disruptive selection is strong enough, speciation results.

## 3.2  SPECIATION

Of all the smaller taxonomic groups specified, races, subspecies, biotypes, species, genera, and families, most are quite arbitrarily defined (Wilson and Brown 1953; Futuyma 1986), and their value is questionable. Unfortunately, estimates of the rate of evolution are often based on the origin and extinction of families or other taxa, so some knowledge of higher systematics is essential. For many biologists, however, one taxonomic category is considered real and nonarbitrary—the **species.**

In any discussion of speciation, it is valuable to have a working concept of species. Perhaps the best known is that of Mayr (1942): "groups of populations that can actually or

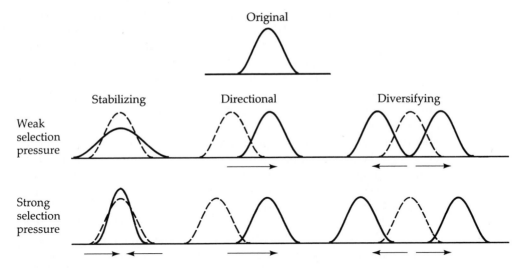

Figure 3.4   *Effects of stabilizing, directional, and diversifying selection upon variation for a quantitative character.*

potentially exchange genes with one another and that are reproductively isolated from other such groups." More recently, it has been noted that for many species with widely separate ranges we have no idea if the reproductive isolation is by distance only or whether there is some isolating mechanism (Donoghue 1985). Thus our judgment of a species is largely based on a surrogate method: morphological criteria. Also bear in mind that asexually reproducing organisms are given names by taxonomists and that paleontologists recognize different temporal portions of the same lineage as distinct—for example, *Homo erectus* and *H. sapiens*—so the above quotation is but a good working definition and is not watertight.

It is clear that the biological species concept is not universally accepted (Levin 1979; Rojas 1992). However, before we can come to the conclusion that a definition of a biological species is only of interest to academics, consider that the definition of species may be critical in conservation. Adoption of different species concepts could alter the way we define and conserve species. This could have a real affect on conservation of biodiversity. If we accept Mayr's definition of species, commonly called the **biological species concept,** or BSC, then present estimates vary from about 5 to 30 million species in the world today (Wilson 1988). Another popular definition of species today is the **phylogenetic species concept** (PSC). This concept depends on the branching, or cladistic, relationships among species or higher taxa. The PSC definition is based on the concept of shared derived characters (Cracraft 1983). Adopting PSC would result in more recognized species than at present (McKitrick and Zink 1988). Many currently recognized species or distinct populations would be elevated to species status

(Photo 3.1 and Fig. 3.5). Populations of species not presently endangered because they are widespread and abundant at some localities might attain endangered status as new species, if split under the PSC system. We should note, however, that the U.S. endangered species law has provision to recognize certain populations of vertebrates as rare even if the remainder of the species is secure (as in the cases of the southern bald eagle and the Florida panther), so this change would only affect plants and invertebrates. It is quite plausible that raising the number of species that are endangered could create even more problems than it solves. There would perhaps be strong public and political backlash against their legal protection. It is also possible that restoration efforts might be hampered because scientists might be more unwilling to move individuals from populations to other populations which they are trying to recover.

Also, Mayr's definition of a species can result in an overly optimistic view of global biodiversity. This is because a species can continue to exist even if many of its populations are destroyed. Those lost populations might represent a decline in biodiversity if they contained unique, genetic traits. A species approach could not tell us that diversity has been lost because the species count remains unchanged. However, using a PSC approach to defining species would reduce this problem. Using the PSC approach, it is clear that we have been losing much more global diversity in the form of populations than is often recognized (Fig. 3.6).

Photo 3.1   *The black racer,* Coluber constrictor, *exists as many different color forms or races throughout the U.S. Shown here is the Northern black racer. Some authorities argue that each race should actually be elevated to the status of a species.*(Z. Leszczynski, Animals, Animals).

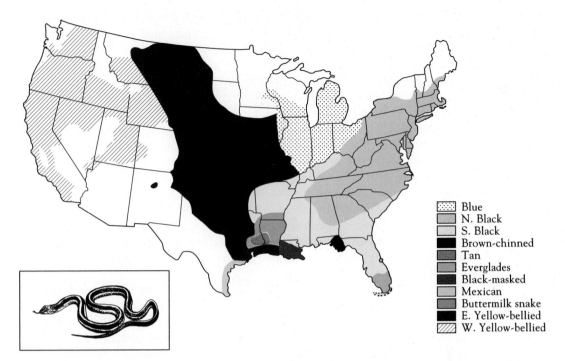

Blue
N. Black
S. Black
Brown-chinned
Tan
Everglades
Black-masked
Mexican
Buttermilk snake
E. Yellow-bellied
W. Yellow-bellied

**Figure 3.5** *Distribution of the various named subspecies of the racer* (Coluber constrictor), *a common snake of North America. Each of the subspecies would probably be designated a full species under the PSC approach (see text).* (Modified from Conant 1975.)

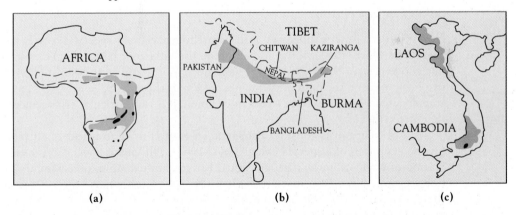

**Figure 3.6** *Present and former distributions of three species of Rhinoceros, showing loss of populational diversity and retreat to a few refuges. The black rhinoceros* (Diceros bicornis), *showing historical distribution (black outline), distribution in 1900 (shaded area), and distribution in 1987 (black areas). The greater one-horned rhinoceros* (Rhinoceros unicornis), *formerly distributed across the shaded area, is now reduced to two populations at Chitwan and Kaziranga reserves. The Javan rhinoceros* (Rhinoceros sondaicus), *showing historical (shaded) and present (black) distributions.* (After G.K. Meffe and C.R.Carroll 1994.)

Mayr was an ornithologist, and his definition seemed to fit birds very well—breeding groups are usually clearly delineated. However, especially in plants, fertile hybrids often form between species that greatly blur species distinctions. Oak trees provide a particularly good example of confusion in species definitions. Oaks often form reproductively viable hybrid populations. That is, oaks from different species interbreed, and their offspring are themselves viable, capable of reproducing with other oaks. For this reason, one might question whether the parental species should be called *species* at all. For example, *Quercus alba* and *Q. stillata* form natural hybrids with 11 other oak species in the eastern United States. It could be asked why, if these oaks cannot tell each other apart, biologists should impose different species names on them. However, Davis (1983) has noted that many examples of viable hybrid formation in nature occur when historically isolated species are brought into contact through climate change, landscape transformation, or transport by humans beyond their historic boundaries. Oaks were absent from the eastern deciduous forests during most of the Pleistocene and in many places their cooccurrence is no older than one hundred generations. There are numerous other examples of fertile hybrids forming when once-isolated species of plants or animals are rapidly thrust into contact.

Calls have often been made to the U.S. Congress to revise the endangered species act to make only full species subject to protection. Protecting subspecies and populations, opponents argue, causes undue economic strain. It is interesting to note that a survey or 492 listings or proposed listings of plants and animals from 1985 to 1991 indicated that 80 percent were full species, only 18 percent were subspecies, and just 2 percent were distinct populations (Wilcove et al. 1993). At the present time, species-level taxonomy is still driving endangered-species legislation.

## *Allopatric Speciation*

How much of a genetic difference is necessary to make a new species? Nobody knows. To paraphrase Mayr (1963), to try to determine the difference between species in terms of nucleotide pairs of DNA would be absurd, much as it would be to compare how different two books were on the basis of the number of letters they use. Species have many genes in common, much as books have words in common; it is the particular arrangement of each that is so critical.

Most evolutionists consider **allopatric speciation** to be the most likely mechanism for the evolution of a species (Mayr 1942, 1963). Allopatric speciation involves separation of populations by a geographical barrier. For example, nonswimming populations separated by a river may gradually diverge because there is no gene flow between them. Alternatively, the upthrusting of mountains often divides populations into many units, among which speciation then proceeds. In an area only 20 × 5 miles on Hawaii, 26 subspecies of land snail, *Achatinella mustelina*, have been recognized, each in a different valley separated from the others by mountain ridges. Some of the best-known instances of divergence among isolated populations are on islands, where a species that is rather homogeneous over its continental range may diverge spectacularly from the continental form in appearance, ecology, and behavior. Darwin's finches have speciated extensively within the Galápagos Archipelago. Could habitat fragmentation from environmental development promote biodiversity by increasing allopatric speciation? Probably not; local extinctions because of resultant small population changes are more likely.

In many situations, the ranges of species that have evolved allopatrically come to overlap, and the species become sympatric once again. The isolating mechanisms between the species, however, have usually continued to evolve separately, and as a result the males and females of the different species are likely to have become incompatible. These isolating mechanisms are a result of divergence, not a cause. Some important isolating mechanisms are outlined in Table 3.1. Among birds and fish, species-specific male coloration seems to be an important isolating mechanism, whereas in insects it is often smell (moths), sound (grasshoppers), or even the correct flight path and flash patterns of lights (fireflies); in frogs and toads, chorusing is important.

## Sympatric Speciation

The alternative to allopatric speciation is **sympatric speciation,** the appearance of new species in an area not geographically separated from other members of the population. Most models of sympatric speciation are highly controversial. Even if some members of a population began to inhabit slightly different parts of a population's range, gene flow between the two groups would probably still be sufficient to prevent speciation. Any slight morphological or behavioral changes in one group would be conveyed to the other. If a single mutation or chromosomal change were to confer complete reproductive isolation in one fell swoop, its bearer would be reproductively isolated. Given the unlikely scenario of the same mutations occurring at the same time in another individual of the opposite sex, and given that these two individuals could find each other, a new species might form. Close inbreeding (self-fertilization or

**TABLE 3.1** *Summary of the most important isolating mechanisms that separate species of organisms.*

A. Prezygotic mechanisms. Fertilization and zygote formation are prevented.

1. Habitat. The populations live in the same regions but occupy different habitats; e.g., different spadefoot-toad species occupy different soil types (Wasserman 1957).

2. Seasonal or temporal. The populations exist in the same regions but are sexually mature at different times, e.g., flowers that bloom in different months (Grant and Grant 1964) or fireflies that mate at different times of the night (Lloyd 1975).

3. Ethological (animals only). The populations are isolated by different and incompatible behavior before mating, e.g., courtship songs of birds or frogs, flash patterns of fireflies.

4. Mechanical. Cross fertilization is prevented or restricted by differences in structure of reproductive structures (e.g., genitalia in animals, flowers in plants).

B. Postzygotic mechanisms. Fertilization takes place, and hybrid zygotes are formed, but these are inviable or give rise to weak or sterile hybrids.

1. Hybrid inviability or weakness. Hybrids of the frogs *Rana pipiens* and *R. sylvatica* do not develop beyond the gastrula stage (Moore 1961).

mating with sibs) may promote the likelihood of such events, and thus sympatric speciation has been proposed for some insect groups such as the Chalcidoidea, parasitic
Hymenoptera in which many individuals develop from a single egg (a process called
*polyembryony*) laid in a host (Askew 1968). Indeed, because insects are themselves so
specious, comprising an estimated 30 million species (Erwin 1982), it is sometimes difficult to imagine allopatric speciation in this order all by means of geographic barriers,
especially given the dispersal abilities of insects (Johnson 1969). In many groups of
insects, closely related species are restricted to different host plants, and it has been
argued that sympatric speciation has occurred there (Bush 1975a,b; Wood and Guttman
1983; Bush 1994). However, in the few cases that have been analyzed, host preference
in insects appears to be controlled by many genes, a situation that would not be favorable to a quick, one-genetic-step method of sympatric speciation (Futuyma and Peterson 1985). Furthermore, even when a species utilizes a new host plant to feed on, it
often still interbreeds with individuals reared on other host plants. Thomas et al.
(1987) showed that the butterfly *Euphydryas editha* has extended its range onto a new
plant, *Lantago lanceolata*, within the last 100 years but that populations still interbreed
with individuals reared on the old host, *Collinsia parviflora*.

How rapidly does speciation occur? The answer, of course, differs for different organisms.
J.B.S. Haldane theorized that species of vertebrates might differ at a minimum of 1,000 loci
and that at least 300,000 generations would be necessary for the formation of new species.
Indeed a great many of the populations isolated for thousands of generations by the
Pleistocene glaciations did not achieve full species status. American and Eurasian sycamore
trees (*Platanus* sp.) have been isolated for at least 20 million years, yet still form fertile hybrids (Stebbins 1950). The selective forces on these two continents have obviously not been
sufficient to cause reproductive isolation between these ecologically general species.
However, several genera of mammals—for example, polar bears (*Thalarctos*) and voles
(*Microtus*)—do appear to have originated relatively recently, in the Pleistocene (Stanley
1979). Many of the Hawaiian species of *Drosophila* flies have arisen in just a few thousand
years, although their generation time is, of course, much shorter than that of mammals. Lake
Nabugabo in Africa has been isolated from Lake Victoria for less than 4,000 years; yet it contains five endemic species of fish (Fryer and Iles 1972). Lake Victoria itself is only
500,000–750,000 years old but harbors about 170 species of cichlid. Again in Hawaii, at least
five species of *Hedylepta* moth feed exclusively on bananas, which were only introduced by
the Polynesians some 1,000 years ago.

## 3.3  RATES OF FORMATION OF NEW SPECIES AND EXTINCTION OF OLD ONES

What are the rates of formation of new species and the rates of extinction of old ones? If
we knew this, we could use the information possibly to decrease current rates of extinction. Usually, the best patterns issue from the fossil record. Marine invertebrates have left
the best fossil records and have been the most intensely analyzed. Sepkoski (1984) documented a steady rise in the number of these families, which reached a plateau in the
Ordovician, suffered a major extinction in the Permian, and have shown a steady increase
in diversity ever since; at least 1,900 families of marine invertebrates are now recognized.

Similar patterns are shown for many other taxa (Fig. 3.7). Some of this increase has been attributed to *provincialization*, the appearance of differentiated regional biotas, but Bambach (1983) has concluded that the increase in community diversity is a consequence of the addition of organisms with new ways of life, for example, new methods of feeding. Patterns for individual families, of course, vary. Diversity in the gastropods and bivalves is staggering and appears to be still on the increase. Other species—the so-called living fossils—coelacanths (*Latimeria chalumnae*), horseshoe crabs (*Limulus*), and ginkgo trees (*Ginkgo biloba*) represent the last members of once-diverse lines. There are five major mass-extinction events, one in each of the Ordovician, Devonian, Permian, Triassic, and Cretaceous periods. By far the most severe was in the late Cretaceous. The causes of these extinctions have been much debated. For the Permian extinction, geologically rapid changes in climate, continental drift, and volcanic activity are probably most important. A single catastrophic event, such as a meteor strike, is probably not to blame. Ordovician extinction appears correlated with a huge global glaciation (the Hirnation glaciation), and the Cretaceous extinction may have been associated with a single catastrophic event—(a meteor strike?)—although this is far from certain. The causes of the Triassic and Devonian

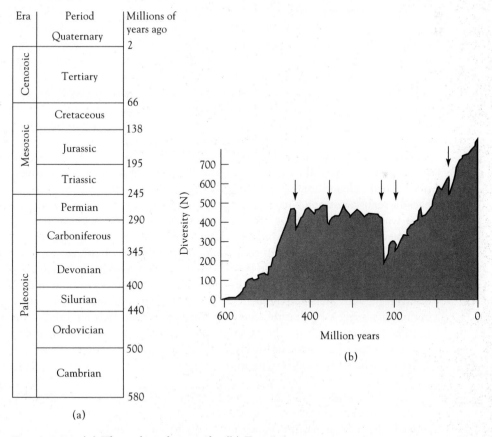

(a)

(b)

Figure 3.7   (a) *The geological timescale.* (b) *Extinction events in marine organisms.* (After Erwin et al. 1987.)

extinctions are not well known. As well as these "major" extinctions, a series of minor extinctions can be noticed with a periodicity of 26 to 28 million years. In sum, these minor extinctions account for more total extinctions than the five major extinctions combined as seen in Fig. 3.7b.

The fossil records for groups other than the marine invertebrates are not so good. Tetrapods have been subject to at least six mass-extinction events since their appearance in the late Devonian, and fishes have experienced eight extinctions since their origin in the Silurian. Most of these events coincide with each other and with those extinctions recorded for marine invertebrates. For example, the five major mass-extinction events outlined above are paralleled by losses of invertebrate diversity. Again, the most significant event is the late Permian, which is the largest recorded extinction both for fishes (44 percent of families disappearing) and for tetrapods (58 percent of families disappearing). The late Cretaceous extinction was far more significant for tetrapods than for other groups: 36 of the 89 families in the fossil record disappeared at this time. These families were mainly confined to three major groups: dinosaurs, plesiosaurs, and pterosaurs. Somewhat surprisingly, most other major vertebrate taxa were very much unaffected. Currently, correlations between the more minor extinctions events in vertebrates and the postulated periodic extinction in marine invertebrates is poor.

For plants, the fossil record does not clearly show the same sudden mass-extinction effect seen in the animal record. For vascular plants, the diversity appears to have increased from the Devonian to the Permian, dropped, and then risen to a plateau that was maintained until the Mesozoic. In the Upper Triassic, angiosperms diversified, a trend that has continued to the present (Niklas, Tiffney, and Knoll 1980; Niklas 1986). The only general similarity with animal extinction events is the end-cretaceous catastrophe, which appears to have had a major influence on the structure and decomposition of terrestrial vegetation and on the survival of species. Perhaps 75 percent of species at this time became extinct.

Is there any overall pattern to the global diversity of all life through time? Does diversity fluctuate around some preordained level, or does it constantly increase until knocked back by some catastrophic natural disaster or extinction event? Population biologists ask the same types of questions about what (if anything) regulates the population densities of modern animals and plants (see Chapter 12), and some analogies can be drawn.

Raup et al. (1973) performed a computer simulation of changes in diversity on the assumptions that a lineage could branch, go extinct, or remain unchanged in a given time period. Extinctions and speciation events were assumed to occur randomly and to occur, on average, with the same frequency. The results mirrored many real historical patterns of diversity, suggesting that the available fossil record is largely a result of random extinctions and speciation events through time. Some particular biological phenomena were not well predicted: the long-term survival of "living fossils" was one; another held that lineages often increase and diversify in the fossil record much more rapidly than they would at random.

Often the origin of new taxa is correlated with the extinction rate (Stanley 1979). The result, of course, is that diversity remains unchanged, leading some to believe that even historically distant communities were saturated with species and that new ones could succeed only in the place of old, extinct ones. Alternatively, it may be more likely that environmental changes affect both processes concurrently. Stanley (1975) assumed that, for newly arisen taxa, the number of species was able to increase in an exponential fashion following the equation $N_t = N_0 e^{rt}$ (see Chapter 6) because there were no competing species to usurp the existing "niche space." In this formula, $r$, the per-capita rate of increase, equals $O - E$, the rate of origin or speciation minus the rate of extinction. If the time $t$ since origin is known from the fossil record, the initial number of species is assumed to be 1, $E$ is calculated from the average life span of fossils, and $N_t$ is the number of species existing, then $O$ could be solved for. Stanley concluded that speciation rates of mammals were higher than those of bivalves (Table 3.2).

Sepkoski (1978, 1979, 1984) has made the analogy between historical diversity and another population model, the logistic (see Chapter 6). This model assumes initial rapid population growth followed by a leveling off at an asymptote as resources for growth become limiting; the result is a **sigmoid curve.** The upper asymptote is depressed if other competing species are present to lower the level of available resources still further. For marine invertebrates in the entire Paleozoic, diversity fits a logistic model. This result implies that, as diversity increases, speciation rates decline and extinction rates rise, possibly because of competition between species. However, one must remember that in many cases a group has diversified only shortly after the demise of another lineage created a potential "empty niche." Thus the crocodilians invaded their present habitats only after the phytosaurs became extinct. The great decline of branchiopods at the end of the Permian was followed by an explosive radiation of clams. In fact, the evidence for this type of event outweighs that for competitive exclusion of one lineage by another (Raup 1984; Jablonski 1986).

It is important to realize that extinction is the rule rather than the exception. Of the 500 million to 1 billion species that have existed during evolutionary time, only perhaps 2 million to 30 million are alive now. For most species in geologic time, and even for some in historical time, very little is known about the immediate causes of extinction (Simberloff 1986b). Predation, parasitism, and even competition can have severe impacts on popula-

TABLE 3.2 *Estimate rates of speciation O, extinction E, and increase in diversity R. (After Stanley 1975.)*

| | $t$ (million years) | $n$ (species) | $r$ | $\bar{r}$ | $e$ | $o$ |
|---|---|---|---|---|---|---|
| | | | | (per million years) | | |
| Bivalvia | | | | 0.07 | 0.17 | 0.24 |
| Veneridae | 120 | 2400 | 0.06 | | | |
| Tellinidae | 120 | 2700 | 0.07 | | | |
| Mammalia | | | | 0.20 | 0.50 | 0.70 |
| Bovidae (cattle, antelopes) | 23 | 115 | 0.21 | | | |
| Cervidae (deer) | 23 | 53 | 0.17 | | | |
| Muridae (rats, mice) | 23 | 844 | 0.29 | | | |
| Cercopithecidae (OW monkeys) | 23 | 60 | 0.18 | | | |
| Cebidae (NW monkeys) | 28 | 37 | 0.13 | | | |
| Cricetidae (mice) | 35 | 714 | 0.19 | | | |

tions of many species, but habitat alteration as a result of climatic change is probably the prime moving force in evolution over geologic time. Over evolutionary time, Van Valen (1973) suggested that within most taxonomic groups, the probability of the extinction of a genus or family is independent of the duration of its existence. Old lineages do not die out more readily than younger ones. For example, among marine invertebrates, the average lifetime of a genus has been 11.1 million years and for Carnivora 8 million years. Within these classifications, the figures vary tremendously, from 78 million years for bivalves to 7.3 million years for ammonites. Again, great care must be exercised in interpreting these records. From early to late geologic times, the number of species described per genus has generally increased. Under the influence of such a trend, because extinction of a family requires extinction of all its species, extinction rates of families will decline even if the probability of extinction of species is constant (Flessa and Jablonski 1985). The same is true of genera.

The assertion that the basal rates of extinction have speeded up at certain times is also a matter of some contention. The so-called periods of "mass extinction" at the ends of five geological eras have been argued to be simply the quantitatively extreme cases in a basal array of extinction rates (Quinn 1983). Raup and Sepkoski (1984) have suggested that mass extinctions occur regularly, with a periodicity of about 26 million years. If they are right, then past adaptations of species provide little preadaptation to extraordinary periodic conditions.

There is some evidence that the survivors of mass extinctions tended to be the more ecologically and morphologically generalized species (Fig. 3.8). Specialization can hamper adaptation to changing conditions, and overspecialization actually can work against a species. Morphologically complex and specialized species tend to occur in more specialized environments and are thought to be more susceptible to local environmental changes. Generalists tend to have a greater breadth of geographic distribution, which appears to be important in enhancing survival.

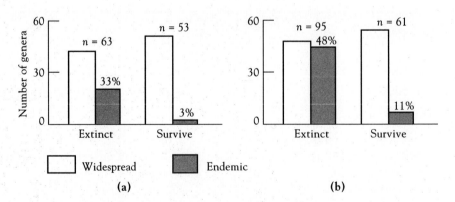

Figure 3.8  *The proportion of genera that survived the extinction event at the end of the Cretaceous period. (a) Bivalves. (b) Gastropods. Among both bivalves and gastropods, the proportion that survived was greater for geographically widespread forms than for genera with narrow geographical distributions (endemics). The histograms plot the percentage survival and extinction of bivalves and gastropods of the Gulf and Atlantic coastal plain of North America, distinguishing those restricted to the region from those that were more widespread.* (Based on Jablonski 1986b.)

The bottom line is that for many extinct lineages there is still no adequate explanation. The horse, *Equus*, evolved entirely in North America, but 8,000 to 10,000 years ago it went extinct there. Yet, when introduced by the Spaniards in 1519, herds became established very quickly. Sometimes, after extinctions, the same morphological forms reappear in the fossil record, derived from new ancestors. More often, the chief impact of extinctions is the obliteration of forms of life whose like never reappears (Strathmann 1978). In summary, it is certainly hard to say why some species die and others persist. Out of the many genera of horselike animals, why did *Equus* alone survive? Was it really structurally better adapted than the others, or was it merely lucky that its habitat persisted?

## 3.4  CAUSES OF RECENT EXTINCTIONS

### Species Life Histories and Correlates with Extinction

Species extinction is a natural process. Fossil records show that the vast majority of species that have ever existed are now extinct, with extinct species outnumbering living species by a factor of perhaps 1,000:1. The current fear is that natural extinction rates are being increased by human activity. Therefore, we should try to determine how human activity increases the chance for species to go extinct.

Species become extinct when all individuals die without producing progeny. They disappear in a different sense when a species' lineage is transformed over evolutionary time, or divides into two or more separate lineages (so-called pseudoextinction). The relative frequency of true extinction and pseudoextinction in evolutionary history is not yet known.

In order to predict what type of human causes are critical in the extinction of wild species, some knowledge of life-history traits that may be correlated with high levels of extinction is desirable. At least seven life-history traits have been proposed as factors affecting a species' sensitivity to extinction (Karr 1991; Lawrence 1991):

1. Rarity. Generally, rare species are more prone to extinction than common ones. This may sound like a truism at best and a platitude at worst. However, it is not as intuitive as is first thought. For example, a very common species might be very susceptible to even the slightest change in climate, whereas rare species, although they may exist at very low numbers of individuals, may be more resistant to climatic change and thus more persistent in evolutionary time as, for example, global warming proceeds. Rabinowitz (1981) has shown that "rarity" itself may depend on three factors: geographic range, habitat breadth, and local-population size. A species is often termed rare if it is found only in one area, regardless of its density there. A species that is widespread but at very low density can also be regarded as rare. Conservation by habitat management is much easier and more likely to succeed for species of the first type than for those of the latter. Schoener (1987) addresses the issue of rarity further.

2. Dispersal ability. Species that are capable of migrating between fragments of habitat, such as between mainland areas and islands, may be more resistant to extinction. Even if one small population goes extinct in one area, it may be "rescued" by immigrating individuals from another population.

3. Degree of specialization. It is often thought that organisms that are specialists; for example, those organisms that can only feed on one type of plant, like panda bears which only feed on a single species of bamboo, are more likely to go extinct. Animals that have a broader diet may be able to switch from one food type to another in the case of habitat loss and are thus less prone to extinction. Plants that can live only in one soil type may be more prone to extinction.

4. Population variability. Species with relatively stable populations, that is, those that are generally maintained at some equilibrium level, may be less prone to extinction than others. For example, some species, especially those in northern taiga ecosystems, show pronounced cycles. Lemmings reach very high numbers in some years, and the population crashes in others. It is thought that these might be more likely to go extinct than others.

5. Trophic status (animals only). Animals occupying higher trophic levels usually have small populations. For example, birds of prey or Florida panthers are far fewer in number than their food items and, as noted above, rarer species may be more vulnerable to extinction.

6. Longevity. Species with naturally low longevity may be more likely to become extinct. Again, this is not as obvious as it first sounds. Imagine two species of birds, one of which lives for 70 or 80 years, like a parrot, and the other which is about the same size but breeds earlier in life, for example at age two or three, and only lives to the age of ten years. The parrot, with its 80-year life span, may be able to "weather the storm" of a fragmented habitat for ten years without breeding. Then it can pick up and breed again when the habitat becomes favorable. Species with naturally low longevity are not able to do this.

7. Intrinsic rate of population increase. Species that can reproduce and breed quickly may be more likely to recover after severe population declines, say, following the introduction of new exotic diseases, than those that cannot. Populations of species that breed only slowly may be more likely to suffer a double setback, for example, a cold winter following a summer when an exotic disease was introduced. Thus, it is thought that those organisms with a high rate of increase—especially small organisms, bacteria, insects, and small mammals—are less likely to go extinct than larger species such as elephants, whales, and redwood trees. The passenger pigeon laid only one egg per year, and this probably contributed to its demise.

## The Causes of Extinction

How long do species last on Earth? The average life span of a species in the fossil record is around 4 million years, which would give, at a very gross estimate, a background extinction rate of four species each year out of a total number of species around 10 million. However, it can be argued that the fossil record is heavily biased toward successful, often geographically wide-ranging species, which undoubtedly have a far longer average persistence time. If background extinction rates were ten times higher than this, extinctions among the 4,000 or so living mammals today would be expected to occur at a rate of one every 400 years, and among birds one every 200 years. It is indisputable that the extinction rate in recent times has been far higher than this.

It is easy to suggest that humans have been the overwhelming cause of recent extinctions, though it is equally easy to suggest that the calculated background rates of extinction are still an underestimate. Nevertheless, the arrival of humans on previously isolated continents—around 50,000 years ago in the case of Australia and 11,000 years ago for

North and South America—seems to coincide with large-scale extinctions in certain taxa. Australia lost nearly all its species of very large mammals, giant snakes and reptiles, and nearly half its large flightless birds around this time. Similarly, North America lost 73 percent and South America 80 percent of their genera of large mammals around the time of the arrival of the first humans. The probable cause is hunting, but the fact that climate changed at around this same time leaves the door open for natural changes as a contributing cause of these extinctions. However, the rates of extinctions on islands in the more recent past seem to confirm the devastating effects on humans. The Polynesians, who colonized Hawaii in the fourth and fifth centuries A.D., appear to have been responsible for exterminating about 50 of the 100 or so species of endemic land birds in the period between their arrival and that of the Europeans in the late eighteenth century. A similar impact probably was felt in New Zealand, which was colonized some 500 years later than Hawaii. There, an entire avian megafauna, consisting of huge land birds, was exterminated by the end of the eighteenth century. This was probably accomplished by a combination of direct hunting and large-scale habitat destruction through burning. On Madagascar, the giant elephant bird, the largest bird ever recorded, 20 species of lemur, most of them larger than any surviving species, and two giant land tortoises have become extinct within the last 1,500 years. In the Caribbean, at least two ground sloths became extinct before Europeans arrived at the end of the fifteenth century. Once again, climate change, particularly progressive desiccation, may have played a part as well. However, it is indisputable that recorded extinctions have increased dramatically in recent years, just as the population of humans has been seen to skyrocket (Figure 3.9).

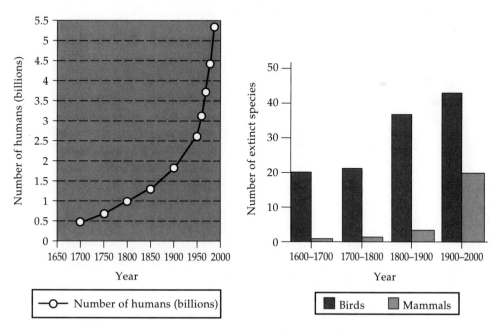

**Figure 3.9** *Population growth and animal extinctions.* (*a*) *Geometric increase in the human population.* (*b*) *Increasing numbers of extinctions in birds and mammals. These figures suggest that, as human numbers increase, more and more living species are exterminated.*

Certain generalized patterns of extinction do emerge on examination of the data. Perhaps one of the most important of these is the preponderance of extinctions on islands over those in continental areas. Although both the mainland and islands have similar over-all numbers of recorded extinctions, there are a lower numbers of species on islands than on continents, making the percentage of taxa extinct on islands greater than on continents (Table 3.3).

The reasons for high extinction rates on islands are perhaps because many species effectively consist of single populations on isolated islands. Adverse factors are thus likely to affect the entire species and bring about its extinction. Also, species on islands may have evolved in the absence of terrestrial predators and may often be flightless. Tameness, flightlessness, and reduced reproductive rates appear to be major contributory factors to species extinction, especially when novel predators are introduced.

When we look at the possible causes of extinction, most often no cause has been assigned (Figure 3.10). Of those causes that have been assigned, introduced animals and direct habitat destruction by humans have been major factors involved in extinctions, being implicated in 17 percent and 16 percent of cases respectively. These are equivalent to 39 percent and 36 percent, respectively, of the extinctions for which causes are known. Hunting and deliberate extermination also contribute significantly (23 percent of extinctions with known cause). Thus, three factors appear of paramount importance in causing extinction. In order of importance, they are:

1. Introduced species
2. Habitat destruction
3. Direct exploitation

We can break up the "introduced species" category a little further. The effects of introduced species can be assigned to either competition from introduced species, predation from introduced species, or disease and parasitism from introduced species. Competition

TABLE 3.3 *Recorded extinctions, 1600 to present, on mainland areas (larger than $10^6 km^2$, the size of Greenland) islands and the ocean.*

| Taxa | Mainland | Island | Ocean | Total | Approximate Number of Species | Percentage of Taxa Extinct Since 1600 |
|---|---|---|---|---|---|---|
| Mammals | 30 | 51 | 2 | 83 | 4,000 | 2.1 |
| Birds | 21 | 92 | 0 | 113 | 9,000 | 1.3 |
| Reptiles | 1 | 20 | 0 | 21 | 6,300 | 0.3 |
| Amphibians | 2 | 0 | 0 | 2 | 4,200 | 0.0 |
| Fish | 22 | 1 | 0 | 23 | 19,100 | 0.1 |
| Invertebrates | 49 | 48 | 1 | 98 | 1,000,000+ | 0.5 |
| Vascular Plants | 380 | 219 | 0 | 599 | 250,000 | 0.3 |
| Total | 505 | 431 | 3 | 936 | | |

may exterminate local populations, but it has not yet been clearly shown to extirpate entire populations of rare species. The best examples of competition-caused extinction occurred long ago in evolutionary time when the land bridge between North and South America formed and species of North American megafauna migrated into South America and vice versa. Niche displacement may have caused some of the extinction of species in both continents. For predation, there have been many recorded cases of extinction. Introduced predators such as rats, cats, and mongooses have accounted for at least 112 of 258 recorded extinctions of birds on islands, or 43.4 percent (Brown 1989). Of the 75 species of birds and mammals that have vanished during the last 300 years or so, predation was a major factor in 25. Parasitism and disease by introduced organisms is also important in causing extinctions. Avian malaria in Hawaii, facilitated by the introduction of mosquitoes, has been thought to have contributed to the demise of many local Hawaiian birds. Similarly, the American chestnut tree and European and American elm trees have been severely impacted by introduced plant diseases, though neither of these has yet become extinct.

In terms of habitat alteration, direct habitat destruction such as deforestation is the prime cause of extinction of species. More subtle alteration of habitat by events such as modified climate due to pollution has not yet been shown to have caused any extinctions. Direct exploitation, however, particularly the hunting of animals, *has* caused many extinctions. Steller's sea cow, a huge species of manateelike mammal, was hunted to extinction on the Bering Straits only 27 years after its discovery in 1740. The dodo was butchered to death on Mauritius soon after Mauritius became a Dutch colony in 1644. In the United States, the passenger pigeon was the most common bird in the entire bird population of North America. Unbelievably, hunting was the primary reason for its eventual extinction by 1900. The Carolina parakeet suf-

Figure 3.10   *The causes of animal extinction, based on knowledge of 484 extinct species.*   (After data from World Conservation Monitoring Centre 1992.)

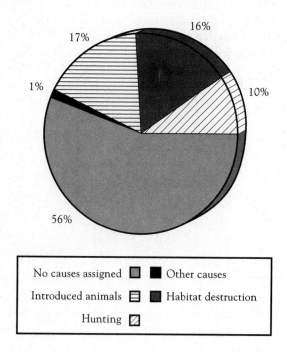

fered the same fate (Photo 3.2). The list of species hunted to extinction goes on and on. Interestingly, in the case of the passenger pigeon, there may not be enough individuals to stimulate synchronous breeding because a number of females were needed to stimulate ovarian development. This is common in some colonial species.

## Threatened Species

Knowing why species have gone extinct helps us to prioritize the threats that are likely to deplete species numbers. This may make it easier to protect threatened species. A threatened species is one that is thought to be at risk of extinction in the foreseeable future.

The growth in public awareness of the problem of depletion and the possible extinction of species perhaps started with the publication of the so-called Red Data Books, a concept founded by Sir Peter Scott during the 1960s. This was an attempt to categorize species at risk according to the severity of the threats facing them and the estimated imminence of their extinction. More recently, the Red Data Books have been compiled on a global basis by the International Union

Photo 3.2    *The Carolina parakeet,* Conuropsis carolineisis, *the only parrot native to the United States, went extinct in 1914, the last individual dying at the Cincinnati Zoo, one month later than the last Passenger pigeon. Hunting and habitat destruction were featured prominently in its demise.* (Runk, Grant Heilman 3PPAC-11A.)

for Conservation of Nature and Natural Resources (IUCN), now called the World Conservation Union. Red Data Books were first compiled for terrestrial vertebrates but have been expanded to include books for plants and invertebrates. The animals' red list has been compiled every two years since 1986 by the World Conservation Monitoring Center in collaboration with the IUCN. In addition to these global books on threatened species, individual countries or areas may issue their own type of red data book. This is because species may be more threatened on a national basis than they are globally. For example, the osprey may be thought of as a rare species in the United Kingdom, but it is common in other parts of the world, for example, coastal areas of the southeast United States. As of 1992, most bird species in the world had been comprehensively reviewed for rarity, about 50 percent of mammal species, probably less than 20 percent of reptiles, 10 percent of amphibians, and 5 percent of fish.

The main categories used for threatened species are as follows:

1. Endangered—taxa in danger of extinction whose survival is unlikely if the causal factors continue to operate. Included are taxa whose numbers have been reduced to a critical level or whose habitats have been so drastically reduced that they are deemed to be in imminent danger of extinction.
2. Vulnerable—taxa believed likely to move into the endangered category in the near future if the causal factors continue operating. Included here are taxa that are decreasing because of overexploitation, extensive destruction of habitat, or other environmental disturbance.
3. Rare—taxa with small world populations that are not at present endangered or vulnerable but are at risk. Taxa in this category usually have very localized distributions, for example, populations that are wholly maintained on small oceanic islands.
4. Indeterminate—taxa that are known to be endangered, vulnerable, or rare, but where there is not enough information to say which of the three categories is appropriate.

## Threats to Species

Most of the causal factors currently threatening species are anthropogenic in nature. These factors include:

1. Habitat loss or modification; causes include pastoral development, cultivation and settlement, forestry, and pollution.
2. Overexploitation for commercial or subsistence reasons, including hunting for meat, for fur, or for the pet trade.
3. Accidental or deliberate introduction of exotic species that may compete with, prey on, or hybridize with native species.
4. Disturbance, persecution, and uprooting, including deliberate eradication of species considered to be pests; this is perhaps most serious for predatory species such as wolves or tigers.
5. Incidental take, particularly the drowning of aquatic reptiles and mammals in fishing nets.
6. Disease, both exotic and endemic, exacerbated by the presence of large numbers of domestic livestock or introduced species.

In many of these cases, individual species are faced by several of these threats simultaneously. Some understanding of the relative importance of these different threat types, as measured by frequency of occurrence, has been estimated from an examination of threats facing the terrestrial mammals of Australia and the Americas and those facing the birds of the world. For mammals, these threats are summarized in Figure 3.11.

Of the 119 species of mammals from these continents that are considered threatened, 75 percent are threatened by more than one factor, and of these, 27 face four or more threats. The major threat, which affects 76 percent of species, is habitat loss and modification. This has a variety of causes, of which the most frequent is cultivation and settlement. Overexploitation affects half the species, the most significant cause being hunting for meat. Introduced predators and competitors affect 18 percent of the threatened species. The most serious other factor is limited distribution, which affects one-quarter of species. An interesting point here is that the factors that threaten endangered species rank differently in importance to those which are known to actually to have caused extinctions, where introduced species rank highest.

Overexploitation affects some animals more than others. Many fur-bearing animals, including chinchilla, giant otter, many species of cats, and some species of monkeys have declined to very low population sizes because their pelts are prized. Valuable timber species, including populations of the West Indies' mahogany in the Bahamas and the caoba (mahogany) of Ecuador, have been severely depleted, and the Lebanese cedar has been reduced to a few scattered remnants of forest. Overexploitation is a more selective threat to species survival than is habitat loss and primarily threatens vertebrates and certain taxa of plants and insects. More specif-

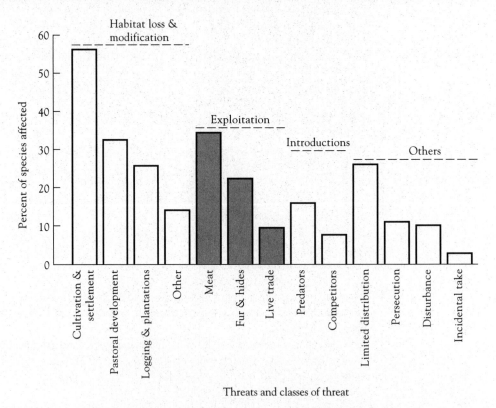

Figure 3.11   *The factors that threaten mammals in Australia and the Americas.*   (World Conservation Monitoring Centre 1992; *Global Biodiversity: Status of the Earth's Living Resources*. Chapman and Hall, London.)

ically, carnivores, ungulates, primates, sea turtles, showy tropical birds, and timber species have been overharvested. Many species of butterflies and orchids have been overharvested for commercial interests, too, and rare plants have been threatened by collectors. Predator control efforts have also significantly reduced the population sizes of many vertebrates, including sea lions, birds of prey, foxes, wolves, various large cats, and bears.

## The Taxonomic Distribution of Threatened Species

In all, some 835 animal species are listed as endangered in the 1990 IUCN Red List, or much less than 0.1 percent of the world's estimated total of more than 1.5 million described animal species. The class with the greatest number of endangered species is fish. Some orders seem to have a disproportionately high number of endangered or threatened species. For example, elephants have two out of two species threatened, manatees and dugongs have four out of four species threatened, and primates, carnivorous cats, and antelopes are also highly threatened with 53 percent, 32 percent, and 31 percent respectively of their constituent species listed. Although these latter three orders combined only contain some 20.6 percent of the world's mammals species, they account for just under half of the listed threatened species and just over half of the endangered species. Vertebrates are probably more vulnerable to extinction than invertebrates because they are much larger and require more resources and larger ranges. For example, the Florida panther population is now extremely low but may have only been about 1,600 animals at its maximum, before humans appeared. On the other hand, many invertebrates have an extremely small range, but populations may still run in the millions. Nevertheless, there probably are many endangered invertebrates, such as insects, but the "higher" orders such as fish and especially birds and mammals are better monitored. On a worldwide basis, mammals are officially acknowledged as the most endangered taxa on Earth, followed by fish, birds, reptiles, amphibians, and invertebrates (Fig. 3.12).

Figure 3.12  *Percent of known species classed as* endangered.

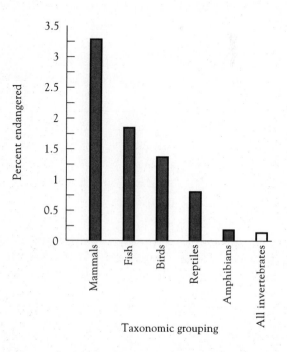

We can also break down the data to show which geopolitical areas contain the most threatened species (Table 3.4). The majority of threatened mammals occur in mainly tropical countries, with highest numbers recorded from Madagascar (53), Indonesia (49), China (40), and Brazil (40). Such countries may have large numbers of threatened animals simply because they have more animals in general. Therefore if the same percentage were threatened in each country, a country having a larger number of animals to begin with would have a higher number threatened. We can obtain some idea of what a country might be expected to have in terms of numbers of threatened species by drawing a graph of that country's area against the number of threatened animals. This is done in Figure 3.13. Madagascar and Indonesia, in particular, have more threatened mammals in relation to country area than would be predicted statistically (points above the line), whereas the United States, for example, has fewer. The patterns for birds are similar to those for mammals, the majority of endangered species being concentrated in Southeast Asia, Mexico, and South America. In comparison, Europe, Africa, Canada, the Middle East, and the Arabian Peninsula have relatively few globally threatened bird or mammal species. It may come as a surprise to find that the United States of America heads the list in terms of the numbers of threatened reptiles, amphibians, and fishes (perhaps because other countries monitor them less well?).

We can also break the data down into the habitat types that contain the highest number of threatened species. This type of data is available for birds of the world and the mammals of Australia and the Americas (Fig. 3.14). The habitat type in which the largest number of threatened species, both bird and mammal, occur is clearly tropical forests. In general, the world's threatened bird species occupy the range of habitat types similar to the threatened mammals. However, there are a couple of noticeable differences. Most important of these are the fact that there are far more threatened birds than mammals on oceanic islands. In fact, the data shown on Figure 3.14 does not include mammals on oceanic islands, but it is known that there are few mammals on such islands, and therefore the number of threatened species on such habitats is likely to be small.

**TABLE 3.4** *The "top-ten" list of countries for different types of endangered species.*

| Mammals | | Birds | | Reptiles | | Amphibians | | Fishes | |
|---|---|---|---|---|---|---|---|---|---|
| Country | Total | Country | Total | Country | Total | Country | Total | Country | Total |
| Madagascar | 53 | Indonesia | 135 | USA | 25 | USA | 22 | USA | 164 |
| Indonesia | 49 | Brazil | 123 | India | 17 | Italy | 7 | Mexico | 98 |
| Brazil | 40 | China | 83 | Mexico | 16 | Mexico | 4 | Indonesia | 29 |
| China | 40 | India | 72 | Bangladesh | 14 | Australia | 3 | South Africa | 28 |
| India | 39 | Colombia | 69 | Indonesia | 13 | India | 3 | Philippines | 21 |
| Australia | 38 | Peru | 65 | Malaysia | 12 | New Zealand | 3 | Australia | 16 |
| Zaire | 31 | Ecuador | 64 | Brazil | 11 | Seychelles | 3 | Canada | 15 |
| Tanzania | 30 | Argentina | 53 | Colombia | 10 | Spain | 3 | Thailand | 13 |
| Peru | 29 | USA | 43 | Madagascar | 10 | Yugoslavia | 2 | Sri Lanka | 12 |
| Vietnam | 28 | Myanmar | 42 | Myanmar | 10 | | | Cameroon | 11 |

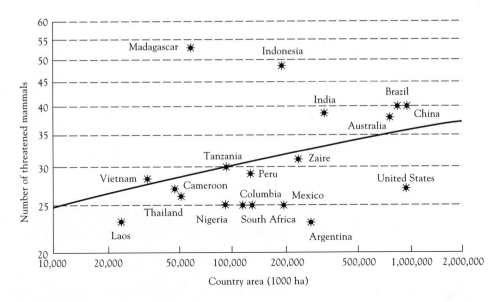

**Figure   3.13**   *Relationship between number of threatened species and country area.* (World Conservation Monitoring Centre 1992.)

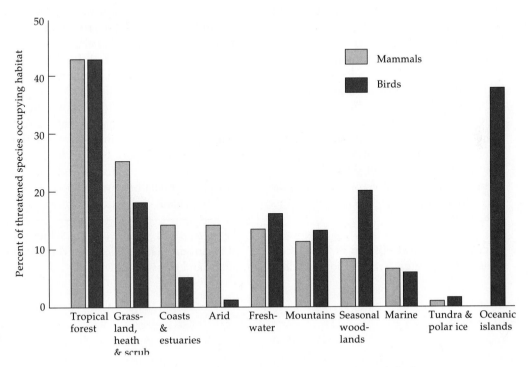

**Figure   3.14**   *Habitat distribution of threatened mammals and birds.* (World Conservation Monitoring Centre 1992.)

Oceanic islands are also known to have very many species of threatened plants. For example, Hawaii has 108 endemic taxa that have already gone extinct, 138 that are endangered, 37 vulnerable, 126 rare, and 9 indeterminate. Hawaii has one of the most distinctive and most threatened floras in the world. Other oceanic islands show similar patterns. St. Helena, in the Atlantic Ocean, has 46 endemic species that are threatened; Bermuda has 14 of 15 endemic species threatened.

Too often, it is difficult to study rare populations to see which of many factors is the most critical to their survival. Zimmerman and Bierregaard (1986) feel that theory will not be as valuable as an informed "naturalist's feel" in answering how to manage the survival of particular species.

## Summary

 1. Some of the best-analyzed examples of natural selection involve industrial melanism—a change in color that takes place in areas that have been heavily polluted. This underscores the fact that observation of practical environmental problems can help us understand theoretical ecological principles. On the flip side, biological indicator organisms can provide a good record of the strength of pollution.

2. Natural selection ultimately results in the formation of races, subspecies, biotypes, and species. An understanding of the seemingly simple idea of a species is in fact quite complex. For example, there is the biological-species concept and the phylogenetic-species concept. The definition of *species* is very important because federal conservation legislation often emphasizes species over other categories.

3. What causes the extinction of species? Such knowledge can be used to decrease current rates of extinction. At least seven life-history traits may affect a specie's sensitivity to extinction: rarity, dispersal ability, degree of specialization, population variability, trophic status, longevity, and intrinsic rate of population increase.

4. In order of importance, the three most important causes of extinction have been introduced species, habitat destruction, and direct exploitation. Introduced predators such as rats and cats have accounted for much extinction of ground-nesting birds and lizards on islands. In more modern times, habitat destruction, such as deforestation, is the prime cause of extinction. Indeed, of the factors that currently threaten living species with extinction, more than three-quarters are threatened by habitat loss, over-exploitation affects half the species, and introduced predators and competitors affect only 18 percent of threatened species.

*Discussion Question:* Which types of organisms deserve priority in conservation efforts and why?

# Group Selection and Individual Selection

I f ecology strives to explain the distribution patterns of plants and animals on Earth, then an understanding of behavior is valuable. Two schools of thought exist concerning the abundance of populations. One, viewed often as the "climate" school, or "density independence," was championed by the Australian biologists Andrewartha and Birch in the 1940s and 1950s. Their view was that most population mortality was ultimately caused by the weather and was unpredictable in its effects. Populations of small insects, called thrips, which Andrewartha and Birch studied on plants, suffered violent changes from the effects of rainfall and other features of the weather. The alternative, often termed the "equilibrium" school or "density dependence," was championed by mathematicians such as Pearl and Lotka in the 1920s and 1930s and by the Australian entomologist Nicholson in the 1950s. The density-dependence school of thought held that populations varied about some mean or equilibrium level. Any variations from that mean were "corrected" by biological phenomena. For example, a bonanza of food that might promote high insect populations, would, in turn, be followed by an abundance of natural enemies, which would reduce pest populations to their mean or equilibrium value. For many years, neither school of thought enjoyed the overwhelming support of evidence, and so it is even today: opinions sway back and forth about which theory is generally held to be more true. We shall return to this theme in the next section, Population Ecology.

Within the density dependence camp, there have been discussions about which factors commonly cause populations to oscillate around the equilibrium value. These factors can be either intrinsic or extrinsic. Extrinsic factors include predators, parasites, and competition with other species—interspecific competition. Intrinsic factors center on control of population size at the level that is imposed by crowding or by competition with conspecifics for resources—intraspecific competition.

Some ecologists have argued that populations self-regulate themselves at a level that can be sustained without the ravages of starvation and disease—that is, at levels below which competition becomes important. In this way, wasteful squabbling and fighting for food is avoided. Nature is neat, tidy, and harmonious. Several pieces of evidence led early ecologists to believe that many populations did self-regulate their numbers. For example, many birds are territorial, and a population in a given area was often below that which could be supported by a given amount of prey. Second, any increase in organisms within a given area resulted in emigration, which again operated to maintain populations about a mean level. Experimental manipulations showed that some of these mechanisms operated at low densities in the absence of food or other resource limitations. Third, many species—in particular, birds—showed high variation in reproductive rate. Some birds fledged more young than others. This was regarded as evidence of the adjustment of fecundity to balance mortality in the population. In the late 1940s, ornithologists David Lack and Alexander Skutch brought the argument of self-regulation versus extrinsic regulation to a head. The argument was precipitated by data that showed that songbirds typically lay a clutch of four to six eggs in temperate regions of North America and Europe and only two or three in the tropics. The trend is general, effecting virtually all groups of birds in all regions of the world. In the next 20 years, the arguments were replayed many times with different data, but the theme was always the same: profusion of young or reproductive prudence?

In 1962 the self-regulation viewpoint was championed by British ecologist V. C. Wynne-Edwards, who articulated the full concept of self-regulation in a book called, *Animal Dispersion in Relation to Social Behavior.* The premise was that most groups of individuals purposely controlled their rate of consumption of resources, and their rate of breeding to ensure that the group would not go extinct: this was known as **group selection.** Individuals in successful groups would not tend to act selfishly. Groups that consisted of selfish individuals would overexploit their resources and die out. In concept the idea of self-regulation or group selection is straightforward, is intellectually satisfying to many, and would seem to represent what nature ought to do—avoid the grizzly clashes and potentially damaging confrontations of competition. Why fight to the death if a contest can be settled amicably or indeed if species never overstepped their limits so that confrontations never occurred? This is perhaps how humans would like to be. In the late 1960s and 1970s, the idea of self-regulation came under severe attack, and some of the data that supported the group-selection camp was equally well explained by **individual selection.**

## 4.1  DRAWBACKS TO THEORIES OF GROUP SELECTION

Attractive though the idea of **group selection** is, it has several flaws:

1. Mutation: Imagine a species of bird in which a pair lays only two eggs and there is no overexploitation of resources. Suppose the tendency to lay two eggs is inherited as a group-selection trait. Now consider a **mutant** that lays six eggs. If the population is not over exploiting its resource, there will be sufficient food for all the young to survive, and the six-egg genotype will become more common very

rapidly. Gene frequencies in the population will change. This process would work for even larger brood sizes, and brood sizes would tend to increase until they became so large that the parents could not look after all their young, causing an increase in infant mortality. Thus, the clutch size in nature evolves so as to maximize the number of surviving offspring. Field studies of great tits in Wytham Woods, England, for example, show a median clutch size of eight to nine eggs, above which adult birds cannot reliably supply sufficient food for all chicks to survive.

2. Immigration: Even in a population in which all pairs laid two eggs and no mutations occurred to increase clutch size, "selfish" individuals that laid more could still migrate in from other areas. In nature, populations are rarely sufficiently isolated to prevent **immigration.**

3. Individual selection: For group selection to work, some groups must die out faster than others. In practice, groups do not go extinct fast enough for group selection to be an important force. Individuals nearly always die more frequently than groups, so individual selection will be the more powerful evolutionary force.

4. Resource prediction: For group selection to work, individuals must be able to assess and predict future food availability and population density within their own habitat. There is little evidence that they can.

Individual selfishness seems a more plausible result of natural selection. Group selection is probably a weak force and is only rarely very important (Maynard Smith 1976a). Any reduction in population sizes from **self-regulation** is likely to come from intraspecific competition, of a selfish nature, in which individuals are still striving to command as much of a resource as they can (see Chapter 8). Indeed we often see animals in nature acting in their own selfish interest. Male lions kill existing cubs when they take over a pride. The proximate cause may be the unfamiliar smell of the cubs. A similar effect, known as the Bruce effect, occurs in rodents, where the presence of a strange male prevents implantation of a fertilized egg or induces abortion in females (Bruce 1966). In the case of lions, the advantage of infanticide for the male lion is that without their cubs, females come into the reproductive condition much faster, in nine months as opposed to 25 if the cubs are spared, hastening the day when males can father their own offspring. A male's reproductive life in the pride is only two to three years before he in turn is supplanted by a younger, stronger male. Infanticide ensures that the male will father more offspring, and the tendency spreads by natural selection (Bertram 1975). There is no advantage of infanticide to the mothers of the killed cubs, but, being smaller, they are powerless to stop the males.

In human society, individuals also rarely act for the good of the group, as illustrated by the "tragedy of the commons" scenario. Where a common pasture is open to public grazing, farmers are likely to overload the land with cattle in an attempt to maximize their own returns. What usually happens is overgrazing, and the area becomes useless to anyone (Hardin 1968). Similar problems of overfishing, overharvesting, and overhunting beset society.

If individual selfishness is more common than group selection, we are obligated to provide explanations for phenomena such as **altruism,** the existence of castes, and even the existence of sex itself, all of which smack of group selection.

## 4.2 ALTRUISM

Although natural selection favors individual rather than group selection, it is still common to see apparent cooperation. Animals of the same species groom one another, hunt communally, and give warning signals to each other in the presence of danger. How can this altruistic behavior be explained by natural selection?

We are not surprised to see a parent working hard to care for its young. All offspring have copies of their parents' genes, so parental care is genotypically selfish. Genes for altruism toward one's young will therefore become more numerous because offspring have copies of those same genes. The "selfish genes" themselves are increasing, by virtue of their effect on behavior and the copies of themselves in the bodies of other individuals. In meiosis, any given gene has a 50 percent chance of going into an egg or sperm. Thus each parent contributes 50 percent of its genes to its offspring. The probability that a parent and offspring will share a copy of a particular gene is a quantity, $r$, called the coefficient of relatedness. By similar reasoning, brothers or sisters are related by an amount $r = 0.5$, grandchildren to grandparents by 0.25 and cousins to each other by 0.125 (Fig. 4.1). It was W.D. Hamilton (1964) who realized the important implication of this relatedness for the evolution of altruism, though the idea was anticipated by two earlier giants in the field, Fisher (1930) and Haldane (1953). Just as gene replication can occur through parental care, so it can by care for siblings, cousins, and other relatives. (Perhaps we should talk about selfish genes, rather than selfish individuals [Dawkins 1989].) Selection for behavior that lowers an individual's own chances of survival and reproduction but raises that of a relative is known as **kin selection.** The conditions under which an altruistic act will spread

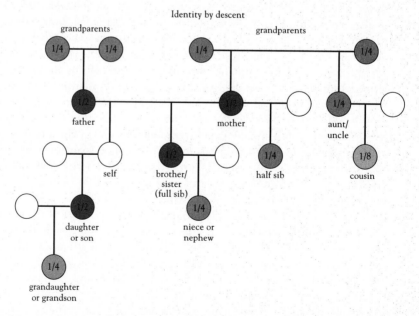

**Figure 4.1** *Degrees of genetic relationship among relatives in diploid organisms. Open circles represent completely unrelated individuals.*

by kin selection can be quantified as follows. If the donor sacrifices $C$ offspring for which the recipient gains $B$ offspring, then the gene causing the donor to act in this way will spread if

$$rB - C > 0$$

where $r$ is the coefficient of relatedness of donor to recipient. Having made these calculations on the back of an envelope in a pub one evening, J.B.S. Haldane reputedly announced that he would lay down his life for two brothers or eight cousins (Krebs and Davies 1993).

## Altruism Between Relatives

Many insect larvae, especially caterpillars, are soft-bodied creatures. They rely on bad taste or poison to deter predators and advertise this condition by bright warning colors. For example, noxious *Datana* caterpillars, which feed on oak and other trees, have bright red-and-yellow stripes (Color insert II, page 3). Of course, a predator has to kill and attempt to eat one of the caterpillars in order to learn to avoid similar individuals in the future. It is of no personal use to the unlucky caterpillar to be killed. However, warningly colored animals often aggregate in kin groups (Table 4.1), so the death of one individual is most likely to benefit its relatives, such as siblings, and its genes will be preserved. Even some solitary species are warningly colored, suggesting a direct benefit as well—if the unlucky larva isn't killed by an attack, it will probably not be chosen again.

Another example of altruism again refers to lions. Lionesses tend to remain within the pride, whereas the males leave. As a result, lionesses within a pride are related, on average, by = 0.15. Females all come into heat at the same time, one individual probably influencing the others' estrous cycles by means of pheromones. A similar phenomenon occurs in humans, where young women living in the same school dormitories may have synchronized menstrual cycles. In lionesses, the result is the simultaneous birth of cubs, and females exhibit the apparently altruistic behavior of suckling other females' cubs. Because the females are related, the selfish-gene hypothesis accounts for this behavior. Similarly, when male lions depart, they may act in concert to take over a pride; each one's genes will be perpetuated both through his own offspring and through those sired by his brothers (Bertram 1976; Bygott, Bertram, and Hanby 1979; Packer et al. 1991).

Although an altruistic act toward two sisters or brothers may seem to be equivalent genetically to a similar act toward an offspring, there are sometimes other ecological or proximate

TABLE 4.1 *Brightly colored species of caterpillars of British butterflies are more likely to be aggregated in family groups than cryptic species. (From Harvey et al. 1983.)*

| | No. species of caterpillars | |
|---|---|---|
| Dispersion | Aposematic | Cryptic |
| Large family groups | 9 | 0 |
| Solitary | 11 | 44 |

factors that tip the scales in favor of the offspring. Young may be more valuable in terms of expected future reproduction, having a higher potential-reproductive output than older siblings or parents. Progeny may also benefit more from a given amount of aid. One insect fed to a nestling contributes more to its survival than the same insect would to the survival of a healthy adult sibling. Young may thus be thought of as superbeneficiaries (West Eberhard 1975).

Not all acts of altruism result in such extremes as suicide and sterility. A common example of altruism is the raising of an alarm call by "sentries" in the presence of a predator. The alarm maker is drawing attention to itself and risking increased danger by its behavior. For some groups such as ground squirrels, *Spermophilus beldingi* (Sherman 1977), individuals near an alarm maker bolt down their burrows, and those close neighbors are most likely to be sisters or sisters' offspring; thus, the altruistic act of alarm calling could be reasoned to be favored by kin selection. On the other hand, sentries can be subordinate individuals who are driven to the edge of the group, where the risk is higher and they are forced to be alert for their own safety.

Working out the exact costs and benefits in a system of kin selection can be a nightmare and has often been a stumbling block in behavioral ecology. To get a precise measure of natural selection, one must be able to calculate an individual's contribution to the gene pool. This might be the contribution of an individual plus 0.5 times its number of brothers and sisters, 0.125 times its number of cousins, and so on. This genetical octopus has been termed **inclusive fitness** by Hamilton (1964). More commonly, behavior or **adaptation** is recognized as beneficial if it shows a closely designed fit to a problem presented by the animal's environment (Williams 1966). At least for the alarm makers, some selfish motive is involved because if a predator fails to catch prey in a particular area, it is less likely to return, and the number of potential attacks over the alarm-maker's lifetime is likely to be reduced (Trivers 1971).

## Altruism Between Unrelated Individuals

Not all acts of altruism are directed to close relatives (Photo 4.1). In more than 200 species of birds and 120 species of mammals, individuals have been recorded helping others to reproduce. When a female olive baboon, *Papio anubis,* comes into estrus, a male forms a consort relationship with her, following her around and waiting for the opportunity to mate. Sometimes an unattached male enlists the help of another to engage the consort male in battle, while the solicitor attempts to mate with the female. On a later occasion the roles are reversed (Packer 1977). This is an example of reciprocal altruism (Trivers 1971). (Reciprocal altruism is common in human society where money is used to mediate its use.) Sometimes, unrelated individuals will occupy the same territories as breeding individuals and help the parents raise offspring by foraging for additional food for the young. Although the helpers may be related to the parents, for example in the well-studied Florida scrub jay (Woolfenden 1975; Woolfenden and Fitzpatrick 1984), sometimes they are not, for example in mongooses (Rood 1978, 1990; Creel and Waser 1991) and dunnocks, *Prunella modularis* (Houston and Davies 1985; Davis et al. 1992). By comparing nest or brood success of breeding pairs with helpers with those whose helpers were removed, Brown et al. (1978, 1982) and Mumme (1992) were able to show that helpers do significantly increase parents' fitness. Again, the motive in most situations seems to be a sort of reciprocal altruism. In these situations, the habitat is usually saturated with breeders, and helpers could probably not obtain a territory for themselves. What they do is help increase ter-

ritory size of a breeding pair; they are later able to carve off a fragment of this territory themselves, or they can take over from the breeding pair after one member dies.

For many predators, reciprocal altruism in the form of social hunting allows bigger game
to be caught. The benefits of a large kill outweigh the cost of having to share the meat
(Caraco and Wolf 1975). Many acts of altruism between unrelated individuals, whether in
defense, attack, or mating, are exhibited by individuals living in social groups.

Of course not all observed behavior can be explained by altruism. Many adaptive explanations
are in fact more the result of an author's ingenuity than natural consequences of the facts. Great
care must be taken to avoid such errors; otherwise, the resultant theories will have no more
substance than Kipling's *Just So Stories* (Gould and Lewontin 1979). Often some individuals are
simply manipulated by others. Cuckoos, for example, have used the devoted parental care of
adult birds to rear their parasitic cuckoo young instead. Some authors have also suggested that in
social insects, some females could actually do better, in terms of **inclusive fitness,** by reproducing themselves rather than being sterile, but they have been manipulated by their parents. Often
more than one factor has to be invoked to explain any apparent pattern in animal behavior.

Photo 4.1    *Grooming behavior in Japanese macaques. You scratch my back, I'll scratch yours, is
an easy to understand maxim. Other types of altruism are much more difficult to explain.*
(Kaufman, Peter Arnold AN-BE-25A.)

As an example, it would be easy to say that the long neck of the giraffe is an adaptation to feeding on high foliage, but it could equally well aid in predator detection. There are often these confounding variables in behavioral ecology, which make it tricky to decide which selective pressures cause a particular trait. Often it may be more than one.

## Castes

Perhaps the most extreme altruism is the evolution of sterile casts in the social insects, in which some females, known as workers, rarely reproduce themselves but instead help others to raise offspring (Photo 4.2).

The differentiation of one species into many different-sized castes in ants can be quite staggering (Fig. 4.2). Of the 263 living genera of ants known worldwide, 44 possess species with caste systems.

The explanation of the peculiar system of castes was thought to lie primarily in the particular genetics of most (at least hymenopteran) social insect reproduction (Hamilton 1967)

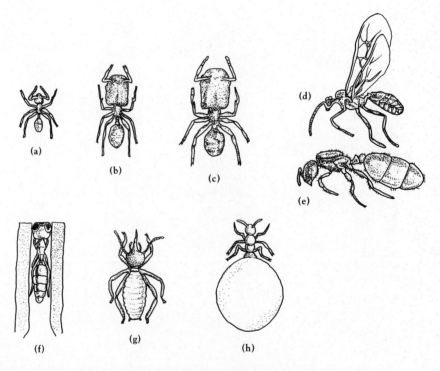

Figure 4.2   *Examples of castes in social insects. The top row shows the female castes and male of the myrmicine ant, Pheidole kingi instabilis.* (**a**) *Minor worker.* (**b**) *Media worker.* (**c**) *Major worker.* (**d**) *Male.* (**e**) *Queen. The bottom row shows various specialized castes in other species.* (**f**) *Soldier of the ant,* Camponotus truncatus, *blocking a nest entrance with its pluglike head which serves as a "living entrance" to the nest.* (**g**) *A sterile caste of the nasute termite,* Nasutitermes exitiosus, *which has a head shaped like a water pistol to spray noxious substances at an approaching enemy.* (**h**) *Replete worker of a* Myrmecocystus *ant, which lives permanently in the nest as a "living storage cask."* (From Wilson 1971.)

Photo 4.2   *In bees, wasps, ants, and termites, sterile castes or workers, like these paper wasps, exhibit the extreme altruistic act of forgoing reproduction.*   (Smith, Photo Researchers, Inc. OU 5589.)

(Fig. 4.3). Males develop from unfertilized eggs and are **haploid.** Male gametes are formed without meiosis, so every sperm is identical. Thus each daughter receives an identical set of genes from her father. Half of a female's genes come from her **diploid** mother, so the total relatedness of sisters is 0.75. Such a genetic system is called **haplodiploidy.** Thus, females are more related to their sisters than they would be to normal offspring. It is therefore advantageous to stay in the nest or hive and to try to produce new reproductive sisters.

The story is complicated a little bit when the interests of the queen are incorporated into the propensity for sterile castes. Queens are equally related to their sons and their daughters; $r = 0.5$ in each case. To maximize her reproductive potential, a queen should therefore produce as many sons as daughters, a 50:50 sex ratio. If she did, then sterile worker females would spend as much

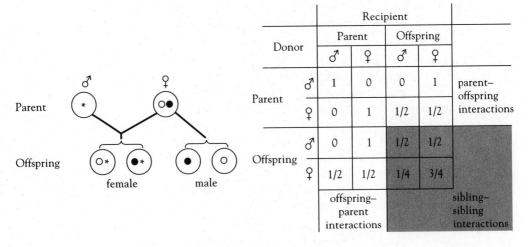

Figure 4.3   *Coefficients of genetic relationship in a haplodiploid mating system.*   (After Ricklefs 1990.)

time rearing brothers (to which they are related only by 0.25) as sisters. The average relatedness of a female to her siblings would then be 0.5, and she would do equally well to breed on her own. From the workers' viewpoint, it is far better to have more sisters, and in this conflict with the queen they appear to have won because in any colony there are more females than males, by a ratio of about 3:1 (Trivers and Hare 1976; but see also Alexander and Sherman 1977).

Elegant though these types of explanations are, they do not provide the whole picture: Large social colonies exist in termites, too, but theirs is not a haplodiploid system, and colony males have only half their genes in common, on average. Furthermore, naked mole rats, mammals living underground in South Africa, have a division of labor based on castes and cooperative broods, but the animals are, of course, **diploid** (Jarvis 1981; Jarvis et al. 1994). There is only one breeding female, the queen; she apparently suppresses reproduction in other females by producing a chemical in her urine that is passed around the colony by grooming after visits to a communal toilet. The other castes perform different types of work in burrowing. One rat chisels away at the face of the burrow, and others shuffle the earth backward toward the burrow entrance. There, another individual kicks it out onto the surface. The earth carriers then return to the face of the burrow by leapfrogging over the crouched earth removers (Fig. 4.4). One caste, the "frequent workers," appears to do most of this work; another, the "infrequent workers," are heavier individuals that do some of the work, and even larger individuals, the nonworkers, rarely work at all. These are often male and may be a reproductive caste. Even before the case of the mole rats was known, Richard Alexander, curator of the Museum of Zoology at the University of Michigan, had suggested that it is the particular life style of animals, not genetics, that promotes eusociality (Alexander 1974). He argued that in a normal diploid organism, females are related to their daughters by 0.5 and to their sisters by 0.5, so it matters little to them whether they rear sibs or daughters of their own. He predicted that mammals could exhibit a castelike society under certain conditions:

1. Where the individuals of the species are confined in nests or burrows.
2. Where food is abundant enough to support a high concentration of individuals in one place.
3. Where adults exhibit parental care.
4. Where there are mechanisms by which mothers can manipulate other individuals.
5. Where "heroism" is possible, whereby individuals give up their lives and by so doing can save the queen.

These factors can immediately account for eusociality in the termites as well. In mole-rat colonies, where the burrows become as hard as cement—a heroic effort by a mole rat effectively stops a predator (commonly a snake) because predators cannot rip open the surrounding substrate. Self-sacrifice by a "worker" does translate into a genetic gain. In mole rats, the queen reigns supreme; all workers and nonworkers, whether male or female, develop teats during her pregnancy—testament to the power of her pheromonal cues. The superabundant food comes in the form of tubers of the plant *Pyrenacantha kaurabassana*, which weigh up to 50 kg and can provide food for a whole colony. Often the tubers remain only half-eaten and can regenerate in time. Why eusociality has developed in this, but not in other, systems of burrowing rodents such as prairie dogs is open to speculation. It could be that movement between mole-rat colonies is effectively zero, promoting intense inbreeding.

Hamilton's genetic-relatedness theory seems even less appealing when one considers that the relatedness of workers in a hymenopteran colony is extremely close only if the colony is formed

Figure 4.4  *Mole rat bucket-brigade.* (Based on Jarvis and Sale 1971.)

by a single queen who has mated once. When a queen mates twice and sperm mixes at random, the average relatedness between sisters is only 0.5. Evidence has come to light that in a variety of hymenopterans, unrelated queens often initiate colonies together, dropping from mating swarms at the same time and engaging in pleiometrosis (cooperative nest digging and egg laying) (see references given by Strassman 1989). Some queens thus give up the opportunity to lay their own eggs, tending instead to the young of a cofoundress. In fire ants, the number of cofoundresses commonly varies from two to five. How is this behavior explained? Females must not be able to predict which one will become the eventual egg layer when they begin a nest together. The benefits of nesting in a group must also be great. Part of the reason for cooperation apparently lies in the fact that many hymenopteran nests or colonies are clumped. Brood raiding by neighboring colonies is common (especially in ants), so attaining a large worker force quickly is critical to colony survival. In fire ants, especially, brood raiding is widespread, and in a study of newly initiated nests, Tschinkel (1990) documented the eventual merger of 80 nests into only two over the course of a month. Although this case was exceptional, brood raiding from four or five colonies over the space of a few hours is common, and the raids go back and forth from nest to nest until one colony eventually ends up with all the brood. Other advantages to large colony size may involve increased defense against predators and parasites and general lower adult mortality.

## Sex

One of the biggest stumbling blocks to the complete acceptance to individual selection over group selection is the existence of sex itself. The traditional explanation to "why have sex?" was based on group selection because it demanded that an individual share its genes with those of another individual when making young. If this did not happen, the species would not evolve and could, a few thousand generations later, be replaced by species that did. For this reason, sexual species were thought to be better off than asexual species. A sexual gene could spread only if it doubles the number of offspring an individual could have, which seems absurd. Imagine that an individual decided to forgo sex and instead pass on all its genes to its own offspring, taking none from its mate. It would then have passed twice as many genes on to the next generation as its rivals had. This would probably put it at a huge advantage because it would contribute twice as much to the next generation and would soon be represented most commonly in the species. Individuals that abandon sex could theoretically outcompete their sexual rivals in passing on genes. But in nature they do not, so sex must be beneficial in some way.

At first, a "lottery" model was popular. The logic here was that an asexual species could produce hundreds of equal quality offspring, but what counts in the evolutionary race is a handful of exceptional offspring. Breeding asexually is like having lots of lottery tickets with the same number. To stand a chance of winning the lottery, you need lots of different tickets. To produce a future president of the United States, one might need a few exceptional offspring, not a lot of "average" offspring. Unfortunately, when scientists looked, there was no correlation between sexuality and ecological uncertainty. Sex should be commonest where, in fact, it is rarest—among highly fecund, small creatures in changeable environments. There, sex is the exception. It is among big, long-lived, slow-breeding creatures in stable environments that sex is the rule.

In the 1980s, sex was linked to a phenomenon known as "Muller's rachet." This theory proposed that sex exists to cleanse a species from accumulated damaging genetic mutations. When mieosis occurs and DNA is copied, some defects tend to accumulate as mutations occur. As successive mutations occur, it is likely that they will add more defects rather than repair previous ones. An analogy is that if you use a photocopier to make a copy of a copy of a copy of a document, the quality deteriorates with each successive copy. Only by using the unblemished original can you regenerate a clean copy. In the 1980s, it is was suggested that sex throws Muller's rachet into reverse. With sexual reproduction, new "perfect" individuals can arise from recombination where the damaging mutations are now absent. Again the field data do not conform to the theory. When two species—say, an aphid that breeds asexually and one that breeds sexually—are competing for the same resource, the sexual population would probably be driven extinct by the asexual population's greater productivity, unless the asexual species' genetic drawbacks appear quickly—and this seems unlikely.

The most recent, and therefore the most accepted theory, is that sex is needed to fight disease. Diseases specialize in breaking into cells, either to eat them, as with fungi and bacteria, or, like viruses, to subvert their genetic machinery for the purpose of making new viruses. To do that, the diseases use protein molecules that bind to other molecules on cell surfaces. There is necessarily a struggle between parasites and their hosts for the use of these binding proteins. Parasites invent new "keys" to use them, and hosts are constantly changing the locks so that parasites are not able to bind with proteins. The logic is that if one so-called lock is common in one generation, the key that fits it will spread like wildfire. Thus, the individuals who have this lock will be quickly killed, and this type of lock will not be common a few generations later. Sexual species have a "library" of locks unavailable to sexual species. Most notoriously variable, or polymorphic, genes are the very ones that affect resistance to disease—the genes for locks. Examples are genes that encode histocompatibility for antigens, proteins found on the surfaces of cells that help the immune system to distinguish "self" from foreign tissue or parasites. There is good evidence that asexuality is more common in species that are little troubled by disease, such as microscopic creatures and arctic or high-altitude plants and insects.

## 4.3  LIVING IN GROUPS

Individual selection can explain the occurrence of altruism and the existence of sex within populations. But another common occurrence in nature is the observation that many species occur in dense shoals, flocks, or herds that would surely only promote intense competition between individuals. If the central concern of ecology is to explain the distribution patterns of plants and animals, we must be able to provide a framework from which to understand the social behavior of animals. If dense congregations promote intense competition, there must be some high, selective advantages of group living to compensate. In today's world, group living may be disastrous in that animals that organize into dense aggregations can easily be harvested by people and reduced to very low numbers. For example, the control of vampire bats in Latin America was greatly facilitated by the bats' habit of communal roosting and grooming (Mitchell 1986). Bats groom for two hours a day, and pesticides applied topically to only a few bats were soon transferred to the others. A ratio of 15 or 16 dead vampires to each one treated was obtained. The annual benefit to farmers in Nicaragua, resulting from increased milk production from host cattle, was U.S. $2,414,158 after vampire control. As annual costs were only $129,750, a favorable benefit-cost ratio of 18.6:1 was obtained.

Guppies, *Poecilia reticulata*, were first discovered in Trinidad in the nineteenth century by the Reverend P.L. Guppy. In their native habitat, guppies live in tighter groups when they are in streams in which predators are more common (Fig. 4.5), suggesting that being in a

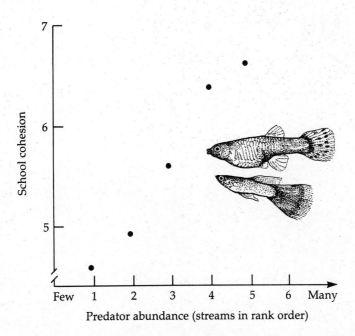

Figure 4.5   *Intraspecific variation in group size may be related to predators. Guppies* (Poecilia reticulata) *from streams with many predators live in tighter schools than those from streams with few predators. Each dot is a different stream, and "cohesion" was measured by a count of the number of fish in grid squares on the bottom of the tank.* (Modified from Seghers 1974.)

group helps an individual to avoid becoming a meal. Group living could reduce predator success in several different ways.

### Increased Vigilance

For many, a predator's success depends on surprise; if a victim is alerted too soon during an attack, the predator's chance of success is low. Goshawks *(Accipiter gentilis)* are less successful in attacks on large flocks of pigeons *(Columba palumbus)* mainly because the birds in a large flock take to the air when the hawk is still some distance away (Photo 4.3). If each pigeon occasionally looks up to scan for a hawk, the bigger the flock, the more likely that one bird will spot the hawk early. Once one pigeon takes off, the rest follow (Fig. 4.6) (Kenward 1978). Of course, cheating is a possibility because some birds might never look up, relying on others to keep watch while they keep feeding. However, at least in groups of Thompson's gazelle, the individual that happens to be scanning when a predator approaches is the one most likely to escape (Fitzgibbon 1989). This tends to dissuade cheating.

### Dilution Effect

Normally, predators take only one prey item per attack. An individual antelope in a herd of 100 has only a 1 in 100 chance of being attacked, whereas a single individual has a 1 in 1 chance. Of course, large herds may well be attacked more, but a herd is hardly likely to attract 100 times more attacks

Photo 4.3   *Why do animals live in groups, when groups may promote competition between individuals? For these snow geese, large flocks may be better able to detect predators, such as the bald eagle shown here just skyward of the flock.* (Schumacher, Peter Arnold AN-BE-50C.)

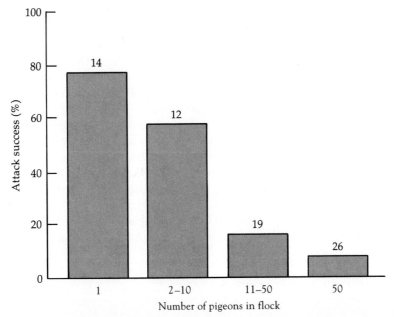

Figure 4.6   *The value of flocks of wood pigeons in detecting the approach of an avian predator, a goshawk. The graph shows the dependence of the goshawk's attack success on the size of the flock.* (Redrawn from Kenward 1978.)

than an individual. In the Camargue, a marshy delta in the south of France, horses in large groups suffer less attacks per horse from biting tabanid flies than solitary horses (Duncan and Vigne 1979).

Associated with herds is a tendency to prefer the middle because predators are likely to attack prey on the edge of the group. Part of the reason may be the difficulty of visually tracking large numbers of prey. Throw two tennis balls to a friend, and the chances are that he or she will drop them both, whereas one ball can be tracked much more easily. A similar phenomenon may operate in predator-prey interaction, explaining why predators take peripheral individuals (Neill and Cullen 1974). It may also be physically difficult to get to the center of a group, with many herds tending to bunch close together when they are under attack (Hamilton 1971). Furthermore, large numbers of prey are able to defend themselves better than single individuals, which usually flee. Nesting black-headed gulls mob crows remorselessly and reduce the success of the crows at stealing gulls' eggs (Kruuk 1964). Natural predators are rare in the Hawaiian Islands, and birds there seldom flock (Willis 1972). Who gets the best spots within the group? Perhaps the older, the more experienced, or the bigger individuals commandeer the best positions. Perhaps the fastest individuals commandeer the positions at the edge of the group because they can flee more rapidly. It is usually easy to come up with more than one reason.

## Group Predation

Living in groups may confer advantages for predators as well as prey. Predatory groups often capture prey that are difficult for a single individual to overcome, either because the prey is too large (lions hunting adult buffalo) or because it is too elusive (killer whales hunting dolphins). For species that feed on large ephemeral food clumps such as seeds or fruits, the limiting factor is often the location of a good site or tree. Once it has been found, there is usually plenty of food. It has been proposed, with good evidence from *Quelea* birds in Africa, that in large roosts successful foragers may be followed by birds that had been previously unsuccessful. This "mutual parasitism" supposes that poor foragers are in some way able to distinguish successful birds (Ward and Zahavi 1973). Brown (1988) has shown that cliff swallows (*Hirundo pyrrhonota*) in southwestern Nebraska nest in colonies that serve as information centers in which unsuccessful individuals locate and follow successful individuals to aerial insect-food resources. Brown was able to factor out the confounding effects of increased ectoparasitism (by nest fumigation) on larger colonies and colony location (by reducing certain colony sizes) to show that increased nestling weight at larger colonies was due to more successful foraging, attributable to more efficient transfer of information among colony residents.

There are additional benefits and further drawbacks to group living (Table 4.2). For example, large groups may attract more parasites, but in small groups individuals may not

TABLE 4.2 *Examples of studies in which possible costs and benefits of group living other than those mentioned in the text have been measured. (After Krebs and Davies 1993.)*

| Hypothesis | Example |
| --- | --- |
| 1. Saving of energy by warm-blooded animals as a result of thermal advantage of being close together. | Pallid bats (*Antrozous pallidus*) roosting in groups use less energy than solitary neighbors. |
| 2. Chance for small species to overcome competitive superiority of a large species by being in a group. | Groups of striped parrot fish (*Scarus croicus*) can feed successfully inside the territories of the competitively superior damselfish (*Eupomacentrus flavifrons*). |
| 3. Hydrodynamic advantage for fish swimming in a school. They save energy by positioning themselves to take advantage of vortices created by others in the group. | Measurements of distances and angles between individuals show that they are not correctly positioned to benefit according to the predictions of the theory. |
| 4. Increased incidence of disease as a result of close proximity to others. | Number of ectoparasites in burrows of prairie dogs (*Cynomys* spp.) increase in larger colonies. Number of ectoparasitic bugs and fleas increases with size of cliff swallow (*Hirundo pyrrhonota*) colonies (Brown and Bomberger Brown 1986). |
| 5. Risk of cuckoldry by neighbors. | In colonially nesting red-winged blackbirds (*Agelaius phoeniceus*), the mates of vasectomized males laid fertile eggs. They must have been fertilized by males other than their mates. |
| 6. Risk of predation on young by cannibalistic neighbors. | In colonies of Belding's ground squirrel (*Spermophilus beldingi*), females with small territories are more likely to lose their young to cannibalistic neighbors than are females with large territories around their burrows. |

be able to find suitable mates. It is clear that conflicting selective pressures operate on group size to determine the eventual number of individuals in a herd or a flock. The operation of just two conflicting variables, competition for food and the presence of predators, on the size of bird flocks is illustrated in Figure 4.7. Increased predation rates increase flock size while increased food levels decrease lock size. Many other variables are likely to influence flock size at the same time.

To explain why some animals live in groups of 50 and others in groups of ten is a little harder. Even if there is a best or "optimal" group size—say seven, neighboring individuals would probably join the group because they could do better than they could alone. Figure

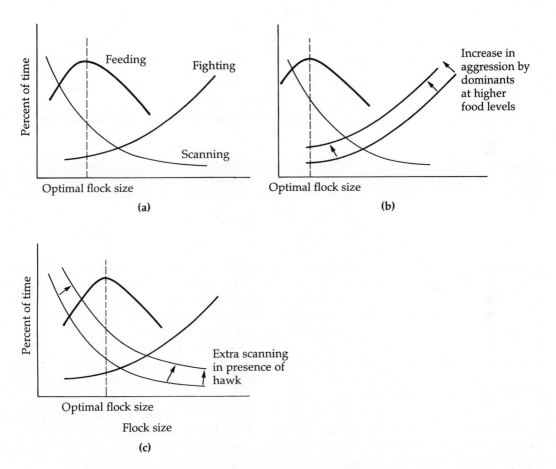

Figure 4.7    *A model of optimal flock size. (*a*) The trade-off between squabbling or fighting and scanning for predators. As flock size increases, birds spend more time fighting and less time scanning. An intermediate flock size gives the maximum proportion of time feeding. (*b*) The effect of an increase in resources on flock size. When food is more plentiful, dominant birds can afford to spend more time attacking subordinates. The optimal flock size for the average bird therefore decreases. (*c*) The effect of an increase in predation on flock size. When predation risk is increased by the flight of a hawk over the flock, the scanning level should go up, and the optimal flock size is increased.* (Modified from Krebs and Davies 1981.)

4.8 illustrates that, for this example, even where the group size becomes 14, individuals would still do better than they would alone (Sibly 1983). Thus, although the optimal group size might be seven, a range of two to 14 would be expected in nature.

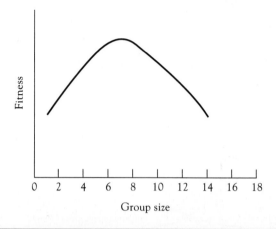

Figure 4.8    *Sibly's model of optimal and stable group size. Each individual joins the group that maximizes its fitness so that the optimal size of seven is not necessarily stable—it will be joined by solitary individuals for example.* (After Sibly 1983.)

Fitness

Group size

## Summary

1. Some ecologists believed that populations of species in nature are maintained at constant or equilibrium levels by group selection—self-regulation by individuals in a population to prevent overexploitation of resources. Attractive though the idea of group selection is, it has several flaws, such as mutation, immigration, and resource prediction. Individual selfishness seems a more plausible result of natural selection.

2. If individual selection is more likely than group selection, how can we explain acts of altruism in nature between individuals? One popular explanation that was invoked was kin selection—selection for behavior that lowers an individual's own chances of survival and reproduction but raises that of a relative that contains some of the same genes. Kin selection was invoked to explain the unusual caste systems of social insects, especially those with a haplodiploid mating system.

3. Eusociality and cooperation are not only found in haplodiploid organisms.

Termites and naked mole rats exhibit a caste system, too. In 1974, Richard Alexander predicted that certain ecological reasons could be sufficient to cause a castelike society. These reasons were confinement to burrows, high food concentrations, parental care of offspring, mechanisms for mothers to manipulate other individuals, and the opportunity for heroism.

4. While ecological or evolutionary reasons can explain the existence of altruism and social behavior, group size can often be seen to be a compromise between large groups that lessen the impact predation and small groups that lessen the impact of competition.

*Discussion Question:* One often reads that a change in the attitudes of humans is necessary to reduce environmental problems such as pollution or the overuse of resources. Such ideas are often based on altruism. This chapter has argued that individual selfishness is much more likely than altruism. Can humans act selfishly and still reduce environmental problems?

# Population
# Ecology

Wild lupines bloom in Acadia National Park, Maine. What is the most important source of mortality for them—competition with other plants, herbivory by insects or vertebrates, or a lack of soil nutrients or moisture? Such questions abound in nature and understanding what limits populations is the province of population ecology. (Till, DRK Photo 118010-0054)

What determines the abundance of species in nature? This question has continued to intrigue ecologists ever since 1957, the date of the famous Cold Spring Harbor Symposium in North America at which the proponents of various views vigorously aired their opinions. The theme of the symposium was the debate over density dependence versus density independence (see Chapter 12). On the one hand, there were the correlations of abundance and population changes of organisms with aspects of weather (Davidson and Andrewartha 1948a,b). There was also good demonstration that intrinsic rate of natural increase of a species could be infleunced by laboratory-controlled temperature and moisture (Birch 1953). On the other hand, there were many laboratory populations that showed a reasonably good fit to a model known as the logistic (which suggests that competition acts so as to slow the growth rates of populations at high densities). There were also some long-term field studies showing population fluctuations but a fairly constant mean level of abundance. Finally, there existed the apparently logical and mathematical argument that without density dependence, populations would fluctuate with increasing amplitude, either going extinct or reaching completely unrealistic numbers.

Since that time, the same types of questions have been posed many times over, but in a more sophisticated manner. Most ecologists accept that there are upper limits within which population densities are constrained, but that still doesn't necessarily mean that density dependence occurs (see Chapter 12). Also, many ecologists are concerned about whether density-dependent factors operate continually or are interrupted by stochastic disturbances. Some argue that disturbance itself is necessary for the stability of some systems (Lewin 1986). If that is true, we need to know the frequency of disturbance and the factors that constitute that disturbance. In order to address fully the density-dependent-versus-density-independent debate, we must examine how populations grow and whether they are affected much by the weather or by other organisms such as competitors, parasites, predators, and—in the case of plants—herbivores. Only after an examination of these factors can we begin to construct a synthesis about which type of population is controlled by which type of factors.

# *Physiological Ecology*

The local distribution patterns of most species are limited by certain physical or **abiotic factors** of the environment such as temperature, moisture, light, pH, soil quality, salinity, and water current. Leibig's law of the minimum, coined by Justus Leibig in 1840, states that the distribution of species will be controlled by that environmental factor for which the organism has the narrowest range of adaptability or control. Often, rather than just one limiting factor, there are many factors, all interacting. For example, forest trees differ in their tolerance to shade. Sun-loving species soon overtop the shade-loving species because of their faster rate of growth in full sun (see Fig. 5.1). However, they cast shade that is too dark for their own seedlings and that inhibits their growth. Light levels below their crowns are still sufficient for the growth of those species that do better in intermediate light levels, and in turn for species the most tolerant of shade. Thus, a stratified canopy may develop with sun-loving species at the highest level, intermediate species below them, and shade-loving species lower still. Plant growth rates differ not only because of shade but also for other resources such as the availability of water and nutrients (nitrogen, phosphorus, potassium, calcium, magnesium, and sulfur). A lack of nitrogen is by far the most critical factor limiting growth of certain species. Species responses to nitrogen availability can be described in much the same way as their response to light. However, species that are tolerant of low light levels are not necessarily also tolerant of low nitrogen availability. Table 5.1 lists some species of eastern deciduous tree forests and their requirements as regards tolerance to shade and nitrogen. The data in the table indicate where each species might be common within a large forested area that contains young/disturbed and old/undisturbed stands. For example, white ash is intermediate for low levels of light and intolerant for lack of nitrogen, and it tends to occur in small disturbances on sites with rich soil. It may now be apparent why there are so many species of tree in a given forest region. Nine distinct types of environment can be delineated by their light and nitrogen availability alone.

Figure 5.1 *Relative rates of tree growth as a function of light availability for sun-loving, intermediate, and shade-loving species.*

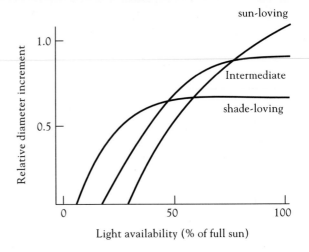

Some species are tolerant of a wide range of environmental conditions (eurytopic), others of only a narrow range (stenotopic), but each functions best only over a limited part of the gradient, and this is termed a species' *optimal range* (Fig. 5.2). It must also be remembered that part of a preferred optimal range may already be occupied by a competitively superior species. In the field, species may not occupy their full ranges, as measured in the laboratory in terms of abiotic factors, because of competition with other organisms (Fig. 5.3) (see also Chapter 7).

By and large, it is the young stages of both plants and animals that are most sensitive to environmental factors. Large trees are more likely to be able to tolerate drought conditions because their roots can penetrate further into the soil and tap lower water levels. Adult animals are often more mobile than juveniles and are better able to find shelter in times of stress.

TABLE 5.1 **Separation of some deciduous tree species in the northeastern United States by tolerance classes for shade and low nitrogen availability.**

| | | Tolerance class for shade | | |
| --- | --- | --- | --- | --- |
| | | *Tolerant* | *Intermediate* | *Intolerant* |
| Tolerance class for low nitrogen availability | *Tolerant* | Closed canopies—poor sites<br>Hickory | Gaps—poor sites<br>White oak<br>Chestnut oak | Open—poor sites<br>Bigtooth aspen |
| | *Intermediate* | Closed canopies—moderate sites<br>Beech<br>Red maple | Gaps—moderate sites<br>Basswood | Open—moderate sites<br>Trembling aspen |
| | *Intolerant* | Closed canopies—rich sites<br>Sugar maple<br>Black gum | Gaps—rich sites<br>White ash<br>Northern red oak | Open—rich sites<br>Tulip poplar |

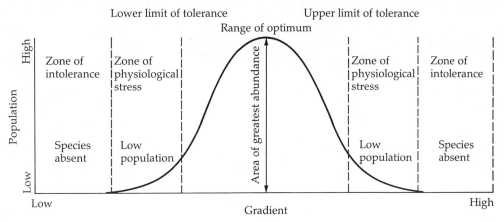

**Figure 5.2** *Organismal distribution along a physical gradient.* (Modified from Cox, Healey, and Moore 1976.)

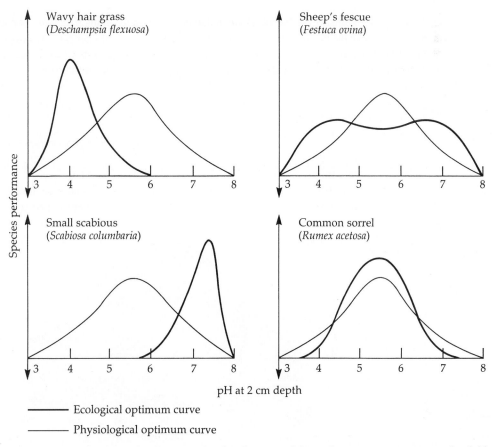

**Figure 5.3** *The difference between the distributions of four plant species growing in the field (ecological optimum curve) and under noncompetitive conditions in controlled laboratory plots (physiological optimum curve).* (Redrawn from Collinson 1977.)

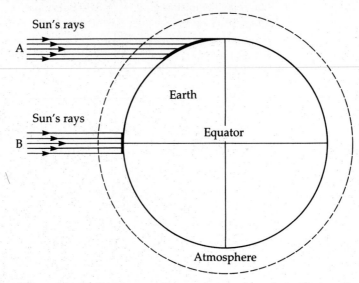

**Figure 5.4**  *Effect of the Earth's shape and atmosphere on incoming radiation. In polar areas the sun's rays strike the Earth in an oblique manner (**A**) and deliver less energy than at tropical locations (**B**) for two reasons: (1) because the energy is spread over a larger surface in **A** and (2) because it passes through a thicker layer of absorbing, scattering, and reflecting atmosphere.*

## 5.1  CLIMATE

There are substantial temperature differentials over the Earth, a large proportion of which are due to variation in the incoming solar radiation. In higher latitudes, the sun's rays hit the Earth obliquely and are thus spread out over more of the Earth's surface than they are in the equatorial regions (Fig. 5.4). More heat is also dispersed in the higher latitudes because the sun's rays travel a greater distance through the atmosphere. The result is a much smaller (40 percent) total annual

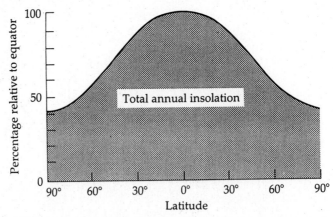

**Figure 5.5**  *Insolation at different latitudes during the year. The amount of solar energy is expressed as a percentage of the amount at the equator.*

insolation in polar latitudes than in equatorial areas (Fig. 5.5). In the summer, increased day length in high latitudes increases insolation, but shorter day length in winter decreases the daily total. The reason is that the Earth's axis of rotation is inclined at an angle of 23.5° (Fig. 5.6); the Northern Hemisphere is treated to long summer days, while the Southern Hemisphere has win-

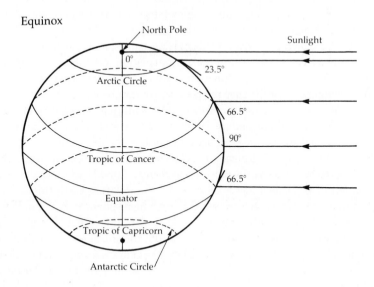

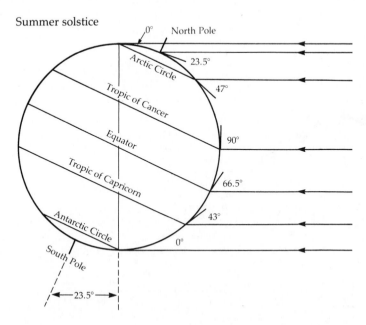

**Figure 5.6** *Effects of the Earth's inclined axis of rotation on amount of insolation. The Earth's axis of rotation is inclined at an angle of 23.5 degrees, which causes increasing seasonal variation in temperature and day length with increasing latitude.*

ter, and vice versa. At the summer solstice in the Northern Hemisphere (June 22), light falls perpendicularly on the tropic of Cancer; on December 22, it shines perpendicularly on the tropic of Capricorn. On March 22 and September 22 (the equinoxes), the sun's rays fall perpendicularly on the equator, and every place on Earth receives the same day length. These effects do not translate into a linear relationship between temperature at the surface and latitude—at the tropics both cloudiness and rain reduce mean temperature, and relatively cloud-free areas beyond this zone increase mean temperature relative to isolation (Fig. 5.7).

Global temperature differentials create winds and drive atmospheric circulation. The first contribution to a classical model of general atmospheric circulation was made by George Hadley in 1735. Hadley proposed that solar energy drove winds which in turn influenced the circulation of the atmosphere. He proposed that the large temperature contrast between the very cold poles and the hot equator would create a thermal circulation. The warmth at the equator caused the surface equatorial air to become buoyant and rise vertically into the atmosphere. As it rose away from its source of heat, it cooled and became less buoyant but was unable to sink back to the surface because of the warm air behind it. Instead, it spread north and south away from the equator, eventually returning to the surface at the poles. From there, it flowed back toward the equator to close the circulation loop. Hadley suggested that on a nonrotating Earth this air movement would take the form of one large convection cell in each hemisphere as shown in Figure 5.8.

When the effect of the Earth's rotation is added, the surface flow becomes somewhat easterly (toward the west). This is a consequence of the the so-called Coriolis effect. In any one revolution of the Earth, any point on the equator circumscribes a greater circle than any point north or south. Thus, the equator of the Earth is moving fastest during rotation. Imagine a rocket fired from the north pole toward the equator. In the time it took the rocket to reach the equator, hours later, that point on the equator may have moved 15° to the east.

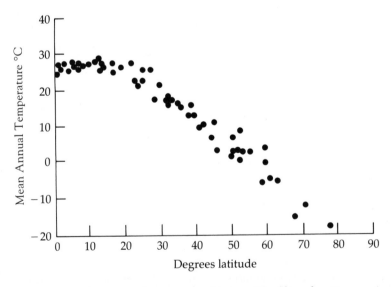

**Figure 5.7** *Mean annual temperature (degrees C) of low-elevation, mesic, continental locations on latitudinal gradient. Note the wide band of similar temperatures between 20 degrees S and 20 degrees N.* (Redrawn from Terborgh 1973.)

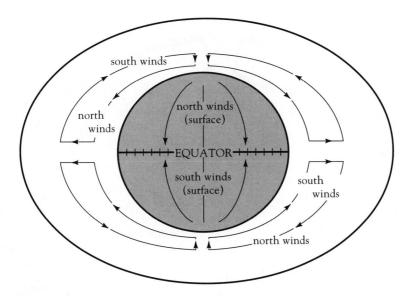

**Figure 5.8** *Simple convective circulation of air on a uniform, nonrotating Earth, heated at the equator and cooled at the poles, according to the scientist George Hadley in 1735.*

Thus, the straight line of the rocket's path effectively becomes curved toward the west. A similar phenomena occurs with winds.

Attractive though this simple theory is, we have to modify it to fit the data. In the 1920s, a three-cell circulation in each hemisphere was proposed to fit the Earth's heat balance (Fig. 5.9).

The contribution of George Hadley is still recognized in that the most prominent of these three cells, the one nearest the equator, is called the Hadley cell. In the Hadley cell, the warm air rising near the equator forms towers of cumulus clouds that provide rainfall, which in turn maintains the lush vegetation of the equatorial rain forests (Fig. 5.10). As the upper flow in this cell moves poleward, it begins to subside in a zone between 20° and 35° latitude. Subsidence zones are areas of high pressure and are the sites of the world's tropical deserts. This is so because the subsiding air is relatively dry, having released all its moisture over the equator. In the subsidence zone, winds are generally weak and variable near the center of this zone of descending air. The region has popularly been called the horse latitudes. The name is said to have been coined by Spanish sailors who, crossing the Atlantic, were sometimes becalmed in these waters and reportedly were forced to throw horses overboard as they could no longer water or feed them.

From the center of the horse latitudes, the surface flow splits into a pole branch and an equatorial branch. The equatorial flow is deflected by the Coriolis force and forms the reliable trade winds. In the Northern Hemisphere, the trades are from the northeast, where they provided the sail power to explore the New World; in the Southern Hemisphere, the trades are from the southeast. The trade winds from both hemispheres meet near the equator in a region that has a weak pressure gradient, the intertropical convergence zone. This region is also called the Doldrums. Here the light winds and humid conditions provide the monotonous weather that may be the basis for the expression "down in the doldrums."

In the three-cell model, the circulation between 30° and 60° latitude is just opposite that of the Hadley cell. Net surface flow is poleward, and because of the Coriolis effect, the winds have

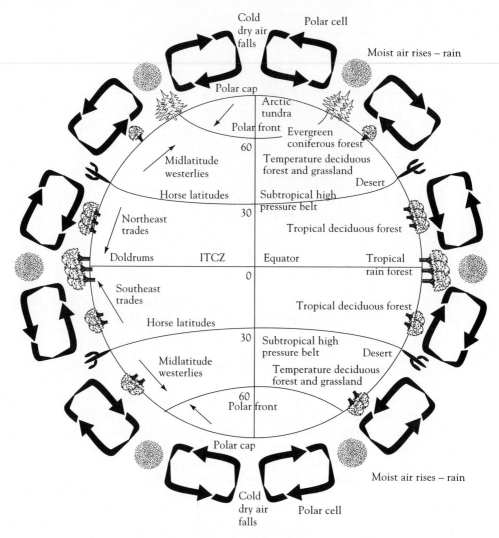

**Figure 5.9** *Three-cell model of the atmospheric circulation on a uniform, rotating Earth heated at the equator and cooled at the poles. The direction of air flow and the ascent and descent of air masses in six giant convection cells determine Earth's general climatic zones. This uneven distribution of heat and moisture over different parts of the planet's surface leads to the forests, grasslands, and deserts that make up the planet's biomass.*

a strong westerly component. These prevailing westerlies were known to Benjamin Franklin, perhaps the first American weather forecaster, who noted that storms migrated eastward across the colonies.

Winds, together with the rotation of the Earth, create currents. The major currents act as "pinwheels" between continents, running clockwise in the ocean basins of the Northern Hemisphere and counterclockwise in those of the Southern Hemisphere. Thus, the Gulf Stream, equivalent in flow to 50 times all the world's major rivers combined, brings warm water from the Caribbean and the U.S. coasts to Europe, the climate of which is correspondingly

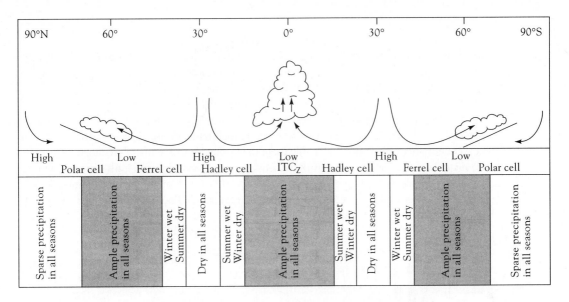

**Figure 5.10**   *Schematic illustration of zonal precipitation patterns.*

moderated. The Humboldt current brings cool conditions almost to the equator along the western coast of South America (Fig. 5.11). The climates of coastal regions may differ markedly from those of their climatic zones; they may never experience frost, and fog is often evident.

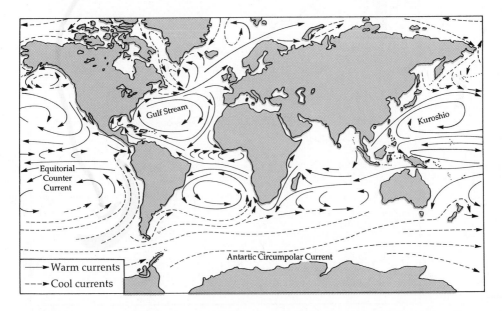

**Figure 5.11**   *Main patterns of circulation of the surface currents of the oceans. In general the major circular gyres in each ocean move clockwise in the Northern Hemisphere and counter-clockwise in the Southern Hemisphere. This pattern results in warm currents along the eastern coasts of continents and cold currents along the western coasts.*

## 5.2 TEMPERATURE

Environmental temperature is an important factor in the distribution of organisms because of its effect on biological processes and the inability of most organisms to regulate body temperature precisely. In plants, temperature is particularly important because cells may rupture if the water they contain freezes. High temperature is critical because the proteins of most organisms denature, that is, are destroyed, at temperatures above 45°C. In addition, few organisms can maintain a sufficiently high metabolic rate at very low or very high temperatures. Some organisms, of course, have extraordinary adaptations to enable them to live outside this temperature range. Penguins can survive the coldest regions. As warm-blooded organisms, mammals and birds can cope with temperature extremes better than many animals because they can regulate their own temperature. But even mammals and birds function best within certain environmental temperature ranges.

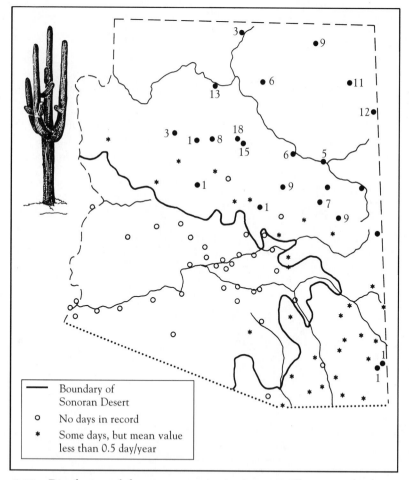

**Figure 5.12** *Distribution of the saguaro cactus in Arizona. There is a close correspondence between the northern and eastern edges of the cactus's range in the Sonoran Desert and the line beyond which it occasionally fails to thaw during the day. The numbers are mean numbers of days per year with no rise above freezing.* (Modified from Hastings and Turner 1965.)

Temperature resistance in plants, though poorly understood, is often critical to their distribution patterns. Under cold conditions, water must be moved outside cell walls or be bound up in such a chemical form that it cannot change to damaging ice, which would rupture the cellular machinery. Injury by frost is probably the single most important factor limiting plant distribution. As an example, the saguaro cactus can easily withstand frost for one night as long as it thaws in the day, but it will be killed when temperatures remain below freezing for 36 hours. In Arizona, the limit of the cactus's distribution corresponds to a line joining places where, on occasional days, it fails to thaw (Fig. 5.12). For some plants, general coldness, not freezing, limits distribution. The northern boundary of the wild madder, *Rubia peregrina*, in Europe coincides with the January 4.5°C isotherm (Fig. 5.13), and it has been suggested that this temperature is critical for the early growth phases of new shoots.

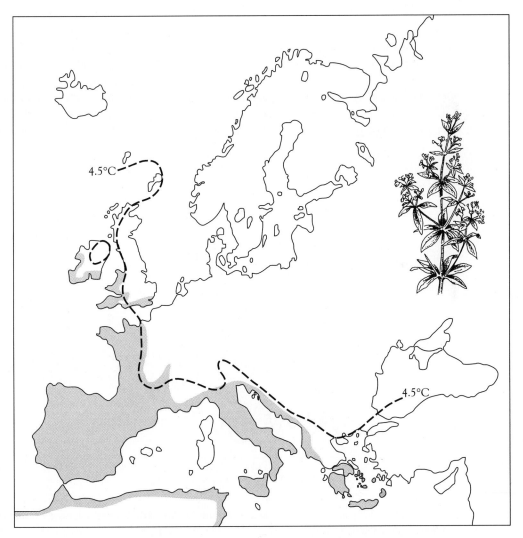

**Figure 5.13**   *The distribution of wild madder* (Rubia peregrina) *in Europe (shaded) and the location of the January isotherm for 4.5 degrees C.*   (Modified from Cox, Healey, and Moore 1976.)

Frost injury has caused losses to agriculture of more than $1 billion annually in the United States and has been considered an unavoidable result of subfreezing temperatures, but genetic engineering is changing the trend. Frost injury is precipitated by the ice-nucleation activity of just five species of bacteria that live on plant surfaces. Recently, the DNA sequences conferring ice nucleation have been identified, isolated, and prevented from working in an engineered strain of one of them, *Pseudomonas syringae* (Lindow 1985). When such a strain is allowed to colonize plants, frost damage is greatly reduced, and plants can withstand approximately 5°C cooler temperatures before frost forms. The promise of this technique for the increase of agricultural yields and the alteration of normal plant-distribution patterns is staggering.

## Global Warming

Because many species are limited in their distribution patterns by global temperatures, there is concern that if global temperatures change, then some species will be driven to extinction or that their geographic range will change and that the geographic location of many centers of agriculture and forestry will be altered. Global warming may be caused by changes of any of a variety of greenhouse gases such as methane, chlorofluorocarbons, nitrous oxides, and carbon dioxide. Of these gases, however, carbon dioxide is thought to be most important. Carbon-dioxide levels have continued to increase because of the

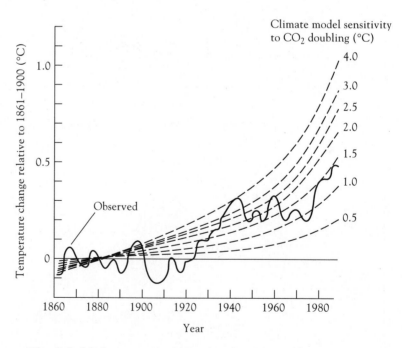

**Figure 5.14**   *Observed global mean temperature changes 1861–1989 compared with predicted values.*  (After Wigley and Raper 1991.)

increase in the burning of fossil fuels, the production of cement, and the increase in global deforestation (see Chapter 20).

Observed global temperature changes from 1860 to 1990 have been small (Fig. 5.14), but the consensus of scientists remains that a warming of between 1.5 and 4.5°C is likely if carbon dioxide or any other greenhouse gas equivalent were to double in the next century. The best guess is that a global warming of 2.5°C will occur under such a scenario (Table 5.2). Global temperature changes will also have indirect effects, notably a decrease in precipitation in some areas but an overall global average increase (Table 5.2).

**TABLE 5.2** *Major equilibrium changes in climate due to doubling* $CO_2$*, as deduced by models. The number of stars indicates the degree of confidence in the predicted change. Five stars indicate virtual certainties; one star indicates low confidence in the prediction. (After IPCC 1990.)*

| Degree of confidence | Predicted change |
|---|---|
| | *Temperature* |
| ***** | The lower atmosphere and Earth's surface warm. |
| ***** | The stratosphere cools. |
| *** | Near the Earth's surface, the global average warming lies between +1.5°C and +4.5°C, with a "best guess" of 2.5°C. |
| *** | The surface warming at high latitudes is greater than the global average in winter but smaller in summer. (In time-dependent simulations with a deep ocean, there is little warming over the high-latitude ocean.) |
| *** | The surface warming and its seasonal variation are least in the tropics. |
| | *Precipitation* |
| **** | The global average increases (as does that of evaporation); the larger the warming, the larger the increase. |
| *** | Increases at high latitudes throughout the year. |
| *** | Increases globally by 3 to 15 percent (as does evaporation). |
| ** | Increases at midlatitudes in winter. |
| ** | The zonal mean value increases in the tropics, although there are areas of decrease. Shifts in the main tropical rain bands differ from model to model, so there is little consistency between models in simulated regional changes. |
| ** | Changes little in subtropical arid areas. |
| | *Soil Moisture* |
| *** | Increases in high latitudes in winter. |
| ** | Decreases over northern midlatitude continents in summer. |
| | *Snow and Sea Ice* |
| **** | The area of sea ice and seasonal snow cover diminish. |

Rapid global warming will increase the number of species that are maladapted to their climatically changed habitats. This has lead some people into a debate over whether humans are needed to remove some species from the field and hold them in captive breeding programs for reintroductions later when the climate reverts back to "normal." Alternatively, translocating wild individuals into potentially more hospitable environments is an option. In this scenario, evolution is usually assumed to be irrelevant because it is thought that most species cannot evolve significantly or rapidly enough to counter climate changes. The anticipated changes in global climate are expected to occur at a rate that will probably be too rapid to be tracked by evolutionary processes such as natural selection. It is also not likely that all species can simply disperse and move north or south into the newly created climatic regions that will be suitable for them. Many tree species take hundreds, even thousands, of years for substantial dispersal via progeny. The tree that has one of the fastest known dispersal rates is spruce, which expanded its range an average of only 20 km every ten years, about 9,000 years ago (Roberts 1988). Because plants can "move" only slowly, the resultant changes in the floras will have dramatic effects on animals, both on the large, biogeographic scale and on the local community scale. The present geographic range of sugar maple and its potentially suitable range under a doubled $CO_2$ regime is shown in Figure 5.15. For the red-cockaded woodpecker, habitat is mature pine and pine/oak forests. If climate change occurred more rapidly than a dispersal rate of these trees, resulting in individuals dying in the sun before individuals can be established in the north, then the red-cockaded woodpecker habitat

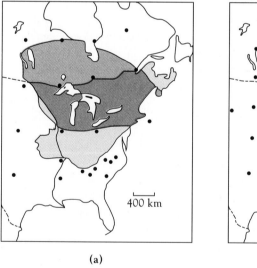
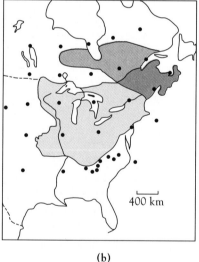

(a)                                              (b)

**Figure 5.15**   *The present geographic range of sugar maple (horizontal lines) and its potentially suitable range under doubled $CO_2$ (vertical lines). Cross-hatching indicates the region of overlap. Note that different models give different predictions: (**a**) uses a scenario from the Goddard Institute for Space Studies. (**b**) uses a scenario derived from the Goddard Fluid Dynamics Laboratory. Grid points are sites of climatic data output for each model.* (From Davis and Zabinski 1992.)

would be very much compressed. If the temperature was to rise fast enough and mature trees died before new established trees reached maturity, then the woodpecker would go extinct.

The odds or true extent of global warming are not known to perfection. Some economists argue that the economic changes following global warming are not sufficient to justify spending lots of money to prevent "minor" global warming. However, if global warming is severe, then clearly we would be justified in taking extreme measures to prevent it. A good analogy is that of a person having a 1 percent chance of catching a cold if a certain activity is performed. A cold is not a severe enough disease that a probability as low as 1 percent would modify anyone's behavior. However, if the cost of the activity is contracting a fatal disease, then a 1 percent probability should be of major concern, and dramatic steps should be taken to avoid the activity (Root 1993).

## 5.3  MOISTURE

Although temperature alone is an important limiting factor, moisture (a combination of water and temperature) has an equally important effect on the ecology of terrestrial organisms. Protoplasm is 85–90 percent water, and without moisture there can be no life. Globally, there is a belt of high precipitation around the equator, that broadly corresponds to the tropics and a secondary peak between latitudes 45° and 55°. Rates of evaporation and transpiration are primarily dependent on temperature, hence the importance of water and temperature combined. Over about one-third of the Earth, evaporation exceeds precipitation—these areas are the deserts.

The distribution patterns of most plants are limited by available water. Some, for example, the water tupelo in the United States, do best when completely flooded and are thus predominant only in swamps. For many plants, the limiting amount of moisture is much lower. In cold climates, water can be present but locked up as permafrost and, therefore, unavailable—a frost-drought situation. A good example of plant limitation by water availability is the **timberline** on most mountain ranges. Alpine timberlines are determined by winter desiccation or frost drought. As one proceeds up a mountain, temperature decreases, rainfall increases, and wind velocity increases. Because temperatures are below freezing during much of the year, available soil moisture decreases. Desiccation is so severe above the timberline as a result of high winds and low temperatures that leaves cannot grow enough in the short summer period to mature, lay down a thick cuticle, and become drought resistant (Tranquillini 1979). Experiments by Hadley and Smith (1986) in southeast Wyoming showed that needle mortality of *Picea engelmannii* conifers was primarily due to winter wind and cuticle abrasion. Sheltering exposed trees from the wind increased needle survival.

Animals face problems of water balance, too, but most can move away from hot, dry, and intolerable environments. Many desert species are small and can hide underground in the heat of the day. Larger animals cannot be accommodated so easily, and because most depend ultimately on plants as food, their distributions are intrinsically linked to those of their food sources. The distributional boundary of the red kangaroo, *Macropus rufus*, in Australia coincides with the 400-mm rainfall contour, because the kangaroos are dependent on arid-zone grasses that are restricted to such low-rainfall areas (Fig. 5.16). In the wake of an extraordinary El Niño event (an irregular increase in water temperature in the eastern Pacific Ocean) in 1982–1983, the rainfall on Isla Genovesa, Galápagos, increased from its normal 100–150 mm during the rainy season to 2,400 mm from November 1982 through July 1983. Plants responded with prodigious growth, and Darwin's finches bred up to eight times, rather than their normal maximum of three (Grant and Grant 1987).

Water type may be important too for freshwater organisms. Such organisms often cannot tolerate the high salinity of the marine environment while, conversely, most marine organisms cannot deal with freshwater conditions.

## 5.4  WIND

Wind can be important as it amplifies the effects of environmental temperature on an organism by increasing heat loss by evaporation and convection (the wind-chill factor). Wind also contributes to water loss in organisms by increasing the rate of evaporation in animals and tran-

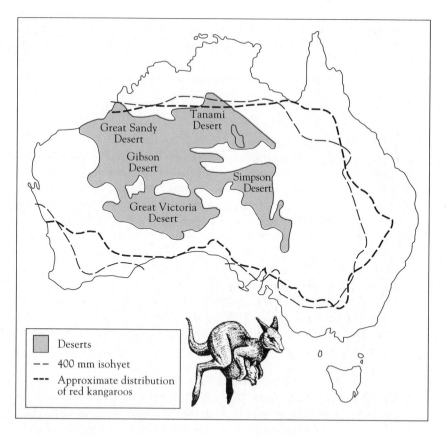

**Figure 5.16**  *Distribution of the red kangaroo in the arid regions of Australia and the 400 mm (15 inch) rainfall line. Red kangaroos are relatively rare in the large desert areas shown.* (Modified from Krebs 1985a.)

spiration in plants. In addition, wind aids in the pollination of many plants by blowing pollen from flower to flower, and it disperses plant seeds, so the strength of the wind is important.

King (1986) has analyzed the growth of tree form and height (using *Acer saccharum*, sugar maple) and its relation to susceptibility to wind damage. Trees should grow as rapidly as possible to escape shade conditions. There is therefore a trade-off between trunk height and diameter—that needed to keep the tree upright. Support efficiency was analyzed in terms of the ratio of actual trunk diameter to the minimum required to keep the tree erect (the stability safety factor). A trade-off was expected between height-growth efficiency (maximized by a minimally designed trunk) and ability to resist storms. The lowest stability safety factors (1.8) were observed in saplings. The maxima were observed in mature canopy trees, which had trunks two to six times the minimum needed to keep trees erect in the absence of winds. However, large trees snap more frequently in high winds than do the more supple saplings. Wind speed is higher in the canopy than closer to the ground, and large trunks are less flexible.

Wind can also be an important mortality factor (Photo 5.1). Fifteen million trees in the south of England perished on October 16, 1987, in the wake of a mighty storm

Photo 5.1   *This huge live oak tree,* Quercus virginiana, *was felled by strong winds in North Florida. Abiotic factors can clearly have a big influence on the population densities of many forms of life.* (Photo by Peter Stiling.)

(Kerr 1988). Records suggest that such winds had not hit the region for at least 300 years, so this may have been a rare event. Others suggest that such severe weather will become more common (Thompson 1988) because the world's climate is changing. Five of England's biggest freezes have come since 1978; the frequency of disastrous hurricanes in the South Pacific, especially over Fiji, has increased from one every twelve years to one every seven years, and six storms were recorded between 1981 and 1985 alone. Woodley et al. (1981) present data to show that hurricanes can absolutely devastate coral reefs. Weather extremes can undoubtedly play a big part in the distributions of plants and animals.

## 5.5  SOIL

Soil characteristics are also of prime importance to plant distributional patterns (Jenny 1980). Nitrogen availability is often crucial. In nitrogen-poor soils, such as bogs or poor sandy areas, only species that can supplement their nitrogen intake survive. Thus roots of alder (*Alnus glutinosa*) have nodules containing bacteria that fix atmospheric nitrogen, enabling the plant to grow in nitrogen-deficient soils. Similarly, bog myrtle, *Myrica gale*, has root nodules that are able to fix nitrogen, and bog myrtle occurs in Scotland in large areas of wet, acid, peat soils. In poor soils of the southeastern United States, some species (like the Venus flytrap and various pitcher plants) supplement their nitrogen intake by trapping insects, whose body fluids they can dissolve and then absorb (Photo 5.2).

## 5.6  FIRE

Distribution patterns of species may be limited by other features, such as chemical pollutants in the air, water and soil, salinity, oxygen content, and pH. Acid rain (see Chapter 20) can radically alter the pH of freshwater lakes and the biotic component within them. Even the frequency of fire can affect the distribution of plants and animals. While it is too lengthy a process to describe examples of distribution limitation by every environmental variable, a discussion of the effects of fire is valuable, particularly as it relates to land-management practices.

Before the arrival of Europeans in North America, fires started by lightning were a frequent and regular occurrence in some areas (Beaufait 1960), for example, in the pine forests of what is now the southeastern United States. These fires, because they were so frequent, consumed leaf litter, dead twigs and branches, and undergrowth before they accumulated in great quantities. As a result, no single fire burned hot enough or long enough in one place to damage large trees—each one swept by quickly and at a relatively low temperature. The dominant plant species of these areas came to depend both directly and indirectly on frequent, low-intensity fires for their existence. The jack pine, *Pinus banksiana*, has

**Photo 5.2**   *Soil quality is another vitally important abiotic factor that affects the distribution and abundance of species. In nitrogen-poor soils, plants that fix their own nitrogen are favored, as are plants which can supplement their nitrogen intake by catching insects. This venus flytrap is native to North and South Carolina.* (Suzio, Photo Researchers, Inc. 2S1490.)

serotinous cones that remain sealed by resin until the heat of a fire melts them open and releases the seeds, and it therefore depends directly on fire for its reproductive success. Much of the rest of the fire-adapted vegetation would be supplanted by other species if fires did not suppress those species periodically (Wade, Ewel, and Hotstetler 1980; Christensen 1981).

Management practice that attempts to maintain forests in their natural state by preventing forest fires completely often has exactly the opposite result. First, trees like the jack pine simply stop reproducing in the absence of fire. Second, species like the longleaf pine and wiregrass that depend on fire to suppress their competitors are soon replaced by species characteristic of other communities. Finally, when a fire does occur, fuel has had a much longer period in which to accumulate on the forest floor, and the result is an inferno—a fire that is so large and burns so hot that it consumes seeds, seedlings, and adult trees, native and competitor alike. Photo 5.3 shows a longleaf pine seedling; part b shows a small, rapidly moving "natural" fire of the sort that prevailed before current management practices were instituted; and part c shows a destructive fire of the sort that results when litter is allowed to accumulate for long periods.

Management practices that prevent fire arise from the mistaken assumption that forests evolved in the virtual absence of fire. Lightning-caused fires are particularly frequent in the southeastern United States, but even in other areas, many more fires are naturally caused than most people believe. For example, in the western half of the United States, nearly half of the yearly average of more than 10,000 fires are thought to be started by lightning (Brown and Davis 1973).

## 5.7 ENVIRONMENT AND ABUNDANCE

Not only does climate play an important role in the global distriubtion of species, it can affect their densities within these ranges, too. The most famous proponents of this idea were the Australian entomologists Davidson, Andrewartha, and Birch who studied densities of small insects, called thrips, in rosebushes. Like many insects, the thrips underwent large fluctuations in density. Davidson and colleagues found that 78 percent of the variation in population maxima was accounted for by variations in weather (Andrewartha and Birch 1954).

Photo 5.3 (a)    *Adaptation to fire. Longleaf pine seedling, showing the dense cluster of long green needles that protects its growing tip from the low temperature, fast-moving fires that suppress its competitors.* (Photo by Dana C. Bryan, Florida Park Service.)

Photo 5.3 (b)   *A "natural" fire in a stand of longleaf pine.*  (Photo by Florida Park Service.)

Photo 5.3 (c)   *The results of unsound management practices—a destructive fire. Natural fires do not burn very hot because they are frequent and not much litter accumulates before each burn. When natural burning is stopped, litter accumulates rapidly; subsequent fires, of whatever origin, quickly get out of control, not only killing the seedlings but leaping into the forest canopy and devastating the mature trees.*  (Photo by E. R. Degginger.)

Enright (1976) suggested that density limitation by climate may be more common than we think. He argued that many ecologists are preoccupied with the effects of competitors, predators, and parasites on populations and that the effect of climate is often overlooked. Biogeographers have described whole arrays of species, or communities, that are restricted to certain climatic zones. Why should we deny that weather and climate can profoundly influence densities of species within their range? Often, abundance declines toward the boundaries of a species range and is correlated with environmental variables. Of course, in the center of a range climatic variables may be less likely to cause observed density differences, if they occur. Again, the issue of scale arises. Over small scales, biotic variables may be important; over larger scales climate may be more important. Even over small scales, the abundance of refuges from climate may be important, and the indirect effects of climate, by reducing foliage quality, may affect the density of herbivores.

## 5.8  ENVIRONMENT AND COMMUNITIES

Environmental factors not only determine the distribution patterns of individual species, but they also play an important part in community richness—the number of species in any given locale can be affected by climate. Currie and Paquin (1987) showed that, of all the environmental data available, evapotranspiration rates (strongly correlated with primary production and hence available energy) are the best predictors of tree species richness in North America (Fig. 5.17).

One can also recognize characteristic assemblages of species for many of the major environmental settings on Earth. For example, tall evergreen trees and lush undergrowth grow rampantly in the tropics, whereas cacti are most prevalent in deserts. Each of the major types of floral and faunal assemblages is referred to as a **biome.** Thus, cooler, drier areas dominated by tall, deciduous trees form the temperate forest biome. A number of major biome types are recognized by ecologists, among which are the tropical rain forest, temperate **forest, desert, grassland, taiga, tundra,** tropical savannah, tropical scrub and seasonal forest, temperate rain forest, and chaparral. In aquatic situations, there are coral reefs, the open ocean, estuaries, freshwater environments, and the intertidal biome. The meteorological conditions necessary for certain biomes and their physical appearances are discussed in the section on communities.

## 5.9  ENVIRONMENT AND EVOLUTION

Finally, although it is generally recognized that most species will not be able to evolve quickly enough to be unaffected by global changes such as global warming, some evolution has been noted over short time periods. Six hundred years ago, all cotton plants were perennial shrubs confined to the frost-free tropics. Gradually, forms were selected to fruit early and produce a sizable crop in the first growing season. Early fruiting varieties were suitable for cultivation in temperate regions, where they could produce cotton before winter. Cold winters and hot summers imposed an annual growth habit, and now all commercial cottons are obligate annuals that can be grown in cold-winter areas and semiarid climates. All this—adaptation to previously lethal environments (Hutchinson 1965)—has been achieved in a maximum of 600 generations. In less than 50 years, the grass *Agrostis tenuis* has evolved populations that live on spoil tips in Great Britain, the areas of mine wastes often rich in such noxious elements as lead, copper, or zinc (Antonovics, Bradshaw, and Turner 1971). In this case, natural selection has favored the very few individuals tolerant to such areas, and these have prospered.

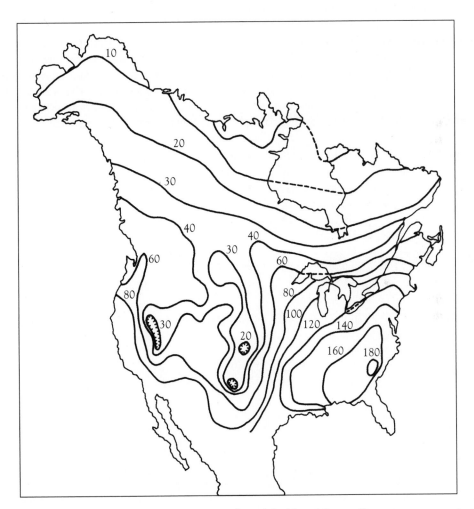

**Figure 5.17**    *Tree species richness in Canada and the United States. Contours connect points with the same approximate number of species per quadrat.*    (Redrawn from Currie and Paquin 1987.)

# Summary

1. The local distribution patterns of most species are limited by certain physical or abiotic factors of the environment such as temperature, moisture, light, pH, soil, salinity, and water current. If such factors can limit where species occur, they probably have a strong influence on population densities within these areas of distributions, too.

2. Perhaps two of the most important environmental variables, in terms of their influence on plant and animal distribution patterns, are temperature and moisture.

3. The influence of humans on abiotic variables may be substantial. Increasing $CO_2$ output from the burning of fossil fuels and increasing deforestation may cause global $CO_2$ levels to increase so much that the world's climate might change slightly (so-called global warming). Other anthropogenic changes include increased $SO_2$ output from fossil-fuel burning, leading to acid rain, acid lakes, and acid streams.

*Discussion Question:*  Do you think abiotic factors, such as soil nitrogen or available water, influence plant growth more than biotic variables, such as the presence of herbivores, or competition from other plants that compete for the same resources?

# 6

# *Population Growth*

W ithin their areas of distribution, plants and animals occur in varying densities. We recognize this pattern by saying an animal is "rare" in one place and "common" in another. Schoener (1987) has shown that, at least for Australian birds, a species we know as rare is in fact more likely to be common in at least one part of its range and rare everywhere else than to be rare throughout its range. Rarity, of course, can be strongly influenced by human activities, especially habitat destruction, but it can also be produced naturally; some species are rare because they are **endemic** to limited areas, such as islands. For example, more than 50 percent of the Canary Islands' plant species, which are about 95 percent endemic, and about 66 percent of the 155 endemic plants on Crete are considered **endangered** (Lucas and Synge 1978). Habitat destruction in these cases adds heavily to the perils of an already restricted range. Rarity can also be the result of the dynamics of food chains (Chapter 16); top predators are normally less abundant than their prey, and only about 1,600 panthers are thought to have occupied the whole state of Florida even before the arrival of Europeans (Cristoffer and Eisenberg 1985).

For more precision, and especially for management purposes, it is desirable to quantify population density and, more precisely, to determine what fractions of a population consist of juveniles and adults. Density is normally calculated for a small area, and total abundance over an entire habitat is estimated from these figures. Apart from pure visual counts of organisms, sampling methods include the use of:

**Traps:** Live traps, snap traps, light traps (for night-flying insects), pitfall traps (for crepuscular species), suction traps, pheromone traps.

**Fecal pellets:** For hare, mice, rabbits, and so on.

**Vocalization frequencies:** For birds or frogs.

**Pelt records:** Taken at trading stations for large mammals.

**Catch per-unit effort:** Especially useful in fisheries, where catch is often given per 100 trawling hours.

**Percentage ground cover:** For plants.

**Frequency of abundance along transects or in quadrants of known area:** For plants and sessile animals, which remain in place to be counted.

**Feeding damage:** Useful for estimating the relative numbers of herbivorous insects.

**Roadside spottings in a standard distance:** Often used in bird counts.

Southwood (1977) provides an exhaustive review of these techniques and many more. From such data, one can estimate not only the density of a population but also the relative frequency of juveniles and adults, larvae and nymphs, or other types of immatures. Great care must be taken in data collection; in particular, one must always consider whether the sampling regime is likely to bias results. For example, Mallet et al. (1987) showed how mark-re-capture techniques strongly influenced butterfly behavior and hence population-size estimates. *Heliconius* butterflies avoid specific sites where they have been handled. Chase (personal communication) has documented how repeated visits to colonies of California gulls in Utah completely upset their behavior, and Spear (1988), in a fascinating article, showed that nesting gulls can recognize and distinguish different individual human investigators, mostly on the basis of facial features!

With data it is possible to construct life tables that show precisely how a population is age structured. Life-table construction is termed *demography*. An accurate measure of age is essential here; the use of size as an indicator of age is tenuous at best and at worst leads to underestimates of juveniles and overharvesting. Basically there are two different types of life table: age-specific (**cohort** analysis) and time-specific (static).

## 6.1  LIFE TABLES

Time-specific **life tables** provide a snapshot of a population's age structure at a given time. These tables were first developed by actuaries in the life insurance business, who had a vested interest in knowing how long people could be expected to live. Time-specific life tables are useful in examining populations of long-lived animals, say herds of elephants, where following a cohort of individuals from birth to death would be impractical. An example of such a time-specific life table, prepared from a collection of skulls of known ages, for Dall mountain sheep (shown in Photo 6.1) living in Mount McKinley (now Denali) National Park, Alaska, is shown in Table 6.1. The values given in the columns of life tables are symbolized by letters:

$x$  = age interval (years)

$n_x$ = the number of survivors at beginning of age interval $x$

$x'$ = age as percent deviation from mean length of life

$d_x$ = number of organisms dying between the beginning of age interval $x$ and the beginning of age interval $x + 1$

$l_x$ = proportion of organisms surviving to beginning of age interval $x$

$q_x$ = rate of mortality between the beginning of age interval $x$ and the beginning of age interval $x + 1$

$e_x$ = mean expectation of life for organisms alive at beginning of age $x$ (the goal of actuaries)

Age intervals can be set at any convenient length—a month, six months, a year, five years. An important point to remember about life tables is that each column is only a different way of expressing the same data. All (except $x$, of course) can be calculated from just one column. For example, $d_x$ can be calculated from two adjacent items in the $n_x$ column ($n_x$ and $n_{x+1}$):

$$n_{x+1} = n_x - d_x$$

A numerical example using the first two $n_x$ values from Table 6.1 is

$$n_1 = n_0 - d_0 = 1,000 - 199 = 801$$

Then, $q_x$ can be calculated from the newly calculated $d_x$ and $n_x$; thus,

$$q_x = \frac{d_x}{n_x}$$

A numerical example using the third row of Table 6.1 is

$$q_2 = \frac{d_2}{n_2} = \frac{13}{789} = 0.0165$$

Photo 6.1    *Dall Mountain sheep, Ovis dalli, in McKinley Park, Alaska. Collections of skulls, together with accurate age estimates of the skulls, permitted construction of an accurate life table for this species.* (Bean, DRK Photo, 116027.)

In turn, $l_x$ is calculated from $n_x$ and $n_0$ (the starting value of $n_x$):

$$l_x = \frac{n_x}{n_0}$$

A numerical example, from the fourth row of Table 6.1, is

$$l_3 = \frac{n_3}{n_0} = \frac{776}{1,000} = 0.776$$

The table also has an extra column purely to facilitate the calculation of $e_x$. For $e_x$, one must first obtain the number of animals alive in each age interval, from $x$ to $x + 1$. This number is known as $L_x$, where

$$L_x = \frac{n_x + n_{x+1}}{2} = \text{number of individuals alive during age interval } x$$

A numerical example, from the fourth and fifth rows of Table 6.1, is

$$L_3 = \frac{n_3 + n_4}{2} = \frac{776 + 764}{2} = 770$$

**TABLE 6.1**  *Time-specific life table for the Dall Mountain Sheep (Ovis dalli) based on the known age of death of 608 sheep dying before 1937 (both sexes combined)[a]. Mean length of life 7.09 years. Data from Murie (1944) are expressed per 1,000 individuals.*

| $x$ | $n_x$ | $x'$ | $d_x$ | $l_x$ | $q_x$ | $T_x$ | $L_x$ | $e_x$ |
|---|---|---|---|---|---|---|---|---|
| 0–1 | 1000 | −100 | 199 | 1.000 | .199 | 7053 | 900.5 | 7.0 |
| 1–2 | 801 | −85.8 | 12 | 0.801 | .015 | 6152.5 | 795 | 7.7 |
| 2–3 | 789 | −71.6 | 13 | 0.789 | .016 | 5357.5 | 776.5 | 6.8 |
| 3–4 | 776 | −57.5 | 12 | 0.776 | .015 | 4581 | 770 | 5.9 |
| 4–5 | 764 | −43.3 | 30 | 0.764 | .039 | 3811 | 749 | 5.0 |
| 5–6 | 734 | −29.1 | 46 | 0.734 | .063 | 3062 | 711 | 4.2 |
| 6–7 | 688 | −14.9 | 48 | 0.688 | .070 | 2351 | 664 | 3.4 |
| 7–8 | 640 | −0.8 | 69 | 0.640 | .108 | 1687 | 605.5 | 2.6 |
| 8–9 | 571 | +13.4 | 132 | 0.571 | .231 | 1081.5 | 505 | 1.9 |
| 9–10 | 499 | +27.6 | 187 | 0.439 | .426 | 576.5 | 345.5 | 1.3 |
| 10–11 | 252 | +41.8 | 136 | 0.252 | .619 | 231 | 174 | 0.9 |
| 11–12 | 96 | +56.0 | 90 | 0.096 | .937 | 57 | 51 | 0.6 |
| 12–13 | 6 | +70.1 | 3 | 0.006 | .500 | 6 | 4.5 | 1.0 |
| 13–14 | 3 | +84.3 | 3 | 0.003 | 1.00 | 1.5 | 1.5 | 0.5 |

[a] A small number of skulls without horns, but judged by their osteology to belong to sheep nine years old or older, were apportioned pro rata among the older age classes.

now

$$e_x = \frac{\sum L_x}{n_x}$$

that is, the sum of all the $L_x$s from age $x$ to the last age, written as $L_\infty$. So for the Dall mountain sheep,

$$e_{10} = \frac{174 + 51 + 4.5 + 1.5}{252} = 0.92$$

In Table 6.1, life expectancy is also shown by $x'$, age as percentage deviation from mean length of life. Doing so permits comparisons between different species, as it would be meaningless to compare $e_x$s between animals that had different average age distributions. The $n_x$ data are commonly plotted against time, age, or the $x'$ data to give a survivorship curve as shown in Figure 6.1. In survivorship curves, introduced to ecology by Raymond Pearl (see Pearl 1928), $n_x$ values are plotted on a logarithmic scale because it is more valuable to examine rate of change than absolute numerical changes. Consider the following:

| Population Size | No. Dying | Log No. Dying | Subsequent Population Size |
|---|---|---|---|
| 1,000 | 500 | 3.00 | 500 |
| 500 | 250 | 2.70 | 250 |
| 250 | 125 | 2.40 | 125 |

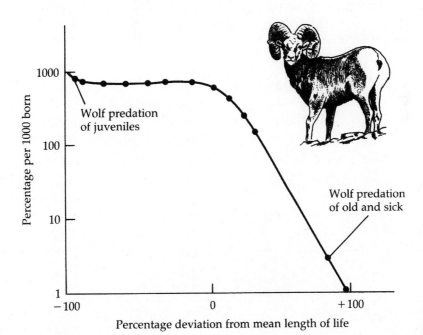

Figure 6.1 *Time-specific survivorship curve for the Dall mountain sheep* (Ovis dalli). (Based on data from Murie 1944.)

Although the numbers lost are certainly different, the rate of change is identical. For the Dall mountain sheep, one might notice that the life expectancy actually increases during the first year of life. The reason is that wolf predation on juveniles is high; once young sheep have passed through this critical stage, their life expectancy is higher than that of newborns. In some life tables, an additional column, $m_x$, is included; this is age-specific fertility. Such tables commonly only refer to females. The product of each entry in the $x$ column and the corresponding entry in the $m_x$ column is the total number of female offspring expected from that part of the population. Theoretically, if all females lived their entire reproductive life, the potential or gross reproductive rate would be maximal and would equal the sum of the $m_x$s. However, in practice some females die prematurely so that the **net reproductive rate,** $R_0$, is then often computed as

$$R_0 = \Sigma l_x m_x$$

**TABLE 6.2.** *Age-specific life table for the 1952–1953 generation of the spruce budworm on the plot G4 in the Green River watershed of New Brunswick. (After Morris and Miller 1954.)*

| Age interval $x$ | No.[a] alive at start of age interval $n_x$ | Factor responsible for mortality | No.[a] dying during age interval $d_x$ | Mortality rate (%) $100q_x$ |
|---|---|---|---|---|
| Eggs | 174 | Parasites | 3 | 2 |
| | | Predators | 15 | 9 |
| | | Other | 1 | 1 |
| | | Total | 19 | 11 |
| Instar I | 155 | Dispersal | 74.4 | 48 |
| Hibernacula | 80.6 | Winter | 13.7 | 17 |
| Instar II | 66.9 | Dispersal | 42.2 | 63 |
| Instar III–VI | 24.7 | Parasites | 8.92 | 36 |
| | | Disease | 0.54 | 2 |
| | | Birds | 3.39 | 14 |
| | | Other | 10.57 | 43 |
| | | Total | 23.42 | 95 |
| Pupae | 1.28 | Parasites | 0.10 | 8 |
| | | Predators | 0.13 | 10 |
| | | Other | 0.23 | 18 |
| | | Total | 0.46 | 36 |
| Moths | 0.82 | | 0.82 | 100 |
| Total for generation (egg to adult) | | | 173.18 | 99.53 |

[a] All numbers expressed per 10 sq. ft. of branch surface.

Values of $R_0$ greater than 1 indicate increasing populations, those less than 1 decreasing populations, and those equal to 1 stationary populations.

Despite the value of time-specific life tables, there are some assumptions that limit their accuracy. Paramount among these is that equal numbers of offspring are born each year. For example, if the rate of mortality of two-year-old Dall mountain sheep were identical to the rate of four-year-old sheep but there were more two-year-old sheep born because of favorable climate in that particular year, then more skulls of two-year-old sheep would be found later on, and a higher rate of mortality of two year olds would be implied. There is often no independent method of estimating birth rates of each age class.

For organisms with short life spans, usually completed within a year, or those with distinct breeding cycles, a snapshot, time-specific life table may not give the correct picture. It will be severely biased toward the juvenile stage common at that moment. In these cases age-specific tables are used, which follow one cohort or generation. Population censuses must be conducted frequently, but only for a limited time (usually less than a year). An age-specific life table for the spruce budworm is shown in Table 6.2 and represented graphically in Figure 6.2. Spruce budworms are larvae of tortricid moths. They excavate the terminal and lateral buds of conifers and are an economically important pest, especially in eastern Canada (Photo 6.2). Eggs are laid in August, and young caterpillars feed until the fall, when they pupate in hibernacula. In the spring, the caterpillars resume feeding and pupate, and adult moths emerge in the early summer to lay new eggs, starting the cycle over again. Various sources of mortality operate at different stages in the life cycle, so, for example, eggs or larvae may be parasitized by small wasps, caterpillars may be eaten by birds or infected by fungi, and so on. In fact, one of the biggest de-

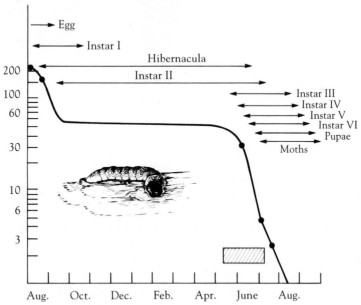

Figure 6.2 *Average survivorship curve for the spruce budworm during the 1945–1960 outbreak in New Brunswick. This insect has one generation per year, and females can lay 200 eggs. Regular census points on the curve are indicated by dots, and the critical large larval stage in early summer is indicated by crosshatching.* (Drawn after Morris 1963.)

Photo 6.2    *Spruce budworms feeding on Jack Pine in Ontario can cause substantial economic loss to planted forests. For this reason there is much impetus to understand what influences their numbers. Because the life cycle is passed so quickly, it is possible to follow a entire cohort of bud-worms, from egg to moth, and construct an age-specific life table.* (Lynch, DRK Photo 147379.)

creases in density is caused by caterpillar dispersal—they balloon away on silken threads to try their luck on new trees. The fact that most do not land on a suitable tree is a big source of mortality.

As a result of both time-specific and age-specific demographic techniques, three general types of survivorship curves can be recognized (Fig. 6.3): Type III, in which a large fraction of the population is lost in the juvenile stages; Type II, in which there is an almost linear rate of loss; and Type I, in which most individuals are lost when they are older. Type I curves are often observed in "higher" organisms, especially vertebrates, that exhibit parental care and protect their young. Type III curves are often exhibited by "lower" organisms, especially invertebrates such as insects; many plants, especially weeds; and marine invertebrates. For example, barnacles release millions of young into the sea, but most drift off and are eaten by predators. Only a few survive and settle in the rocky intertidal (although, once there, they show excellent survivorship). Examples of a steady decline in survivorship, a Type II curve, are hydra and many birds.

Many other interesting divisions can be made on the basis of life-history characteristics. For example, some organisms, like salmon, and plants, like bamboo and yucca, are *semelparous*. They reproduce once only and die. This condition is in contrast to *iteroparous*, or repeated, reproduction. Even among iteroparous organisms, strategies vary considerably. Weedy plants reproduce quickly and die, presumably before they are replaced by other colonists. Trees, however, often delay reproduction. The reasons for these sorts of differences in demographic processes are not always clear.

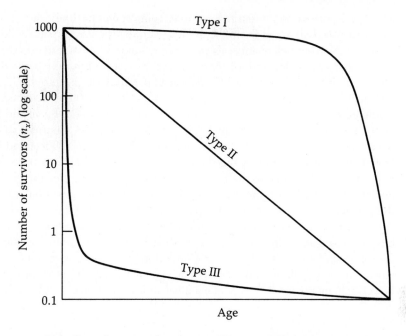

Figure 6.3   *Hypothetical survivorship curves. Type I includes man, mammals, and higher animals, often with parental care of young. Type II includes some birds and some invertebrates such as Hydra. Type III includes some insects and many lower organisms with pelagic juvenile stages such as benthic invertebrates, molluscs, oysters, and fishes.*

When a species can invade and successfully reproduce in a habitat, many factors contribute to its eventual population size. Such factors commonly include relationships with other organisms in the form of competition, predation, and herbivory. Before the effects of these selective pressures on population size can be examined, however, it must be known how populations grow naturally in the absence of these pressures.

## 6.2   DETERMINISTIC MODELS

### Geometric Growth

A population released into a favorable environment will begin to increase in numbers. Consider a **univoltine** insect that has one feeding season and a lifespan of one year. The population size at time $t + 1$ is given by

$$N_{t+1} = R_0 N_t$$

where $N_t$ is the population size of females at generation $t$; $N_{t+1}$ is the population size of females at generation $t + 1$; and $R_0$ is the net reproductive rate.

Clearly, much depends on the value of $R_0$: when $R_0 < 1$, the population goes extinct; when $R_0 = 1$, the population remains constant; and when $R_0 > 1$, the population increases. When

$R_0 = 1$, the population is often referred to as being at **equilibrium,** where no changes in pop-ulation density will occur. Even if $R_0$ is only fractionally above 1, population increase is rapid (Fig. 6.4). Northern elephant seals were nearly hunted to extinction in the late nineteenth century because of demand for their blubber. About 20 surviving animals were found off Mexico on Isla Guadalupe in 1890, and the population was protected. The actual growth of the elephant-seal population and recolonization of old habitats matched well the growth predicted when $R_0 = 2$ and generation time was eight years (Le Boeuf and Kaza 1981). Pre-dicted numbers were about 80 in 1906 and 40,960 in 1978. Censuses showed that actual population numbers were 125 in 1911 and 60,000 in 1977 (Le Boeuf and Kaza 1981). The growth of protected, annually breeding populations recovering from overexploitation in the past often provide some of the best fits of actual population-growth curves to this kind of model. The growth of some exotic species introduced into novel habitats also seem to fit the geometric pattern. The rapid expansion of rabbits after their introduction into South Australia in 1859 is a case in point. Sixteen years later, rabbits were reported on the west coast, having crossed an entire continent, more than 1,100 miles, despite the efforts of Australians to stop them by means of huge, 1,000-mile-long fences (Fenner and Ratcliffe 1965). The global human population also seems to fit a geometric pattern (see Chapter 18). In general, however, in more than 200 cases of mammalian introductions, most have failed to grow in an

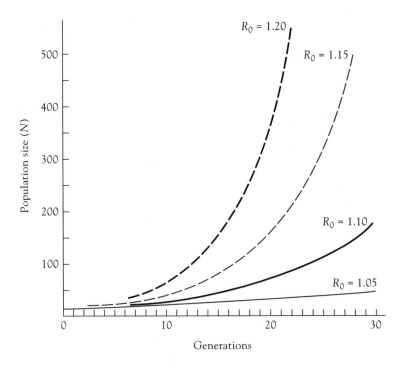

Figure 6.4 *Four examples of geometric population growth, discrete generations, constant repro-ductive rate.* $N_0 = 10$.

explosive or geometric pattern of increase (De Vos, Manville, and Van Gelder 1956). Of course, the Earth is not overrun with animals or plants, usually because $R_0$ decreases at high densities because of an increase in death rates and decrease in birth rates brought about by food shortage or epidemic disease. The value of $R_0$ usually changes as resources run short. For humans, this doesn't seem to have happened yet. However, the affect of changing reproductive rates are usually discussed with reference to a different set of equations that follow below.

For many species, including bacteria, fungi, insects, invertebrates, and many plants and animals in warm climates, reproduction occurs not seasonally but year round, and generations overlap. For such species the rate of increase is best described by

$$\text{rate of increase} \quad \frac{dN}{dt} = rN = (b - d)N$$

where $N$ = population size, $t$ = time, $r$ = **per-capita rate of population growth,** $b$ = instantaneous birth rate, and $d$ = instantaneous death rate.

In this equation, $r$ is analogous to $R_0$ but is actually related to it by the equation

$$r \simeq \frac{(\ln R_0)}{T_c}$$

where $T_c$ is the generation time (Southwood 1976)

Again, the result is a curve of **geometric increase** similar to that in Figure 6.4.

## Logistic Growth

For many species, resources become limiting as populations grow. The intrinsic rate of natural increase decreases as these resources are used up. Thus, a more appropriate equation to explain population growth under these conditions is

$$\frac{dN}{dt} = rN \left( \frac{K - N}{K} \right) \quad \text{or} \quad \frac{dN}{dt} = rN \left( 1 - \frac{N}{K} \right)$$

where $K$ is the upper asymptote or maximal value of $N$, commonly referred to as the carrying capacity of the environment at the equilibrium level of the population. In essence this equation means:

$$\begin{array}{ccccc}
\text{rate of increase} & & & & \text{population size times} \\
\text{of population} & = & \text{rate of population} & \times & \text{unused opportunity} \\
\text{per unit time} & & & & \text{for population growth} \\
\\
& = & r & \times & N \left( \frac{(K - N)}{K} \right)
\end{array}$$

When this type of growth is represented graphically, a **sigmoidal,** S-shaped, or so-called *logistic curve,* results (Fig. 6.5). This equation was first described by Verhulst (1838) and was derived independently by Pearl and Reed (1920) to describe human population growth in the United States.

Logistic growth entails many important assumptions. Five of the most important are listed below:

1. The relation between density and rate of increase is linear.
2. The effect of density on rate of increase is instantaneous.
3. The environment (and thus $K$) is constant.
4. All individuals reproduce equally.
5. There is no immigration or emigration.

For many laboratory cultures of small organisms, such assumptions are easily met. Thus, early tests of these models using laboratory cultures of yeast or bacteria suggested that they were valid (Gause 1934). For field populations of larger animals, assumptions are not so easily met.

1. In nature, *each* individual added to the population probably does not cause an incremental decrease to $r$.
2. In nature, there are often time lags, especially in species with complex life cycles. For example, in mammals, it may be months before pregnant females give birth.
3. In nature, $K$ may vary seasonally or with climate.
4. In nature, often a few individuals command many matings.
5. In nature, there are few barriers preventing dispersal.

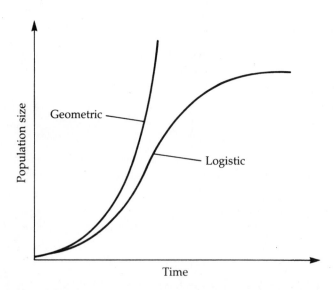

Figure 6.5 *Geometric or exponential growth in an unlimited environment and logistic growth in a limited environment.*

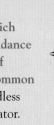

Is mutualism a potent force which affects the distribution and abundance of species? There are a variety of mutualisms in nature that are common and important. *(Upper left)* Cloudless sulfur on thistle, plant and pollinator. (Peter Stiling.) *(Middle)* Beauty berry *Callicarpa americana*, seeds ready to be dispersed. (Peter Stiling.) *(Lower left)* Coral *Millepora* sp. Dinoflagellates often enter into a mutualism with corals where they provide photosynthate to their hosts. (Roessler, Animals Animals, 620159.) *(Lower right)* Cleaner fish and host. This moray looks as though it might at any moment cross the line from mutualist to predator. (Wu, Peter Arnold, AN-BE-70C.)

**M**ore mutualism. *(Upper left)* Staghorn lichen on western white pine. Lichens are a mutualistic association of fungi and algae. (Levin, Animals Animals, 612539.) *(Lower left)* Mycorrhizae are a mutualistic association of fungus and root tissue which benefits both species. Here the root systems of a lemon tree grown with and without mycorrhizae are shown at 4½ months. (Runk/Schoenberger, Grant Heilman, CR2-80B.) *(Upper right)* Commensalism. Black rhino and egrets at the Ngorongore crater, Tanzania. The egrets benefit from this association because the rhino dislodges insect prey for the egrets from the grass. The rhino is unaffected. (Murphy, Animals Animals, 423630-M.) *(Lower right)* Ants tending aphids on a milkweed plant. The aphids secrete a sugar-rich substance called honey dew which the ants eat. In return, the ants protect the aphids from marauding natural enemies. (Robinson, Photo Researchers, Inc., 5J 5043.)

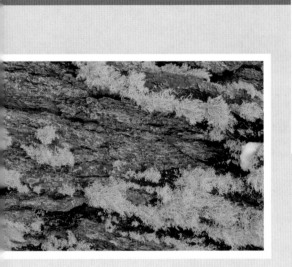

There are many lines of evidence to suggest competition is frequent and important in nature. (*Upper*) Competition may be intraspecific, between members of the same species, as between these California poppies on Portal Ridge, Antelope Valley, California (Ulrich, DRK Photo, 164211.) (*Lower*)…or interspecific, as between Iceland poppies, baby's breath, and cornflower in a meadow garden. (Lefever/Grushow, Grant Heilman, CO4X-65B.)

Competition continued. *(Upper left)* Consumptive competition over resources is common in nature. (Lynn, Photo Researchers, Inc., 2A1776.) *(Upper right)* Pre-emptive competition, over space, is common among intertidal organisms such as barnacles. Here barnacles compete for space on the body surface of a California gray whale. (Kolar, Animals Animals, 408301M.)

*(Lower left)* Plants commonly exhibit overgrowth competition, where certain species are denied light as in this tropical forest. (Dimijian, Photo Researchers, Inc., 621642.) *(Lower right)* Allelopathy. Purple sage, *Salvia dorrii*, is surrounded by bare ground, although Indian Paintbrush can be seen growing among it. Very little work has been done on allelopathy, and chemicals which may be detrimental to one species may not have an effect on others. (Duffurrena, Grant Heilman, CO5SAD11A.)

Thus, it is not surprising that there are few good examples of population data fitting the logistic. For example, in 1911 reindeer were introduced onto two islands, St. Paul and St. George, in the Bering Sea off Alaska. About 20 reindeer were introduced onto each island, both of which were completely undisturbed, having no predators and no hunting pressure (Fig. 6.6). The St. George population reached a low ceiling of 222 in 1922 and then subsided to a herd of about 40. The St. Paul population grew enormously, to about 2,000 in 1938, but then crashed to eight animals in 1950. There appeared to be no ecological differences between the islands and no reason for the differences in population growth observed. Neither population had fit the pattern of logistic growth (Klein 1968). Even Pearl and Reed's (1920) own prediction on human growth in the United States has not held true. On the basis of census data taken from 1790 to 1910, they projected an asymptote of 197 million to be reached around the year 2060. The census data for 1920 to 1940 fit the curve well, but since then the population has increased geometrically rather than logistically, resulting in projections of 260–350 million in the year 2025. To be fair to them, immigration rates have greatly increased and now contribute the same number of new people to the U.S. population as births. At least 500,000 immigrants enter the country each year legally, and probably 1 million illegally (more than 1.5 million were caught along the U.S.–Mexican border in 1986 [Brewer 1988]).

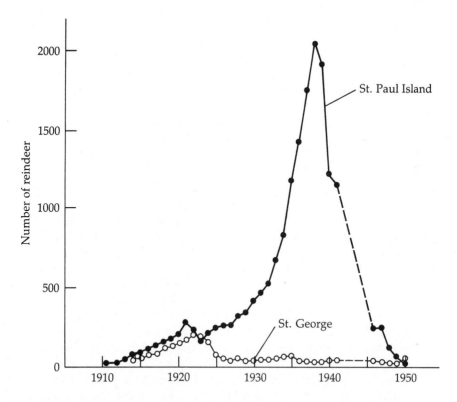

Figure 6.6 *Reindeer population growth on the Pribilof Islands, Bering Sea, from 1911, when they were introduced, until 1950.* (Redrawn from Scheffer 1951.)

## Time Lags

If there is a time lag of length $\tau$ between the change in population size and its effect on population growth rate, then the population growth at time $t$ is controlled by its size at some time in the past, $t - \tau$. If we incorporate the time lag into the logistic growth equation

$$\frac{dN}{dt} = rN\left(1 - \frac{N_{t-\tau}}{K}\right)$$

then population growth is affected by the length of the time lag $\tau$ and the response time of the population, which is inversely proportional to $r$. It follows that populations with quick growth rates have short response times $(1/r)$. The ratio of the time lag $(\tau)$ to the response time $(1/r)$, or $r\tau$, affects population growth. If $r\tau$ is small, $< 0.368$, the population increases smoothly to the carrying capacity (Fig. 6.7). If $r\tau$ is large, $> 1.57$, the population enters into a stable limit cycle, rising and falling around $K$ but never settling on the equilibrium value (May 1976a). We can think of a population at just below the carrying capacity having enough food for every individual to reproduce. After reproduction, however, few of the resulting juveniles will command enough resources to breed and the population will crash again. In this scenario, territorial species might be less likely to undergo limit cycles because territory holders will usually always command enough resources to breed. On the other hand, invertebrate species that often scramble for resources many be more likely to undergo a limit cycle. When values of $r\tau$ are intermediate, the population undergoes oscillations that dampen with time until $K$ is reached.

In stable limit cycles, the period of the cycle is always about $4\tau$, so a population with a time lag of one year (annual breeders) may expect to reach peak densities every four years. This may explain the observation of a four-year cycle in high-latitude annually breeding mammals, such as lemmings.

For species with discrete generations, the logistic equation becomes

$$N_{t+1} = N_t + rN_t\left(1 - \frac{N_t}{K}\right)$$

Although we know $r\tau$ controls population growth, with discrete generations, the time lag is always 1.0; thus the value of $r$ alone controls the dynamics. Again, if $r$ is small, $< 2.0$, the population generally reaches $r$ smoothly. At values of $r$ between 2 and 2.449, the population enters a stable two-point limit cycle with sharp "peaks" and "valleys" rather than smooth ones (Fig. 6.8 top). We can imagine that with continuously breeding species, there is more chance for these rough edges to be "smoothed" out, but with annual breeders, there is less chance for this to happen. Between $r$ values of 2.449 and 2.570, more-complex limit cycles result (May 1976a). One illustrated in Figure 6.8b has two distinct peaks and valleys before it starts to repeat. At values of $r$ larger than 2.57, the limit cycles break down and the population grows in complex, nonrepeating patterns, often known as "chaos." It is important to note that *chaos* doesn't mean "random," although random population growth and chaos certainly look the same. The important point is that with chaos, the same chaotic population-growth patterns would repeat every time the model is run, as long as the specified values of $N_0$, $r$, and $K$ are the same. With a "messy" data set, it is sometimes hard to know if a population is behaving in a chaotic manner or if random or stochastic changes are driving the dynamics. The effects of these random or stochastic changes are explained a little more below.

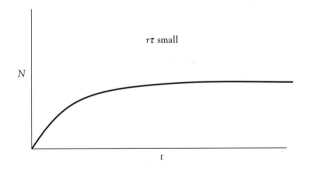

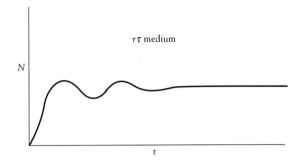

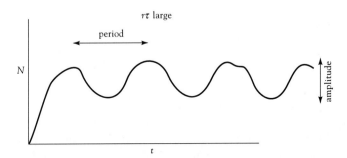

**Figure 6.7**  *Logistic growth curves with a time lag for species with overlapping generations. Growth depends on the value of rτ, the product of the intrinsic rate of increase and the time lag. In this figure, rτ increases from top to bottom. At a certain value of rτ, the population oscillations become stable (stable limit cycle) and no longer converge on the carrying capacity. (After Gotelli 1995.)*

## 6.3 STOCHASTIC MODELS

It is obviously unrealistic to expect animals in real life to behave exactly like numbers in an equation. Part of the reason is the genetic variability between individuals that causes some females to produce more offspring, on average, than others or some animals to be more re-sistant to climatic stress, predator pressure, and other factors. Given this variability, termed *stochasticity*, can we ever hope to predict population processes accurately or to incorporate such variability into a model?

**Stochastic models** of population growth are based largely on probability theory. Rather than having exactly two offspring, one might assume that each female has a 0.5 probability

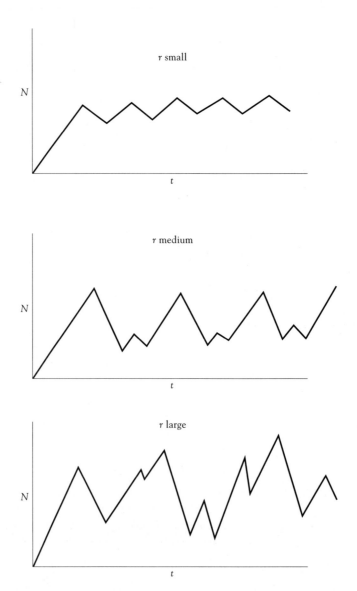

Figure 6.8 *Limit cycles and chaos for species with discrete generations. The discrete logistic growth curve is effected by the size of r. Values of r increase from top to bottom in this figure, and the number of points in the limit cycle also increases. In the end, a chaotic patterns results that is difficult to distinguish from randomness.* (After Gotelli 1995.)

of giving birth to two offspring, a 0.25 chance of producing three progeny, and a 0.25 chance of producing one.

For the geometric **deterministic model** discussed previously, if $R_0 = 2$ and $n = 5$, then

$$
\begin{aligned}
N_{t+1} &= R_0 N_t \\
&= 2 \times 5 = 10
\end{aligned}
$$

For a stochastic model, a coin can be flipped to mimic the probability of the outcome, where tails/heads or heads/tails = two offspring, two tails = one offspring, and two heads = three offspring:

|  | Outcome of Trial | | | |
| --- | --- | --- | --- | --- |
| Parent | 1 | 2 | 3 | 4 |
| 1 | 2 | 3 | 3 | 2 |
| 2 | 3 | 1 | 1 | 1 |
| 3 | 3 | 1 | 2 | 2 |
| 4 | 1 | 1 | 3 | 3 |
| 5 | 3 | 1 | 1 | 1 |
| Total population in next generation | 12 | 7 | 10 | 9 |

Some of the outcomes are above the expected value of 10, and some are below. If this technique is continued, a frequency histogram can be constructed (Fig. 6.9).

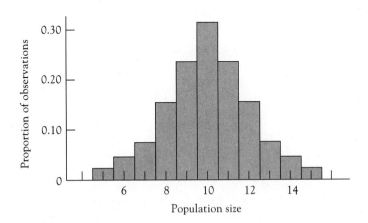

Figure 6.9 *Stochastic frequency distribution for size of a female population after one generation. In this case, probability of having two female offspring = 0.5, probability of having three female offspring = 0.25, and probability of having one female offspring = 0.25.*

Stochastic models can also be developed for logistic growth. Again, such a model is best explained by reference to the corresponding geometric equation, where

$$dN/dt = (b - d)\ N$$

and if $b = 0.5$, $d = 0$ (often true in new populations), and $N_0 = 5$, then

$$N_t = N_0 e^{rt} = 8.244$$

With a stochastic model, the probability of one individual not reproducing in one time interval must be calculated as $e^{-b} = 0.6065$, and likewise the probability of one individual reproducing once in one time interval is

$$1 - e^{-b} = 0.3935$$

Then for $N = 5$, the chance that no individuals will reproduce is $0.6065^5 = 0.082$. Similar calculations for other combinations eventually produce a frequency histogram, from which a population-growth curve can be constructed (Fig. 6.10).

If death occurs in the population, there will be a chance that the population will become extinct. If birth rate is greater than death rate, then

$$\text{probability of extinction} \ = \ \left(\frac{d}{b}\right)^{N_0} \ \text{as time} \rightarrow \infty$$

Thus, if $b = 0.75$, $d = 0.25$, and $N_0 = 5$, then

$$(d/b)^{N_0} = 0.0041$$

but if $b = 0.55$, $d = 0.45$, and $N_0 = 5$, then probability of extinction = 0.367.

The larger the initial population size and the greater the value of $b - d$, the more resistant to extinction the population becomes. In reality $b - d$ is often zero, so $(d/b) = 1.0$ as time $\rightarrow \infty$. In other words, extinction is a certainty for a population given a long enough time span and is likely to occur more quickly for a small population. Fischer, Simon, and Vincent (1969) believed that probably 25 percent of the species of birds and mammals that have become extinct since 1600 may have died out naturally. Such stochastic effects are particularly important when the conservation of small populations of

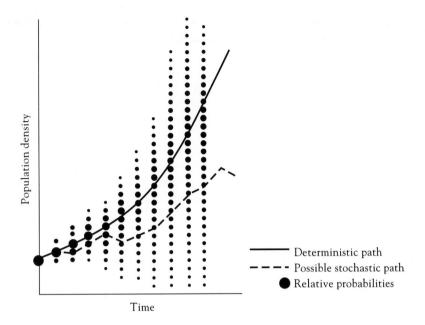

Figure 6.10 *Stochastic model of geometric population growth for continuous, overlapping generations.* (Redrawn from Krebs 1985a.)

rare species is considered. For example, Schaffer and Samson (1985) have predicted that if $N_e$ (effective population size) = 50 for grizzly bears, demographic stochasticity alone would cause extinction on average once every 114 years. A model of the spotted owl (*Strix occidentalis caurina*) suggests that demographic stochasticity is more likely than genetic factors to extinguish local subpopulations over the short term of decades (Simberloff 1986b).

Stochastic models introduce biological variation into population growth and are much more likely to represent what is happening in the field. The price paid is complicated mathematics as many new factors must be incorporated, such as the probability that a predator will kill a certain number of individuals or that there will be enough food available. Carrying capacity, $K$, may vary seasonally or randomly. For example, the number of insect prey available to birds may vary from spring to summer and may also be depressed by unusual cold snaps or wet weather. In general, the more variable the environment, the lower the population size. Stochastic models become more important as population sizes become smaller. If all populations were in the millions, one could throw away stochastic models—deterministic ones would do. These days, of course, populations of many mammals, except humans, tend toward the thousands rather than the millions.

Although some readers may have by now despaired of ever producing a robust model of population growth that incorporates all the necessary factors, some closing generalizations can be made about population growth:

1. There is a strong correlation between size and generation time in organisms ranging from bacteria to whales and redwoods (Fig. 6.11).
2. Organisms with long generation times have lower per-capita rates of population growth (Fig. 6.12).
3. Therefore, larger animals have lower rates of increase, $r$ (Fig. 6.13). For any given size, warm-blooded animals, homeotherms, have a higher rate of increase than heterotherms, which, in, turn are more fecund than unicellulars.

Such generalizations are particularly important to conservation efforts because they underline how long it takes for populations of large animals to rebound after ecological disasters or for large trees to reappear after forest clear-cutting. They may also point to reasons that larger species are more prone to extinction by people than are smaller species.

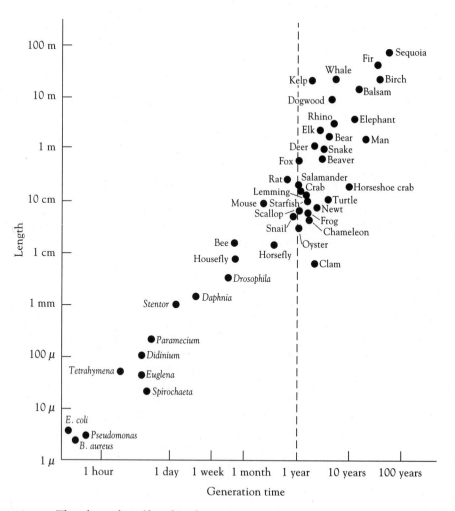

**Figure 6.11**   *The relationship of length and generation time, on a log-log scale, for a wide variety of organisms.*   (Redrawn from Bonner 1965.)

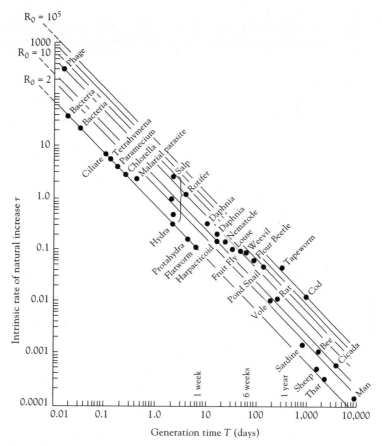

Figure 6.12 *The relationship of the intrinsic rate of natural increase, r, and generation time, with diagonal lines representing values of $R_0$ from 2 to $10^5$, for a variety of organisms.* (Redrawn from Heron 1972.)

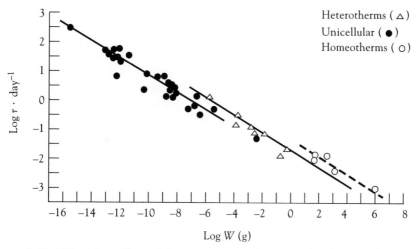

Figure 6.13 *The relationships of the intrinsic rate of natural increase to weight for various animals.* (Redrawn from Fenchel 1974.)

# Summary

1. For longer-lived organisms, time-specific life tables provide a snapshot of a population's age structure at a given time. Such tables often contain information on the number of organisms dying at any given age interval, the proportion surviving, and the mean expectation of life for organisms at any age interval. Data in most any column can be calculated from any other column.

2. For organisms with short life spans, usually completed within a year, age-specific life tables, which follow one cohort or one generation, are used.

3. Three types of survivorship curve can be recognized: Type III, where most individuals are lost as juveniles (invertebrates, plants); Type II, in which there is an almost linear rate of loss (some birds), and Type I where most individuals die after they have reached sexual maturity (mammals).

4. Population growth may be discussed under the banners of deterministic models or stochastic models. In deterministic models, rates of population increase are given precisely determined values. In stochastic models, variation is allowed around this value—as there would be in nature. For small populations, stochastic events are important, and stochastic models are therefore often used in conservation theory. Stochastic models are more difficult to develop than deterministic ones. For large populations, deterministic models give good approximations of population growth.

5. In both deterministic and stochastic population growth, where environments are favorable, populations may undergo geometric or exponential growth, which results in J-shaped population-growth curves.

6. In the real world, there are often limits set on population growth. Such limits might be exhaustion of space or resources or limitation by waste production. As a result, most population-growth curves are logistic and result in an S-shaped growth curve, where the upper level or carrying capacity, $K$, represents the total number of individuals that can exist in an area.

7. Population growth curves are not always smooth in trajectory. They may oscillate about the carrying capacity or exhibit what appears to be random fluctuations. Such fluctuations may actually be the result of time lags and may result in chaotic growth curves—curves that look as if they are generated by random fluctuations but are actually the result of precise time lags and rates of reproduction.

## Discussion Question:

Human populations appear to be growing geometrically. Do you think there will be a limit to human population growth? If so, what will it be, and will it be set by: space, food, water, or pollutants? What would happen to human population growth if the carrying capacity, $K$, and the net reproductive rate, $r$, are not constant but vary with time?

# Mutualism and Commensalism

O rganisms do not exist alone in nature but instead co-occur in a matrix of many species where the interactions in the table below are possible.

| Nature of Interaction | | Species 1 | Species 2 |
|---|---|---|---|
| Mutualism* | (Chapter 7) | + | + |
| Commensalism* | (Chapter 7) | + | 0 |
| Herbivory | (Chapter 10) | + | − |
| Predation | (Chapter 9) | + | − |
| Parasitism* | (Chapter 11) | − | − |
| Allelopathy | (Chapter 8) | − | 0 |
| Competition | (Chapter 8) | − | − |

+ = positive effect; 0 = no effect; − = deleterious effect.

* Examples of a symbiotic relationship, in which the participants live in intimate association with one another.

Herbivory, predation, and parasitism all have the same general effects, a positive effect on one population and a negative effect on the other. Competition affects both species negatively. **Mutualism** and **commensalism** are less commonly discussed in ecology but are tied together with **parasitism,** under the banner of **symbiotic** relationships. In symbiotic relationships, the partners in the association live in intimate association with one another; they are always found in close proximity (Saffo 1992). The effects of mutualism and commensalism are different from those of parasitism, however, and are better discussed separately here.

## 7.1 MUTUALISM

In mutualistic arrangements (reviewed by Boucher 1985), both species benefit. The color insert shows members of some mutualistic relationships. In mutualistic pollination systems, both plant and pollinator (insect, bird, or bat) benefit, one usually by a nectar meal and the other by the transfer of pollen. In one extraordinary case, male euglossine bees visiting orchids in the tropics do not collect nectar or pollen but are rewarded instead with a variety of floral fragrances, which they modify to attract females (Williams 1983). The tightness of the relationship in some pollination systems is underlined by the phenomenon of *buzz pollination* (Erickson and Buchmann 1983). Certain flowers whose anthers open through pores at the top shed their pollen when subjected to vibrations emanating from the buzzing of bee wings. To ensure that some pollen falls on its target bee below, the pollen is negatively charged. As the pollen rains down, it is attracted electrostatically to the bees, which tend to have positive charges.

Both parties also benefit in mutualistic fashion from seed dispersal when fruits are eaten by frugivorous birds, bats, or other mammals; the consumer receives a meal, and the plant receives an effective means of progeny **dispersal.** The lack of plant mobility has made many of them dependent on animals for pollination and seed dispersal. Temple (1977) has argued that the tree *Calvaria major* on the island of Mauritius has produced no seedlings for the past 300 years because its seeds do not germinate unless they have first passed through the digestive system of the now-extinct dodo. However, Witmer (1991) exploded this myth by noting that (a) seeds can germinate without abrasion in bird guts and (b) some living trees that are less than 300 years old exist, whereas the dodo went extinct in the 1660s, more than 300 years ago. Corals are mutualists, too. The animal polyps contain unicellular algae. Some people even view the association of humans with domestic animals or crops as a mutualism.

Several recent articles (Cherif 1990; Keddy 1990) have commented that mutalistic interactions are not covered in sufficient detail in modern ecology texts. This is a contrast to ecological textbooks of the 1920s–1940s where positive interactions were hypothesized to be important driving forces in communities (Clements, Weaver, and Hansen 1926; Allee et al. 1949). However, even in the ecological literature, the frequency of research articles on mutualism (14 percent) is much less than that for such other interaction types as competition (31 percent), predation (24 percent), or herbivory (30 percent) (Bronstein 1991). So textbooks may merely reflect the state of the ecological literature. This low level of representation could be because many mutualism studies have been descriptive and have focused on particular adaptations or life-history characteristics of organisms and not on theory, whereas studies on other interactions focus on the interaction itself and the mechanisms involved. Bertness and Hacker (1994) also suggest that studies of mutualism are not often experimental and are paid little attention by theorists. There are few appropriate models for mutualistic interactions. We could incorporate the positive effect of one species on the other by modifying the Lotka–Volterra equations such that

$$\frac{dN_1}{dt} = r_1 N_1 \left( \frac{K_1 - N_1 + \alpha N_2}{K_1} \right)$$

$$\text{and} \quad \frac{dN_2}{dt} = r_2 N_2 \left( \frac{K_2 - N_2 + \beta N_1}{K_2} \right)$$

where $\alpha N_2$ and $\beta N_1$ are the positive effects of species 2 on species 1 and species 1 on species 2, respectively. (See Chapter 8 for fuller discussion of similar modifications to the Lotka-Volterra equations for competition theory.) Such modifications often lead to unrealistic solutions in which both populations increase to unlimited size. We could allow each species to increase the carrying capacity of the other but place a limit on the interaction such that $\alpha \beta < 1$. Again, the models are unstable and often lead to extinction of one species (May 1976a). However, some authors have suggested that is an accurate finding: few obligate mutualisms in nature are very stable in the face of environmental change.

## 7.2 POLLINATION

Pollination studies are booming in modern ecology. Nearly 45 percent of all studies of mutualism involve pollination systems (Bronstein 1991). This may be partly because of the great diversity of apparently tight, **coevolved**, and interesting systems to study and partly because money is available to study them. More than 90 crops in the United States alone are pollinated by insects. One of the finest studies in obligate mutualism involves figs and fig wasps as studied by Janzen (1979a). More than 900 species of *Ficus* exist, and virtually every one must be pollinated by its own species of agaonid wasp. The fig that we actually eat has an enclosed inflorescence containing many flowers. A female wasp enters through a small opening, pollinates the flowers, lays eggs in their ovaries, and then dies. The progeny develop in tiny galls and hatch inside the fig. Males hatch first, locate female wasps within their galls, and thrust their abdomens inside the galls to mate with them. The males then die without ever having left the fig. Females collect pollen from the fig and then leave to search out new figs in which to lay their progeny. A similar mutualistic relationship occurs between yucca plants and yucca moths (Addicott 1986). Both mutualisms are highly coevolved. The distribution of each species of yucca or fig is controlled by the availability of its pollinator and vice versa. In the late nineteenth century, Smyrna figs were introduced into California, but they failed to produce fruit until the proper wasps were introduced to pollinate them.

It is interesting that such highly coevolved systems arose because on superficial examination, the needs of the plants and those of the pollinators seem to conflict sharply. From the plant's perspective, an ideal pollinator would move quickly among individuals but retain a high fidelity to a plant species, thus ensuring that little pollen is wasted as it is inadvertently brushed onto the pistils of other plants. The plant should provide just enough nectar to attract a pollinator's visit. From the pollinator's perspective, it would probably be best to be a generalist and to obtain nectar and pollen from flowers in a small area, thus minimizing energy spent on flight between patches. This casts doubt on whether such relationships are truly mutualistic in nature or whether both species are actually "trying to win" in an evolutionary arms race. One way in which the plants encourage the pollinator's species fidelity is by sequential flowering through the year of different plant species and by synchronous flowering within a species (Heinrich 1979). There are cases where both flower and pollinator try to cheat. In the bogs of Maine, the grass pink orchid (*Calopogon pulchellus*) produces no nectar, but it mimics the nectar-producing rose pogonia (*Pogonia ophioglossoides*) and is therefore still visited by bees. Bee orchids have even gone so far as to mimic female bees; males pick up and transfer pollen while trying to copulate with the flowers. So effective are the stimuli of flowers of the orchid genus *Ophrys* that male bees prefer to mate with them even in the presence of real female bees! Conversely, some *Bombus* species cheat by biting through the petals at the base of flowers and robbing the plants of their nectar without entering through the tunnel of the corolla and picking up pollen.

Finally, it is interesting to speculate about why ants, usually the most abundant insects in a given area, are so rarely involved in pollination. One reason might be that the subterranean nesting behavior of many ants exposes them to a wide range of pathogenic fungi and other dangerous microorganisms, to which they respond by producing large amounts of antibiotics. These antibiotics inhibit pollen function (Beattie et al. 1984; Beattie 1985). Peakall, Beattie, and James (1987) demonstrated that ants without metapleural glands and, therefore, without these secretions do successfully pollinate orchids in Australia.

## 7.3   DISPERSAL SYSTEMS

Mutualistic relations are highly prevalent in seed-dispersal systems of plants. Studies on these dispersal systems are also highly prevalent in the mutualism literature with 38 percent of studies focusing on dispersal. In the tropics, some fruits are dispersed by birds that are strictly frugivorous. These fruits provide a balanced diet of proteins, fats, and vitamins (Proctor and Proctor 1978). In return for this juicy meal, birds unwittingly disperse the enclosed seeds, which pass unharmed through the digestive tract. Some plants, instead of producing highly nutritious fruits to attract an efficient disperser, simply produce abundant mediocre fruit in the hope that some of it will be eaten by generalists. Fruits taken by birds and mammals often have attractive colors—red, yellow, black, or blue; those dispersed by nocturnal bats are not brightly colored but instead give off a pungent odor to attract the bats. (In contrast, because birds do not have a keen sense of smell, fruits eaten by them are generally odorless.)

In general, the relationships are not as obligately mutual as are plant-pollinator systems because seed dispersal is performed by more generalist agents. Nevertheless, a wide array of adaptations exists; one has only to look at the impressive specialization of parrot beaks, strong and sharp to crack and peel fruits, to see that the mutualistic relationship between plant and seed disperser is strong in this case. Some bizarre strategies also exist; for example, in the floodplains of the Amazon, fruit- and seed-eating fish have evolved that disperse seeds (Goulding 1980). Microbes appear to be good at cheating the plant in this system, for they will readily attack the fruit without dispersing it. Janzen (1979c) has suggested that microbes deliberately cause fruit and other resources such as carcasses to "rot," reserving it for themselves, by manufacturing ethanol and rendering the medium distasteful to vertebrate consumers. Animals eating such food are selected against because they become drunk and are easy victims for predators. As a countermeasure, some vertebrates have "learned" evolutionarily to tolerate these microorganisms and even to use them in their own guts to digest food. Alternatively, they often possess the enzyme alcohol dehydrogenase to break down alcohol.

## 7.4   OTHER EXAMPLES OF MUTUALISM

Numerous other types of mutualism exist, although most are not commonly studied. On coral reefs, "cleaner" fish nibble parasites and dead skin (which might otherwise cause disease) from their customer fish at specific cleaning stations (Ehrlich 1975). Such systems often leave their participants open to cheaters. Saber-toothed blennies of the genus *Aspidontus* bear a striking resemblance to the common cleaner wrasse, *Labroides dimidiatus*. Instead of performing a cleaning function, however, the blenny bites chunks out of the customers. Saber-tooths are protected from attack by their resemblance to the cleaners, though customers, in time, learn to avoid the cleaning stations that blennies frequent.

One of the oldest ideas about mutualisms is that they are more common under harsh physical conditions when neighbors buffer one another from limiting physical stresses (Hay 1981; Wood and Del Moral 1987; Bertness 1987).

Bertness and Hacker (1994) have described an interesting mutualism in the salt marshes of New England where only a few plant species exist, each generally in distinct zonation patterns (Fig. 7.1) according to tolerances of soil salinities and waterlogging. In southern New England, terrestrial marsh borders are dominated by the perennial shrub *Iva frutescens* (marsh elder), whereas at the seaward side needle rush, *Juncus gerardi*, dominates. In between the two is a region of needle rush mixed with stunted *Iva*. Bertness and Hacker experimentally removed *Juncus* from around some *Iva* plants and found that soil salinities doubled and soil oxygen levels decreased. The photosynthetic rate of *Iva* went down and, fourteen months later, the *Iva* were dead. *Juncus* neighbors clearly have strong positive effects on adult marsh elders by reducing the soil salinity and enabling *Iva* to survive at places in the salt marsh where, alone, they would die. Next, Bertness and Hacker moved *Iva* and *Juncus* together into three different habitat types: low marsh (high salt stress), transition zone, and high marsh (low salt stress). In the low marsh, the biomass of both species was high, suggesting a mutualistic association in regions of high stress. However, in the high marsh, where conditions were relatively benign, plants did not do better; in fact, they did worse as competition ensued. This type of experiment, predicting when and where mutualisms will occur, will be of much value to ecologists in the future.

## Protection

In terrestrial systems one of the most commonly observed mutualisms exists between ants and aphids. Aphids are fairly helpless creatures, easy prey to marauding ants. Yet in general,

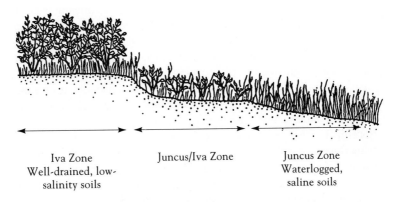

Iva Zone
Well-drained, low-
salinity soils

Juncus/Iva Zone

Juncus Zone
Waterlogged,
saline soils

Figure 7.1 *Schematic diagram of the terrestrial border of a typical New England salt marsh.*

ants tend to farm aphids like so many cattle. The aphids secrete honeydew, a sticky exudate that is rich in sugars and that the ants enjoy. In return, ants protect aphids from an array of predators, such as syrphid larvae, such parasites as braconid wasps, and other competing insects, by vigorously attacking them (Fowler and MacGarvin 1985).

## Obligatory Mutualism

In some cases, the mutualistic relationship is so tight that neither participant could exist without the other and is called *obligatory mutualism*. Such is the case for many lichens, which are combinations of algae (which provide the photosynthate) and fungi. The "lichenized" fungi include within their bodies and near the surface a thin layer of algal cells, forming only 3 to 10 percent of the weight of the thallus body. Of the 70,000 or so species of fungi, 25 percent are lichenized. Lichenized forms occur in deserts, in alpine regions, and across a wide range of habitats. Nonlichenized fungi are usually restricted to being parasites of plants or animals or to being involved in decomposition. Many ruminants shelter symbiotic bacteria in their guts, which break down plant tissue to provide energy for their hosts; cellulose is otherwise indigestible for mammals (Hungate 1975). Likewise, the roots of most higher plants (except the Cruciferae) are actually a mutualistic association of fungus and root tissue—the mycorrhizae. The fungi require soluble carbohydrates from their host as a carbon source (up to 40 percent of the photosynthate produced) , and they supply mineral resources, which they are able to extract efficiently from the soil, to the host (Harley and Smith 1983; Lynch 1990). The relationship between systemic fungal endophytes and vascular plants, usually thought to be a parasitic infection, has also been viewed as mutualism (Clay 1988). The fungi are thought to aid their hosts in defense against herbivory.

## 7.5  COMMENSALISM

In commensal relationships, one member derives benefit while the other is unaffected. Such is the case when sea anemones grow on hermit-crab shells. The crab is already well protected in its shell and gains nothing from the relationship, but the anemone gains continued access to new food sources. The same benefits accrue to members of a phoretic relationship (**phoresy**), in which the association involves the passive and more-temporary transport of one organism by another, as in the transfer of flower-inhabiting mites from bloom to bloom in the nares of hummingbirds (Colwell 1973). Some of the most numerous examples of commensalism are provided by plant mechanisms of seed dispersal. Many plants have essentially cheated their potential mutualistic seed-dispersal agents out of a meal by developing seeds with barbs or hooks to lodge in the animals' fur rather than their stomachs. In these cases, the plants receive free seed dispersal, and the animals receive nothing. This type of relationship is fairly common; most hikers have been plagued by "burrs" and "sticktights." Sometimes, these barbed seeds can cause great discomfort to the animal in whose fur they become entangled. Fruits of the genus *Pisonia* (cabbage tree), which grows in the Pacific region, are so sticky that they cling to bird feathers. On some islands, birds and reptiles can become so entangled with *Pisonia* fruits that they die.

# Summary

1. Organisms do not exist alone but instead co-occur with many species where many different interactions such as mutualism, commensalism, herbivory, predation, parasitism, and competition are possible. Mutualism, commensalism, and parasitism are all symbiotic relationships where the partners live in close association with one another. However, in mutualism and commensalism, there are no negative effects of one species on the other, so discussions of these phenomena are usually treated together.

2. In mutualisms, both species benefit. A common example is pollination, where plants benefit from the transfer of pollen and where the pollinator, often an insect, gains a nectar meal. In seed-dispersal systems, the plant provides a fruit meal for birds and mammals and, in turn, benefits from dispersal of seeds into new areas.

3. Obligatory mutualisms are mutualisms in which the species cannot live apart. Examples include lichens, a mutualism between fungi and algae; mycorrhizae, an association between fungae and plant roots; and the association between fungal endophytes and the plant leaves in which they live.

4. In commensalisms, one member derives benefit while the other is unaffected. One of the most common examples of such phenomena is phoresy, the passive transport of one organism by another.

*Discussion Question:* At least six types of mutualism have been recognized in nature: those that deter predation; increase the availability of prey or resources; feed on (or compete with) a predator; increase the competitiveness of one partner in the mutualism with other species outside the mutualism; decrease the vigor of a competitor; or feed on (or compete with) a competitor (Addicott and Freedman 1984). Discuss examples of these. Where might mutualisms be more common: in more-stressful temperate areas or in the more-stable tropics?

# CHAPTER 8

# *Competition and Coexistence*

For many biologists, implicit in Darwin's theory of natural selection was a view of nature "red in tooth and claw" in which species scrambled to outcompete each other and leave the most offspring. How true a picture is this?

It is first worthwhile to be specific about what types of competitive event may occur in nature. **Competition** may be **intraspecific** (between individuals of the same species) or **interspecific** (between individuals of different species). Competition can also be characterized as scramble competition or contest competition. In scramble (**resource**) competition, organisms compete directly for the **limiting resource,** each obtaining as much as it can. Under severe stress, for example when fly maggots compete in a bottle of medium, few individuals can command enough of the resource to survive or reproduce. Such competition is most evident between invertebrates. In **contest** (interference) competition, individuals harm one another directly by physical force. Often, this force is ritualized into threatening behavior associated with territories. In these cases, strong individuals survive and take the best territories, and weaker ones perish or at best survive under suboptimal conditions. Such behavior is most common in vertebrates.

## 8.1 MATHEMATICAL MODELS

Mathematically, competitive interactions can be described by equations derived independently by Lotka (1925) in the United States and by Volterra (1926) in Italy, commonly called the Lotka-Volterra equations. For two species growing independently,

$$\frac{dN_1}{dt} = r_1 N_1 \left( \frac{K_1 - N_1}{K_1} \right)$$

$$and \quad \frac{dN_2}{dt} = r_2 N_2 \left( \frac{K_2 - N_2}{K_2} \right)$$

where $r$, $N$, and $K$ represent the same variables introduced in Chapter 6, namely per-capita rate of population growth, population size, and carrying capacity.

Another term must be introduced to allow for the effect of each population on the other. In most cases, individuals of one species are larger than those of the other, and a conversion factor is needed to convert species 2 into units of species 1, such that

$$N_1 = \alpha N_2$$

where $\alpha$ is the conversion factor. Thus, in the diagram below one individual of species 2 uses the same amount of resources as four individuals of species 1.

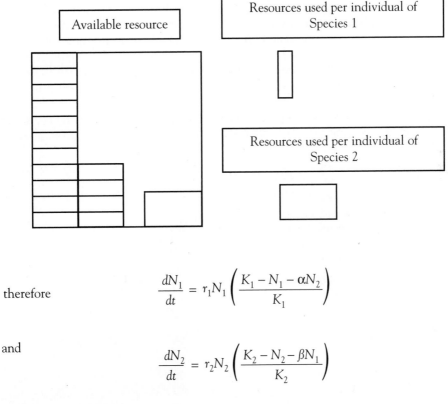

therefore

$$\frac{dN_1}{dt} = r_1 N_1 \left( \frac{K_1 - N_1 - \alpha N_2}{K_1} \right)$$

and

$$\frac{dN_2}{dt} = r_2 N_2 \left( \frac{K_2 - N_2 - \beta N_1}{K_2} \right)$$

where $\beta$ is the conversion factor to convert $N_1$ into units of $N_2$.

Such relationships can be expressed graphically (Fig. 8.1). Population growth of $N_1$ continues to the carrying capacity of the environment $K_1$ in the absence of $N_2$. If there are $K_1/\alpha$ individuals of $N_2$ present, no population growth of $N_1$ is possible. Between these two extremes are many combinations of $N_2$ and $N_1$ at which no further growth of $N_1$ is possible. These points fall on the diagonal $dN_1/dt = r_1 = 0$, which is often called the *zero isocline*. Population growth of $N_2$ can be represented by a similar diagram. Combining the two figures and adding the arrows by vector addition illustrates what happens when the species co-occur (Fig. 8.2). Essentially there are four possible outcomes: species 1 goes extinct; species 2 goes extinct; either species 1 or species 2 goes extinct, depending on the initial densities; or the two species **coexist.**

A drawback to the Lotka-Volterra model of competition is that no mechanisms are specified that drive the competitive process. Tilman (1982, 1987) criticized this approach and emphasized that we need to know the mechanism by which competition occurs. Knowing the mechanism will enable better predictions of the outcome. Tilman began by considering the responses of two competing plant species to environmental variables, say nitrogen and light again. As with the Lotka-Volterra models, we can draw zero-growth isoclines for both species, this time based on their responses to light and nitrogen (Fig. 8.3). If light levels are too low, a species will not grow; above a certain light level, growth proceeds. The same happens for nitrogen levels. A sort of all-or-nothing response is envisaged. We can do the same for the second species. By superimposing the two zero-growth isoclines, we can determine the outcome if the two species coexist in the same habitat (Fig. 8.4). Once again, four different outcomes are possible. In the first instance, species A will always need more of both resources than species B, and species B wins out while species A goes extinct. In the second instance, the roles are reversed. In the two remaining cases, the zero-growth isoclines cross and there may be an equilibrium point. To determine what happens in these other two scenarios, an extra piece of information is needed: the consumption curves of each species and the ratio of existing resources, that is, the position of the resource supply point.

**Figure 8.1**    *Changes in population size of species 1 when competing with species 2. Populations in the shaded area will increase in size and will come to equilibrium at some point on the diagonal line. Along the diagonal, $dN/dt = 0$.*

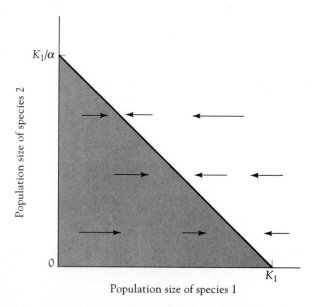

Population size of species 1

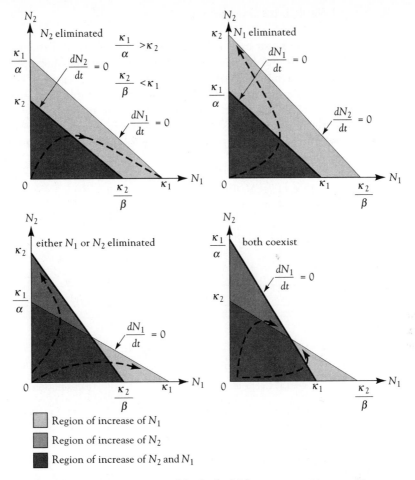

Figure 8.2 *The four consequences of the Lotka-Volterra competition equations.*

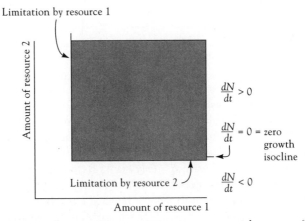

Figure 8.3 *The response of an organism to variations in two essential resources (like nitrogen and light for plants). The thick line represents the zero-growth isocline. Above this line in the shaded area, the population will increase in size. Below this line in the white area, the population will decline.* (After Tilman 1982.)

The resource supply point can be located in any position, 1 through 6, in Figure 8.4. In region 4 in Figure 8.4 (c) and (d), Species A is limited more by resource 2, and Species B is limited more by resource 1. If species A consumes relatively more of resource 1 than species B, the equilibrium point is unstable, and one or other of the species will go extinct (Fig. 8.4c). On the other hand, if species B consumes relatively more of resource 1 than species A, there is a stable equilibrium point, as illustrated in Figure 8.4d. In this situation, each species consumes more of the resource that lim-

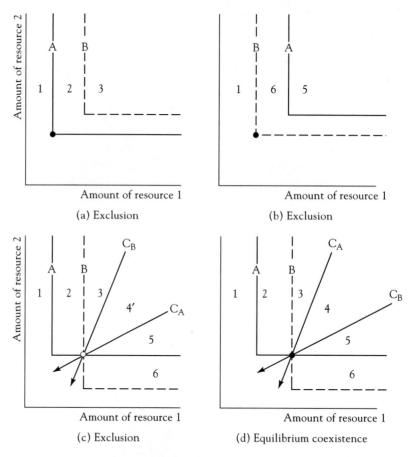

Figure 8.4   *Tilman's model of competition for two essential resources. The zero-growth isoclines for species A and B are shown, along with the consumption rates for each species ($C_A$ and $C_B$). For all four cases, the regions are labeled as follows: 1 = neither species can live; 2 = only species A can live; 3 = species A wins out; 4 = stable coexistence; $4^1$ = unstable coexistence, one or the other species wins; 5 = species B wins; 6 = only species B can live. • = stable equilibrium, o = unstable equilibrium. In (c) and (d), species B is limited more by resource 1 and species A by resource 2. In (c), species A consumes relatively more of resource 1 than species B and wins out in competition. In (d), each species consumes more of the resource that more limits its own growth and the species comes to a stable equilibrium with both species coexisting.* (From Tilman 1982.)

its its own growth and the system equilibrates at the intersection point of the two zero-growth isoclines. Thus, competitors may coexist depending on their resource utilization. Figure 8.5 illustrates what the actual population densities of species A and B look like, given scenario (d) in Figure 8.4. Tilman tested this model by growing two species of diatoms, *Asterionella formosa* and *Cyclotella meneghiniana,* in chemostats under controlled rates of nutrient supply. *Asterionella* required higher levels of silicon, and *Cyclotella* higher levels of phosphorus. In Tilman's experiments, there were four levels of supply: low phosphate, high silicate to high phosphate, low silicate. As predicted by his theory, Tilman could get either coexistence or one or the other species to go extinct by varying the nutrient supply.

## 8.2  LABORATORY STUDIES OF COMPETITION

One must ask whether mathematical formulations represent real biological systems. One of the first and most important tests of these equations was performed in 1932 by a Russian microbiologist, who studied competition between two species of yeast, *Saccharomyces cervisiae* and *Schizosaccharomyces kephir* (species renamed since 1932) (Gause 1932). Alone, both species grew according to the logistic curve; the asymptote reached was a function of ethyl alcohol concentration. Ethyl alcohol is a by-product of sugar breakdown under anaerobic conditions and can kill new yeast buds just after they separate from the mother cell. In cultures where the two yeasts grew together, population densities were lower than they were under single-species conditions (Fig. 8.6). From these data, Gause was able to calculate that $\alpha$ = 3.15 and $\beta$ = 0.44; that is, 1 volume of *Saccharomyces* = 3.15 volumes of *Schizosaccharomyces.* Because alcohol is the limiting factor, Gause argued that he could determine $\alpha$ and $\beta$ by measuring alcohol production of the two yeasts, which turned out to be 0.113 percent EtOH/cc yeast for *Saccharomyces* and 0.247 percent for *Schizosaccharomyces.* Thus, $\alpha$ = 0.247/0.113 = 2.18, and $\beta$ = 0.113/0.247 = 0.46. The values of $\alpha$ and $\beta$ obtained from the Lotka-Volterra equations were indeed in general agreement with those obtained independently by a physiological method.

In the late 1940s, Thomas Park and his students at the University of Chicago began a series of experiments examining competition between two flour beetles, *Tribolium confusum* and *Tribolium castaneum.*(Photo 8.1) *Tribolium confusum* usually won, but in initial experiments the beetle cultures were infested with a sporozoan parasite, *Adelina,* that killed some beetles, particularly individuals of *T. castaneum.* In these early experiments (Park 1948), *T. confusum* won in 66 out of 74 trials because it was more resistant to the parasite. Later, *Adelina* was removed, and *T. castaneum* won in 12 out of 18 trials. Most important, with or without the parasite, there was no absolute victor; some stochasticity was evident. Park then began to vary the abiotic environment and obtained the results shown in Table 8.1. It was evident that competitive ability was greatly influenced by climate; each species was a better competitor in a different microclimate. However, single-species rearings in a given climate could not always be relied on to predict the outcome of mixed-species rearings (examine the entry for cold-moist climate). Later, it was found that the mechanism of competition was largely predation on eggs and pupae by larvae and adults. Park then varied the aggressive or cannibalistic tendencies of the beetles by selecting different strains; he obtained different results according to the strain of each beetle used (Table 8.2) although again the results could

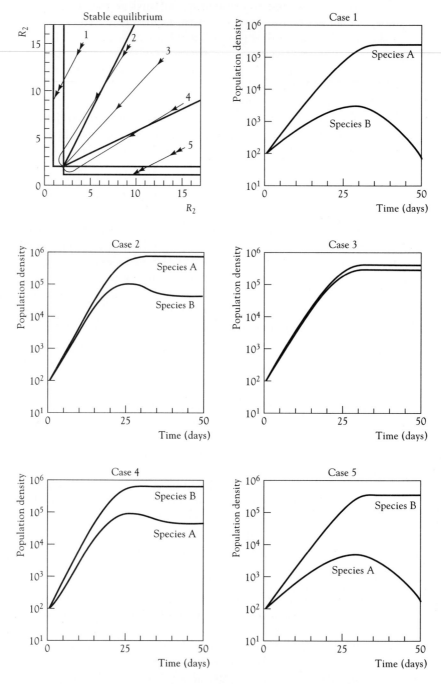

**Figure 8.5** *The outcome of competition for Tilman's model for five different resource supply points for Figure 8.4, equilibrium coexistence. The resulting five population curves for the two species show the time course of competition.* (From Tilman 1982.)

not always be predicted from the particular strains used. However, Park had demonstrated a complete reversal of competitive outcome as a function of temperature, moisture, parasites, and genetic strains.

## 8.3  COMPETITION IN NATURE

What of systems in nature where far more variability exists? One view holds that competition in nature is rare because by now, of all potential competitors, one has displaced the other. An alternative view holds that competition is a common enough force in

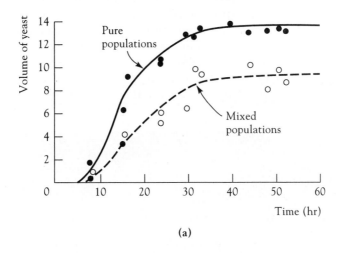

**(a)**

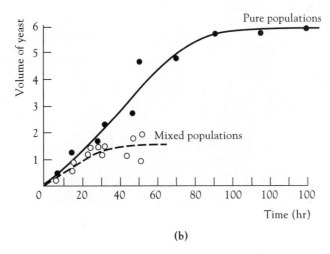

**(b)**

**Figure 8.6 (a)**  *Growth of population of the yeast Saccharomyces in pure cultures and in mixed cultures with Schizosacchoromyces.* (After Gause 1932.) **(b)** *Growth of population of the yeast Schizosacchoromyces in pure cultures and in mixed cultures with Sacchoromyces.* (After Gause 1932.)

nature to be a major factor influencing evolution. A third alternative is that predation and other factors hold populations below competitive levels. The question is important in applied situations—for example, in biological-control campaigns—because it is vital to know whether releasing one natural enemy against a pest is likely to be more effective than the release of many, where competition between enemies might reduce their overall effectiveness. Circumstantial evidence suggests that fewer natural enemies become established where many are released (Ehler and Hall 1982); yet sometimes this phenomenon does little to reduce overall effectiveness of control (Ehler 1979). Competitive effects between plants are also often thought to be of paramount importance in influencing crop yields, and many applied ecologists immediately assume all plants compete if resources are limiting (for example, Reynolds 1988). Again, this is important in applied ecology because while agronomists may strive to reduce competition, entomologists argue that more than one crop is valuable to encourage a wide variety of natural enemies and thus so reduce insect pest densities. The most direct method of assessing the importance of competition is to remove individuals of one species A and to measure the responses on species B (Wise 1981). Often, however, such manipulations are difficult to make outside the laboratory. If individuals of species A are removed, what's to stop them migrating back into the area

Photo 8.1   *In the 1940s, confused flour beetles, Tribolium confusum, were used as a model system by University of Chicago researcher Thomas Park to study competition.* (Degginger, Photo Researchers, Inc. 7W1467.)

TABLE 8.1 *Results of competition between the flour beetles* Tribolium castaneum *and* T. confusum. *(After Park 1954.)*

| Temp. °C | Relative Humidity % | Climate | Single species numbers | Mixed Species (% wins) | |
|---|---|---|---|---|---|
| | | | | T. confusum | T. castaneum |
| 34 | 70 | Hot-moist | conf=cast | 0 | 100 |
| 34 | 30 | Hot-dry | conf>cast | 90 | 10 |
| 29 | 70 | Temperate-moist | conf<cast | 14 | 86 |
| 29 | 30 | Temperate-dry | conf>cast | 87 | 13 |
| 24 | 70 | Cold-moist | conf<cast | 71 | 29 |
| 24 | 30 | Cold-dry | conf>cast | 100 | 0 |

TABLE 8.2 *Results of competition experiments between the flour beetles* Tribolium castaneum *and* T. confusum.[a] *(After Park et al. 1964.)*

| T. castaneum strain | T. confusum strain | No. castaneum wins | No. confusum wins |
|---|---|---|---|
| CI | bI | 10 | 0 |
| | bII | 10 | 0 |
| | bIII | 10 | 0 |
| | bIV | 10 | 0 |
| CII | bI | 1 | 8 |
| | bII | 0 | 10 |
| | bIII | 0 | 10 |
| | bIV | 4 | 6 |
| CIII | bI | 0 | 10 |
| | bII | 0 | 9 |
| | bIII | 0 | 10 |
| | bIV | 0 | 10 |
| CIV | bI | 9 | 1 |
| | bII | 9 | 0 |
| | bIII | 9 | 1 |
| | bIV | 8 | 2 |

[a] Cannibalistic tendency of strain goes low——> high as CI——> CIV and bI——> bIV.

of removal? If cages are used to stop immigration of species A, emigration of species B is prevented, and the numbers of species B may go unnaturally high (the so-called cage effect). Three of the most cited examples of competition in nature involve barnacles, parasitic wasps, and chaparral shrubs.

Two barnacles, *Chthamalus stellatus* and *Balanus balanoides*, dominate the British coasts. Their distribution on the intertidal rock faces are often well defined (Fig. 8.7). Joseph Connell (1961) showed that *Chthamalus* could survive in the *Balanus* zone when *Balanus* was removed. In nature, *Balanus* grows faster on rocks of the middle intertidal zone, squeezing *Chthamalus* out. The limits of *Balanus* distribution are determined in the upper zone by desiccation and in the lower zone by predation and competition for space with algae. *Chthamalus* is more resistant to desiccation than *Balanus* and is normally found only high on the rock face. Thirty-four years later, in 1988, Connell repeated his competition experiments in the same area of the Scottish coast and once more observed strong evidence for competition (Connell, personal communication).

Pest control is big business in the United States, and for some biological control projects, there is much labor available in the form of trained "scouts" to survey an area for evidence of successfully reproducing released enemies. Thus, when three parasitic wasps of the genus *Aphytis* (Photo 8.2) were introduced into southern California to help control the red scale (*Aonidiella aurantii*), an insect pest of orange trees, there were unprecedented amounts of data on the results. *Aphytis chrysomphali* was introduced accidentally from the Mediter-

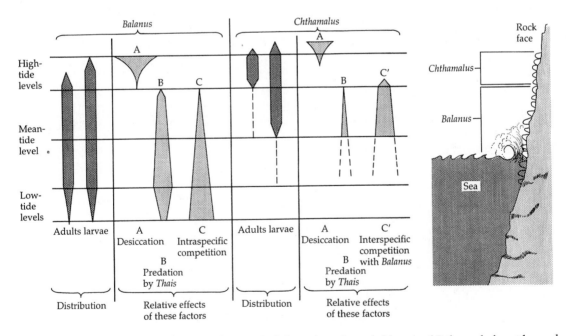

Figure 8.7   *Intertidal distribution of adults and newly settled larvae of* Balanus balanoides *and* Chthamalus stellatus *at Millport, Scotland, with a diagrammatic representation of the relative effects of the principal limiting factors.* (Modified from Connell 1961.)

Photo 8.2   *What is a parasitoid?—a parasite that lays its egg in another organism and whose resultant larva kills its host. Here a parasitic wasp lays its egg in a Dactynotus aphid. Parasitoids can be very effective agents in biological control; the interaction of three species of Aphytis wasp parasitoids in California provided a good example of competition in the field.*   (Kuhn, DRK Photo 140701.)

ranean in 1900 and became widely distributed. In 1948, A. *lignanensis* was introduced from south China and began to replace A. *chrysomphali* in many areas (Debach and Sundby 1963), such as Santa Barbara County and Orange County:

|  | Percent of Individuals | |
| --- | --- | --- |
|  | A. *chrysomphali* | A. *lignanensis* |
| Santa Barbara County |  |  |
| 1958 | 85 | 15 |
| 1959 | 0 | 100 |
| Orange County |  |  |
| 1958 | 96 | 4 |
| 1959 | 7 | 93 |

In 1956–1957 another species, A. *melinus*, was imported from India and immediately displaced A. *lignanensis* from the hotter interior areas:

| | Percent of Individuals | |
| --- | --- | --- |
| | A. ignanensis | A. melinus |
| Coastal, Santa Barbara County | | |
| 1959 | 100 | 0 |
| 1960 | 95 | 5 |
| 1961 | 100 | 0 |
| Interior, Santa Barbara County | | |
| 1959 | 50 | 50 |
| 1960 | 6 | 94 |
| 1961 | 4 | 96 |

The mechanism by which competitive displacement occurred was that female A. *melinus* could use smaller scale insects as hosts and could lay a higher proportion of "female" eggs in them than the other two species (Luck and Podoler 1985). Thus, A. *melinus* preempted most scales as hosts before the other parasites could use them. There are probably very few other examples of such well-documented competitive displacement in nature. However, the caveat here is that all species were exotic and may have been expected to compete more than native species that have evolved together over millions of years.

Plants are often thought to suffer more from competition than do animal populations because plants are rooted in the ground and cannot move to escape competitive effects. In southern California chaparral, grassland shrubs such as the aromatic *Salvia leucophylla* and *Artemisia californica* are often separated from adjacent grassland by bare sand 1 to 2 m wide. Volatile terpenes are released from the leaves of the aromatic shrub; these inhibit the growth of nearby grasses (Muller 1966). Some plants, such as black walnut, *Juglans nigra*, produce similar chemicals (here, juglone) from their roots, which leach into the soil, killing neighboring roots (Massey 1925). This phenomenon is termed *allelopathy*; the action of penicillin among microorganisms is a classic case. In many cases, such allelopathic chemicals are toxic to some competitors but not to others. Competitive interactions between plants are of paramount importance in *agroforestry*, a relatively new concept in which crops are grown under forest cover so that the land will yield both food and timber. Many species of tree, especially *Eucalyptus* in tropical regimes, are not suited to the practice of agroforestry because of their adverse effects on plants growing beneath them (Young 1988).

Other good but more anecdotal instances of competitive displacement in nature include the deliberate introductions of animals into areas for economic gain and the effects of habitat alteration. Although such examples do not represent well-conceived scientific experiments because of the economic impacts involved, more observations over a larger scale were available than for most scientific experiments. The introduction to Gatun Lake, Panama, of the cichlid fish *Cichla ocellaris* (a native of the Amazon) is thought to have led to the elimination of six of the eight previously common fish species within five years (Zaret and Paine 1973; see also Payne 1987, Chapter 21). Also, the introduction of the game fish *Micropterus salmoides* (largemouth bass) and *Pomoxis nigromaculatus* (black crappie) led to the diminution of local fish and crab populations. Similarly, after construction of the Welland Canal linking the Atlantic Ocean with the Great Lakes, much of the native fish fauna was displaced by the alewife (*Alosa pseudoharengus*) through competition for food (Aron and Smith 1971). Introduced African dung beetles were successful in reducing the numbers of pest flies that competed for dung in Australia. A fuller review of the affects of introduced species is given in Chapter 21.

Despite the difficulties of demonstrating competition, it is valuable to know how frequent it might be in nature. Two reviews of the literature attempted to answer this question. Connell (1983) reviewed 72 studies on active competition as reported in the literature. Competition was found in 55 percent of the 215 species, and in 40 percent of 527 experiments involved. Connell suggested that his result appears logical if one takes the following view. Imagine a resource set with, say, four species distributed along it. Then if only adjacent species competed, competitive effects would be expected in only three out of the six species pairs (50 percent) (Fig. 8.8). Of course, the mathematics would be drastically different according to the number of species on the axis. For any given pair of adjacent species, however, competition would be expected, and indeed Connell found that, in studies of single pairs of species, competition was almost always reported (90 percent), whereas in studies involving more species, the frequency was only 50 percent.

In a parallel but independent review of 150 field experiments, Schoener (1983) reported competition in more than 90 percent of 164 studies and in 75 percent of the species studied. Why the difference between Connell's and Schoener's studies? It is because of slightly

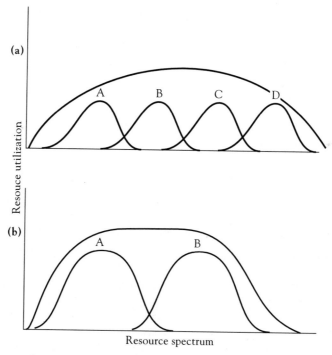

Figure 8.8    (**a**) *Resource supply and utilization curves of 4 species, A, B, C, and D along a resource gradient. If competition occurs only between species with adjacent resource utilization curves, then competition will be expected between A and B, B and C, and C and D—three of the six possible pairings. (**b**) When only two species utilize a resource set, competition would nearly always be expected between them.*

different samples of studies, methods of analysis, and perhaps predispositions of the authors. Connell, perhaps believing predation to be a more important force in nature, was more rigorous in what he accepted as a satisfactory experiment. For example, Hairston (1989) points out that at least one of the experiments accepted as evidence of competition by Schoener did not meet the necessary requirements of experimental design.

Both reviews may well overrepresent the actual frequency of competition in nature because of some common flaws:

1. "Positive" results demonstrating a phenomenon (here competition) may tend to be more readily accepted into the literature than "negative" results demonstrating patterns indistinguishable from randomness (Connell 1983).
2. Scientists do not study systems at random; those interested in competition may well choose to work in a system where competition may be more likely to occur.

On the other hand, the reviews may fail to reveal the true importance of competition because

1. By now most organisms have evolved to escape competition and the lack of fitness it may confer.
2. Competition may only occur in certain "crunch" years where resources are scarce (Wiens 1977). Nevertheless, this competition is severe enough to structure the community. If a crunch year occurs only one in five years and a researcher does experiments in any of the other four, competition may go undetected.

Despite these drawbacks some general patterns are evident from Connell's and Schoener's work if one assumes there are no taxonomic biases in reporting the frequency of competition (Table 8.3). Folivorous insects (leaf feeders) and filter feeders (such as clams) showed less competition than plants, predators, scavengers, or grain feeders. Marine intertidal organisms tended to compete more than terrestrial ones, and large organisms more than small ones. Some patterns like this might be expected; for example, given limited intertidal space, it would not seem odd to detect competition for space between sessile organisms.

TABLE 8.3 *Percentage of experimental studies showing interspecific competition. (After Connell.)*

| | Terrestrial | | Marine | | Freshwater | | Total | |
|---|---|---|---|---|---|---|---|---|
| | No. Exp. | % | No. Exp. | % | No. Exp. | % | No. Exp. | % |
| Plants | 205 | 30 | 31 | 68 | 2 | 50 | 238 | 35 |
| Herbivores | 45 | 20 | 13 | 69 | 0 | — | 58 | 31 |
| Carnivores | 36 | 11 | 5 | 60 | 3 | 67 | 44 | 20 |
| Total | 286 | 26 | 49 | 67 | 5 | 60 | 340 | 32 |
| Invertebrates | 57 | 16 | 37 | 32 | 0 | — | 94 | 22 |
| Vertebrates | 47 | 23 | 10 | 90 | 3 | 67 | 60 | 37 |

Seeds and grains also provide a limiting but very important nutrient-rich resource for desert granivores. Brown and co-workers (Brown et al. 1986; Heske, Brown, and Mistry 1994) provide good evidence that all members of the grain-feeding guild from rodents and birds to ants compete for this resource.

Lawton (1984) argued that there is much evidence of vacant **niche** space on the plants of the world. As evidence, he showed that bracken fern (*Pteridium aquilinum*) in Europe has a large array of chewing, sucking, mining, and galling insects in a wide range of habitats, but in the United States and especially in Papua, New Guinea, whole guilds (for example, gall formers) are missing (Fig. 8.9). With such vacant niches available, insects cannot be expected to compete so fiercely. The fact that so many introduced insects, other animals, and plants have become established and thrive when introduced into new countries and novel habitats suggests that few niches in natural ecosystems are filled. The implication is that many empty ecological niches exist, but it is hard for a human observer to tell what constitutes available niches independently of the species that occupy them. For example, could vampire bats later evolve in Africa to take advantage of the big game there? Could sea snakes, present now in the tropical parts of the Indian and Pacific oceans, evolve in the Atlantic region?

Lawton and McNeill (1979) also suggested that insects often "lie between the devil and the deep blue sea," that is, between a huge array of predators and parasites on the one hand and a deep blue sea of abundant but low-quality food on the other. As a result, they could scarcely become abundant enough to compete. Schoener himself had noted a dearth of competitive effects between **herbivores,** as compared to those between plants. Thus, plants and perhaps other groups with a lack of control from the next **trophic level** above could be reasoned to compete more. The foundations of this argument had been laid years earlier by Hairston, Smith, and Slobodkin (1960), who had essentially argued that the "earth is green" and that the phytophagous insects that could eat this greenery must therefore be held in check by animals from the next trophic level up. Herbivores, of course, constitute a huge fraction of the Earth's biota—more than 25 percent of the Earth's species are phytophagous insects alone—so such patterns must be taken seriously. We will return to this theme later, in Chapter 10 which compares the strength of competition with other mortality factors. Interestingly, the latest review (Denno, McClure, and Ott 1995) suggests that, in the mid-1980s to early 1990s, competition in insects has actually been found more frequently than in previous decades. An examination of 193 pair-wise species interactions showed interspecific competition in 76 percent of the cases. Perhaps more modern methods have been able to detect competition more easily. Clearly, we may have to reevaluate our position on competition in general, and insect competition in particular, as more modern data became available.

Finally, it is valuable to examine the precise mechanisms of the competitive process. Schoener (1985) divided these mechanisms into six categories:

1. Consumptive or exploitative—using resources
2. Preemptive—using space
3. Overgrowth—one species growing over another and blocking light or depriving the other of a resource.
4. Chemical—by production of toxins
5. Territorial—behavior or fighting in defense of space
6. Encounter—transient interactions directly over specific resources.

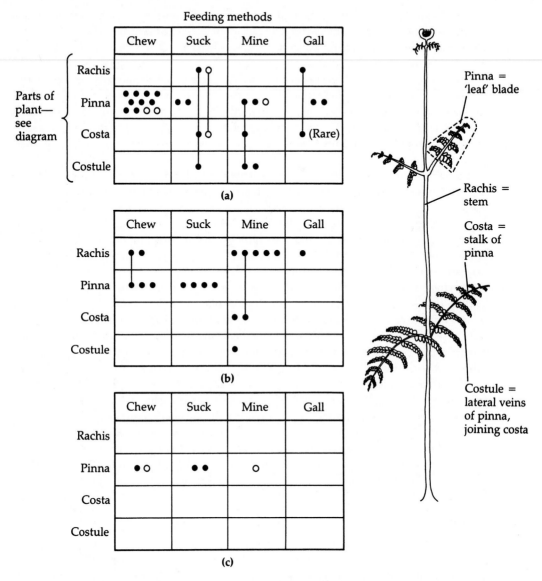

**Figure 8.9** *Feeding sites and feeding methods of herbivorous insects attacking bracken (Pteridium aquilinum) on three continents. (**a**) Skipwith Common in northern England. (**b**) Hombron Bluff, a savannah woodland in Papua New Guinea. (**c**) Sierra Blanca in the Sacramento Mountains of New Mexico and Arizona. Each bracken insect exploits the frond in a characteristic way. Chewers live externally and bite large pieces out of the plant; suckers puncture individual cells or the vascular system; miners live inside tissues; and gall-formers do likewise but induce galls. Feeding sites are indicated on the diagram of the bracken frond. Feeding sites of species exploiting more than one part of the frond are joined by lines. Closed circles represent open and woodland sites; open circles represent open sites only. (Drawn after Lawton 1984.)*

Exploitative competition is by far the most common, occurring in 71/188 = 37.8 percent of cases (Table 8.4). This has led some observers to underscore the frequency of indirect effects in nature because, in exploitative competition, species only interact via a third species, which is the shared-food source. Again, we will return to indirect effects later. Some of Schoener's findings about mechanisms are easy to interpret—preemptive and overgrowth competition appear among sessile space users, primarily terrestrial plants and marine macrophytes and animals living on hard substrates.

Territorial and encounter competition occur among actively moving animals, especially birds and mammals. Chemical competition occurs among terrestrial plants; toxins become too dilute in aquatic systems.

A final tidbit from both Connell's and Schoener's studies is that most often only one member of a species pair responded to the addition or removal of individuals of the other. The logic here is that such asymmetric competition should be expected; the superior competitor will probably be more strongly limited by some other factor—environmental tolerances or predators. Usually, the larger organism has the competitive advantage (Persson 1985).

## 8.4  COEXISTENCE

Active competition may not always lead to competitive displacement. It is conceivable, for example, that in areas of overlap, species change their lifestyles or feeding habits so that competition is minimized. If species do compete in nature, the important question is perhaps not how much competition goes on but how similar can competing species be and still live together. This question has received more attention in ecology than any other single topic (Schoener 1974), but Lewin (1983) suggests it has led ecologists into futile works and blind alleys.

TABLE 8.4 *Mechanisms of interspecific competition in experimental field studies. (After Schoener 1985.)*

| | Mechanism | | | | | | |
|---|---|---|---|---|---|---|---|
| Group | Consumptive | Preemptive | Overgrowth | Chemical | Territorial | Encounter | Unknown |
| Freshwater | | | | | | | |
| Plants | 0 | 0 | 1 | 1 | 0 | 0 | 0 |
| Animals | 13 | 1 | 0 | 1 | 1 | 5 | 2 |
| Marine | | | | | | | |
| Plants | 0 | 6 | 4 | 1 | 0 | 0 | 0 |
| Animals | 9 | 10 | 6 | 0 | 7 | 6 | 0 |
| Terrestrial | | | | | | | |
| Plants | 28 | 3 | 11 | 7 | 0 | 1 | 9 |
| Animals | 21 | 1 | 0 | 1 | 11 | 15 | 6 |
| Total | 71 | 21 | 22 | 11 | 19 | 27 | 17 |

## Theories Based on Morphology

In a seminal paper entitled "Homage to Santa Rosalia, or why are there so many kinds of animals?" G. Evelyn Hutchinson (1959) looked at size differences, particularly in feeding apparatus, between congeneric species when they were **sympatric** (occurring together) and **allopatric** (occurring alone) (Table 8.5). (The conceptual basis and explicit discussion of this approach had actually been laid out by Julian Huxley 17 years earlier in 1942 [Carothers 1986]).

Ratios between characters studied when species were sympatric ranged between 1.1 and 1.43, and Hutchinson tentatively argued that the mean value of 1.28 could be used as an indication of the amount of difference necessary to permit coexistence at the same trophic level but in different niches. Some authors extrapolated that ratios of between 1.3 and 2.0 indicated sufficient differences to permit coexistence because weight varies to the third power of length and $1.3^3 = 2.2$. Hutchinson's idea came under heavy fire because

1. In a large series of examples purporting to support this hypothesis, statistical analysis showed no more differences between species than would occur by chance alone (Simberloff and Boecklen 1981).
2. Size-ratio differences have too loosely been asserted to represent the ghost of competition past (Connell 1980) when, in fact, they could have evolved for other reasons.
3. Biological significance cannot always be attached to ratios, particularly those of structures not used to gather food: ratios of 1:3 have been found to occur between members of sets of kitchen skillets, musical recorders, and children's bicycles (Horn and May 1977).
4. Maiorana (1978) argued that, in such cases, these values may simply reflect something about our perceptual abilities.
5. Ratios of between 1:1 and 2:2 often result when things are lognormally distributed in nature (see Chapter 14) and have small variances (Eadie, Broekhoven, and Colgan 1987).

Needless to say, followers of Hutchinson's ideas have been swift to rebut some of these ideas. Losos, Naeem, and Colwell (1989) argued that the Simberloff–Boecklen statistics had

**TABLE 8.5** *Size relationships, the ratio of the larger to the smaller dimension, between congeneric species when they are sympatric and allopatric. (From Hutchinson 1959.)*

| Animals and Character | Species | Measurement (mm) when | | Ratio when | |
|---|---|---|---|---|---|
| | | sympatric | allopatric | sympatric | allopatric |
| Weasels | *Mustela nivalis* | 39.3 | 42.9 | 1.28 | 1.07 |
| (skull) | *M. erminea* | 50.4 | 46.0 | | |
| Mice | *Apodemus sylvaticus* | 24.8 | | 1.09 | |
| (skull) | *A. flavicollis* | 27.0 | | | |
| Nuthatches | *Sitta tephronota* | 29.0 | 25.5 | 1.24 | 1.02 |
| (culmen) | *S. neumayer* | 23.5 | 26.0 | | |
| Darwin's | *Geospiza fortis* | 12.0 | 10.5 | 1.43 | 1.13 |
| Finches | *G. fuliginosa* | 8.4 | 9.3 | | |
| (culmen) | | | | | |

deficiency of statistical power. In other words, they erred too strongly on the side of accept-
ing a null hypothesis of no effect of competition when such an hypothesis was false. The
paper also refutes the claim of Eadie, Broekhoven, and Colgan that the effects of competition cannot be distinguished from the affects of lognomal distributions. Doubtless, variations
of these arguments will continue to sway back and forth in the future. It is noteworthy, however, that even one of the chief protagonists in the debate, Daniel Simberloff, has recently
published articles that support the idea of character displacement in mammals (see, for example, Dayan and Simberloff 1994).

There are some authors who have tried to use natural experiments to support the validity of Hutchinson's ratios. In the primeval forests of Canada, four indigenous parasitoids attacked the wood-boring siricid larva *Tremex columba*. Each had a different ovipositor length
and laid an egg only when the ovipositor was fully extended. The ovipositor can be regarded
as a food-provisioning apparatus for the larva; each species layed eggs in *Tremex* cocoons at
a different depth in logs.

In the 1950s, a fifth species, *Pleolophus basizonus*, was introduced into the area in an attempt to control another pest, the European sawfly. Its ovipositor length was intermediate
between those of two of the existing species:

Even before the introduction of *P. basizonus*, the first three species were tightly packed
but still maintained the minimum separation of about 1:1 noted by Hutchinson. However, when *P. basizonus* was introduced, strong competition ensued. *Pleolophus basizonus* or
either of the two other species could have been displaced. In fact, M. *aciculatus* and M. *indis-
tinctus* were forced out of the more favorable high-host-density sites (Price 1970) (Fig. 8.10).

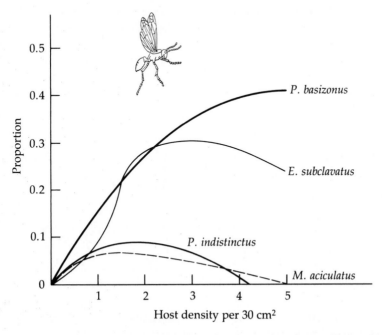

Figure 8.10 *The response of parasitoids to increasing host density illustrated by the change in the proportion of each species in the total parasitoid complex. As the host density increases, competition between parasitoids becomes more severe. Note that as the introduced* Pleolophus basizonus *increases,
the first species to be suppressed are those closest in ovipositor lengths.* (Redrawn from Price 1970.)

Besides separation in size, species may also differ in their use of a particular set of resources, such as food or space. Consider three species normally distributed on a resource set (Fig. 8.11), where $K(x)$ is the resource availability of $x$ or carrying capacity, $d$ is the distance between abundance maxima, and $w$ represents one standard deviation, approximately 68 percent of the area on $x$ one side of the curve. It has been argued mathematically that, if $d/w < 1$, species cannot coexist; if $d/w$ is $< 3$, there will be some interaction between species; and if $d/w > 3$, species coexist harmoniously (see Southwood 1978). The problem is that species abundancies are often not normally distributed.

Where the resource has a discontinuous distribution or occurs in distinct units, like leaves on a shrub, resource utilization can be illustrated graphically as in Figure 8.12. The niche breadth of a species can then be quantified by Levins's (1968) formula:

$$\text{niche breadth} = \frac{1}{\sum\limits_{i=1}^{s} p_i{}^2 \; (S)}$$

where $p_i$ = proportion of species found in the $i$th unit of a resource set of $S$ units, such that $B_{max} = 1.0$ and $B_{min} = 1/S$. Proportional similarity between species is then given by

$$PS = \sum_{i=1}^{n} p_{mi}$$

where $p_{mi}$ is the proportion of the less-abundant species of the pair in the $i$th unit of a resource set with $n$ units. Finally, the niche overlap of one species with another is represented by $\alpha_{ij}$ where

$$\alpha_{ij} = \sum_{h=1}^{n} p_{ih} \, p_{ij} \; (B_i)$$

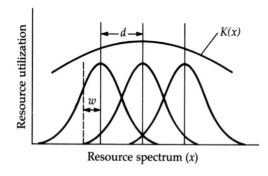 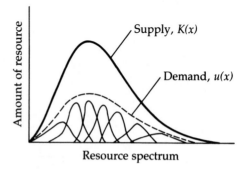

**Figure 8.11**   *Theoretical resource-utilization relationships.* (**a**) *The "simplest case" of three species with similar (and normal) resource-utilization curves.* $d$ = *distance apart of means,* $w$ = *standard deviation of utilization, and* $d/w$ = *resource separation ratio.* (**b**) *The more typical case with varying resource-utilization curves broadest in the region of fewer resources and less interspecific competition.* ([a] *Modified from May and MacArthur 1972;* [b] *redrawn from Pianka 1976.*)

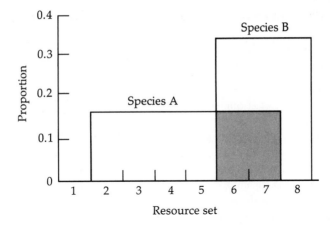

Figure 8.12 *Hypothetical distributions of a species, A, with a broad niche and a species, B, with a narrower niche, on a resource set subdivided into eight resource units. The species have the same proportional similarity (shaded zone), but species A overlaps B more than B overlaps A.*

where $p_{ih}$ and $p_{ij}$ are the proportion of each species in the *i*th unit of the resource set. Niche overlaps calculated in this way have been used as competition coefficients in classical Lotka-Volterra equations, although a safer method is to measure the effect of one species on another experimentally.

In accordance with Hutchinson's ideas, PS values of less than 0.70 have been taken to indicate possible coexistence, and those greater than 0.70, competitive exclusion. However, species may differ not only along one resource axis, but along many, such as food, temperature, and moisture. For two resource axes, proportional similarity indices can be combined—proportional similarity values of 0.8 and 0.6 on two axes combine to give an overall PS of 0.48. Theoretically, coexistence would be permitted in cases where combined PS values $0.7 \times 0.7 = 0.49$ or less. Such analyses become more and more complex and less biologically meaningful as new axes are included.

Perhaps the most serious criticism of both $d/w$ and PS treatment is that resource axes identified by the researcher as important may not accurately reflect limiting resources for organisms. Can sweep-net samples of insects on foliage reliably indicate food availability for birds? Furthermore, faced with the apparent contradiction that many ecologically similar species coexist with no apparent differences in biology, many researchers would argue that the correct niche dimensions had not yet been examined.

It is worth noting that, in many situations, competing species have been found to differ hardly at all in morphology and yet still are found together. Two reasons have been proposed. First, in the presence of high levels of predation, competitively **dominant** species are likely to be selected by predators over less-abundant prey. Thus, good competitors will probably never be able to eliminate poor competitors totally if predation occurs; this is the idea of predator-mediated coexistence. Second, it is important to realize that many real populations in nature exist not in closed systems but in open areas where migration is possible and where there is good connectance between populations. Caswell (1978) has theorized that, given good connectance between areas, immigration into an area of competitively inferior species, from areas where they do well, will be sufficient to maintain reasonable population sizes of both competitors for an indefinite time. Thus, if predators open up resources in an environment by killing members of the competitively dominant species, high connectance between populations means that competitively inferior species may first appear there by immigrating from other areas.

## r *and* K *Selection*

One of the most popular concepts to come out of competition theory is the idea of the **r–K continuum** (MacArthur and Wilson 1967). Not all organisms are well suited to compete with others; some are better able to live in more hostile environments, often in a competitive vacuum. There is a continuum of reproductive strategies that encompass so-called r-selected species at one end and K-selected species at the other. r-selected species are fugitives with a high rate of per-capita population growth, r, but poor competitive ability. An example is a weed that quickly colonizes vacant habitats (such as barren land), passes through several generations, and then disappears or is competitively excluded by individuals of more K-selected species. K-selected species compete well but tend to increase more slowly to the carrying capacity, K, of the environment. Some of the most important biological differences between these strategies are detailed in Table 8.6. It is interesting to note that biological control—the control of pests by natural enemies (usually insect parasitoids or predators)—has proved more successful on pests that are closer to the K type than to the r type (Southwood 1977) (Table 8.7). Conway (1976) has proposed different control techniques for pests that lie toward either the r or the K end of the spectrum (see Table 8.8).

More recently, alternatives to the r and K continuum have been proposed. Gill (1974) suggested a three-way classification scheme with r, K, and α strategies, the last being characterized by high competitive ability. For plants, Grime (1977, 1979) proposed the R, C, and S strategies, where R strategists (ruderals) are adapted to cope with habitat disturbance (especially

---

TABLE 8.6 *Some of the correlates of r and K selection. (After Pianka 1970.)*

|  | r selection | K selection |
|---|---|---|
| Favorable climate | Variable and/or unprdictable; uncertain. | Fairly constant and/or predictable; more certain |
| Mortality | Often catastrophic, nondirected, density-independent | More directed, density-dependent |
| Survivorship | Often type III | Usually types I or II |
| Population size | Variable in time, nonequilibrium; usually well below carrying capacity; communities saturated; no recolonization;portions thereof unsaturated; ecologic vacuums; recolonization each year. | Fairly constant in time, equilibrium; at or near carrying capacity of the environment; communities saturated, no recolonization necessary |
| Intra- and interspecific competition | Variable, often lax | Usually keen |
| Selection favors | 1. Rapid development | 1. Slower development |
|  | 2. High reproductive rate | 2. Great competitive ability |
|  | 3. Early reproduction | 3. Delayed reproduction |
|  | 4. Small body size | 4. Larger body size |
|  | 5. Single reproduction | 5. Repeated reproduction |
| Length of life | Short, usually less than one year. | Longer, usually more than one year |
| Leads to | Productivity | Efficiency |

**TABLE 8.7 *Cases of biological control of insects by imported natural enemies, grouped according to habitat characteristics on an r–K continuum. (After Southwood 1977.)***

| Habitat type | Cases attempted | | Cases graded "complete control" | |
|---|---|---|---|---|
| | No. | % | No. | % |
| *r* end | | | | |
| Cereals and forage crops | 6 | 4 | 0 | 0 |
| Vegetables | 21 | 14 | 0 | 0 |
| Sugarcane, cotton, pasture | 19 | 12 | 2 | 8 |
| Trees | 107 | 70 | 23 | 92 |
| *K* end | | | | |

man-made); C strategists (competitors) are adapted to live in supposed highly competitive environments such as the tropics; and S strategists (tolerators) are adapted to cope with severe abiotic environmental parameters. Finally, Greenslade (1983) proposed the *r, K,* and A strategies, where A species are adapted to tolerate adverse environmental conditions. Useful though each of these schemes is, MacArthur and Wilson's original concept remains the rock on which each is based. Sometimes, however, the MacArthur–Wilson foundation seems a little shaky. Despite the apparently broad array of support for the *r* and *K* concept from a wide variety of taxa, on closer examination the actual empirical evidence for this idea is difficult to assess. Different authors have used the terms *r* and *K* selection too loosely and in different senses (Parry 1981). It has thus become an "omnibus" term (Milne 1961), a term with such different intuitive definitions as to be ambiguous. For many people, *r* and *K* selection is what Hardin (1957) termed a *panchreston*—something that can explain almost anything.

**TABLE 8.8 *Principal control techniques for different pest strategies. (After Conway 1976.)***

| Technique | *r* pests | Intermediate pests | *K* pests |
|---|---|---|---|
| Pesticides | Early wide-scale applications based on forecasting | Selective pesticides | Precisely targeted applications based on monitoring |
| Biological control | ——— | Introduction of and/or enhancement of natural enemies | ——— |
| Cultural control | Timing, cultivation, sanitation, and rotation | ——— | Changes in agronomic practice, destruction of alternative hosts |
| Resistance | General, polygenic resistance | ——— | Specific monogenic resistance |
| Genetic control | ——— | ——— | Sterile mating technique |

## *Summary*

 1. Competition may be intraspcific (between individuals of the same species) or interspecific (between individuals of different species). It may also be viewed as scramble competition for a limiting resource, or contest competition where individuals compete directly with one another.

2. The first attempts to model how competing species interacted were originated by Lotka (1925) in the U.S. and by Volterra (1928) in Italy. In a two-species interaction, four outcomes are possible: species 1 goes extinct; species 2 goes extinct; either species 1 or 2 goes extinct, depending on starting conditions; or both species coexist.

3. Thomas Park's laboratory studies of competition between four beetles showed that the outcome of competition could be changed by environmental conditions, by the presence or absence of natural enemies, and by the genetic strain of the competitors involved.

4. In nature, there are watertight experimental studies that show competition between different type of organisms, such as between barnacles, between parasitic wasps, and between chaparral plants.

5. Reviews on the frequency of competition in nature demonstrate that competition between species has been found in 55 percent to 75 percent of the species involved.

6. Schoener (1985) recognized at least six mechanisms of competition: consumptive (exploitative), preemptive, overgrowth, chemical, territorial, and encounter. Exploitative competition is by far the most common, occurring in at least 37.8 percent of cases.

7. It has frequently been argued that competition is minimized and that species can coexist, if they utilize different resources. What is the amount by which competitors can overlap in resource utilization but still coexist? Hutchinson (1959) suggested that a ratio of 1:1.3 in the morphological sizes of feeding apparatus was necessary. Other authors have studied resource use more directly and have suggested proportional similarity values of no more than 70 percent are necessary.

8. From competition theory has sprung the idea of the $r–k$ continuum. $r$-selected organisms are poor competitors but have high per-capita population growth rate ($r$). Such species are often highly dispersive species that colonize new habitats before being outcompeted by $K$-selected species. $K$-selected organisms are good competitors and often exist in mature habitats where they can outcompete most species. They become more common as the carrying capacity of an area is approached.

*Discussion Question:* Which type of competition would you expect to be the more important in nature—intraspecific or interspecific? Does intraspecific competition have to be weak in gregarious species? If interspecific competition between species can cause species to partition resources or to have different morphological characteristics, why wouldn't intraspecific competition cause even greater character displacement or resource partitioning between individuals of the species?

# Predation

$S$everal types of predation can be recognized:

**Herbivory:** Animals feeding on green plants

**Carnivory:** Animals feeding on herbivores or other carnivores

**Parasitism:** (a) Animals or plants feeding on other organisms without killing them and (b) parasitoids, usually insects, laying eggs on other insects, which are subsequently totally devoured by the developing parasitoid larvae

**Cannibalism:** A special form of predation, predator and prey being of the same species

Although herbivory and parasitism could be viewed widely in the context of predation, each has special characteristics that tend to separate it out into a special subdivision of predation, and each subject will be discussed separately. Herbivory, for example, is often nonlethal predation on plants. Many leaves or parts of leaves of a plant can be eaten without serious damage to the "prey." In the animal world, on the other hand, predation generally means death for the prey. Parasitism, too, is a unique type of event, in which one individual prey is commonly utilized for the development of one predator. In the special case of insect herbivores on plants, there is great debate about whether such insects can be called plant *parasites* because they seem to fit such a definition fairly well (Price 1980). Carnivory embodies the traditionally held view of predation, and cannibalism can be seen as a special case of it. The degree of lethality of these interactions on their hosts/prey and the intimacy with them are summarized in Figure 9.1.

## 9.1  STRATEGIES TO AVOID BEING EATEN: EVIDENCE FOR THE STRENGTH OF PREDATION AS A SELECTIVE FORCE

In response to predation, many prey species have developed strategies to avoid being eaten. Edmunds (1974), Owen (1980), and Waldbauer (1988) have reviewed many case studies, which include the following.

- *Aposematic, or warning, coloration:* Aposematic coloration advertises a distasteful nature. Lincoln Brower (1970) and coworkers (Brower et al. 1968) showed how inexperienced blue jays took monarch butterflies, suffered a violent vomiting reaction, and learned to associate forever afterward the striking orange-and-black barred appearance of a monarch with a noxious reaction. The caterpillar of this butterfly gleans poison from its poisonous host plant, a milkweed. Many other species of animals, especially invertebrates, are also warningly colored, for example, ladybird beetles. Caterpillars of many Lepidoptera are bright and conspicuous, too, because

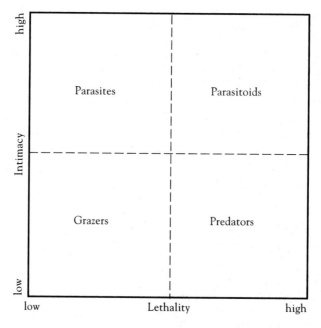

**Figure 9.1** *Possible interactions between populations. Lethality represents the probability that a trophic interaction results in the death of the organism being consumed. Intimacy represents the closeness and duration of the relationship between the individual consumer and the organism it consumes.* (After Pollard 1992.)

being noxious is their main line of defense—being soft-bodied, they would otherwise be very vulnerable to predators.

- *Crypsis and catalepsis:* **Crypsis** and **catalepsis** are camouflage and the development of a frozen posture with appendages retracted. This is another method of avoidance of detection by invertebrates. For example, many grasshoppers are green and blend in perfectly with the foliage on which they feed, exhibiting **apatetic coloration.** Often, even leaf veins are mimicked on grasshopper wings. Stick insects mimic branches and twigs with their long slender bodies. In most cases, these animals stay perfectly still when threatened because movement alerts a predator. Crypsis is prevalent in the vertebrate world, too. A zebra's stripes supposedly make it blend in with its grassy background, and the sargassum fish even adopts a grotesque body shape to mimic the sargassum weed in which it is found. Perhaps first prize should go to the chameleon, whose skin tones can be adjusted to match the background on which it is resting.
- *Mimicry:* Though many organisms may try to blend into their background, mimicking the foliage or background around them, some animals mimic other animals instead. For example, some hoverflies mimic wasps. Several types of mimicry can be defined.

**Mullerian:** The convergence of many unpalatable species to look the same, thus reinforcing the basic distasteful design, as for example with wasps and some butterflies (Muller 1879). (See color insert II, page 1.)

**Batesian** (after the English naturalist Henry Bates [1862], who first described it): Mimicry of an unpalatable species by a palatable one or of dangerous species (coral snakes) by innocuous ones (pseudocorals). Wickler (1968) has documented many cases in which flies, especially hoverflies of the family Syrphidae, are striped black-and-yellow to resemble stinging bees and wasps (see color insert II, page 1). Also, some butterflies painted to mimic distasteful models were recaptured more frequently than those painted to mimic palatable forms, suggesting the great survival value of mimicry (Sternberg, Waldbauer, and Jeffords 1977). Until very recently, monarch and queen butterflies, themselves aposematically colored, were considered models for mimic viceroy butterflies. However, David Ritland (Ritland and Brower 1991; Ritland 1994) showed that all three species were unpalatable and thus coexisted as a Mullerian mimicry complex.

**Aggressive:** In this case, the body coloration permits individuals not so much to escape predation as to be better predators themselves. Many praying mantises mimic flowers so as to entrap insect prey. This strategy is shared by the crab spiders. Certain bottom-dwelling ocean fish also mimic the substrate so as to get in closer proximity to their prey. In extreme cases, such as with anglerfish, parts of the body are modified to act as lures for prey. Some species also use light-emitting organs to attract prey. On land, aggressive mimicry is practiced by female fireflies of the genus *Photuris* (Lloyd 1975). Normally, males respond to the species-specific light flashes of their females and move toward them. *Photuris* females mimic the flashing patterns of females of other species to lure the males of those species and eat them.

- *Intimidation displays:* An example of intimidation display is a toad that swallows air to make itself appear larger. Frilled lizards extend their collars when intimidated to have the same effect.

- *Polymorphisms:* **Polymorphism** is the occurrence together in the same population of two or more discrete forms of a species in proportions greater than can be maintained by recurrent **mutation** alone. Often, this phenomenon takes the form of a color polymorphism; if a predator has a preference or **search image** for one color form, usually the commoner (Tinbergen 1960), then the prey can proliferate in the rarer form until this form itself becomes the more common (Cain and Sheppard 1954b) (so-called **apostatic selection,** Clark 1962). Stiling (1980) advocated just this type of choice of prey, by a visually searching **parasitoid,** to maintain the difference between two distinct color morphs, orange and black, in some leafhopper nymphs of the genus *Eupteryx,* though Stewart (1986a,b) also implicated thermal **melanism** as an important agent of selection in some species; that is, black morphs occur in cooler climates because they heat up faster in the sun. Owen and Whiteley (1986) have pointed out that in many species the form of the polymorphism is such that *every* individual is slightly different from all others. This is true in brittlestars, butterflies, moths, echinoderms, and gastropods. They suggest that such a staggering variety of form thwarts predators' learning processes, and they suggest the term *reflexive selection* for this type of phenomenon.
- *Phenological separation of prey from predator:* Fruit bats, normally nocturnal foragers, are active by day and at night on some small, species-poor Pacific islands such as Fiji. Wiens et al. (1986) suggest that the fruit bats are constrained elsewhere to fly only at night by the presence of predatory diurnal eagles.
- *Chemical defenses:* One of the classic defenses involves the bombardier beetle (*Bradinus crepitans*) as studied by Tom Eisner and coworkers (Eisner and Meinwald 1966; Eisner and Aneshansley 1982). These beetles possess a reservoir of hydroquinone and hydrogen peroxide in their abdomens. When threatened, they eject these chemicals into an explosion chamber, where they mix with a peroxidase enzyme. The resultant release of oxygen causes the whole mixture to be violently ejected as a spray that can be directed at the beetle's attackers. Many other arthropods, such as millipedes, have chemical defenses too. This phenomenon is also found in vertebrates, as people who have had a close encounter with a skunk can testify.
- *Masting:* Masting is the synchronous production of many progeny by all individuals in a population to satiate predators and to allow some progeny to survive (Silvertown 1980). It is commonly documented in trees, which tend to have years of unusually high seed production. A similar phenomenon is exhibited by the emergence of 17-year and 13-year cicadas. It is worth noting in this context that both 13 and 17 are prime numbers, so no predator on a shorter multiannual cycle could repeatedly use this resource.

How common is each of these defense types? Brian Witz (1989) surveyed 354 papers that documented antipredator mechanisms in arthropods, mainly insects (Table 9.1) of 555 predator/prey interactions. By far, the most common antipredator mechanism was chemical defense, noted in at least 46 percent of the examples—*at least* because many categories were not mutually exclusive; for example, aposematic coloration is also usually coupled with noxious chemicals.

It is obvious that predation constitutes a great selective pressure on plant and animal populations. Despite the impressive array of defenses, predators still manage to survive by eating individuals of their chosen prey, often by circumventing the defenses in some way. The co-

TABLE 9.1 *Antipredator mechanisms in arthropods. (After Witz 1989.)*

| Mechanism and rank | Example of mechanism | % frequency |
| --- | --- | --- |
| 1. Chemical | Reflexive bleeding, toxic chemicals (especially beetles) | 46 |
| 2. Fighting | Stinging (especially wasps), biting, kicking | 11 |
| 3. Crypsis | Camouflage, especially caterpillars | 9 |
| 4. Escape | Running away, flying | 8 |
| 5. Mimicry | Batesian and Mullerian | 5 |
| 6. Aposematism | Warning coloration | 5 |
| 7. Intimidation display | Posturing | 4 |
| 8. Dilution | Masting, satiation | 4 |
| 9. Mutualism | Defense by other organism | 3 |
| 10. Armor | Spines, thorns | 2 |
| 11. Acoustic | Loud noise, e.g., grasshopper | 2 |
| 12. Feigning death | | 1 |

evolution of defense and attack can be seen as an ongoing evolutionary arms race. According to Dawkins and Krebs (1979), the prey are always likely to be one step ahead. The reason is what they termed the *life-dinner principle*. In a race between a fox and a rabbit, the rabbit is usually faster because it is running for its life, whereas the fox is running "merely" for its dinner. A fox may still reproduce, even if it does not catch the rabbit. The rabbit never reproduces again if it loses. This arms race has been argued to be run not only between predators and prey but also between parasite and host and between plant and herbivore. In the latter case, the race often proceeds by the production of toxins by the host and detoxifying mechanisms by the predator. Usually, however, at least for plants and herbivorous insects, phylogenetic relationships among herbivores do not correspond well to those of their host plants (Futuyma 1983; Mitter and Brooks 1983). The result is that at each defensive turn by a plant (or other host), it is as likely to run into a new set of enemies as it is to escape the old ones.

## 9.2  MATHEMATICAL MODELS

What effect has the predator on its prey population? The answer depends on many things, including prey and predator density and predator efficiency. Rosenzweig and MacArthur (1963) modeled predator-prey dynamics using a graphical method. First, they assumed that, in the absence of predators, the prey population increases exponentially according to the Lotka-Volterra formula

$$\frac{dN}{dt} = rN$$

The rate of removal of prey increases with an increase in encounter rate by predators dependent upon the number of predators $(C)$ and the number of prey $(N)$. The rate of removal is also dependent upon a "searching efficiency" or attack rate of predators, termed $a$. The consumption of prey by predators is then $a'CN$, so

$$\frac{dn}{dt} = rN - a'CN$$

The rate of growth of the predators is given by

$$\frac{dC}{dt} = fa'CN - qC$$

where $q$ is a mortality rate based on starvation in the absence of prey and $fa'CN$ is predator birth rate based on $f$, the predator's efficiency at turning food into offspring.

The properties of this model can be investigated by location of zero isoclines (regions of zero growth, neither positive nor negative). In the case of the prey, when

$$\frac{dN}{dt} = 0, \quad rN = a'CN$$

or

$$C = \frac{r}{a'}$$

Because $r$ and $a'$ are constants, the prey zero isocline is a line for which C, the number of predators, is itself a constant (Fig. 9.2a). Along this line, prey neither increase nor decrease in abundance.

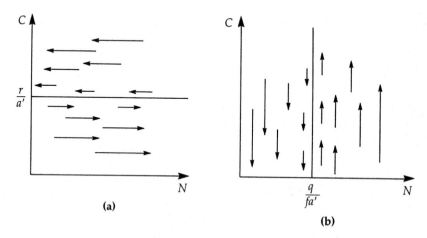

(a)

(b)

Figure 9.2   A Lotka-Volterra type predator-prey model. (**a**) The prey zero isocline, with prey (N) increasing in abundance at lower predator densities (low C) and decreasing at higher predator densities. (**b**) The predator zero isocline, with predators increasing in abundance at higher prey densities and decreasing at lower prey densities. (Redrawn from Begen, Harper, and Townsend 1986.)

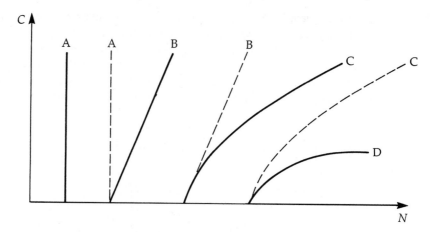

**Figure 9.3**   *Predator zero isoclines of increasing complexity, A to D. A is the Lotka-Volterra isocline. B shows that more predators require more prey. C shows that the consumption rate is progressively reduced by mutual interference among predators. D shows that predators are limited by something other than their food.* (Redrawn from Begon, Harper, and Townsend 1986.)

For the predators, when

$$\frac{dC}{dt} = 0, \quad fa'\,CN = qC$$

or

$$N = \frac{q}{fa'}$$

The predator-zero isocline is also a straight line, one along which $N$, the number of prey, is constant (Fig. 9.2b). An assumption is that if there are enough prey, a population of predators will increase and that if there are not enough prey, they will starve. This is obviously a gross oversimplification. First, larger populations of predators require larger populations of prey to maintain them, so the zero isocline for predator growth should slant to the right (line B, Fig. 9.3). As predator density increases, so will mutual interference between predators, who will spend more time fighting one another and less time tackling prey. For zero growth, then, more realistically, there must be even more prey (line C, Fig. 9.2). Finally, at the highest densities of prey, it seems that the rate of growth of predators will be limited by something other than prey availability—say, social constraints on territory size or the availability of burrows or nest sites—so that the zero isocline will appear as in line D in Figure. 9.2. Remember that at predator-prey combinations to the left of this line, predator numbers decrease, whereas to the right of it they increase.

The refinement of the prey-zero isocline is dependent on two concepts. The first is that the recruitment rate of prey into the population is high when $N$ is low but low when $N$ is high and near the carrying capacity of the environment. The logic is that strong competition for resources at levels of $N$ near $K$ severely reduces recruitment rate; hence, the recruitment curve is semicircular (Fig. 9.4 top). The rate of growth of the prey population is also dependent on the level of predation, and different predation rates are represented by the steepness of the dashed lines in Figure 9.4 top. At each of the points where the consumption curve crosses the recruitment curve, the rate of population growth of the prey is zero, and these pairs of densities can be directly transferred to give the zero isocline for growth of the prey (Fig. 9.4 bottom).

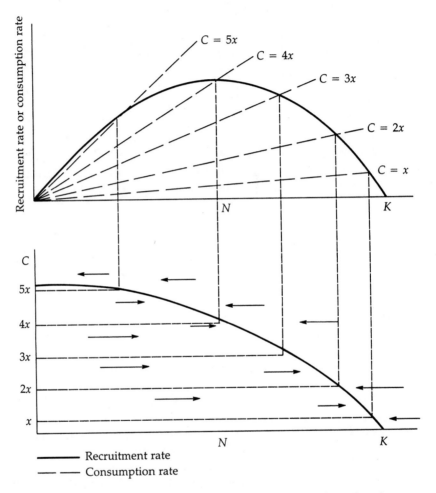

Figure 9.4  *Refinement of prey zero isocline. The solid line in the top figure describes variation in prey recruitment rate with density. The dashed lines in the upper figure describe the removal or consumption of prey by predators. There is a family of dashed curves because the total rate of consumption depends on predator density: increasingly steep dashed curves reflect these increasing densities. At the points where a consumption curve crosses the recruitment curve, the net rate of prey increase is zero (consumption equals recruitment). Each of these points is characterized by a prey density and a predator density, and these pairs of densities therefore represent joint populations lying on the prey zero isocline in the bottom figure. The arrows in the lower figure show the direction of change in prey abundance. (Redrawn from Begon, Harper, and Townsend 1986.)*

The effect of different densities of predators and prey can now be assessed by combination of the two isoclines (Fig. 9.5). The highest levels of predation translate into large oscillations of predator and prey (i), sometimes so violent that one or both species goes extinct. Predators in this case are relatively efficient and abundant. Less-efficient predators (ii) give rise to small and often damped oscillations in the system. At the lowest levels of predation, the system is highly stable, but only low levels of predators are supported (iii). The problem with this analysis is that it is almost impossible to distinguish this type of predator-prey cycling from population fluctuations resulting from environmental factors such as bad weather. Furthermore, predator-prey isoclines can change dramatically with the incorporation of such factors as refuges from predators and the ability of predators to switch back and forth between multiple species of prey. The predictive value of these models seems to decrease as their complexity increases.

How do the data on predator and prey abundances in the field compare with these types of graphical models? Predators of the Serengeti plains of eastern Africa (lions, cheetahs, leopards, wild dogs, and spotted hyenas) seem to have little impact on their large-mammal prey (Bertram 1979). Most of the prey taken are injured or senile and are likely to contribute little to future generations. In addition, most of the prey sources are migratory, and the predators—residents—are more likely to be limited in numbers by prey who also are resident in

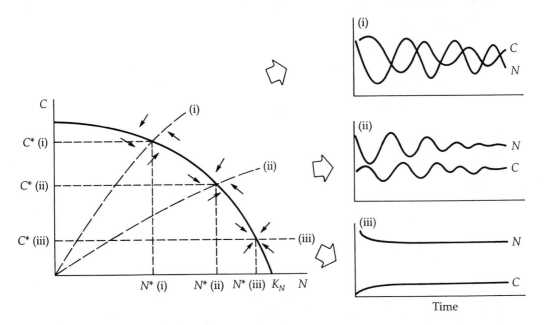

Figure 9.5  *A prey zero isocline with self-limitation, combined with predator zero isoclines with increasing levels of self-limitation: (i), (ii), and (iii). C* is the equilibrium abundance of predators, and N* is the equilibrium abundance of prey. Combination (i) is least stable (most persistent oscillations) and generally has most predators and least prey: the predators are relatively efficient. Less efficient predators (ii) give rise to a lowered predator abundance, an increased prey abundance, and less persistent oscillations. Strong predator self-limitation (iii) can eliminate oscillations altogether, but C* is low and N* is close to $K^N$.* (Redrawn from Begon, Harper, and Townsend 1986.)

the dry season when migratory ungulates are elsewhere. Pimm (1979, 1980) has argued that the importance of predation is dependent on whether the system is "donor controlled" or "predator controlled." In a donor-controlled system, prey supply is determined by factors other than predation, so removal of predators has no effect. Examples include consumers of fruit and seeds, consumers of dead animals and plants, and intertidal communities in which space is limiting. In a predator-controlled system, the action of predator feeding eventually reduces the supply of prey and their reproductive ability. Removal of predators in a donor-controlled system is obviously likely to have little effect, whereas in a predator-controlled system such an action would probably result in large changes in abundance.

Many predator-prey systems appear stable; others appear to fluctuate dramatically. Determining which mechanism works in which situation is not easy. For example, in the Arctic, there are two groups of primarily herbivorous rodents—the microtine varieties (lemmings and voles) and the ground squirrels. Ground squirrels exhibit the strongly self-limiting behavior of aggressive territorial defense of burrows (Batzli 1983). Their populations are remarkably consistent from year to year. On the other hand, the microtines are renowned for their dramatic population fluctuations. Thus, even in the same habitat, results for different species vary dramatically. One series of population fluctuations analyzed in great detail is that of the Canada lynx and snowshoe hare.

The Canada lynx (*Lynx canadensis*) eats snowshoe hares (*Lepus americanus*) and shows dramatic cyclic oscillation every nine to eleven years (Fig. 9.6). Charles Elton analyzed the records of furs traded by trappers to the Hudson's Bay Company in Canada over a 200-year period and showed that a cycle has existed for as long as records have been kept (Elton and Nicholson 1942). This cycle has been interpreted as an example of an intrinsically stable predator-prey relationship (Trostel et al. 1987), but Keith (1983) has argued that it is winter food shortage and not predation that precipitates hare decline. He showed that heavily grazed grasses produce shoots with high levels of toxins, making them unpalatable to hares.

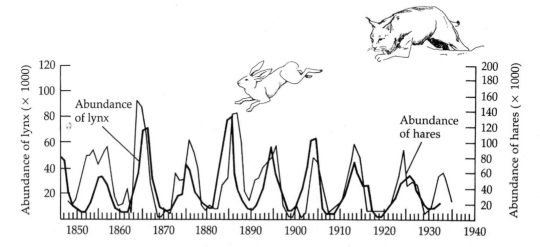

Figure 9.6 *The apparently coupled oscillations in abundance of the snowshoe hare (Lepus americanus) and Canada lynx (Lynx canadensis) as determined from numbers of pelts lodged with the Hudson's Bay Company.*

Such chemical protection remains in effect for two to three years, precipitating further hare decline. Predators, he argued, simply exacerbate population reduction. Thus, although lynx cycles depend on snowshoe-hare numbers, hares fluctuate in response to their host plants. Subsequently, Smith et al. (1988) showed that, although food quality greatly affects hare **biomass,** most hares die of predation, not starvation. However, death due to predation is greatly exacerbated by poor quality of hares, which is of course greatly affected by food quality; thus, there seems to be a good deal of common ground between the predation and starvation camps. Most recently Sinclair et al. (1993) suggested a correlation between the frequency of sunspots, the level of herbivory on white spruce by snowshoe hares, and hare fur records stretching back 200 years.

## 9.3  EVIDENCE FROM EXPERIMENTS AND INTRODUCTIONS

Perhaps the best way to find out whether predators determine the abundance of their prey is to remove predators from the system and to examine the response. One of the best examples involves dingo predation on kangaroos in Australia (Caughley et al. 1980). The dingo, *Canis familiaris dingo*, is the largest naturally occurring carnivore in Australia and an important predator of imported sheep. Dingoes have been intensively hunted and poisoned in sheep country—southern and eastern Australia—and long fences (some up to 9,600 km) extend to prevent them from recolonizing areas, providing a classic experiment in predator control. The result has been a spectacular increase, 166-fold, of red kangaroos where the dingoes have been eliminated in New South Wales (Fig. 9.7), over their density in south Australia, where dingoes have not been molested.

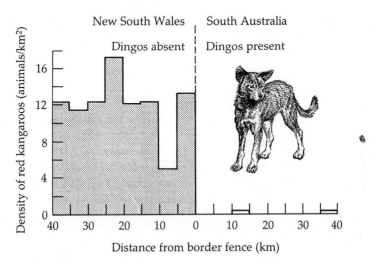

**Figure 9.7**   *Density of red kangaroos on a transect across the New South Wales-South Australia border in 1976. The border is coincident with a dingo fence that prevents dingoes from moving from South Australia into the sheep country of New South Wales.* (Redrawn from Caughley et al. 1980.)

Emus (*Dromaius novaehollandiae*) are also more than 20 times more abundant in dingo-free areas. Dingoes are also frequent predators of feral pigs in tropical Australia (Newsome 1990). In Cape York, northern Queensland, there is a gross shortage of young pigs less than two years old on the mainland where there are dingoes. On neighboring Prince of Wales Island where dingoes are absent, recruitment is considerable (Fig. 9.8).

Other important exotic animals in Australia are European foxes and feral cats. Both can do damage to domestic livestock and are subject to eradication by shooting. In areas where these predators were shot, numbers of rabbits, also exotic in Australia, increased (Fig. 9.9). Where rabbits increase, valuable rangeland may become overgrazed. The effects of predators on their prey is clearly a subject of interest to farmers as well as biologists.

Another striking example of predation pressure has been provided by an inadvertent introduction by humans. Marine sea lampreys (*Petromyzon marinus*) live on the Atlantic coast of North America and migrate into freshwater to spawn. Adult lampreys feed by attaching themselves to other fish, then rasping a hole, and finally sucking out the body fluids. The passage of the lamprey to the upper Great Lakes was presumably blocked by Niagara Falls before the Welland Canal was built in 1829. The first sea lamprey was found in Lake Erie in 1921,

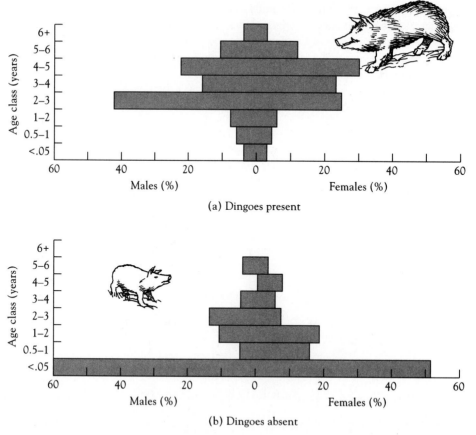

Figure 9.8  *Contrasting population structures of feral pigs where dingoes are* (**a**) *present and* (**b**) *absent in tropical northern Australia.* (After Newsome 1990.)

in Lake Michigan in 1936, in Lake Huron in 1937, and in Lake Superior in 1945 (Applegate 1950). Lake trout catches decreased to virtually zero within about 20 years of lamprey invasion (Fig. 9.10). Control efforts have been applied since 1956 to reduce the lamprey population, and attempts are being made to rebuild the Great Lakes fishery (Christie 1974).

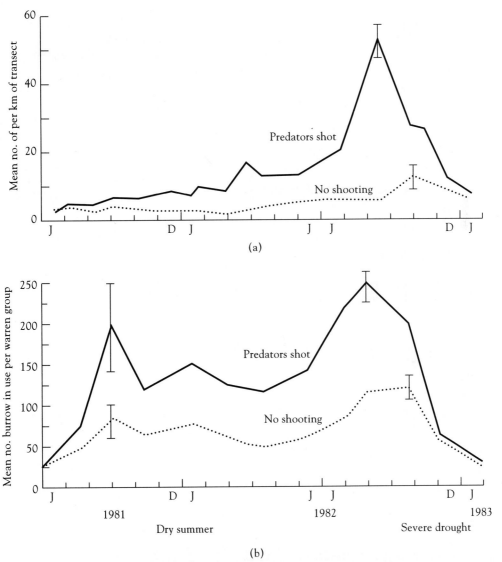

Figure 9.9   *Accelerated increase in rabbits with removal of European foxes and feral cats in an Australian field experiment. (**a**) Comparison of counts of rabbits per kilometer along transects where predator populations were continually shot (solid line) or left intact (dotted line). (**b**) Comparison of burrow use in warren groups where predator populations were continually shot (solid line) or left intact (dotted line). The error bars represent one standard error on either side of the mean. (After Newsome et al 1989.)*

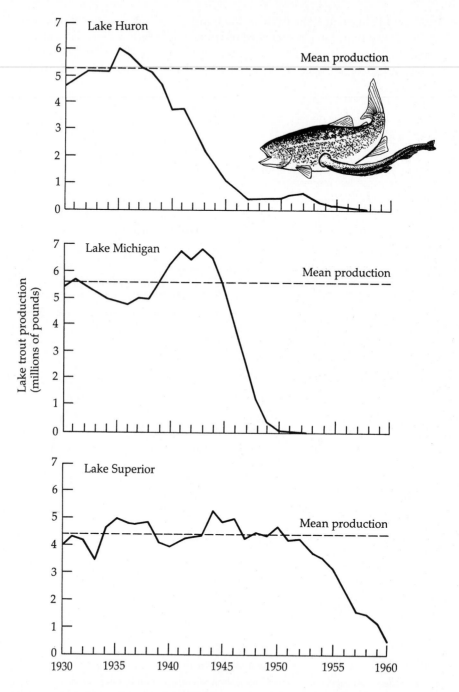

**Figure 9.10** *Effect of sea lamprey introduction on the lake trout fishery of the upper Great Lakes. Lampreys were first seen in Lake Huron and Lake Michigan in the 1930s and in Lake Superior in the 1940s.* (Redrawn from Baldwin 1964.)

## 9.4 REVIEWS AND NATURAL SYSTEMS

Most of the examples discussed so far have focused on introduced species. Effects of such introductions are often very strong, but what of totally natural predator-prey systems where predator and prey have evolved together for long periods? In 1903, lions were shot in Kruger National Park, South Africa, to allow numbers of large prey to increase. Shooting ceased in 1960, by which time wildebeest (*Connochaeter taurinus*) had increased so much that human culling was instigated from 1965 to 1972.

For many years, the moose population on Michigan's Isle Royale, a 45-mile long island enjoyed a wolf-free existence. Then, in 1949, during a particularly hard winter, a pair of Canadian wolves were able to colonize the island, walking across frozen Lake Superior. In 1958, wildlife biologist Durwood Allen of Purdue University began to track wolf and moose numbers. The wolf population peaked at 50 individuals in 1980, and then in 1981 it took a severe nosedive (Fig. 9.11). Wildlife ecologist Rolf Peterson of Michigan Technological University has followed this population in the most recent years. The wolf population has continued to decline. Only four pups were born in 1992 and 1993, all to the same female in one wolfpack. The other two packs on the island are down to just a pair of wolves each. Many observers feel the total wolf population is on its way to extinction. As for the moose population, it has reached a record level of about 1,900 moose in 1993. This seems good evidence that predation does have a strong effect on natural populations.

Why did the wolves decrease in numbers? First of all, there was a narrow genetic base to begin with. Restriction enzyme analysis of the wolves mitochondrial DNA turned up just a single pattern, indicating the wolves were all descended from a single female. They had only about

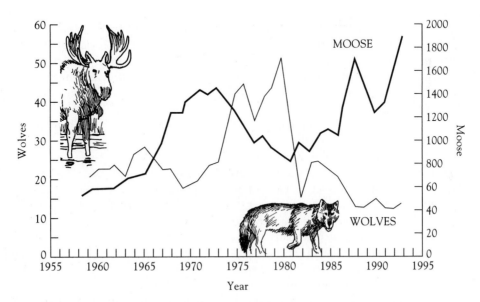

**Figure 9.11**   *The effects of wolf predation on moose numbers in Isle Royale. As the wolf population declined, the number of moose went up.*   (After Mlot 1993.)

half the genetic variability of the mainland wolves. Second, there was evidence of a deadly canine virus in 1981. This was probably a result of a parvovirus outbreak in nearby Houghton, Michigan, which was carried to the island on the hiking boots of visitors (Mlot 1993).

In 1985 Andrew Sih and colleagues (Sih et al. 1985) surveyed 20 years (1965–1984) of seven ecological journals for field experiments concerned with predation. Their survey yielded 139 papers involving 1,412 comparisons. Virtually every report, 132/139, or 95 percent, showed some significant effects. At least one comparison existed in 85.5 percent, where prey showed a large response to predator manipulations. Can we conclude that in most cases predators influence the abundance of their prey in the field? Probably, although the same caveats remain as in the reviews of field experiments investigating competition, that is, under reporting of negative results, biased views of investigating scientists, and so on.

Finally, it is worth noting that studies on the effects of predators are particularly timely now, given the desire of certain conservation groups to reintroduce large predators into certain areas. The U.S. Fish and Wildlife Service would like to reintroduce the wolf into Yellowstone National Park and to stabilize its numbers in Montana and Minnesota, the only states other than Alaska to possess viable populations (Mitchell 1994). Cattle ranchers are fearful that wolves would decimate their herds. There is also considerable argument as to whether wolves constitute a large predation pressure on other prey such as caribou herds and should be controlled in times of declining caribou numbers (Bergerud 1980).

## *Summary*

1. Is predation common in nature? The existence of the following phenomena suggest that it is: aposematic coloration, crypsis and catalepsis, Batesian and Mullerian mimicry, intimidation displays, polymorphisms, and chemical defenses.

2. The existence of prey refuges, mutual interference between predators, specific territory sizes, and the ability of predators to feed on more than one prey type make it very difficult to accurately predict or model how populations of predators and prey interact.

3. Accidental and deliberate introductions of predators in different parts of the world, have often had profound effects on populations of native prey. Although this suggests that predators can have important regulatory effects on prey, this data is not from "natural" systems and should be treated with caution.

4. Evidence from natural systems, where both predators and prey are native, and from a 1985 review by Andrew Sih, suggest that virtually every study that has looked for significant effects of predators on prey has found it. This means that predation is not a casual or unimportant force in nature but is frequent, with often strong effects.

*Discussion Question:* Should ranchers be concerned about the reintroduction of large predators like wolves or panthers? Do sea lions, otters, or dolphins decrease the stock of fish available for people who fish? Would the number of deer available for hunters be the same in the presence of large predators as it would in the absence of them? Do predators control herbivore populations or do they merely take weak and sickly individuals?

# Herbivory

Plants appear to present a luscious green world of food to any organism versatile enough to attack and use it. Why couldn't more of this food source be exploited? After all, plants cannot even move to escape being eaten.

There are three possible reasons that more plant material is not eaten. First, natural enemies—predators and parasites—might keep **herbivores** below levels at which they could make full use of their resources. This idea is preferred by some ecologists, such as Lawton and Strong (1981) who studied insect herbivores. Second, herbivores may have evolved mechanisms of **self-regulation** to prevent the destruction of the host plant, perhaps ensuring food for future generations. But this argument relies on group selection, and, as we saw in Chapter 2, this is unlikely to be true. Third, the plant world is not as helpless as it appears—the sea of green is in fact tinted with shades of noxious chemicals and armed with defensive spines and tough cuticles.

## 10.1 PLANT DEFENSES

An array of unusual and powerful chemicals is present in plants, such as nicotine (an alkaloid) in tobacco, morphine and caffeine (also alkaloids), mustard oils, terpenoids (in peppermint and catnip), phenylpropanes (in cinnamon and cloves), and many others. A teaspoon of mustard should be enough to convince anyone of the potency of these chemicals. Such compounds are unusual—they are not part of the primary metabolic pathways of plants. They are therefore referred to as **secondary chemicals** and are thought to be synthesized specifically as a deterrent to herbivores.

199

Before this aspect of plant defense is discussed, it is important to note an alternative viewpoint—that secondary plant substances are merely the waste products of plant metabolism. It must be remembered that excretion is a necessary part of metabolism even for plants, but it must take a fundamentally different form (that is, storage) than it does in animals. Some authors (for example, Muller 1970) argued that it is only coincidental that these waste metabolites have an effect on herbivores.

Ehrlich and Raven (1964) were the first to crystallize the notion that secondary plant compounds evolved specifically to thwart herbivores. They proposed that most secondary compounds are produced only at a metabolic cost to the plant, not as energy-free by-products. This is the prevalent view among plant-insect ecologists today. Rhoades (1979) has formulated a general defense theory based on the idea that such plant compounds are costly to produce:

1. Higher herbivory levels lead to more defenses.
2. Higher costs of defense lead to fewer defenses.
3. More defenses are allocated to the most valuable tissues.
4. Environmental stress may lessen the availability of energy for defensive mechanisms.
5. Defense mechanisms are reduced when enemies are absent and increased when plants are attacked. This is known as the *theory of induced defense*.

## Types of Defensive Reactions

Defensive reactions in plants can be classified into several main types, of which the most commonly used are quantitative and qualitative.

*Quantitative defenses* are substances that gradually build up inside the herbivore as it eats and that prevent digestion of food. Examples are tannins and resins in leaves, which may occupy 60 percent of the dry weight of the leaf (Feeny 1976). In the case of tannin, the compound is not toxic in small doses, but it has cumulative effects. Tannins (the compounds in many leaves, such as tea, that give water a brown color) act by binding with proteins in insect-herbivore guts. The more leaf the herbivores ingest, the more difficult it is for them to digest it. Feeny (1970) was the first to document such a defense (by oaks against externally feeding caterpillars). Zucker (1983) proposed that there are two main classes of tannins, each with differing biological functions. *Hydrolyzable tannins* inactivate the digestive enzymes of herbivores, especially insects, whereas *condensed tannins* are attached to the cellulose and fiber-bound proteins of cell walls, thereby defending plants against microbial and fungal attack. Some authors have found that other insect herbivores are not much affected by quantitative defenses (Coley 1983; Karban and Ricklefs 1984; Faeth 1985; Mauffette and Oechel 1989), but the evidence that tannins affect vertebrate herbivory is stronger. Cooper and Owen-Smith (1985) showed that for browsing ruminants in Africa (kudus, impalas, and goats), palatability of 14 species of woody plants was clearly related to leaf contents of condensed tannins. The effect showed a distinct threshold; the browsers rejected all plants that contain more than 5 percent condensed tannins. Cooper and Owen-Smith suggest that the reason is that ruminants depend on microbial fermentation of plant cell walls for part of their energy needs. For grey squirrels, acorns with higher concentrations of tannins are less preferred than acorns with lower concentrations (Smallwood and Peters 1986).

*Qualitative defenses* are, essentially, highly toxic substances, very small doses of which can kill herbivores. These compounds are present in leaves at low concentrations, like 1 to 2 percent of dry weight. Examples include alkaloids and cyanogenic compounds in leaves.

Atropine, produced by deadly nightshade (*Atropa belladonna*), is a most potent poison. Of course, the plant must store many of these poisons in discrete glands or vacuoles or in latex or resin systems in order not to poison itself. Some compounds are stored as precursors and only become toxic when they are metabolized by the herbivore. For example, fluoroacetate found in certain Dichapetalaceae is metabolized by herbivores to fluorocitrate, a potent inhibitor of Krebs-cycle reactions (McKey 1979).

A good review of theory and pattern in plant defense allocation is given by Zangerl and Bazzaz (1992). The two defense strategies, qualitative and quantitative, are correlated with plant "apparency" (Feeny 1976; Rhoades and Cates 1976; Chew and Courtney 1991). Apparent plants are long-lived and always apparent to the herbivores (for example, oak trees). Their defenses are thought to be mainly of the quantitative kind, effective against specialist herbivores with a long history of association with these $K$-selected plants. Unapparent plants are weeds, which are ephemeral and unavailable to herbivores for long periods. Their defenses are thought to be mainly qualitative, guarding against generalist enemies. Thus, trees nearly all contain digestibility-reducing compounds, and weeds contain the cheaper-to-make toxins. The value of these terms is of dubious value, however. An "unapparent" plant is not likely to be unapparent to the herbivores that specialize in finding it by chemical cues and eating it. Apparency probably best reflects the ability of human searchers to find plants. Nevertheless, the terms have spawned a great deal of research.

Cates and Orians (1975) set out to test the idea that early and late successional plants (see Chapter 15) might be differentially palatable to generalist herbivores. They found, using a slug as a test animal, that early successional plants were preferred. Is this result so surprising, though, given that the slug is an early successional animal, presumably predisposed to feeding on early successional vegetation? Quite the reverse was found to be true when late successional herbivores were used. For instance, gypsy moths, *Lymantria dispar*, prefer the leaves of trees and other late successional plants to those of weeds and other perennials (Bernays 1981). Coley (1988) studied growth, herbivory, and defenses in 41 common tree species in a lowland Panamanian rain forest. Species with long-lived leaves had significantly higher concentrations of immobile defenses such as tannins and lignins. She also found that trees with lower growth rates generally suffered higher herbivory.

Other defense mechanisms consist of the following:

- **Mechanical defenses:** Plant thorns and spines deter vertebrate herbivores, if not invertebrate ones. In Africa, in the presence of a large guild of vertebrate herbivores, much of the vegetation is thorny and spinose (Cooper and Owen-Smith 1986). Many neotropical plants are also armed in this fashion, for, although now absent, large browsers were abundant in this area until recently (Janzen and Martin 1982).
- **Failure to attract:** Some plants may stop herbivory by failing to attract herbivores. They do so by lacking a certain chemical attractant that the herbivore uses as a cue.
- **Reproductive inhibition:** Some plants, for example firs (*Abies* spp.), contain insect hormone derivatives that, if digested, prevent successful metamorphosis of insect juveniles into adults (Slama 1969). In this way herbivory in the future is lessened by a decrease in the herbivore's reproductive output.
- **Masting** (see also Chapter 9): The synchronous production of progeny, seeds, in some years satiates herbivores, permitting some seed to survive. Nilsson and Wästljung (1987) compared seed predation on beeches (*Fagus sylvatica*). In mast years, 3.1 percent of seeds were destroyed by a boring moth; in nonmast years, this

figure was 38 percent. Vertebrate predation of seeds was 5.7 percent in mast years but 12 percent in normal years.

Some plants defend themselves against herbivores by enlisting the help of other animals. Such a relationship can, of course, be seen as **mutualism.** A very common example is that plants attract ants by providing sugary nectar secreted from extra-floral nectary glands (Barton 1986; Smiley 1986). African *Barteria* and neotropical *Cecropia* trees have hollow stems in which ants maintain populations. The ants are obligate occupiers of the trees; in return for food and shelter, they protect the trees from other herbivores and from encroaching vines, both of which they bite to death (Janzen 1979b). Schupp (1986) demonstrated the benefits to juvenile *Cecropia* by removing ants. The experimental plants suffered more damage from nocturnal herbivorous Coleoptera than unmanipulated controls. The *Acacias* of Central America have hollow thorns, which provide homes for extremely aggressive *Pseudomyrmex* ants, which also kill other herbivores and chop away encroaching vegetation (Janzen 1966). In numerical terms, the effects of ants can be quite substantial. Schemske (1980) found that seed production of *Costus woodsonii* in Panama was reduced by 66 percent where ants were excluded. In England, Skinner and Whittaker (1981) used exclusion techniques to show that the wood ant, *Formica rufa*, was able to reduce herbivory from 8 percent of the leaves to 1 percent. Not all ant attendance is beneficial to trees, however, because many ants simply farm sap-sucking Homoptera and effectively protect their populations, a behavior that is of little use to the plant.

Of course, some defensive schemes backfire on the plants. Some chemicals that are toxic to generalist insects actually increase the growth rates of adapted specialist insects, which can circumvent the defense or actually put it to good use in their own metabolic pathways. Danaid (monarch) butterflies are attracted to milkweed plants that contain cardiac glycosides. These substances are vertebrate heart poisons, and cattle will not eat the milkweeds, but monarch butterflies can assimilate these poisons and use them in their own bodies as a defense against their own predators, advertising their distastefulness with bright colors (Brower 1969). With decreased rates of predation, monarch caterpillars may actually cause more herbivory than they would if they were acceptable to predators.

### Induced Defenses

Plants do not necessarily keep their tissues permanently suffused with defensive and deadly chemicals. There is much evidence that chemicals are produced only as they are needed. The initiation of herbivore attack is usually sufficient to start the metabolic pathways of defense grinding. This defensive tactic is known as *induced defense* (Schultz and Baldwin 1982; Karban and Carey 1984; Edwards and Wratten 1985; Fowler and Lawton 1985; Faeth 1988; Rausher et al. 1993). For example, Rhoades (1979) removed 50 percent of the leaves of ragwort, *Senecio jacobaea*, and detected a 45-percent increase in leaf alkaloids and N-oxidases in the undamaged leaves. Some of these effects can persist for long periods. Haukioja (1980) found that leaves of a birch tree were still poor-quality food for a moth three years after the tree had been damaged. This phenomenon may help to explain the severe oscillations of many forest insects. Sheep fertility was reduced six weeks after aphid attack on the alfalfa the sheep fed on had increased the production of its estrogen mimic, coumestrol (Schutt 1976).

Facultative defenses are not confined to chemical mechanisms. For example, the prickles on cattle-grazed *Rubus* plants were longer and sharper than those on ungrazed individuals nearby (Abrahamson 1975). Browsed *Acacia depranolobium* trees in Kenya have longer thorns than unbrowsed ones (Young 1987). Stinging nettles, *Urtica dioica*, exhibit increased density of stinging trichomes after herbivore damage (Pullin and Gilbert 1989). On holly trees, *Ilex aquifolium*, in Britain, lower leaves, subject to grazing, are heavily armed with spines; upper leaves, free from herbivory, are not (Crawley 1983). Interestingly enough, though the same phenomenon is apparent on American holly, *Ilex opaca*, Potter and Kimmerer (1988) regard it as an ontogenetic phenomenon rather than a defense against browsers. They suggest that the poor nutritional quality of holly and the high concentrations of saponins are more important as deterrents of vertebrate herbivores and that fibrous leaf edges deter invertebrate herbivores, which are unlikely to be affected by spines. Williams and Whitham (1986) have even argued that leaves infested with sessile insects (for example, gall makers and leaf miners) are abscised prematurely as a defense against herbivory; the insects are killed as the leaf senesces on the forest floor. Stiling and Simberloff (1989) examined such a situation on oak trees in northern Florida. Although it is true that leaves infested with leaf miners do abscise earlier than noninfested leaves and that premature abscission kills miners inside the leaf, leaf abscission is not likely to be a complex, induced defense. It is more likely that premature abscission is a simple wound response on the part of the tree to rid itself of damaged leaves, possibly to avoid infection (Stiling and Simberloff 1989). For abscission to work as an induced defense in this situation, miners would have to show a high fidelity to a host tree; otherwise the tree would simply be reinfected by miners from neighboring trees. There is little fidelity in the leaf miner–oak tree system.

Though induced defenses undoubtedly occur in nature, their effectiveness as a deterrent is open to speculation and is probably less than that of permanent chemical defenses. Fowler and Lawton (1985) reviewed much of the work on induced defenses and concluded that as a result of such chemicals, only small changes, generally less than 10 percent, occurred in such things as larval development time or pupal weights. Other studies have shown that certain insects even benefit by feeding on previously damaged plants (Myers and Williams 1984, 1987; Niemelä et al. 1984).

## 10.2  MATHEMATICAL MODELS

Crawley (1983) has provided a series of models of plant-herbivore interactions. The most simple of these assumes that there is an upper limit, a carrying capacity, $K$, for a population of plants. The rate of change of a population of plants is given by

$$\frac{dV}{dt} = A - B$$

where $A$ is the gains, $B$ is the losses, and $V$ is plant abundance. Similarly, for the herbivores

$$\frac{dN}{dt} = C - D$$

where $C$ and $D$ are gains and losses and $N$ is herbivore numbers.

It is assumed that, in the absence of herbivores, plant populations increase exponentially such that gains, $A$, $= aV$ where $a$ is the plant's intrinsic rate of increase. Losses for plants,

$B_1 = bNV$ where $b$ is the feeding rate of herbivores. For the herbivores, $C = cNV$ where $c$ describes the numerical response of herbivores, and $D = dN$ where $d$ is herbivore death rate. The assumption is that, in the absence of plants, herbivores starve and their numbers decline exponentially. Assuming an upper carrying capacity, $K$, and following a basic Lotka-Volterra type of logistic model, then $A = aV (K - V)/K$. The other elements, $B$, $C$, and $D$, remain the same. Therefore,

$$\frac{dV}{dt} = \frac{aV(K - V)}{K} - bNV$$

and

$$\frac{dN}{dt} = cNV - dN$$

At equilibrium both $dV/dt$ and $dN/dt$ are zero; there is no population change. The equilibrium plant density $V^*$ and equilibrium herbivore abundance $N^*$ can then be solved for because $A = B$ and $C = D$. Therefore,

$$V^* = \frac{d}{c}$$

and

$$N^* = \frac{a(K - (d/c))}{bK}$$

The effect of each of the parameters $a$, $b$, $c$, and $d$ is best assessed by graphical techniques. Each parameter can be made to vary while the others remain fixed. The effect of increasing the plant's intrinsic rate of growth, $a$, stabilizes and increases herbivore equilibrium density but has no effect on equilibrium plant abundance (Fig. 10.1): essentially, the faster the plants grow, the faster the herbivores eat them up. When $b$, the feeding rate of herbivores, is increased, there is a lower herbivore equilibrium, but the size of the equilibrium plant population is unchanged (Fig. 10.2). The more efficient an herbivore is at food gathering, the rarer the herbivore is because fewer herbivores can be supported by a given level of plant production. The numerical response of herbivores is described by $c$; the higher the value of $c$, the more herbivores can be borne per unit feeding. This is essentially the efficiency with which herbivores turn food into progeny. The higher this efficiency, the lower the equilibrium population of plants because an efficient herbivore turns all production into herbivores (Fig. 10.3). Perhaps more important is the effect of $c$ on stability. When plant equilibrium density is low, much below $K$ (the carrying capacity), the population is under lax control and tends to increase exponentially when herbivore numbers decrease. Thus, the lower the equilibrium plant population, the lower the stability of plant and herbivore numbers. The effects of herbivore death rate, $d$, are exactly opposite to those of $c$. Increasing herbivore death rate increases **stability** because it leads to an increase in equilibrium plant numbers. Finally, plant carrying capacity, $K$, is critical to the stability of these models. If $K$ is high, then populations may be under lax control even where equilibrium plant numbers are quite high. An increase in $K$ could lead to an increase in the number of herbivores, too. However, equilibrium plant density $V^*$ is determined solely by $c$ and $d$. Thus increasing $K$ reduces the relative level of plant equilibrium populations, and control of plant numbers is lax. This effect explains why high values of $K$ decrease the stability of

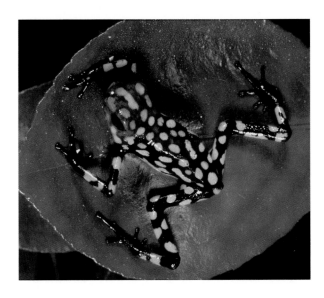

Predation: Is nature red in tooth and claw? The color patterns and behaviors of many animals suggest predation is of vital importance. (*Upper left*) Aposematic or warning coloration. *Dendrobates histronicus*, a poison dart frog. (Dr. E. R. Degginger, 248.) (*Upper right*) If predation were not important, why would this katydid mimic leaves? (Laval, Animals Animals, 4G447S-2.) (*Lower left*) The monarch butterfly, Batesian model or Mullerian mimic. (Degginger, Animals Animals, 520202-1.) (*Lower right*) Batesian mimicry: a harmless hoverfly, *Epistrophe balteata*, imitating a harmful wasp. (Vock/Okapia, Photo Researchers, Inc., 6Z0719.)

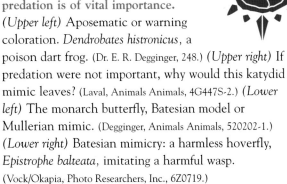

Predation continued. (*Upper left*) *Photuris pennsylvanicus*, a firefly or lightning bug, has a bioluminescent organ in its abdomen. Each species has a specific flash pattern to attract mates. (Runk/Schoenberger, Grant Heilman, CH5-67-62.) (*Above*) Fireflies gathered over this farmyard in Iowa are visible as faint streaks of green. Some females mimic the signals of other species so that they can feed on the "mates" they attract. This could be thought of as aggressive mimicry, allowing a predator better access to its prey. (Kent, Photo Researchers, Inc., 7R2777.) (*Lower left*) Bluff. This frilled lizard, *Chlamydosaurus kingii*, from Australia, tries to surprise potential predators by extending the frill around its neck to appear larger. (Uhlenhut, Animals Animals, 544996-R.) (*Lower right*) Chemical defense is used by some animals like the bombadier beetle, the skunk, or in this case, the octopus, which releases a cloud of ink. (Wu, Peter Arnold, AN-BE-15.)

Many features of plants suggest that herbivory is important in nature. (*Above*) As these spiny oak-worm caterpillars feed on water-oak leaves, their digestion will become less efficient as tannins bind to substances in their guts and prevent digestion. (Peter Stiling.) (*Upper right*) Bittersweet, *Solanum dulcamara* (also sometimes called deadly nightshade) defends itself against herbivores by being mildly poisonous. An infusion of juice from the European deadly nightshade, *Atropa belladonna*, was formerly dropped in women's eyes causing dilation of the pupils to produce a "wide-eyed" look—hence bella-donna, beautiful lady. (Dr. E.R. Degginger, 260.) (*Middle*) Spines on the desert cactus *Echinocereus mohavensis* in Joshua Tree National Monument, California, protect it from herbivory. (Gerlach, Animals Animals, 632235.) (*Lower right*) Some "thorns" are actually insects mimicking thorns, like these membracid treehoppers in South Florida. (Kolar, Animals Animals I-6078.)

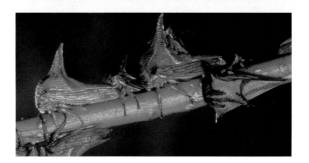

**H**erbivory continued. *(Upper left)* Thorns on *Acacia collinsii,* in Costa Rica, provide homes for *Pseudomyrmex ferruginea* ants, whose aggressive behavior further deters any herbivores. (Dimijian, Photo Researchers, 7Q0700.) *(Upper right)* Inside this frothy mass on a stinging nettle leaf lives a cercopid, an insect which feeds on xylem fluid. The cercopid kicks up the excreted xylem fluid into a frothy mass, deterring predators. (Peter Stiling.) *(Lower left)* Leaf abscission kills endophytic insects within leaves, like these leafminers. (Peter Stiling.) *(Lower right)* The array of spines on the Io caterpillar are testament to strong pressure on it by natural enemies. If the plant can delay insect development long enough, a parasite or predator may kill the caterpillar, thus doing the plant a favor. Tri-trophic interactions involve the influence of plants on natural enemies of their herbivores. (Peter Stiling.)

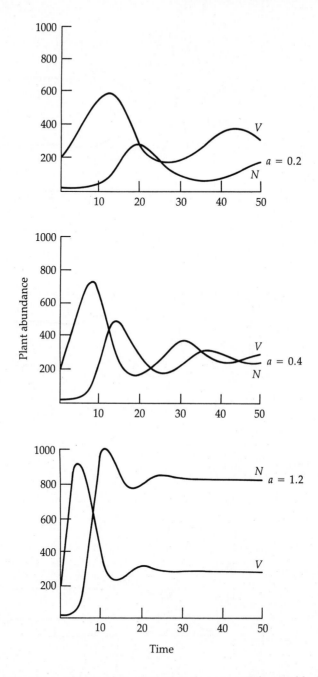

**Figure 10.1**    *The effect of the plant's intrinsic growth rate a on plant and herbivore abundance. Increasing a from 0.2 to 1.2 increases stability and increases herbivore equilibrium density but has no effect on equilibrium plant abundance. Other parameters are* b = 0.01, c = 0.001, d = 0.3, K = 1,000. *As in all these types of models, the curve* V *represents plant abundance, and the curve* N *represents herbivore numbers (in arbitrary units).*  (Redrawn from Crawley 1983.)

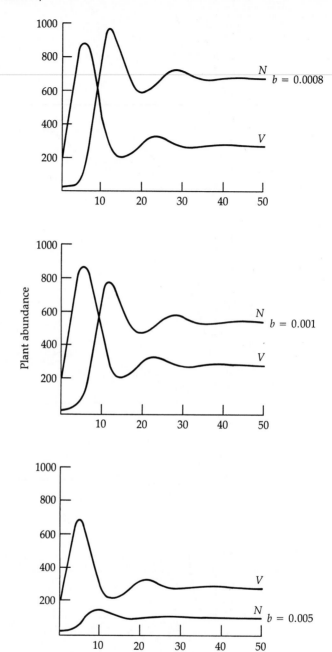

**Figure 10.2** *Resource-limited plants—the effect of herbivore searching efficiency b. Increasing b from 0.0008 to 0.005 reduces herbivore equilibrium density but has no effect on stability or on equilibrium plant abundance. Other parameters are a = 0.5, c = 0.001, d = 0.3, K = 1,000.* (Redrawn from Crawley 1983.)

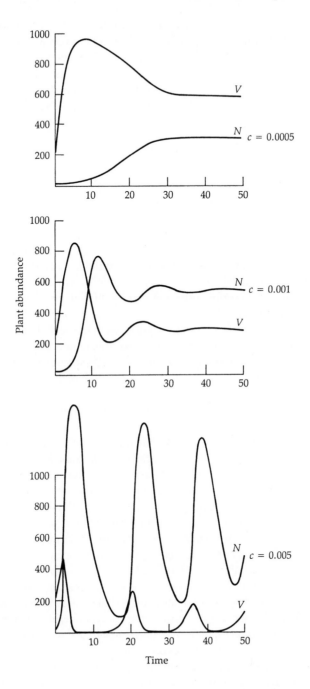

**Figure 10.3**   *The effect of herbivore growth efficiency* c. *Increasing* c *from 0.0005 to 0.005 reduces equilibrium plant abundance and thereby reduces stability but increases equilibrium herbivore numbers. Other parameters are* a = 0.5, b = 0.001, d = 0.3, K = 1,000. (Redrawn from Crawley 1983.)

a system (Fig. 10.4). This is what Rosenzweig (1971) called the "paradox of enrichment" and may account for the unstable population cycles of herbivorous insects in the vast boreal forest of the Northern Hemisphere, where the carrying capacity for evergreen trees is enormous.

Crawley has concluded that plants have a much more important impact on herbivores than herbivores have on the dynamics of plants. Mammals are especially prone to food limitation, though insect herbivores are probably more influenced by predation, parasitism, and disease. Plant death rates themselves, Crawley suggests, are largely determined by competition with other plants and by self-thinning.

## 10.3   THE FIELD FREQUENCY OF HERBIVORY

Despite the impressive array of defenses in their arsenals, plants do not have things all their own way in the plant-herbivore interaction. Herbivores can detoxify many poisons by four chemical pathways: oxidation, reduction, hydrolysis, and conjugation (Smith 1962). Oxidation occurs in mammals in the liver and in insects in the midgut. It is brought about by a group of enzymes known as mixed-function oxidases (MFOs). Conjugation, often the critical step in detoxification, involves the uniting of two harmful elements into one inactive and readily excreted product. Given that herbivores can circumvent plant defenses in certain situations, what is their measured effect on plant populations in the field? There are numerous studies showing how herbivores reduce plant growth, flowering, reproduction, and survival (Fay and Hartnett 1991; Fox and Morrow 1992). There is even good evidence that long-term suppression of insect herbivores increases the production and growth of roots (Cain, Carson, and Root 1991). However, on average, no more than 10 percent of net primary productivity seems to be taken by herbivores and about 90 percent by decomposers, in most natural systems (Crawley 1983) (see also Chapter 17). In a review of 93 cases of herbivory in terrestrial systems, an average of 7 percent was found to be consumed (Pimentel 1988). It must be remembered, of course, that such figures mask large and important variations. For example, the larch budmoth may take less than 2 percent of the net production of forest trees in some years but 100 percent in others. Also, workers interested in herbivory would probably not choose to study a plant that suffers very little herbivory; thus, even 7 percent could be an overestimate.

Sometimes, damage from insect feeding causes wilting or allows disease to cause loss of more plant biomass. Grasshoppers feeding on needlerush, *Juncus roemerianus,* in salt marshes feed in the middles of the tall, narrow leaves so that even though only the middle of the leaf is actually digested, the top half is cut off and added to the litter layer (Parsons and de la Cruz 1980). This author has observed that regions of *Quercus geminata* leaves that are distal to *Stilbosis* leaf mines often turn brown and senesce, even though the miners do not damage them directly.

Ultimately, the best way to estimate the effects of herbivory on plant populations is to remove the herbivores and examine subsequent growth and reproductive output. Some of the best evidence on the impact of herbivores on plants comes from the biological control of weeds. Following its importation from the Americas in 1839, the prickly pear cactus, *Opuntia stricta,* became a serious pest in Australia, occupying by 1925 more than 240,000 km$^2$ of once-valuable rangeland (Fig. 10.5). After some initial imports of insects that failed to control the growth of the cactus, *Cactoblastis cactorum,* a moth, was introduced from South America in 1925 (Osmond and Monro 1981). By 1932, the original stands of prickly pear had collapsed under the onslaught of the moth larvae. Despite a small resurgence of prickly pear in 1932–1933, *Cactoblastis* has dev-

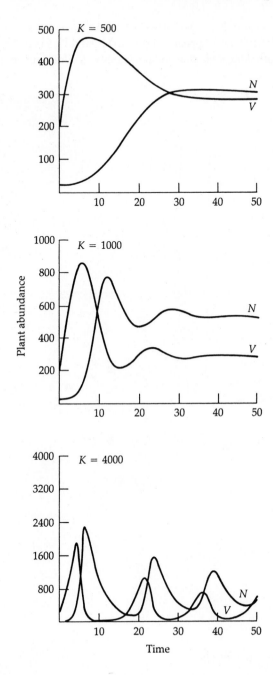

Figure 10.4   *The paradox of enrichment. Increasing the carrying capacity of the environment for plants K from 500 to 4,000 increases herbivore equilibrium density but has no effect on plant equilibrium (note scale changes on y axis). The main effect of increasing K is to reduce stability. Other parameters are a = 0.5, b = 0.001, c = 0.001, d = 0.3.* (Redrawn from Crawley 1983.)

astated prickly pear populations ever since, and the cactus is now confined to isolated areas. "Before" and "after" views appear in Photo 10.1a and Photo 10.1b. Similar sucess stories were reported in South Africa, Hawaii, and the Caribbean. Unfortunately, the moth has now invaded

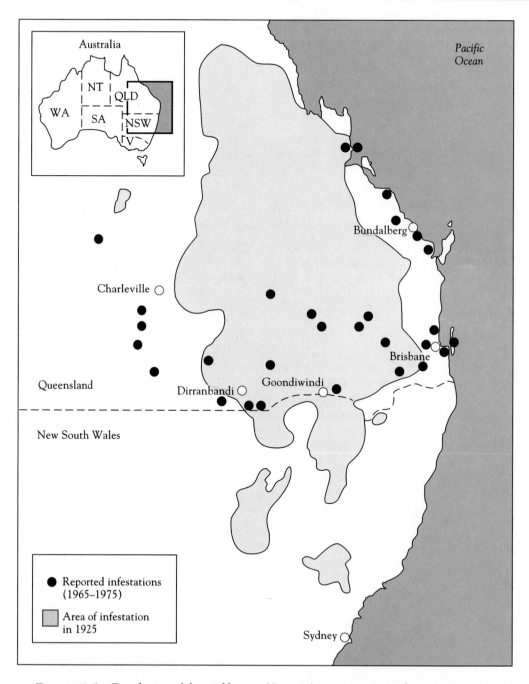

Figure 10.5  *Distribution of the prickly pear (Opuntia) in eastern Australia in 1925 at the peak of infestation and modern areas of local infestation.* (Redrawn from White 1981.)

Photo 10.1 (a)   *The strength of herbivory as illustrated by a terrestrial example of biological control. Rangeland in Australia infested with Opuntia cactus.*

Photo 10.1 (b)   *The same site after introduction of a cactus-eating moth, Cactoblastis.* (Photos courtesy of Commonwealth Scientific and Industrial Research Organization of Australia.)

south Florida from the Caribbean and is damaging some rare and endemic cacti there (Johnson and Stiling 1995). This is important because it is one of the relatively few examples of biological control that has unintended side effects.

There have been several other successes in the biological control of weeds by natural enemies, and these tend to overshadow the probably more numerous failures. Klamath weed (*Hypericum perforatum*), a pest of pastureland in California, was controlled by two French beetles (Huffaker and Kennett 1959). As illustrated in Photo 10.2a and Photo 10.2b, floating fern, *Salvinia molesta*, choked a lake in Australia and was controlled by the weevil *Cyrtobagus singularis*, introduced from Brazil, where the fern is native (Room et al. 1981). Alligatorweed was controlled in Florida's rivers by the so-called alligatorweed beetle, *Agasicles hygrophila*, from South America, and hopes are high that water hyacinth can be controlled biologically, too (Buckingham 1987). On the other side of the coin, large numbers of insects have been introduced to control *Lantana camara*, an introduced weed in Hawaii. Very few have had any impact on the growth of the plant, though its spread might have been slowed.

In successful cases of biological control of weeds, all the plants have been alien, and most were perennials. No native weeds and very few annual weeds have yet been controlled by insects. Annuals can produce huge amounts of seed even when heavily infested by insects. No seed-eating insect has yet controlled a weed plant. The weevil *Apion ulicis* was introduced into New Zealand to control gorse, *Ulex europaeus*, and has become one of the most abundant in-

Photo 10.2 (a)   *An aquatic example of biological control. Lake Kabufwe, Papua New Guinea, choked with the floating fern Salvina in October 1983.*

sects in New Zealand. Unfortunately, although the insect eats up to 95 percent of the gorse seed produced every year, there has been no appreciable effect on plant numbers (Miller 1970).

In a natural setting, removal of herbivores from their host plants has been done less commonly and with mixed results. Karban (1982) used this technique to demonstrate reduction in growth of apple trees due to insect infestation, and he detected a similar effect in scrub oak, *Quercus ilicifolia* (Karban 1980). Australian *Eucalyptus* trees with insects removed were almost 100 percent taller than control trees after three years of this treatment (Fox and Morrow 1992). Increased growth in herbivore removal experiments has also been shown for sand-dune willow trees (Bach 1994). Meyer (1993) was able to show, by selective removals, that xylem-feeding spittlebugs had more of an effect on goldenrod (*Solidago atissima*) growth than either leaf-chewing beetles or phloem-feeding aphids. However, we are not yet in a position to know which type of herbivores have the greatest effects on which type of plants. One of the strongest effects was shown by Gibbens et al. (1993) who excluded rabbits from browsing rangeland plants in New Mexico for more than 50 years. By the end of this period, the basal area of some plants were 30-fold greater in the exclusion treatments than in the controls.

The most serious effects of herbivory may be on reproductive output. Crawley (1985) removed herbivores from oaks in Britain by spraying techniques. Though unsprayed trees

Photo 10.2 (b)   *The same lake clear of the weed in November 1984, after release of the herbivorous weevil Cyrtobagus salvinae.* (Photos by P.M. Room, courtesy of Commonwealth Scientific and Industrial Research Organization of Australia.)

lost only 8 to 12 percent of their leaf area, the sprayed trees consistently produced from 2.5 to 4.5 times the number of seeds produced by unsprayed plants. Waloff and Richards (1977) found almost three times more seed on broom bushes sprayed for insect control than on unsprayed ones. Once again, however, results tend to be mixed. Karban (1985) could find no effect of periodical cicada nymphs on acorn production by *Q. ilicifolia* in New York State. In instances where an exotic herbivore is introduced in the absence of its enemies, the results are much more dramatic. Bermuda cedar, *Juniperus bermudiana*, was virtually wiped out by an introduced scale insect, *Lepidosaphes newsteadi* (Bennett and Hughes 1959; Cock 1985). Young trees in forestry plantations can also suffer large mortalities when they are girdled by introduced goats, rabbits, sheep, or squirrels. In times of abnormally high densities, other animals can have these same effects. For example, in 1958 only 24 percent of mature trees surveyed in an African *Terminalia glaucescens* woodland were dead, but in 1967 after elephant densities were boosted by immigration beyond the carrying capacity of the region, almost 96 percent of the trees were dead (Laws, Parker, and Johnstone 1975). Crawley (1983) has provided many other examples of the impact of herbivory.

In most cases, herbivores cause subtle alterations of growth rates of stems and roots, rather than outright death of the plant. Flower, seed, and fruit production can also be influenced, though it is likely that predators of fallen seed and fruit are more important in a scheme of plant fitness. In this respect, herbivores can be seen as successful **parasites** because they do not kill their hosts—they merely reduce the growth rate. In an agricultural setting, of course, there are many estimates of losses of crops to herbivores. Damage is often severe enough not only to justify control but to cause economic hardship to agriculturalists (May 1977; Pimentel et al. 1980; Barrons 1981). Even though plants are rarely killed outright, the effects of even minor damage on crop yields can be substantial. It is clear, however, that control measures are often initiated when populations of pests are so small that significant damage is unlikely to occur (Pimentel et al. 1980).

## Beneficial Herbivory?

Some authors have argued that herbivory can be beneficial to plants (see McNaughton 1986; Crawley 1987; Owen and Weigert 1987). The rationale is that plants are stimulated to regrow after damage—they end up overcompensating, growing even more than they would have had they not been damaged. The result is often more seed production from more vegetative plant parts. Simberloff, Brown, and Lowrie (1978) noted that the action of isopod and other invertebrate root borers of mangroves tended to initiate new prop roots at the point of attack (Fig. 10.6). More prop roots meant greater stability of mangroves against wave and storm action, so root herbivory could in fact be beneficial. However, in a review of the 20 papers most commonly cited as evidence for beneficial herbivory, Belsky (1986) found fault with the logic, experimental design, or statistics of nearly all of them. Even newer papers that purport to support the beneficial-herbivory theory are usually fraught with methodological or technical errors (Belsky 1987). In addition, Strauss (1988) has pointed out that very carefully designed experiments involving measurements of plant size before and after herbivory are needed because herbivores themselves naturally choose larger plants, which might be expected to show more growth than would stunted plants, even after herbivory. Because of the economic damage supposedly done by herbivores, especially insects, it is not

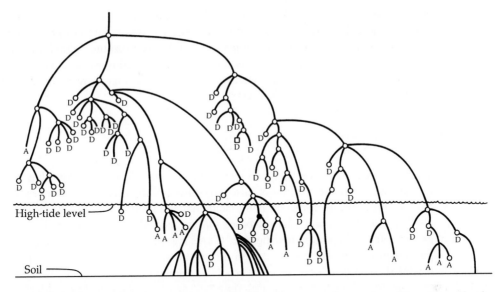

Figure 10.6   *Branching pattern for a single Rhizophora mangle root from Clam Key, Florida. A = alive; D = dead; open square = bored by* Ecdytolopha *sp.; open circle (above water) = bored by unknown insect; open circle (below water) = bored by* Sphaeroma terebrans; *shaded circle = bored by* Teredo *spc.* (Redrawn from Simberloff, Brown, and Lowrie 1978.)

trivial to assess whether or not defoliation is beneficial. Mattson and Addy (1975) have tried to model forest growth with and without insect herbivores. They examined two situations. In the first, aspen was defoliated by forest tent caterpillars. Forest tent caterpillars begin to infest a forest slowly, reach a peak in numbers, remain there for three or four years, and then subside. Stemwood production in the years of peak infestation is much reduced, but foliage production increases to compensate for insect defoliation. Within roughly ten years after the infestation, the biomass production was identical to that of unaffected forests. In the short term, the caterpillars reduced wood production, but in the long term they had no major effect.

In a second example, balsam fir was defoliated by spruce budworms. These larvae actually kill mature trees aged 55 to 60 years, but they leave young trees largely alone. The saplings grow quickly after their parents are killed, and a resurgence of the forest, from its juveniles or saplings, occurs. The end result is the same as in the previous case: in the short term, there is a considerable effect on wood production, but there is not in the long term. Production rates in the young forest remain elevated above that of the mature forest for 15 years because in a mature woodland, most trees have passed their rapid-growth phase. The role of foresters in this cycle is not at all clear (Holling 1978). Perhaps an effective action would be to harvest those trees in the center of budworm outbreaks because they would be killed in any case and because many larvae would be removed with them.

In herbivory on grasses, the placement of the meristem is important . In most species, the meristem is down low, safe from herbivores and protected in the basal leaf sheaths. The few species that produce elongated vegetative shoots are very vulnerable to grazing. Because growth is very slow from axillary buds, which must take over after the meristem is eaten, such plants tends to be outcompeted by species whose meristems survive. The inflorescences of most species, however, must extend upward into the air to ensure pollination of the flowers. Because certain species die only after flowering, heavy herbivory or constant clipping by a lawnmower can, by preventing flowering, virtually ensure their immortality. Grasses in sports stadiums may effectively live forever.

## 10.4 THE EFFECTS OF PLANTS ON HERBIVORES

There is much evidence that herbivores themselves select the plants that are the most nutritionally adequate in terms of nitrogen content of the tissue (Mattson 1980; Scriber and Slansky 1981; White 1984) or amino-acid concentration of the sap (Brodbeck and Strong 1987). Iason, Duck, and Clutton-Brock (1986) showed how red deer fed preferentially on grasses defecated upon by herring gulls (*Larus argentatus*). Where the number of gull droppings increased, so did the vegetation nitrogen content. For birds, Watson, Moss, and Parr (1984) showed how food enrichment affects numbers and spacing behavior of red grouse. In some cases, however, such correlations between host-plant quality and herbivore density are present, but observed population patterns of herbivores are more dependent on other phenomena, such as predation or parasitism (Stiling, Brodbeck, and Strong 1982). There seems to be no easy way to predict when herbivore densities are controlled by the quality of their hosts. When food quality declines, many herbivores, especially vertebrates, respond simply by feeding at a higher rate or for a longer time. However, this will often result in increased exposure time to predators and parasites and higher enemy-induced mortality rates (Loaker and Damman 1991). On the reverse side, deaths of herbivores due to depletion of food plants are witnessed very infrequently (Crawley 1983), perhaps because herbivores can leave an area of poor food availability, but they cannot easily escape the weather or an outbreak of disease. The relatively few examples of mass starvation due to overexploitation of plants come mainly from studies of insects that habitually undergo periodic outbreaks or from cycles of Arctic rodents.

Herbivores are not only influenced by plant quality; they are affected by host defenses, many of which are genetically inherited. Herbivore densities can also be affected by the weather, as noted by Davidson and Andrewartha's (1948a,b) studies on thrips in Australia. Sometimes, there is an interaction between host genotype and environment such that at some sites plants are resistant to herbivores, but at other sites they are susceptible. Presumably the abiotic environment, soil type, water levels, or nutrient levels influence whether or not plants can maintain their resistance to herbivores. Which of these factors is most important? Rick Karban (1992), in an excellent review of the effect of plant variation on herbivorous insects, showed that more than 80 percent of studies that had used experiments to test for the presence of plant genetic variation on insect herbivore densities found it. This is a high percentage. However, Karban also reviewed studies that compared the effects of plant genotype to other factors and found that in 17/30 cases (56.7 percent), plant genotype explained less variation in herbivore numbers than did other factors, such as yearly variability. For example, Stiling and Rossi (1995) showed that environmental conditions affected the population densities of all eight major herbivores on coastal plants in Florida more than did plant genotype. This kind of result reminds us of the ideas of Davidson, Andrewartha, and Birch who suggested that environmental factors such as the weather may be of paramount importance in determining herbivore abundance.

# Summary

 1. A variety of plant defenses are testament to the strength and frequency of herbivory in nature. There are chemical defenses such as nicotine in tobacco and caffeine in tea and mechanical defenses such as spines and stinging hairs. Plants may also contain insect hormone mimics that disrupt insect molts. Other plants enter into mutualisms with ants that attack and remove herbivores in return for shelter and food (extra floral nectaries).

2. Chemical defenses can be subdivided into qualitative and quantitative defenses. Quantitative defenses gradually build up inside herbivore guts and prevent food digestion. The more foliage that is eaten, the worse the situation becomes for the herbivore. Examples are tannins and resin in leaves. Qualitative defenses are toxic compounds, such as cyanogenic compounds in leaves, which are lethal in small doses. A good example is Atropine, produced by deadly nightshade.

3. Because chemical defenses are energetically costly to produce, most are allocated only to the most valuable tissues. Also, some defense mechanisms are only initiated following herbivory. These are known as induced defenses.

4. Mathematical models have led some ecologists to conclude that plants have a much more important impact on herbivores than herbivores have on the dynamics of plants. Reviews suggest that, on average, between 7 and 10 percent of plant tissue is consumed by herbivores. Of course, this masks much important variation, and there are many systems where there are periodic outbreaks of herbivores, such as locusts in the tropics or moth larvae on conifers in boreal zones.

5. Biological control projects have shown that many exotic weeds that have undergone population explosions in the absence of their native herbivores in foreign countries can be brought under control when the native herbivore is reunited with its host. Some experiments which have removed native herbivores from native plants have also shown dramatic effects of herbivory. Thus in some systems, the effects of herbivores are known to b substantial.

## Discussion Question: Are grazing mammals or insects likely to be the most important herbivores? Which types of plants might suffer more herbivory? Are chemical defenses more likely to be found in temperate or tropical plants, in desert species or wetland species? Would your conclusion hold true for other types of defenses?

# Parasitism

When one organism feeds off another but does not normally kill it outright, the predatory organism is termed a **parasite** and the prey a **host.** Some parasites remain attached to their hosts for most of their lives, like tapeworms, which remain inside the host's alimentary canal. Others, such as ticks and leeches, drop off after prolonged periods of feeding. Mosquitoes remain attached for relatively short periods. By this definition, many species of phytophagous insects are parasitic upon their "host" plants. Still, there remain many problems of definition. Should organisms that feed off more than one individual without killing them be known as parasites or predators? For example, saber-tooth blennies (*Plagiotremus*) on the Great Barrier Reef dash out and bite chunks out of fish hosts/prey that swim by. Should the large ungulates of the Serengeti plains—wildebeest, zebra, and the like—be known as parasites? Although they feed off more than one individual host grass, the grass is not killed and will grow back later. Should we retain the term *parasite* for organisms that remain in intimate contact with their hosts? Mosquitoes develop as larvae in a nonparasitic manner in pools of water, and the adults only come into contact with hosts for short periods. Rhinoceroses live on top of their food supply for their entire lives. What about parasites of insects? Many of these develop as internal parasites of caterpillars or other immature stages. In these cases, the host almost never survives, and the term **parasitoid** is used to refer to these parasites, each of which uses only one host but invariably kills it. Even in this case, further gradation between parasitoid/parasite and predator is evident, as when an egg parasitoid hatches from a host egg and has to devour several more in the clutch before it is mature (Askew 1971). May and Anderson (1979) have tried to distinguish two types of parasites—microparasites, which multiply within their hosts, usually within the cells (bacteria and viruses), and macroparasites, which live in the host but release infective juvenile stages outside the host's body. For most microparasite infections, the host has a strong immunological response. For macroparasitic infections, the response is short lived, the infections tend to be persistent, and hosts are subject to continual reinfection.

Despite these problems of definition, the biology of host-parasite relationships has a rich history of interesting, **coevolved,** and complex life-history patterns. Parasites on animals include those of interest to the conventional parasitologist—viruses, bacteria, protozoa, flatworms (flukes and tapeworms), thorny-headed worms (Acanthocephala), nematodes, and various arthropods (ticks, mites, and so on). Parasitoids from the parasitic Hymenoptera and Diptera are of more interest to the entomologist and biological-control specialist. Such parasitoids are often arraigned against other insects and may contribute more than 70 percent of the insect fauna (Price 1980). Because about 75 percent of the known global fauna consists of insects, then at least 50 percent of the animals on Earth might be considered parasitic. When the other large groups of parasites are considered—nematodes, fungi, viruses, and bacteria—it is clear that parasitism is a very common way of life. A free-living organism that does not harbor several parasitic individuals of a number of species is a rarity. The frequency of human infection by parasites is staggering. There are 250 million cases of elephantiasis in the world and more than 200 million of bilharzia, and the list goes on and on.

## 11.1  DEFENSES AGAINST PARASITES

The defensive reactions developed by hosts to resist parasites are nearly as impressive as those to combat predation:

- **Cellular defense reactions:** These reactions particularly are found in insect larvae as a defense against parasitoids, where eggs of the parasitoid are "encapsulated" or enclosed in a tough case rendering them inviable (Salt 1970).
- **Immune responses in vertebrates** (Cox 1982): These responses are the vertebrate body's defense against the parasitic microbes that cause disease in humans and animals. Phagocytes may engulf and digest small alien bodies and encapsulate and isolate larger ones. For microparasites, the host may develop a "memory" that may make it immune to reinfection.
- **Defensive displays or maneuvers:** These actions are intended to deter parasites or to carry organisms away from them. For example, gypsy moth pupae spin violently within their cocoons to deter pupal parasites (Rotheray and Barbosa 1984), and syrphid larvae often drop to the ground from the foliage they forage on to escape parasites (Rotheray 1981, 1986).
- **Grooming and preening behavior:** This behavior is found in mammals and birds, respectively, to remove ectoparasites (Struhsaker 1967; Kethley and Johnston 1975).

## 11.2  THE SPREAD OF DISEASE

The dynamics of microparasites, the spread of disease, have been modeled by May (1981) and Anderson (1982). Instead of examining basic reproductive output, $R_o$, of a parasite, scientists more usually observe $R_p$, the average number of new cases of a disease that arise from each infected host. The reason is that in **epidemiology,** the study of the spread of disease, the number of infected hosts is the most important factor, not the number of parasites. The transmission threshold, which must be crossed if a disease is to spread, is therefore given by the condition $R_p = 1$. For a disease to spread, $R_p$ must be greater than 1, and for a disease to die out, it must be less than 1. The term $R_p$ is influenced by

$N$, the density of susceptibles in the population

$B$, the transmission rate of the disease (a quantity correlated with frequency of host contact and infectiousness of the disease)

$f$, the fraction of hosts that survive long enough to become infectious themselves

$L$, the average period of time over which the infected host remains infectious

The value $R_p$ is related to these factors by the equation $R_p = BNfL$.
Two generalizations can thus be made.

1. As $L$, the period of the host's life when it is infectious, increases, $R_p$ increases. Some hosts remain infectious long after they are dead. This is especially true in plant parasites, which leave a residue of resting spores.
2. If diseases are highly infectious (have large $B$s) or are unlikely to kill their hosts (have large $f$s), $R_p$ increases. An efficient parasite therefore keeps its host alive.

By rearranging the above equation, we can obtain the critical threshold density $N_T$ (where $R_p = 1$), the number of infected hosts needed to maintain the parasite population:

$$N_T = \frac{1}{BfL}$$

Now, if $B$, $f$, or $L$ is large, $N_T$ is small. Conversely, if $B$, $f$, or $L$ is small, the disease can only persist in a large population of infected hosts. Cockburn (1971) has provided some interesting medical and anthropological evidence to back up these ideas, at least for humans. Measles, rubella, smallpox, mumps, cholera, and chicken pox, for example, probably did not exist in ancient times (Black 1975) because the hunter-gatherer populations were small—bands of 200–300 persons at most. These bands were too small to constitute reservoirs for the maintenance of infectious diseases of the types described. In a small population, there should be no infections such as measles, which spreads rapidly and immunizes a majority of the population in one epidemic. Measles only occurs endemically in human populations larger than 500,000. Instead, typhoid, amoebic dysentery, pinta, trachoma, or leprosy were probably the common afflictions, diseases for which the host remained infective for long periods of time. Malaria and schistosomiasis would still have been very prevalent because of the presence of outside vectors to serve as additional reservoirs. Paradoxically, civilization has increased the kinds and frequencies of diseases suffered by humans by enlarging the source pools and by domestication of certain animals. Most modern diseases have arisen because of intimate association with animals and their viruses (Foster and Anderson 1979). Smallpox, for example, is very similar to the cowpox virus, measles belongs to the group containing dog distemper and cattle rinderpest, and human influenza viruses are closely related to those found in hogs. AIDS is similar to a virus found in monkeys in Africa.

For parasites that are spread from one host to another by a vector (for example, an insect), the life-cycle characteristics of both host and vectors become important in the calculation of $R_p$. The appropriate equation is:

$$R_p = \beta^2 \; \frac{N_v}{N_w} \; f_v f_n L_v L_n$$

$N_V$ and $N_h$ represent the density of vector and host respectively, for example, mosquito and man or aphid and tomato plant. $f_v$ and $f_h$ represent the fraction of infected vectors and hosts that survive to become infectious themselves. $L_v$ and $L_h$ represent the periods of time over which vectors and hosts remain infectious. $\beta$ is the effective transmission rate, the rate of mosquito biting or aphid feeding that leads to infection. It appears as a squared term because infection occurs both to and from the host.

Another important feature of the dynamics of vector-borne parasites is that the transmission threshold ($R_p = 1$) is dependent on a ratio,

$$\frac{N_v}{N_L} = \frac{1}{\beta^2 \, f_v \, f_L \, L_v \, L_h}$$

For a disease to establish itself, and spread, the ratio of vectors to hosts must exceed a critical level—hence, disease control measures usually aim directly at reducing the numbers of vectors and are aimed indirectly at the parasite. Insecticides are used to kill aphids and mosquitos, which transmit virus diseases of crops and malaria, rather than directing chemicals at the parasite. Of course, this is not always true; for example, yellow fever was eradicated in the United States by inoculation rather than by extinction of all mosquitos.

One of the diseases causing most concern in the United States recently is Lyme disease, a disease caused by a spirochete that is transferred from animals to humans by the bite of hard-bodied ticks. The disease came to public attention in the 1980s and has been increasing ever since. The disease is not new—it has been known in Europe for more than a hundred years—but it is only in the 1980s and 1990s that it has become prevalent in the United States. More than 40,000 cases were reported from 1982 to 1991.

There are two species of tick, one in the eastern United States, and one in the western United States. Both have many possible hosts, consisting of more than a hundred species of mammals, birds, and lizards. Adults of both species of tick feed preferentially on large and, to a lesser extent, medium-sized mammals. One of the problems in the increase of the disease appears to be related to land-use patterns and limited hunting programs that have allowed one of the main hosts, white-tailed deer, to increase. Most evidence suggests that disease will become more prevalent and the ticks will continue to expand their range and become more numerous.

## 11.3  MODELS OF HOST-PARASITOID INTERACTIONS

Because the vast majority of parasites are actually parasitoids attacking insects, and because insect pests are such economically important organisms in agriculture, other population models of parasite-host relations have centered on parasitoid-host interactions. In many ways, these are similar to predator-prey relationships because only a single host is killed.

Early models of parasitoid-host interactions centered on the **numerical response** of a population of parasitoids to host density. This contrasts with a functional response, which concerns the behavior of individual parasitoids to prey density. The first model was developed by the Australian entomologist Nicholson in the 1930s, who proposed that the success of randomly searching parasitoids would be limited not by their own egg supply but by their ability to find hosts (Nicholson 1933; Nicholson and Bailey 1935). The average area that one parasitoid searched in its lifetime was deemed to be constant and was termed the area of discovery, $a$, as distinct from the area actually traversed. Thus, if 0.3 of the total habitat area is traversed by a parasitoid, the area of discovery may be only 0.254 because some ground may be covered twice. In other words, it may be expected that only 25.4 percent of the hosts will be parasitized instead of 30 percent. If this model is to be tested, $a$ must be calculated because measuring it is usually quite impractical. If $N$ is the total number of hosts, $P$ the number of searching parasitoids, and $S$ the proportion of hosts not parasitized, then

$$a \ = \ \frac{1}{P} \ log_e \ \frac{N}{S}$$

Once a is known, the proportion of hosts parasitized can be predicted from the number of searching parasites:

$$logN_{(n+1)} \ = \ logN_{(n)} \frac{aP_n}{2.3} \ + \ logF$$

where $N_{(n)}$ and $N_{(n+1)}$ represent successive host populations, and $F$ the host reproductive rate. The outcome of this model is usually to produce increasing oscillations in the populations of both species, and laboratory tests have often been in good agreement with the theory (Fig. 11.1) (see also Debach and Smith 1941). Nicholson was aware that increasing oscillations did not often occur under natural conditions, except for some species of alpine insects—pests of coniferous forests that show regular oscillations of outbreak densities over a number of years (Klomp 1966). Nicholson envisaged a fragmentation of large host populations into small ones at high densities and, thus, a proliferation of subpopulations. Although this sort of interaction does occur in Australia between the moth *Cactoblastis cactorum* and its food supply, the prickly pear, once again such events have not been commonly documented in the field.

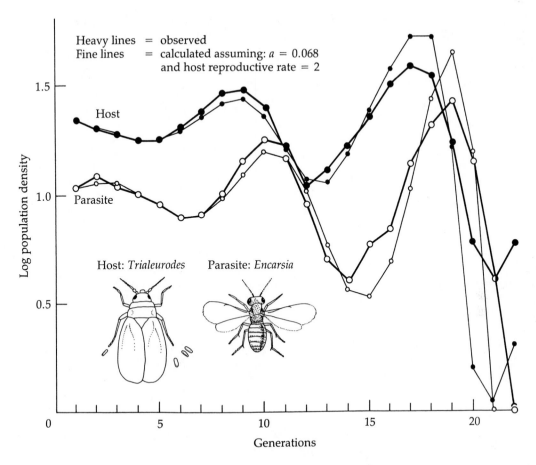

Figure 11.1   *Observed and calculated results of an interaction between Encarsia, a parasitoid, and its host, the greenhouse white-fly, Trialeurodes. Model calculated on basis of a constant area of discovery of 0.068 and a host reproductive rate of 2.*   (Redrawn from Varley, Gradwell, and Hassell 1973.)

The relationships between area of discovery and parasite density were investigated by M.P. Hassell and G.C. Varley (1969), who proposed

$$\log a = \log Q - m \log P$$

where $Q$ is the "quest constant" or area of discovery when the parasite density is one, and $m$ is the mutual interference constant. The outcome of some models based on quest theory are completely different from those of a Nicholsonian model because they include *interference*.

The stability of the quest models increases with greater values of $m$ as shown in Fig. 11.2. More important, there is a wide range of values for $Q$ and $m$ that allow the coexistence of

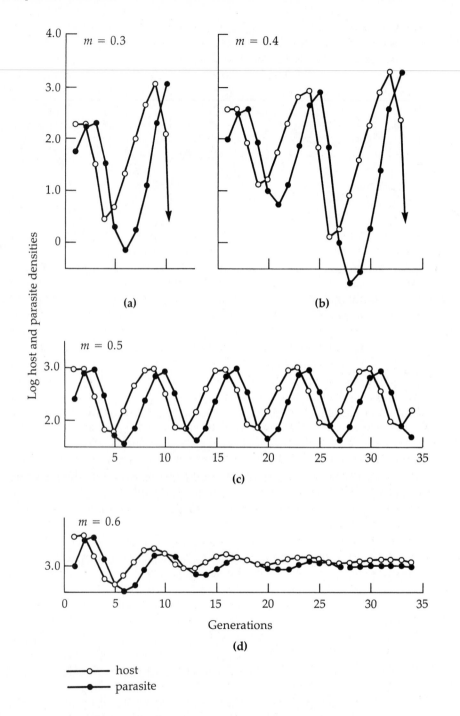

Figure 11.2 *Population models showing the increasing stability as the mutual interference con-stant m is increased from 0.3 in (a) to 0.6 in (d).* (Redrawn from Hassell and Varley 1969.)

two or more parasite species on a single species of host, a very common phenomenon in nature, yet one not accounted for in Nicholson's models.

Despite the added realism of newer models, there remain some serious flaws, the most important of which is the assumption of random search by predators and parasites. There is little evidence that random search is generally prevalent. Most parasitoids are attracted by the scent of their prey, or they remain in areas where they have previously been successful. In either case, the searching population will tend to aggregate in areas of high host density. Unfortunately, nonrandom search is difficult to incorporate into models, though field studies give the impression that it tends to increase the stability of the system.

## 11.4  THE EFFECTS OF PARASITES ON HOST POPULATIONS

Once again, the best way to find out the effect of parasites on the population abundances of their hosts is to remove the parasites and to reexamine the system. This has rarely been done, probably because of the small size and unusual life histories of many parasites, which makes them difficult to exclude. Also, as pointed out earlier, parasites may not always kill; they may merely impair the health of their hosts. This makes their effects even more difficult to gauge. However, inoculations which combat disease are known to save many lives in humans and indeed in domestic animals. Agricultural sprays also reduce crop losses to disease.

Evidence from natural populations also suggests that parasites do have a substantial impact on their hosts. In North America, chestnut blight has virtually eliminated chestnut trees (Photo 11.1). In Europe and North America, Dutch elm disease has devastated elms. In Italy, canker has had severe effects on cypress. The population dynamics of bighorn sheep in North America are dominated by a massive mortality resulting from infection by the lungworms *Protostrongylus stilesi* and *P. rushi*. This parasite predisposes the animals to pathogens causing pneumonia. A fetus can become infected through the mother's placenta, and mortality in lambs can be enormous (Hibler, Lange, and Metzger 1972). The lungworm-pneumonia complex is regarded as one of the most influential mortality factors in many sheep populations with reported mortalities of up to 50–75 percent (Uhazy, Holmes, and Stelfox 1973). Sometimes, epidemics can be even more severe when the parasite is less specific.

Rinderpest, caused by a virus, has at least 47 natural artiodactyl hosts (Scott 1970), most of which occur in Africa. The disease is usually fatal in buffalo, eland, kudu, and warthog and less fatal in bushpig, giraffe, and wildebeest. Other species, such as impala, gazelle, and hippopotamus, appear to suffer little. A major epidemic swept through Africa in 1896, leaving vast areas uninhabited by certain species, and even in the 1970s distribution patterns reflected the impact (May 1983). For example, zebra were exterminated in an area of the Elizabeth National Park in Uganda, and they still had not recolonized the area by 1954 (Pearsall 1954). Because wildlife was eliminated, other parasites were affected, too. Tsetse flies became absent from large areas of Africa south of the Zambesi River (Stevenson-Hamilton 1957), Therefore, large areas became free from trypanosomiasis—sleeping sickness—a disease borne only by the tsetse fly. One parasite, rinderpest, thus had a severe impact on the pattern of life in an area. Furthermore, in the absence of tsetse flies, humans and cattle could move in, supplanting wildlife even further. Prins and Weyerhaeuser (1987) described two recent and major epidemics in wild mammals in a national park in Tanzania, east Africa. An anthrax outbreak lasted almost a year and killed more than 90 percent of the

Photo 11.1    *American chestnut. This species of tree was common in eastern United States deciduous forests but was essentially eliminated by an introduced fungal disease, chestnut blight. Only a few individuals remain.*  (Photo from National Archives, no. 95-G-250527.)

impala population, and a rinderpest outbreak of just a few weeks killed some 20 percent of the buffalo. The conclusion is that epidemics have a more severe impact on these populations than does predation. This is a disconcerting finding for conservation biology, for it means that certain populations on small reserves could be wiped out by disease unless recolonization is encouraged.

Another interesting situation exists in North America. The usual host of the meningeal worm *Parelaphostrongylus tenuis* is the white-tailed deer, *Odocoileus virginianus*, which is tolerant to the infection. All other cervids and the pronghorn antelope are, however, potential hosts, and in these species the worm causes severe neurological damage, even when very small numbers of the nematode are present in the brain. This differential pathogenicity of *P. tenuis* makes the white-tailed deer a potential competitor with other cervids because they cannot survive in the same area with white-tails. The deleterious effects of the parasite probably include direct mortality, increased predation, and reduced resistance to other disease. The activities of humans have altered the normal distribution pattern of the white-tailed deer. As northern forests were felled, the deer expanded their range from a stronghold in the eastern United States, eventually coming into contact with moose. In Maine and Nova Scotia, white-tailed deer have replaced moose as the major cervid, and they have also replaced mule deer and woodland caribou in some parts of their ranges. Whether or not reintroduction of caribou into regions now occupied by white-tailed deer is possible because of the action of parasites is now debatable (Schmitz and Nudds 1994). So, apart from direct mortality from parasites, competitive interactions between populations can be mediated by the action of parasites. This phenomenon is similar to that discussed in Chapter 8, in which Park (1948) compared competing populations of flour beetles with and without the parasite *Adelina triboli*. It is clear that parasites can have direct affects on the species they parasitize and indirect effects on other species. The subject of indirect affects is addressed later, but another interesting example concerns the parasites of the red grouse in England (Hudson, Dodson, and Newborn 1992). More birds are killed by foxes as their burdens of the caecal nematode *Trichostrongylus tenuis* increase, suggesting that parasites increase susceptibility of red grouse to predation.

Cornell (1974) has argued that distributional gaps between bird species, where apparently favorable habitat exists, are maintained by the capacity of vectors to travel between populations. The rationale is that each population has a pathogen to which it is adapted but the other species is not. This concept has led to the idea that populations might compete for parasite-free space. Because the same type of phenomenon might operate for predators, the idea of enemy-free space is more widely circulated. Crosby (1986) argues that the Old World diseases smallpox, measles, typhus, and chicken pox were so devastating to peoples of the New World that the Old World invaders, who had a limited measure of immunity, found subjugation of the people much easier.

## 11.5  PARASITES AND BIOLOGICAL CONTROL

Not all parasites are seen as detrimental by humans. Many are used as an effective line of defense against insect pests of crops, although only about 16 percent of classical biological-control attempts qualify as economic successes (Hall, Ehler, and Bisabri-Ershadi 1980), so we cannot abandon chemical control as yet. The theory of pest control by means of the release of parasitoids is so far not very advanced. Huffaker and Kennett (1969) have suggested five necessary attributes of a good agent of **biological control:**

General adaptation to the environment and host

High searching capacity

High rate of increase relative to the host's

General mobility adequate for dispersal

Minimal lag effects in responding to changes in host numbers

Although these attributes seem necessary for a good control agent, they are clearly not sufficient. So far, the application of biocontrol agents has been carried out by a hit-or-miss technique, rather than by a sound biological method. Some authors consider that this trial-and-error method probably makes best economic sense, given the high cost of research into the biology of natural enemies (van Lenteren 1980). Others have recommended new techniques, for example presenting novel parasite-host associations, as the most likely avenue for control where hosts have not had the opportunity to evolve complex defenses against these parasites from foreign lands (Hokkanen and Pimentel 1984). Arguments still rage as to whether it is better to introduce one parasite at a time or many. The problem is that, if more than one enemy is introduced, competition between parasites could ensue, lessening the overall level of control. Ehler and Hall (1982) provided evidence from a world review of 548 control projects to show that this phenomenon could be a problem; the more parasites released, the lower the rate of establishment, although this analysis was disputed by Keller (1984). It is probably fairest to say that in this, as in so many ecological situations, the jury is still out. Stiling (1990) reviewed the factors affecting success in biological control. The factor of greatest importance was the climatic match between the control agent's locality of origin and the region where it was to be released. This result stresses the value of studies in physiological ecology and that climatic variation is of vital importance in affecting biotic relations (Chapter 8). This was underscored by a separate analysis of reasons for biological failures (Stiling 1993) where reasons related to climate (34.5 percent) were more common than any other type of reason, including competition or parasitism by native insects.

It is noteworthy that the introduction of parasites for biological control can be a risky business for other, nontarget species and should be avoided unless strictly necessary. Howarth (1983), for example, lamented the reduction of native Hawaiian lepidopterans, partially due to wasp species introduced for biological control. He called for a more narrowly focused release effort rather than a hit-or-miss campaign. Such a concern is even more important when release of insect enemies to control weeds is considered. In this case, stringent host-specificity tests are performed to ensure introduced insects will not turn to feed on valuable crops, even in times of starvation. Nevertheless, *Cactoblastis* moths, the agents which so successfully controlled the pest cactus *Opuntia* in Australia, have recently arrived in Florida and are decimating native cacti there (Johnson and Stiling 1995).

The release of genetically engineered parasites is a subject of even more concern. Australians are trying to use generically engineered viruses to control exotic pest mammals. When the British colonized Australia, they brought with them the English rabbit and the European red fox, both in the mid-1800s. The rabbits were for food, and the fox was to be hunted. Not long after these introductions, people realized they had made a mistake: the rabbits began to breed like rabbits, and the predatory foxes turned their attention to the native animals. So far, they have been implicated in the extinction of 20 species of local marsupials (Morrel 1993). The Cooperative Center for Biological Control of Vertebrate (VBC) pest populations (a government-and-university consortium) plans to release genetically designed viruses that will sterilize most foxes and rabbits by tricking the female's immune system into attacking male sperm. Some of the problems are that a live vector might get into other species and that it has the potential to travel abroad. Any travel of rabbits from Australia back to Europe could facilitate movement of the disease. This could have serious implications back in the native habitat of foxes and rabbits. On the other hand, if the project is successful and safe, it could provide a model for wiping out pests in other fragile, threatened habitats. For example, the U.S. Department of Agriculture is considering using the technique to control feral pigs which are wreaking havoc in Hawaii. In New Zealand, introduced possums are causing trouble.

One of the key aspects of these programs is the simultaneous sterilization of both species. For example, because the rabbits are the main item on the foxes menu, any sudden population crash in the rabbits might force the foxes to seek alternative food, the likely choices being small, endangered marsupials. Clearly, getting the rate of spread of the diseases to be similar in foxes and rabbits would be the optimal solution.

# Summary

 1. The true definition of a parasite is problematic. Parasites may include many species that feed on plants as well as more "traditional" parasites like tapeworms, leeches and parasitoids. Parasitism is undoubtedly an extremely common way of life with perhaps 50 percent of all animals considered to be parasitic.

2. The presence of various defenses against parasites, such as the immune response in vertebrates, is testament to the importance of parasitism in nature.

3. Mathematical models suggest that efficient parasites are likely to keep their hosts alive as long as possible to facilitate transmission to other hosts. Such diseases include leprosy, typhoid, and amoebic dysentery. Because human populations are now so large, diseases that used not to be common, such as measles and cholera, flourish with so many potential hosts. Finally, the intimate association of humans with animals has allowed the crossover of animal viruses to people. Smallpox in humans is derived from cowpox in cattle. AIDS originated in monkeys to people.

4. The huge influence of introduced diseases, such as chestnut blight and dutch elm disease in America and rinderpest in Africa, are testament to the severe effects parasites can have on host populations, sometimes driving them close to extinction.

5. Parasites of insects can often be used as control mechanisms against agricultural and forestry pests. This is called biological control. Finding the attributes of successful biological control gents is a valuable ecological endeavor.

*Discussion Question:* Which types of plants and animals might be expected to have more species of parasites and suffer higher rates of parasitism? Why can we eradicate some diseases, such as yellow fever, through vaccinations, while we have not been able to eradicate other diseases, such as malaria? Can we ever expect chemical pesticides to be replaced entirely by biological control? If not, why not?

# The Causes of Population Change

I f parasitism, predation, competition, and weather can all affect the population densities of living organisms, which effect is most important? This is a very difficult question to answer for many reasons. First of all, it is necessary to determine the mean population density or equilibrium value then determine which mortality factor causes most change in population density. This may sound simpler than it actually is. The longer you observe a population, the more variable it seems to be (for example, see Fig. 12.1). As such, it is almost impossible to know what the equilibrium value is. Large variation in population density has been shown to exist for many species of farmland birds, woodland birds, insects, and mammals (Pimm and Redfearn 1988). The existence of such a general phenomenon casts doubt over whether many populations have an equilibrium. If they do, it would be hard to know what it was without decades, even centuries of data. Secondly, while comparing the strengths of mortality factors is one of the most important pursuits in ecology, few scientific papers are broad enough in scope to be able to do this. Instead, most papers focus one or two mortalities. Thirdly, some authors have argued that species may exist not as a single population but as a series of metapopulations.

## Metapopulations

A metapopulation is a series of small, separate, populations united together. In this scenario, even if the individual populations go extinct, perhaps because of natural enemies or because of food shortage, other populations survive and they supply dispersing individuals who recolonize "extinct" patches. Even if the population that supplied the colonists itself becomes extinct, it will be recolonized again later. In such a scenario, there is no population mean or equilibrium level, as extinctions may happen at any time. Metapopulations are

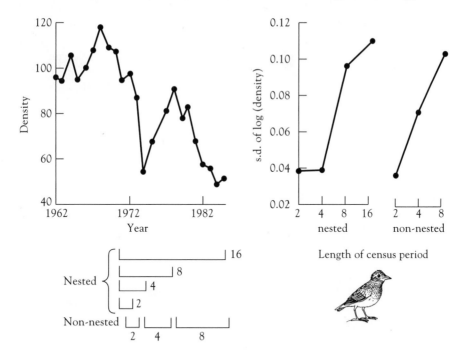

**Figure 12.1**    *Population densities and the change in variability with the length of census.* (**a**) *Density of the skylark (Alauda arvensis) in English farmlands is plotted against year; the scale is relative and set to 100 in 1966.* (**b**) *For the same population, the standard deviations of the logarithms (SDL) of density are plotted against the period over which the calculation was made. For both nested years (2, 4, 6, and 16) or nonnested years (2, 4, and 8) SDL increases with period. Pre-1970 densitites were ignored because before 1970 many bird populations were recovering from a crash in 1963 following a hard winter. Inclusion of these data would have made the increase in SDL even more marked.* (After Pimm and Redfearn 1988.)

viewed as sets of populations persisting in a balance between local extinction and colonization. Persistence depends on factors affecting extinction and colonization rates such as interpatch distances, species dispersal abilities, and number of patches. Harrison (1991) reviewed the empirical literature to identify situations that fitted this description well. There were few situations that fitted a classical metapopulation scenario. More common were three related situations (Fig. 12.2):

1. Core/satellite or source/sink or mainland/island metapopulations in which persistence depended on the existence of one or more extinction-resistant populations, usually large patches, that constantly supplied colonists to small, peripheral patches that often went extinct.
2. Patchy populations in which dispersal between patches or populations was so high that colonists always "rescued" populations from extinction. The system was then effectively a single extinction-resistant population.

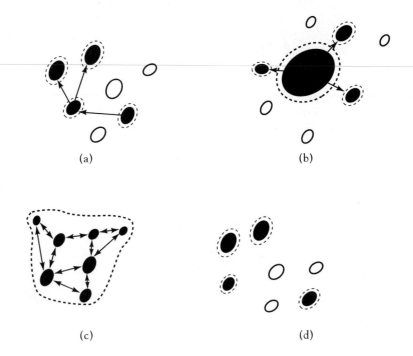

(a)

(b)

(c)

(d)

**Figure 12.2**    *Different kinds of metapopulations. Closed circles represent habitat patches; filled = occupied; unfilled = vacant. Dashed lines indicate the boundaries of "populations." Arrows indicate migration (colonization). (**a**) Classic metapopulation. (**b**) Core-satellite metapopulation (common). (**c**) Patchy population. (**d**) Nonequilibrium metapopulation (differs from a in that there is no recolonization) often happens as part of a general regional decline.* (After Harrison 1991.)

   3. Nonequilibrium metapopulations, in which local extinctions occurred in the course of species' overall regional decline. Many rare species conform to this scenario with habitat fragmentation reducing population density. Lack of dispersal between populations effectively eliminates a true metapopulation scenario.

## 12.1 KEY-FACTOR ANALYSIS

One of the best techniques for empirically comparing the importance of the effect of predators, parasites, or other factors on the size of field populations is key-factor analysis (Morris 1959). In this method, population density is expressed on a logarithmic scale and plotted against time because population changes are more easily visible in log plots. The killing power, or $K$ value, of each source of mortality is then given by

$$\log N_t - \log N_{(t+1)}$$

where $N_t$ is the density of the population before it is subjected to the mortality factor (say, microsporidian disease), and $N_{(t+1)}$ is the density afterwards. The generation mortality $K$ is then defined as

$$K = k_1 + k_2 + k_3 + k_4 + k_5 + \ldots$$

when each mortality factor, or $k$ value, acts in a specific way, for example, as parasitism of eggs or larval predation.

The $k$ value that most closely mirrors overall generation mortality ($K$) is then termed the *key factor*. More precisely, individual sources of mortality or $k$ values can be plotted on the y axis against total mortality, $K$, and the key factor is then the source of mortality with the biggest correlation coefficient, $r$, with $K$ (Podoler and Rogers 1975). In oak winter moths in England (Fig. 12.3), a species probably subject to the most comprehensive key-factor analysis ever done (Varley, Gradwell, and Hassell 1973), the key factor is overwintering loss (Fig. 12.4). Young winter-moth larvae emerge in the spring and feed on the newly developing foliage. To find fresh leaves, they often disperse by ballooning away on silken threads to new trees. This is obviously a chancy business, and many larvae are lost; hence the high $k$ values. Overwinter loss is a key factor in many animal populations because of severe climatic stress. Key factors can usually be detected only by analysis of many generations; in **univoltine** animals, this analysis may take many years.

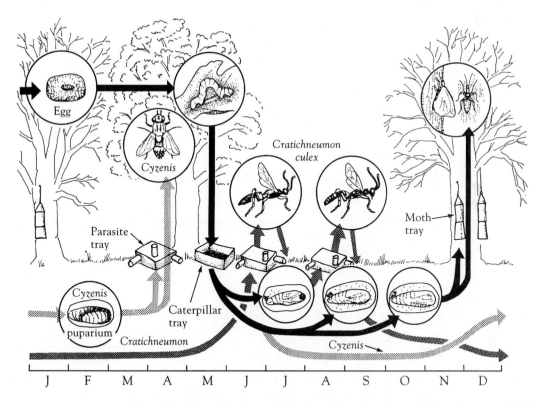

**Figure 12.3**   *Life cycle of the oak winter moth and sampling methods. Adult female winter moths were counted in moth traps on the tree trunks; their larvae were counted in the caterpillar trays into which they fell when prepupal. Larvae of the parasite Cyzenis were counted by dissection of the fallen caterpillars, and adult Cyzenis and Cratichneumon were counted upon emergence from the soil into the parasite traps.* (Redrawn from Varley 1971.)

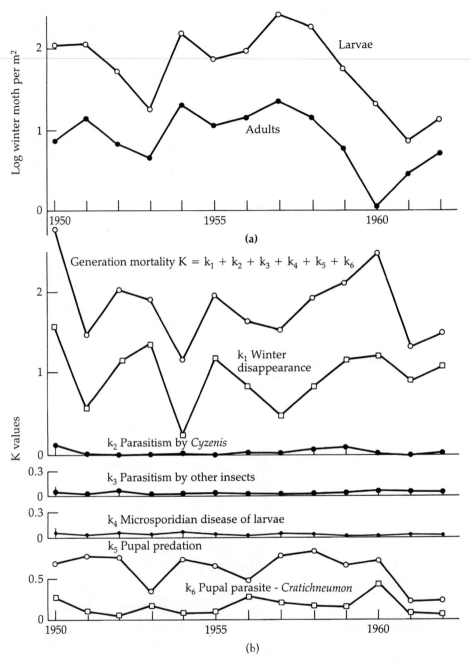

**Figure 12.4** *Key-factor analysis of winter moth population changes. (a) Winter moth population changes expressed as generation curves for larvae and for adults. (b) Changes in the mortality, expressed as k values, showing that the biggest contribution to changing the generation mortality K comes from changes in $k_1$, winter disappearance.* (Redrawn from Varley, Gradwell, and Hassell 1973.)

For other animals, key factors are many and varied. There are few generalizations that can be made as to which type of key factors operate on which types of population (Table 12.1). Even for related species, such as the insects, there is no key factor of overiding importance (Stiling 1988b). Often key factors cannot be precisely linked to specific mortality agents. Census data are only good enough to determine key factor phases in lifestyles, for example, juvenile mortality or adult mortality. However, in terms of strength of mortality factors on populations, regardless of whether they are key factors, the effects of natural enemies (frequency 48%) have been deemed as important as all other mortalities combined: weather, competition, and plant induced, at least for insects (Cornell and Hawkins 1995).

## 12.2  DENSITY DEPENDENCE

Over winter mortality in oak winter moths tends to disturb population densities away from the mean. Are there any compensatory mechanisms that tend to return populations to a mean value? One way in which population densities might be regulated about a mean value is by some negative feedback process, commonly called a **density-dependent factor.** Density dependence can be determined from a plot of the $k$ values of each source of mortality against the logarithm of the density of the life stage on which it acts. If a positive slope results and mortality increases with density, then the $k$ factor is tending to affect less of sparse popu-

**TABLE 12.1** *Key factors and density-dependent factors for a variety of plants and animals. (From Podler and Rogers 1975; Stubbs 1977; and Stiling 1988.) Some species have no obvious density-dependent or key factor, other species may have more than one density-dependent factor. Different species tend to have different types of density-dependent or key factors.*

| Organism | Key factor | Density-dependent factor |
|---|---|---|
| Sand-dune annual plant | Seed mortality in soil | Seedling germination |
| Colorado potato beetle | Emigration of adults | Larval starvation |
| Tawny owl | Reduction in egg clutch size from maximum | Losses of birds outside the nesting season |
| African buffalo | Juvenile mortality | Adult mortality |
| Partridge | Chick mortality | Shooting of adults! |
| Great tit | Loss of birds outside the breeding season | a. Variation in clutch size <br> b. Hatching success |
| Broom beetle | Larval mortality on foliage | a. Larval mortality in ground <br> b. Survival of adults |
| Grass-mirid insect | No obvious key factor | No obvious density-dependent factor |
| Cabbage-root fly | Reduction in egg production | Pupal parasitism and predation |

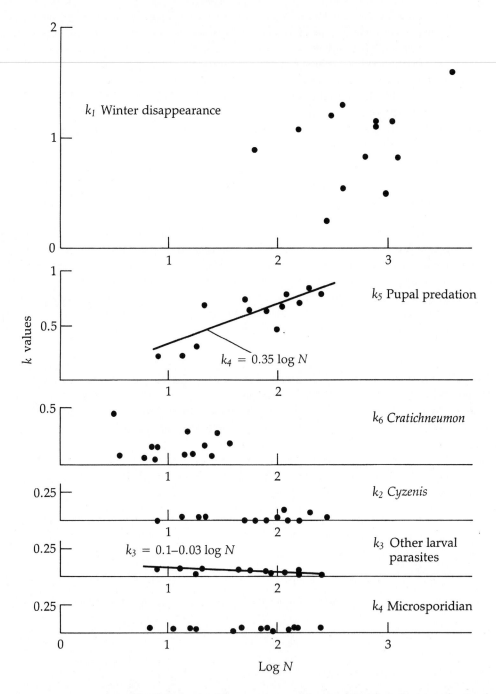

**Figure 12.5**    K *values for the different winter moth mortalities plotted against the population densities on which they acted.* $k_1$ *and* $k_6$ *are density independent and vary quite a lot;* $k_2$ *and* $k_4$ *are density independent but are relatively constant;* $k_3$ *is weakly inversely density dependent; and* $k_5$ *is quite strongly density dependent.* (Redrawn from Varley, Gradwell, and Hassell 1973.)

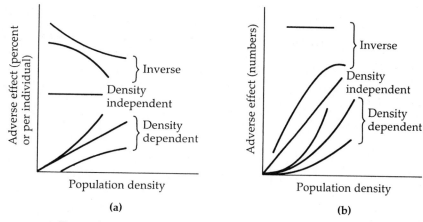

**Figure 12.6**  *Types of response to changes in population density.* (**a**) *Expressed as a percentage response to increasing population.* (**b**) *Expressed as the numbers of individuals affected with increasing populations.*

lations and more of dense ones and is clearly acting in a density-dependent manner. In the winter-moth example, pupal predation of overwintering moth pupae in the ground (by beetles) is the density-dependent factor (Fig. 12.5). This is also true in Canada, where the winter moth was introduced in the 1930s (Roland 1988).

Density dependence can also be detected where adverse effect, expressed as percent or raw numbers, is plotted against population density (Fig. 12.6). In such plots, factors that appear not to change with density and thus do not contribute to population regulation are termed **density independent.** Those sources of mortality that decrease with increasing population size are called **inversely density dependent.**

In some situations, the effect of density-dependent factors on a population is delayed by one or two generations. In such cases, when percent mortality is plotted against host density and the points are joined in a time series, the **time lag** results in a counterclockwise spiral (Fig. 12.7). While most studies have searched for density-dependent regulation, very few have examined the frequency of time-lagged density dependence. Which is more common, delayed density dependence or density dependence? Turchin (1990) evaluated the evidence for delayed density dependence in 14 forest insects and found eight to show evidence of delayed density dependence while only three showed direct density dependence.

Which factors tend to act in a density-dependent manner? For many herbivores, plant quality or availability may limit population densitites. For others, natural enemies may be more important. We can examine this question with reference to insects, the most specious group on Earth. Bernays and Graham (1988) contend that for most insects, generalist natural enemies are most important. However, a group of nine articles in *Ecology* (1988, volume 4) that followed Bernays and Graham's article contested this point and stressed the importance of host quality. Furthermore, Stiling (1987) showed that the percentage of field studies that detected density-dependent parasitism is quite low—on the order of 25 percent (see also Walde and Murdoch 1988). Stiling (1988) went on to examine density dependence by any mortality factor in insect populations, using data sets for 58 species. Again, density dependence was quite low—it was detected in only about half of the studies. However, Mike Hassell and his colleagues (1989) showed that many of the data sets Stiling used may have

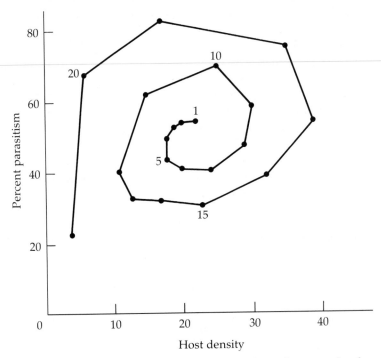

Figure 12.7    *An example of a delayed density-dependent relationship using the observed results from Figure 11.1.* (Redrawn from Varley, Gradwell, and Hassell 1973.)

been of insufficient duration to detect density dependence, even if it were present. They noted that the more generations that were studied, the more frequently density dependence was detected. Solow and Steele (1990) theorized that up to 30 generations would need to be studied before density independence could be rejected with a high probability. This has led to much debate about the number of generations needed and the type of test used to detect density dependence (Vickery and Nudds 1991; Holyoak and Crowley 1993; Wolda and Dennis 1993). While valuable, Hassell's study is disconcerting because it means only long-term studies are useful in the search for density dependence. Such studies are likely to be time consuming, difficult to perform, and probably difficult to get funding for. How many scientists would perform a study for 30 years and admit to finding no density dependence?

The density-dependent factor that is most important is probably different for populations of different species at different times. Stiling's (1988) review of the life tables of 58 species of insects showed that no single process could be regarded as a regulatory factor of overriding importance. Different sources of mortality were important in different systems, even within the insects. By way of illustration, Karban (1989) examined the effects of different sources of mortality on three herbivores of the seaside daisy, *Erigeron glaucus*. For spittlebugs, *Philaneus spumarius*, damage caused by caterpillars increased rates of desiccation and mortality of nymphs. Thus, competitive effects were strongest for this interaction. For the caterpillars, *Platyphilia williamsii*, predation by savannah sparrows, *Passerculus sandwichensis*, was the strongest effect, and protecting caterpillars from sparrows by enclosing colonies under chicken-wire cages greatly increased survival. Finally, with the third herbivore, the thrips *Apterothrips secticornis*, host-plant chemical effects were of greatest importance in determining survival (Karban 1987). The factor that was most important for each herbivore species did not interact with other biotic factors and was different for each herbivore. One of the few generalizations to emerge (Stubbs 1977) has been that more than 80 percent of density-dependent factors act on juvenile stages for *r* selected animals, whereas for *K* selected animals the figure is only 15 percent. Instead, reduced fecundity seemed to act more frequently in animals of permanent habitats.

Perhaps more accurate than observational data are experimental approaches: manipulating various mortality factors in well-replicated factorial experiments. Morris (1992) used factorial experiments to assess the effects of natural enemies (ladybird beetles and syrphid flies), interspecific competitors (chewing flea beetles), and plant quality (water availability) on the abundance of an aphid, *Aphis varians*, on fireweed plants (*Epibolium angustifolium*). While varying water availability and removing flea beetles had little affect, colonies protected from natural enemies increased prodigously. In another study that experimentally manipulated more than one mortality factor, Worthen, Mayrose, and Wilson (1994) found that survivorship of flies feeding on mushrooms was affected more by direct and interactive effects between predation and microclimate (soil moisture content) than by competitive interactions with other flies. Ant predation was important, but only on wet soils. Experimental studies like these, which compare mortality effects, are comparatively few. Review papers of experimental manipulations are even rarer. One of the few review papers that attempted to compare the frequency of different mortality factors from field experiments was that of Sih et al. (1985). They found only 17 studies that manipulated both the densities of competitors and predators. Both competition and predation were found to occur on an equal basis and to be equally important. However, 17 studies is not very many to look for patterns. There have been more reviews that have looked just at predation or just at competitions. For example, in a review of the frequency of predation, the overall frequency of predation was very high, more than 80 percent (Sih et al. 1985). This was much higher than the frequency of competition in nature—40 to 50 percent of all studies (Connell 1983). Could the reviews of competition and of predation be different in methodology? Sih et al. reevaluated the 72 studies of Connell and scored them for competition, using the same methods they'd used in their review of the frequency of predation. Competition was found in 38.4 percent of all comparisons (n = 594), still very much lower than the frequency of predation, 59.9 percent of 1,412 comparisons.

Even though predation may be more important than competition, both are undoubtedly important. As Strong (1988) has summed it up, no single factor along the gamut from plant chemistry to abiotic influences can be ruled out even for a minority of cases. A complex of influences participates in the coactions and regulations of herbs and insect herbivores, and probably other organisms, too. By way of illustration, Sinclair's (1986) data on the causes of population limitation for snowshoe hares did not support either of two separate hypotheses, **resource** limitation and **self-regulation** through social behavior. Results corroborated a multifactor hypothesis—that is, that social behavior causes differential survival through weight loss only when food is limited. Such questions are important to game managers as well as to theoretical ecologists. The population of Thompson's gazelles in the Serengeti National Park, Tanzania, declined by almost two-thirds over a 13-year period, from 660,000 in the early 1970s to 250,000 in 1985. Predation, interspecific competition, and disease have all been implicated (Borner et al. 1987). Ecology, it might seem, is entering a period of pluralism, in which simplistic, one-dimensional explanations of population phenomena are not sufficient.

## Density Vagueness

An alternative notion of population regulation has sprung from the work of Steve Pacala, Mike Hassell, and Bob May (1990). Strong (1986) had suggested that even if a trend of density dependence was observed in nature, the actual scatter of the points around the line of best fit was often substantial—there was much variance in real biological data. Real relationships were more often "density vague" than density dependent (Fig. 12.8). Such a relationship was often taken to mean that nothing interesting was going on. Pacala and his colleagues suggested that this "disorder" was actually a source of order. Consider a population of parasitoids attacking different populations of host insects. If rates of parasitism vary sufficiently from one patch of hosts to another, some patches of hosts can be counted on to escape attack and thereby sustain the host population into the next generation. Pacala and his colleagues proposed what they called the "$CV^2 > 1$." This rule states that, for a broad suite of simple models, if the square of the coefficient of variation (the scatter in the figures) among host patches is larger than one, there is enough variability to stabilize otherwise unstable host-parasitoid interactions. Data from 34 studies (21 different interactions) showed that $CV^2$ was greater than one in about a third of the interactions.

Finally, there may be specific biological reasons why mortalities do not operate in a density-dependent fashion. For example, a searching parasitoid may not always oviposit in a density-dependent fashion, laying more eggs in dense concentrations of caterpillars; she may save some of her eggs to oviposit elsewhere, even in a seemingly suboptimal place. The reason may be that some local catastrophe could occur in the best area, wiping out all her progeny. For example, a flock of birds could also be attracted to a dense congregation of caterpillars

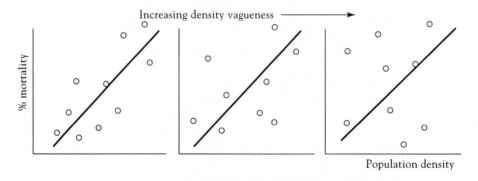

Figure 12.8   *Types of density-dependent relationships likely to be found in real-world populations. An increasing scatter of points makes it difficult to determine if there is a relationship or what it is.* (After Strong 1984.)

and eat every one. To avoid losing all her eggs in such a catastrophe, the wasp oviposits in a few solitary caterpillars in out-of-the-way places. Ferguson and Stiling (1995) showed that parasitism of coastal aphids was not density dependent and that predatory ladybird beetles often devastated local populations. This phenomenon has become known as *spreading the risk* (den Boer 1968, 1981). Spreading the risk may provide a good explanation for the apparently random patterns that are so often observed in the field. Furthermore, even if they wanted to, animals may not always behave in a density-dependent fashion in the face of a huge array of conflicting pressures. Their life-history patterns may be adaptations to survival, not maximization of fitness. Searching parasitoid wasps may not oviposit in a density-dependent manner because of the presence of a concentration of predatory spiders that preys on them.

## Spatial Phenomena

Analysis of population change is not only restricted to examination of long series of population data through time. Similar problems exist on a spatial scale as well as on a temporal one. For example, variations of population densities of herbivorous insects commonly exist from host plant to host plant in a great many different systems. Africa's abundant large herbivores are very heterogeneously distributed, both geographically and regionally. Within a region, some localities contain dense animal concentrations, although areas nearby may be virtually unoccupied (McNaughton 1988). In agricultural settings, some fields of crops may suffer heavy losses to insects while other fields are not so heavily attacked. There is a clear impetus to explain this spatial variation in population density. The reasons may be very similar to those behind temporal variation with one important exception. Climate-related changes may be important in temporal variation but are unlikely to be significant in plant-to-plant, field-to-field, or tree-to-tree variation in herbivore density.

Stiling and colleagues have investigated the differences in densities of a leaf-mining moth, *Stilbosis quadricustatella,* among different individual trees of sand live oak, *Quercus geminata.* Levels of infestation range from less than 1 percent of the leaves mined on some trees to more than 70 percent on others. The advantages of studying spatial phenomena in this system are many. The main ones are that individual leaf miners are restricted to feeding in the leaf tissue between the surfaces of just one leaf and that leaf mines themselves leave excellent records of the fates of the miners. As Photo 12.1 shows, larvae that emerge successfully cut crescent-shaped holes in the lower surface of the mine and drop to the forest floor to pupate; parasitoids leave circular holes in the upper surface of the leaf; and predators rip open the mine in dramatic fashion. In addition, leaves can be ground up and analyzed for quality by means of assays for amino acids and nitrogen levels. Levels of parasitism, predation, and host quality differ between trees, but none correlates positively or negatively with miner densities between trees (Fig. 12.9). Variation in leaf abscission rates may be why some trees are preferred over others (Stiling, Brodbeck, and Simberloff, 1991). When a leaf abscises, it falls to the forest floor and dries up. As it dries, any leafminer inside it dries out too and dies. Leafminers cannot relocate between leaves. Some trees have a higher propensity to abscise leaves than other trees. Just as with temporal studies, the reasons for population variation among different spatial areas are likely to vary with the system. For thrips on seaside daisies in California, variation is due mainly to variation in the particular genetic makeup of the plants (Karban 1987). Are there any generalizations to be made about the reasons for spatial variation, or are causal factors different for each system?

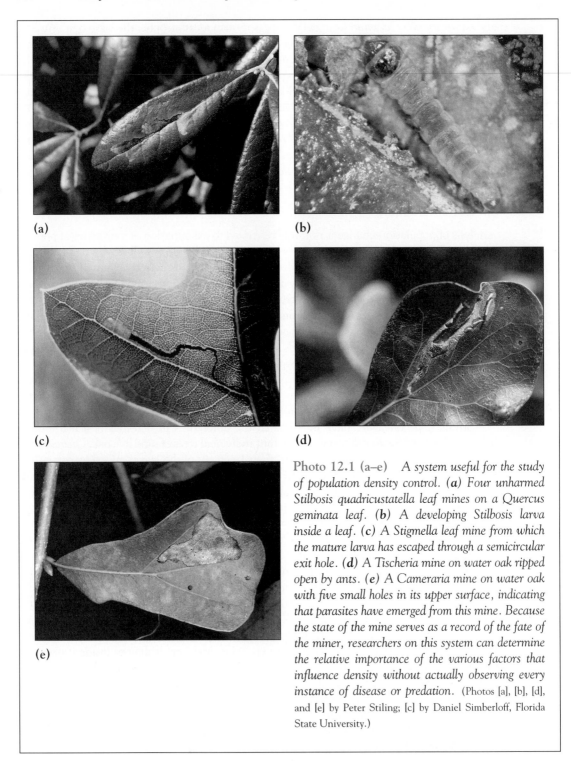

(a)

(b)

(c)

(d)

(e)

Photo 12.1 (a–e)   *A system useful for the study of population density control.* (**a**) *Four unharmed Stilbosis quadricustatella leaf mines on a Quercus geminata leaf.* (**b**) *A developing Stilbosis larva inside a leaf.* (**c**) *A Stigmella leaf mine from which the mature larva has escaped through a semicircular exit hole.* (**d**) *A Tischeria mine on water oak ripped open by ants.* (**e**) *A Cameraria mine on water oak with five small holes in its upper surface, indicating that parasites have emerged from this mine. Because the state of the mine serves as a record of the fate of the miner, researchers on this system can determine the relative importance of the various factors that influence density without actually observing every instance of disease or predation.* (Photos [a], [b], [d], and [e] by Peter Stiling; [c] by Daniel Simberloff, Florida State University.)

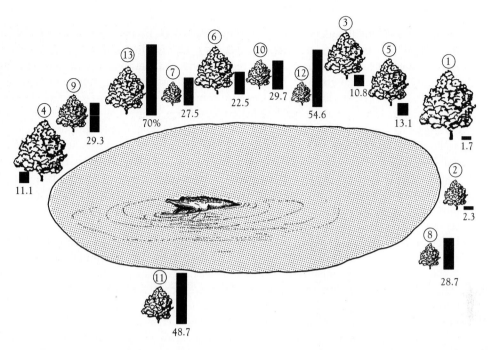

Figure 12.9   *Tree-to-tree (Quercus geminata, sand live oak) variation and percentages of leaves mined by larvae of a moth, Stilbosis quadricustatella, around a lake in northern Florida. Numbers are code numbers for individual trees, and percentages and bars reflect percentage and intensity of mining on leaves.*

Schoener (1986) has attempted to provide some generalities about sources of mortality by ranking communities along certain environmental and biotic axes. For example, he has listed six environmental factors: severity of physical stress, trophic position in food chain, resource input (from closed, a lake, to open), spatial fragmentation (fragmented to continuous), long-term climatic variation (high to low), and partitionability of resources (low, bare rock, to high, a complex leaf or prey of different sizes). To characterize organisms, Schoener has developed another set of axes: body size (small to large), recruitment, generation time (short to long), individual motility (sessile to mobile), and number of life stages (low to high, for example, holometabolous insects). He has proposed that, at some time in the future, it will be possible to describe communities in terms of such axes, with the result that, for "similia-communities" (his definition) (say, short-lived ephemeral marine algae), similar processes would shape the population densities. In this case, the algae are less affected by competition than by predation (herbivory).

## 12.3   CONCEPTUAL MODELS

Many different models have been proposed to describe the types of mortality factors that should be most important in which systems. Among the first was Hairston, Smith, and Slobodkin's

(1960) idea (often called HSS) that because the Earth appears "green," herbivores have little impact on plant abundance. They suggested that this is because herbivores are ordinarily limited by their predators, not their food supply. The implication was that plants, being abundant, endure severe competition for resources but that herbivores, suffering high rates of mortality from natural enemies, do not compete. Natural enemies themselves, being limited only by the availability of their prey, also compete. These authors (Slodbodkin, Smith, and Hairston 1967) later reformulated their arguments to include only consumers of producing tissue (that is, not granivores, nectarivores, or frugivores), only nonintroduced species and only terrestrial systems. Perhaps because of these restrictive caveats, it is the more general HSS hypothesis that has become entrenched in the literature. Oksanen, Fretwell, and colleagues (1981) (OF) proposed that the strength of various types of mortalities varies with the type of system involved—particularly as a function of primary productivity. Thus, for very simple systems with low primary productivity such as in Arctic tundra, productivity is so low that few herbivores exist. Plants are resource limited (competition). As productivity increases, some herbivores can be supported, but there are too few herbivores to support carnivores. In the absence of carnivores, levels of herbivory can be quite high. Plant abundance becomes limited by herbivory, not competition. The abundance of herbivores, in the absence of carnivores, is limited only by competition for limiting plant resources. As primary productivity increases still more, so carnivores can be supported, and there are three trophic levels—this is the HSS scenario. Finally, as productivity increases still further, secondary carnivores might be supported, and these in turn would depress numbers of carnivores, which in turn would increase levels of herbivory and lessen competition between plants. Sound confusing? The scheme is outlined in Table 12.2. The importance of different mortalities in this scheme is linked to the number of links in the food chain, or, more precisely, to primary productivity.

Support for the OF hypothesis comes from freshwater systems, low productivity terrestrial systems, and Schoener's (1989) studies on Bahama Islands. In the latter case, strengths of mortalities at different trophic levels were found to differ according to the number of trophic levels present. On some islands that had few higher predators (lizards and spiders), levels of herbivory were greater than on islands where these organisms were present in large numbers.

**TABLE 12.2** *Effects of number of trophic levels on major mortality factors in natural systems.*

| Taxa | Plants only | Plants and herbivores | Plants, herbivores, and carnivores | Plants, herbivores, carnivores, and secondary carnivores |
|------|-------------|-----------------------|-----------------------------------|----------------------------------------------------------|
| Plants | Competition | Herbivory | Competition | Herbivory |
| Herbivores | | Competition | Predation | Competition |
| Carnivores | | | Competition | Predation |
| Secondary carnivores | | | | Competition |

An alternative to both HSS and OF was suggested by Menge and Sutherland (1976) (MS), who postulated that biotic complexity decreases with increasing environmental stress. Thus in stressful habitats, herbivores have little effect because they are rare or absent and plants are therefore affected mainly by environment stress. There is little herbivory or predation of herbivores. In habitats of moderate stress, there is little herbivory, but plant densities are affected by competition. In benign environments, there are many herbivores, and herbivory controls plant abundance, not competition or environmental stress. As a good example of the MS hypothesis in action, consider salt marshes. These are among the most productive ecosystems in the world, equivalent in productivity to tropical forests or coral reefs, but they are stressful, being inundated constantly by the tide. Here levels of herbivory are low (see Chapter 17) as predicted by the MS hypothesis. In a comparison of the MS and HSS hypotheses, Sih et al. (1985) examined 25 papers on terrestrial plants and herbivores and found that only two gave support to HSS. Furthermore, Menge and Farrell (1989) reviewed experimental studies in marine ecosystems and found that the MS model was well supported. It is interesting to note that Hairston (1991) has raised objections to the validity of many of the studies reviewed by Sih et al. (1985), and Sih (1991) has replied to these criticisms. It is interesting to read firsthand the arguments of both authors.

## Top-down and Bottom-up Effects

The HSS, OF, and MS models can all be categorized under the banner "top-down effects" because the effects of predators are important at least some of the time. Another popular view is the "bottom-up" theory. In the "bottom-up" scenario, nutrients are argued to control plant numbers, which in turn control herbivore densities which in turn control carnivore numbers, and so on. One of the strongest advocates of this position was T.C.R. White (1978, 1984), who argued that the availability of nitrogen in plants was critical to insect herbivore abundance. Hawkins (1992) has also argued how host attributes are critical in affecting parasitoid abundance in host-parasitoid food chains. Finally, Hunter and Price (1992) argued that "bottom-up" forces must logically be the most important. The removal of higher trophic levels leaves low levels present—although greatly modified. However, the removal of primary producers leaves no system at all!

Needless to say, there is a middle ground. Some authors argue that both bottom-up and top-down affects are important. McQueen et al. (1989) showed how trophic cascades that are produced by top-down forces in lakes attenuate before reaching the plants, implying the importance of bottom-up effects for lower trophic levels. Strong (1992) noted that mortalities that cascade from the top of the food chain downward—trophic cascades—often do not get all trophic levels below "wet." In a similar vein, Mittlebach, Osenberg, and Liebold (1988) have suggested that because predators often require different resources as juveniles and adults, it prevents them from effectively tracking resources of just one type; hence, the effects of predators will not cascade throughout the system. Other suggestions for the importance of colimitation by predators and natural resources are summarized in Table 12.3. A diagrammatic comparison of the top-down and colimitation models is presented in Fig. 12.10.

TABLE 12.3 *Views on the relative importance of top-down and bottom-up regulation in food webs in decreasing order of the relative strength attributed to top-down forces. (After Powers 1992.)*

### Top Down

*Menage and Sutherland 1976:* Most trophic levels below the top are potentially predator limited. Physical disturbance shorten food chains.

*Hairston, Smith, and Slobodkin 1960:* Predators regulate herbivores, releasing plants to attain densities at which they become resource limited. Herbivores are predator limited; plants and predators are resource limited.

*Fretwell 1977, 1987; Oksanen et al. 1981:* Food chains can have fewer or more than three trophic levels. Top trophic levels and those even numbers of steps below them are resource limited; trophic levels odd numbers of steps below the top are predator limited.

### Co-Limitation by Predators and Resources

*McQueen et al. 1989:* Trophic cascades produced by top-down forces in limnetic lake food webs attenuate before reaching plants.

*Getz 1984; Arditi and Ginzburg 1989:* Interference among predators prevents their efficient exploitation of resources, so that prey populations, though reduced by exploitation, can increase with increases in their own resources.

*Mittelbach et al. 1988:* Predators require different resources as juveniles than as adults. This decoupling prevents predator populations from efficiently tracking resources when increases involve food of only one predator life history stage.

*Leibold 1989:* Control of prey by consumers diminishes after initial exploitation shifts community dominance to less edible species.

*Sinclair and Norton-Griffiths 1979:* Starvation-weakened prey become more vulnerable to predation or disease.

*Sih 1982; Mittelbach 1988; Power 1984a:* Prey in spatial refuges from predation become more food limited.

### Bottom-Up Limitation

*White 1978:* Plants are not appreciably limited by herbivores except when unusually stressed (for example, by drought). All trophic levels are potentially limited by availability of food resources.

In order for us to fully evaluate top-down and bottom-up models, we need to know how commonly species from one trophic level can affect species more than one level away. Usually, when one species affects another in this manner, it does so indirectly. How common and potent a force are such indirect effects?

## 12.4  INDIRECT EFFECTS

Indirect effects embrace a wide variety of similar phenomena, many of which are similar but bear different labels: apparent competition, facilitation, some mutualisms, cascading effects, tritrophic level interactions, higher-order interactions,, and nonadditive effects. *Indirect effects* are often defined as "how one species alters the effect that another species has on a third". Thus, three is often considered the minimum number of species required for indirect effects. However, species interactions can be mediated by a nonliving resource just as well as through a living one, so that in some instances indirect effects can be detected with just two species. Alteration of water quality in temporary ponds by early colonizing frogs may affect the subsequent success of other species after the first species is no longer present (Alford and Wilbur 1985). Some authors prefer to call such phenomena "priority effects" or "historical effects" rather than indirect effects. Much competition in nature is the result of an indirect effect too because resource exploitation, rather than direct interference, is the mechanism involved.

Indirect affects are probably very common in nature. If we accept that there are five major classes of interaction (competition, predation, herbivory, parasitism, and mutualism), so there are at least $5 \times 4/2 = 10$ classes of indirect interaction. Examples follow.

1. The mutualism of cleaner fishes and the fishes they clean occurs because of parasitism. The cleaner fish actually rid the customer fish of parasites.
2. The mutualism of acacia plants and ants occurs because of herbivory. Ants reduce levels of herbivores.

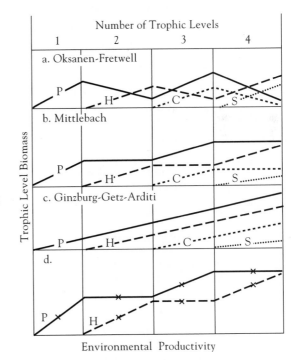

**Figure 12.10**  *Patterns of trophic level biomass accrual expected along environmental productivity gradients under pure top-down models (Oksanen and Fretwell) (a, b) and joint control by predators and resources (Arditi-Ginzburg and Getz models (c). P represents primary producers, H herbivores, C primary consumers that eat herbivores, and S secondary carnivores that eat primary carnivores. In (d), crosses represent positions along the gradient of hypothetical biomass samples taken for producers and herbivores. The positive covariance of consumers and resources predicted by the third model could be mistakenly inferred from a pattern that in reality was stepped if regions where populations plateau are undersampled, and if transitions between n and n + 1 trophic levels are undetected. (After Powers 1992.)*

3. Predators may often reduce the abundance of competitors below the level where competition is important.

4. Less commonly, the risk of predation in some habitats compels prey species to feed and compete in the same habitat (true of coral reef fish).

5. Parasitized or sick individuals are often more subject to predation.

6. Apparent competition. Two species might mistakenly be construed as competitors when their negative interactions are actually the consequence of an (unconsidered) shared predator or parasite (Holt 1977). In the absence of the predator/parasite, no competition would be detected. Individuals weakened by parasites may lose in competitive interactions (Park's work on *Tribolium* beetles and the parasite *Adelina*).

7. Darwin's own classic work showed that grazing animals increase plant species diversity by reducing dominant competitors.

8. Facilitation. Shoots of the common reed *Phragmites australis* when damaged by the stem-boring moth *Archanara geminipunctata* develop narrow side shoots. Shoots are also attacked by the gall-making fly *Giraudiella inclusa*. More shoots mean more available sites for *Giraudiella* galls (Tscharntke 1989).

9. Nonadditive effect. Galls of *Giraudiella* are parasitized by eight parasitoid species. Some parasitoids are most common in large galls. Birds preferentially peck open large galls and feed on the insects inside. There is therefore a nonadditive effect of bird predation on gall midge larvae bacause many parasitoids are also eaten (Tscharntke 1992).

10. Cascading effect. *Pheidole megacephala* ants on the plant *Pluchea indica* in Hawaii tend scale insects, *Coccus viridis* (Bach 1991). They remove predatory coccinellid larvae and decrease parasitism rates of scales. As a result, honeydew accumulation on plants increases, resulting in greater colonization by sooty mold and greater rates of leaf death and abscission. The cascading effect is also known as top-down effect: Tree growth on Isle Royale in Michigan increased when wolf predation on herbivores reduced herbivory (McLaren and Peterson 1994); and the tritrophic effect: Insectivorous birds increase growth of white oak trees in Missouri through consumption of leaf-chewing insects (Marquis and Whelan 1994).

11. Outbreaks of disease on herbivores may lead to reductions in grazing pressure and increases plant abundance (Dobson and Crawley 1994).

## Detecting Indirect Effects

Bender, Cuse, and Gilpin (1984) proposed experiments designed to distinguish direct effects from indirect ones. In "pulse" experiments, a species is removed only once, and any change in population density of the remaining species reflects the direct effects of the removed species. In "press" experiments, removals are maintained over several generations of the target species. Any resultant effects are argued to the result of both direct and indirect effects. This assumes that indirect effects take more time to appear. The differences between the two types of treatment reflect the extent of indirect effects.

## Indirect Effects Compared to Direct Effects

There are very few studies that attempt to compare the strengths of direct and indirect effects. Paine (1992) has made a comparison of direct and indirect effects by painstakingly removing individual species in a marine intertidal food web in the Pacific northwest. He found very little evidence for strong indirect effects and more evidence for strong direct effects of predation. However, Strauss (1991), in a review of indirect effects, marshaled much evidence to stress their importance, and Menge (1995) suggested that indirect effects accounted for about 40 percent of the changes in community structure resulting from manipulations in intertidal communities. In this case, as in many others, there is no general consensus of opinion. But now the separation between population ecology and community ecology starts to blur. Many of the affects of mortalities have wide ramifications throughout food webs and whole assemblages of species. With this in mind, we will now begin to study, in more detail, these assemblages, often known as communities.

# Summary

1. Do population densities commonly vary around mean or equilibrium values, or do they fluctuate widely? Often, it is hard to tell. The longer a population is observed, the more it seems to vary.

2. Which factors affect population densities the most? There can be two kinds of effects. Those that perturb populations away from mean levels can be thought of as key factors and identified by a technique known as key factor analysis. The key factors for plants and animal are many and varied, and there seems to be no generalization as to which key factors are important for which types of organism.

3. Factors that act so as to return populations to equilibrium levels are called density-dependent factors. Although there are many statistical problems inherent in the detection of density dependence, once again there are few generalizations as to which

factors (such as predators, parasites, or disease) act most frequently in a density-dependent fashion.

4. There have been several different types of models that have been proposed to describe the types of mortality factors that should be most important and in which system they are. The Hairston, Smith, and Slobodkin model (HSS) was among the first. It suggested that because the Earth is "green" with plants, herbivores must have little impact on plant abundance. This is so because predators keep herbivore numbers down. Thus, predators undergo severe competition for prey (herbivores); herbivores are limited by predators, and plants, which do not suffer much herbivory, undergo strong competition for resources.

5. The HSS model was the jumping off point for many other models, the most well known of which are models developed by Oksanen, Fretwell, and their associates (OF).

*Continued on page 250*

The OF model suggests that the frequency of competition and predation in a food web varies with productivity and the number of trophic levels in that web.

6. An alternative to both HSS and OF was suggested by Menge and Sutherland (MS) who postulated that biotic complexity decreases with increasing stress. Thus, in polar habitats the climate is so severe that predators and herbivores are absent and the few plants that survive are limited by environmental stress. In benign environments such as tropical forests, herbivory and predation are important.

7. The HSS, OF, and MS models can be categorized under the banner of "top-down" models, with the effects of predators being potentially important. In contrast, "bottom-up" models suggest that populations of all species are ultimately affected by plant populations. Large plant populations permit the growth of herbivore populations and natural enemies. Removal of predators or herbivores still leaves plant populations intact. Removal of plant populations eliminates all other populations.

*Discussion Question:* There are clearly a plethora of ecological effects on populations from direct effects such as predation, competition, parasitism, herbivory, and mutualism to various types of indirect effects. Can we ever hope to erect a framework whereby we can predict which effects are important to which species? How could such theories best be tested?

# Community Ecology

More rain in the tropics means more plants. But why are there more species in the tropics and not just more individuals of one species? Contrast the vegetation of the desert island of San Pedro Martir, Baja, with just one species of cactus, with the species-rich jungle vegetation of Southeast Asia. Community ecology attempts to address questions like these. (Photos by E.R. Degginger)

If we are to preserve biodiversity on Earth, what should we preserve? Areas with the most numerous assemblages of species? Areas with species most different than those from other areas? How exactly should we measure diversity? Why are there many more different species in tropical forests than in temperate forests? Why are there generally a few common species and lots of rare species in a given area? Why do we see weeds and herbs in an old field gradually being replaced by shrubs and trees? How are populations of different species linked together to form food webs, and how do energy and nutrients flow between members of that web? Community ecology attempts to answer these questions. Thus, community ecology deals with populations not in isolation but together as a whole in the natural areas where they interact. Communities can occur on a wide range of scales and can be nested—the tropical-forest community encompasses the community living in the water-filled recesses of bromeliads, which in turn encompass the microfaunal communities of cellulose-digesting insects' guts. Once an **association** between species has been recognized and a community identified, we can describe the basic type of community present, determine the **trophic** structure (who eats whom), and determine the relative **biomass** of individual components. We can also count the number of species present—the abundance of each species—and try to come up with an index of diversity.

The understanding of communities is particularly important in a wide variety of disciplines. For example, modern agriculture, where emphasis today is on the integration of pasture, trees, and livestock (so-called *agroforestry*), especially in tropical regions. Problems associated with particular facets of the community may develop; for example, in tropical Asia, cattle can cause damage to young trees, and more important, their dung can serve as a breeding place for rhinoceros beetles, one of the major pests of coconut (Reynolds 1988). Adjusting the species mix for optimum yields is a complex problem (Young 1988). In conservation, land managers often strive to maximize biodiversity, so a thorough understanding of this concept is necessary. In restoration ecology, practioners are keen to know if, following the replanting of areas with natural vegetation, the restored area will recruit animals and begin to function like a natural community.

Some ecologists view community ecology as a poor excuse for a science (Schrader-Frechette and McCoy 1993), arguing that communities are no more than the sum of their individual components. They claim that a grassland prairie community is simply a collection of populations of several species of grasses, each of which has the same environmental requirements. These grasses are fed upon by an array of insect and mammalian herbivores. Others view the integration of populations as unique communities with special properties, in much the same way that salt has unique characteristics (taste, for example) and does not simply combine the attributes of sodium and chlorine. Such unforeseen characteristics of a community are often termed *emergent properties*. Life itself is an emergent property. In this view, some members of a community are thought to change their habitat subtly and fractionally, enabling other species to exist. For example, in a log community, the first invaders attack and weaken the wood slightly, facilitating the entry of other organisms, though both can be found at the same time and can be thought of as part of the same community. The existence of emergent properties may be important in applied situations because human interventions may not only alter the population biology of certain species but may also disrupt these higher-order interactions.

# The Main Types of Communities

The concept of the **community** is valid over a wide range of scales: we can refer to the community of a rotting log, the aquatic community of a lake, or the desert community. Large-scale communities are referred to as **biomes** and include **tropical forest, temperate forest, grassland, desert, tundra,** coral reef, coastal shelf, and open ocean. The meteorological conditions that produce different biome types, together with a rudimentary map of their global distribution patterns, are shown in Figures 13.1 and 13.2. Each biome type shows general features characteristic of the community. Thus, a tropical-forest community may show a distinct vertical structure, whereas a single population of one plant species would not. Most communities do, in fact, show a vertical structuring. Leaves are very efficient at intercepting light—little passes through them. In low light, leaves are arranged in a virtual monolayer, and not enough light passes through to permit other plants to grow underneath. In high-intensity light, more light is reflected from leaf to leaf, and more penetrates between the leaves, so some leaves grow beneath the upper canopy to utilize this light. This phenomenon has been more rigorously approached by Horn (1971), who measured the approximate numbers of layers of leaves in an oak-hickory forest in New Jersey and found numerical agreement with these ideas. In canopy trees, the mean number of layers of leaves was 2.7, in understory trees 1.4, in shrubs 1.1, and in ground cover 1.0. Light, it must be remembered, is, in times of abundant water and nutrients, a **limiting resource,** and one might expect leaves to be optimally placed to intercept it. This limitation often means that the successful plants are those that grow a little taller than their neighbors. Height of leaves in a canopy is usually greater where more herbaceous cover exists. Plants on barren ground suffer no competition for light and remain short. There are trade-offs, of course, in growing tall because more energy must be devoted to stem support (Givnish 1982). Aquatic communities also show distinct vertical differentiation. For

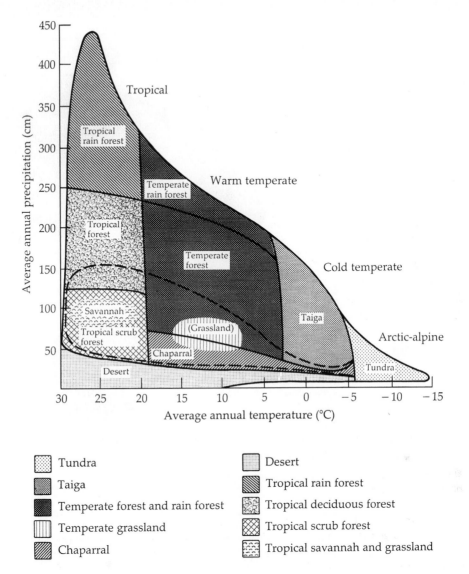

**Figure 13.1**  *Temperature and precipitation conditions that give rise to the world's major biomes. Dashed line represents the area where soil type and fire frequency determine whether woodlands or grasslands occur in the area.*  (Modified from Whittaker 1975.)

example, lakes and oceans can be divided into **epilimnion, thermocline,** and **hypolimnion** on the basis of temperature or into **euphotic** and noneuphotic zones on the basis of light availability (see Chapter 13 section 9).

As the complexity of the vegetational structure of a community increases, so often does the animal diversity in terms of numbers of species. Erwin (1982, 1983) has argued that the richest diversity of organisms on Earth exists in the canopies of tropical rain forests.

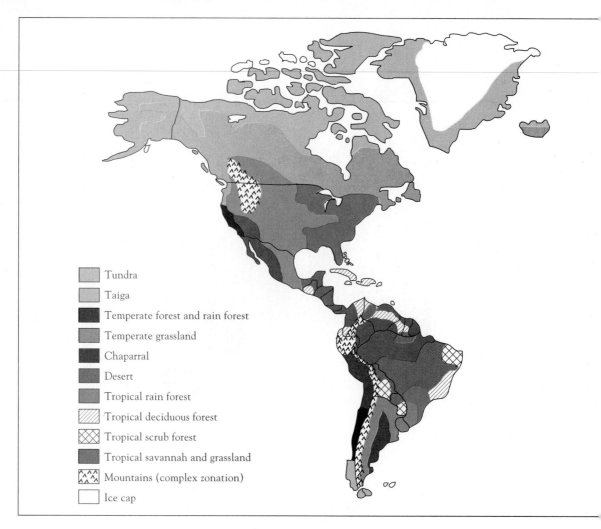

**Figure 13.2**   *Distribution of the major biomes throughout the world.*   (Modified from Odum 1971.)

Communities can also differ in their seasonality. It is often assumed that tropical communities are aseasonal and that members are able to reproduce year-round. Although it is true that there are no hot and cold spells in tropical environments, seasonality is often based on wet and dry periods. Even in the tropics, there are distinct wet and dry seasons; indeed, few forests exist in the world on which approximately the same amount of rain falls in every month of the year. The main effect of seasonality in the tropics is that leaves are generally shed quickly and replaced in the dry season (although some plants produce leaves at a constant rate all year). Thus, although some leaves are available year-round for generalist foliage feeders, leaf supply still changes seasonally. Young succulent leaves appear at specific times, and their appearance is reflected by a distinct seasonality of many monophagous trop-

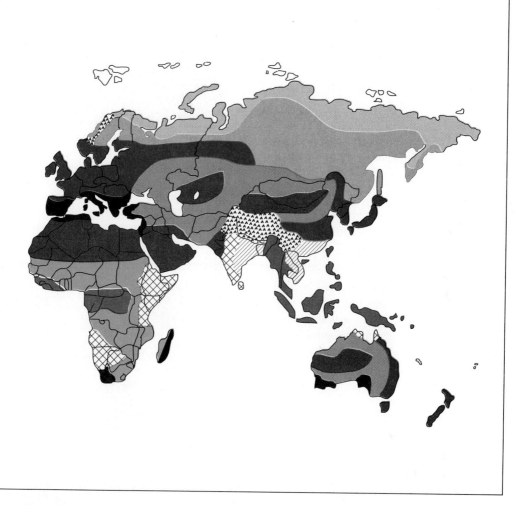

Figure 13.2   *(continued)*

ical insects (Wolda 1983; Wolda and Broadhead 1985). Wolda (1986, p. 93) has stated, "All available information suggests that tropical animals do not differ from temperate zone species in terms of temporal stability." Flowering is also seasonal for individual species, although each species may flower at a different time in the year, enabling pollinators—hummingbirds and insects—to exist year-round. Tropical forests are never a blaze of color as temperate forests are in the spring, and the paucity of blooms can be quite a shock to the first-time visitor to the tropics.

Terrestrial biomes are often named after the predominant vegetation, like tropical forest biome. However, each biome is also characterized by specific microorganisms, types of fungi, and common animals that are adapted to these particular environments. Grasslands, for example, are more likely than forests to be populated by large grazing mammals. The actual species composition throughout a biome may vary slightly from one location to another, but the similarities are more noticeable than the differences.

## 13.1  TROPICAL FOREST

The most species-rich and complex communities on Earth are tropical **rain forests,** an example of which is shown in Photo 13.1. Studying these jungles is quite difficult, as is eloquently expressed by Sanderson (1945, p. 124):

> The principal difficulty encountered in studying a jungle is that you can never see it. This may sound absurd. An average jungle is about 100 feet tall and is shaped like a much-flattened and inverted saucer so that its edges slope down all around to meet the earth. When you enter it you become lost. You can't see the wood for the trees, and you often can't see the

Photo 13.1   *Tropical rain forest on the southwest coast of Costa Rica near Golfo Dulce. Tropical Forest has distinct vertical layers that provide niches for animals living on the top of the canopy, in the mid-layers, on epiphytes, and on the forest floor. Leaves and fruits exist year-round permitting a rich array of specialist feeders.*
(Dimijian, Photo Researchers, Inc.)

Biomes
Rain forrest
savanna
Desert
grasslang/ Prarie
Dicidouous forest / Temp.
coniferous forest / TAIGA
Tundra
mountain
Marine / Fresh

trees for the creepers, lianas, and epiphytic green plants that grow all over them. If you fly over a jungle in an airplane you see even less, for nothing but a gently undulating, bumpy green mat is unfolded beneath you. If you climb a tree, you still do not gain any real conception of the jungle generally, for ants, leaves, hummingbirds, and a riot of tangled vegetation obscure the view and bother you.

These forests are generally found in equatorial regions where annual rainfall exceeds 240 cm a year and the average temperature is more than 17°C. Thus, neither water nor temperature is a limiting factor. Surprisingly, soils in such areas can be fairly poor and yet still support a luxuriant vegetation. Much of the "goodness" is **leached** out by heavy rainfall. There is no rich humus layer as there is in temperate systems; fallen leaves are quickly broken down and nutrients returned to the vegetation, where most of the mineral reserves are locked up. Consequently, cleared tropical forestland does not support agricultural practices well.

Tropical forests cover much of northern South America, Central America, western and central equatorial Africa, and some of Madagascar, southeast Asia, and various islands in the Indian and Pacific oceans. The total amounts to a land area of about 3,000 million ha, 23 percent of the world total (Bunting 1988). The human population in these areas is about 20 percent of the world total. The diversity of species in tropical forests is staggering, often reaching more than 50 tree species per ha; indeed the record for most tree species in an area alternates back and forth between southeast Asia and South America as different areas are censused. Gentry (1988) recorded 283 tree species in 1 ha of Peruvian rain forest. Sixty-three percent of the species in a 1 ha plot were represented by a single tree, and there were only twice as many individuals as species.

Rain-forest trees are often smooth barked and have large oval leaves narrowing to "drip-tips" at the apex so that rainwater drains quickly before **photosynthetic** efficiency is impaired. Many trees have shallow roots with large buttresses for support (Warren et al. 1988). The tallest trees reach heights of 60 m or more and emerge above the tops of lower trees, which interdigitate to form a closed canopy. Little light penetrates this canopy, and the understory is often sparse. Tropical rain forests are also characterized by **epiphytes,** air plants that live perched on trees and are not rooted in the ground. Bromeliads are common epiphytes in New World forests. Climbing vines or lianas are also common.

Animal life in the tropical rain forests is also diverse; insects, reptiles, amphibians, and birds are well represented. Because many of the plant species are widely scattered in tropical forests, it is a more risky operation for plants to rely on wind to be pollinated or for wind to disperse their seed. This means that animals are more important in dispersing fruits and seeds. Many plants rely on mutualistic interactions with animals to deliver pollen. As many butterflies can be found in a single rain forest as occur in the entire United States—500–600 species. Tropical rain forests are the great reservoirs of diversity on the planet; as many as half the species of plants and animals on Earth live in them. The island of Trinidad, off northwest Venezuela, is only a few hundred square miles in area, yet can boast the same number of butterflies as the United States, largely as a result of the presence of tropical rain forest. Bright protective coloration and mimicry are rampant. Large mammals, however, are not common, though monkeys may be important herbivores. Though the genealogy of the major species in rain forests is different in different parts of the world, many species converge to a similar body form because they are adapted to a similar lifestyle (Fig. 13.3).

## 13.2  TEMPERATE FOREST

Temperate forest is the type of forest with which many people in the United States and Europe are most familiar (Photo 13.2). It occurs in regions where temperature falls below freezing each winter, but not usually below −12°C, and annual rainfall is between 75 and 200 cm. Large tracts of such habitat are evident in the eastern United States, east Asia, and western Europe. Commonly, leaves are shed in the fall and reappear in the spring, though there are exceptions. In the Southern Hemisphere, evergreen *Eucalyptus* forests occur in Australia, and large stands of southern beech, *Nothofagus*, occur in southern South America, New Zealand, and Australia. Species diversity is much lower than in the tropics; one or two tree species are dominant in a given locality—for example, oaks, hickories, or maples are usually dominant in the eastern United States. Many herbaceous plants flower in spring before the trees leaf out and block the light (Heinrich 1976), though even in the summer the forest is usually not as dense as in tropical situations, so there is often abundant ground cover. Epiphytes and lianas are rare. Soils are richer because the annual leaf drop cannot be as quickly decomposed. With careful agricultural practices, soil richness can be preserved, and as a result agriculture flourishes. Like the plants, animals are well adapted to the vagaries of the climate, and many mammals hibernate during the cold months. Birds migrate, and insects enter **diapause,** a condition of dormancy passed usually as a pupa (though sometimes as an egg, larva, or adult instead). The reptile fauna, dependent on solar radiation for heat, is somewhat impoverished. Mammals include wolves, bobcats, foxes, bears, and mountain lions.

## 13.3  DESERTS

**Deserts** are biomes that suffer from water deficit. They are found generally around latitudes of 30°N and 30°S, between latitudes of tropical forests and temperate forests or grasslands. More than one-third of the Earth's surface is occupied by these hot, dry regions. One reason for the locations of deserts is the movement of winds in the Earth's atmosphere. Warm, moist air rises over the equator; as it cools, rain is produced, and the air rises again, moving north or south of the equator. This rising, raining, and movement continue to about 30° latitude, where the now dry and relatively cold air begins to sink groundward. It warms by compression and produces a downward flow of warm dry air at about 30° north and south of the equator. Such patterns of circulation, in which the drier air then filters back to the equator, are known as *Hadley cells.* They produce deserts at characteristic latitudes, including the Sahara of north Africa, the Kalahari of southern Africa, the Atacama of Chile, the Sonoran of the southwest United States (shown in Photo 13.3), the Gobi of central Asia, and the Simpson of Australia.

Deserts are characterized by two main conditions, lack of water (less than 30 cm per year) and usually high daytime temperatures. However, cold deserts do exist and are found west of the Rocky Mountains, in eastern Argentina and in much of central Asia. Lacking cloud cover, all deserts quickly radiate their heat at night and become cold. The degree of aridity is reflected in the ground cover. In true deserts, plants cover 10 percent or less of the

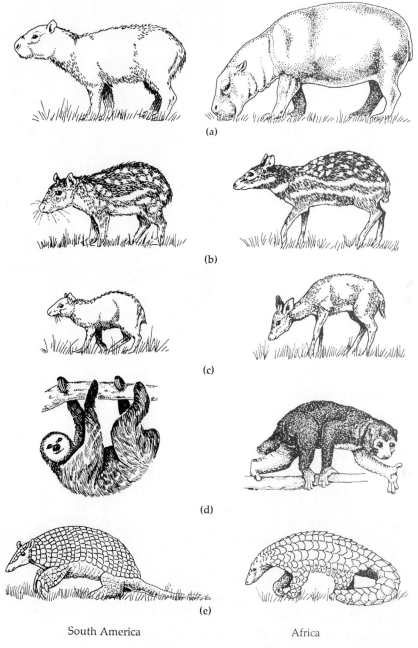

South America                                        Africa

**Figure 13.3**   *Convergence in mammals of tropical rain forests in South America and Africa.* (**a**) *Capybara and pigmy hippopotamus.* (**b**) *Paca and African chevrotain.* (**c**) *Agouti and royal antelope.* (**d**) *Three-toed sloth and Bosman's potto.* (**e**) *Giant armadillo and pangolin. The first three South American animals are rodents, and the first three from Africa are ungulates, but convergence in shape and body size is still strong.* (Modified from Ehrlich and Roughgarden 1987.)

Photo 13.2 *Temperate forest, like this Douglas fir forest at Cathedral Grove, Vancouver Island, British Columbia, has sparser vegetation than its tropical counterpart, with fewer layers between the forest canopy and the ground. Lower temperatures mean a slower rate of decay so fallen leaves and branches remain on the ground longer, resulting in a richer humus layer in the oil.* (Krasemann, Peter Arnold 36837.)

Photo 13.3 *Sonoran Desert, Arizona. The prominent plants include the tall, columnar saguaro cactus,* Canegiea gigantea; *the green spray-like ocotillo,* Fouquieria splendens; *and smaller cholla cacti,* Opuntia sp. (Photo by Peter Stiling.)

soil surface; in semiarid deserts, such as thorn woodlands and some grasslands, they cover 10 to 33 percent. Only rarely do deserts consist of completely lifeless sand dunes, but such places do exist: in the Atacama Desert of western Chile, no rainfall has ever been recorded.

Three forms of plant life are adapted to deserts: (1) the annuals, which circumvent drought by growing only when there is rain; (2) succulents, such as the saguaro cactus (*Carnegiea gigantea*), and other barrel cacti of the southwestern deserts, which store water; and (3) the desert shrubs, such as the spraylike ocotillo (*Fouquieria splendens*), which have short trunks, numerous branches, and small thick leaves that may be shed in prolonged dry periods. As a strategy against water-seeking herbivores, many plants have spines or an aromatic smell indicative of chemical defenses, although the physical structure of the desert plants—their few leaves and sharp spines—is probably also linked to water conservation and heat load. The above-ground parts of desert plants are more widely spaced than those of their forest counterparts because their roots are longer and occupy greater areas to ensure maximum water-gathering potential (Fig. 13.4). Typical perennial plants of the desert are the succulent and thorny cacti of the Western Hemisphere and succulent thorny members of the milkweed (Asclepiadaceae) and spurge (Euphorbiaceae) families in African deserts. In North America, the creosote bush (*Larrea*) is widespread over the southwestern hot desert, and sagebrush (*Artemisia*) is more common in the cooler deserts of the Great Basin.

**Figure 13.4** *Illustration of the regular spacing that allows desert plants to maximize their water uptake after rains.*

Seed-eating animals, such as ants, birds, and rodents, are common in deserts, feeding on the numerous small seeds produced by the plants. Reptiles are numerous because high temperatures permit these cold-blooded animals to maintain their body temperature. Lizards and snakes are important predators of seed-eating animals. Like the plants, desert animals have also evolved many ways of conserving water, like dry excretion (uric acid and guanin), heavy-wax "waterproofing" in insects, and generally crepuscular habits and burrow living in the day.

Irrigated deserts can, because of the large amount of sunlight, be extremely productive for agriculture, though large volumes of water must flow through the system or detrimental salts may accumulate in the soil because of the rapid evaporation rate. Desert civilizations that harnessed the flow of such rivers as the Tigris, the Euphrates, the Indus, and the Nile dominated early human history. Unlike tropical forests, deserts seem to be expanding under human influence because overgrazing, faulty irrigation, and the removal of what little hardwood exists all speed up desertification. The Sahel region, a narrow low-rainfall band south of the Sahara whose name is derived from the Arabic word for "border," is often argued to be a case in point. The acacia tree, ubiquitous in many arid zones and useful as firewood and forage, was common around the Sudan capital, Khartoum, as recently as 1955; by 1972, the nearest trees were 90 km south of the city.

## 13.4  GRASSLAND

**Grasslands** occur in the range between desert and temperate forest in which the rainfall, between 25 and 70 cm, is too low to support a forest but higher than necessary to support desert life forms. Alternatively, in the view of some ecologists (Bragg and Hurlbert 1976; Kucera 1981), the extensive grasslands of central North America, Russia, and parts of Africa are zones between forest and desert in which fire and grazing animals have worked together to prevent the spread of trees (Photo 13.4). Often, forest removed from nearby areas is replaced by grassland. From east to west in North America and from north to south in Asia, grasslands show differentiation along moisture gradients. In Illinois, with about 80 cm annual rainfall, tall prairie grasses about 2 m high such as big bluestem (*Andropogon*) and switchgrass (*Panicum*) dominate, whereas along the eastern base of the Rockies, 1,300 km to the west, where rainfall is only 40 cm, shortgrass prairies exist, rarely exceeding 0.5 m in height and consisting of buffalo grass (*Buchloë*) and blue grama (*Bouteloua*). Similar gradients occur in South Africa (the veldt) and in Argentina and Uruguay (the pampas). In some grasslands, there may be just sufficient rainfall to support isolated trees. For example, African savannas contain isolated acacia trees, and the same is true in South America and Australia

Nowadays, little original grasslands remain. Prairie soil is the richest in the world, having twelve times the humus found in a typical forest soil. Historically and where the grasslands remain, large mammals are the most prominent members of the fauna; examples are bison (buffalo) and pronghorn antelope in North America, wild horses in Eurasia, large kangaroos in Australia, and a diversity of antelopes, zebras, and rhinoceroses in Africa, as well as their associated predators (lions, leopards, cheetahs, hyenas, and coyotes). Burrowing animals such as gophers and mole rats are also common.

Photo 13.4   *In between the areas of desert and temperate forest are the grasslands—vast areas of treeless plain such as this prairie pasture in Nebraska. Some of the largest mammalian herbivores and their predators exist on prairies and savannahs.* (Grant Heilman, Grant Heilman CNB-92B.)

## 13.5  TAIGA

North of the temperate-zone forests and grasslands lies the biome of coniferous forests, known commonly by its Russian name, **taiga.** Most of the trees are evergreens or conifers with tough, narrow leaves, needles that may persist from three to five years. Spruces (*Picea*), firs (*Abies* and *Pseudotsuga*), and pines (*Pinus*) dominate, but some deciduous species such as aspens, alders, larches, and willows may occur in disturbed areas or along water courses. Many of the conifers have conical shapes to reduce bough breakage from heavy loads of snow. As in tropical forests, the understory is thin because of the dense year-round canopies; soils are also poor and acidic because of the slow decay of fallen needles. Snakes are rare, and few amphibians exist. Insects are strongly periodic but may often reach outbreak proportions on the dense wood and foliage available in times of climatic relaxation. Mammals such as bears, lynxes, moose, beavers, and squirrels are heavily furred. The taiga is famous for cyclic population patterns, of which the abundances of hares and lynxes are a well-known example. These patterns have given rise to the notion that diversity and stability are linked in ecological systems (Chapter 14). In the Southern Hemisphere, little land area occurs at latitudes at which one would expect extensive taiga to exist.

## 13.6 TUNDRA

The **tundra** is the last major biome, occupying 20 percent of the Earth's surface but existing only in the Northern Hemisphere, north of the taiga. It is often treeless, as precipitation is generally less than 25 cm per year. What little water that does fall is locked away for a large part of the year as **permafrost**. Summer temperatures are only 5°C, and even in the long summer days the ground thaws to less than 1 m in depth. Midwinter temperatures average –32°C. Vegetation occurs in the form of fragile, slow-growing lichens, mosses, sedges, and occasional dwarf trees such as willow, which grow close to the ground. In some places, vegetation cannot exist, and desert conditions prevail because so little moisture falls. Because permafrost is impenetrable, water drainage is inhibited and surface water lies in shallow lakes and ponds on the surface of the Earth in the summer. The anaerobic (oxygenless) conditions of the water-logged soil and the low temperatures have major effects on nutrient cycling. Organic matter cannot completely decompose, and it often accumulates in soggy masses called peat. Animals of the arctic tundra have adapted to the cold by living in burrows or having good insulation. Many birds, especially shorebirds and waterfowl, migrate. The fauna is much richer in summer than in winter. Many insects spend the winter at immature stages of growth, which are more resistant to cold than the adult forms. The larger animals include such herbivores as musk oxen and caribou in North America, and reindeer in Europe and Asia, as well as the smaller hares and lemmings. Common predators include arctic fox, wolves and snowy owls, and polar bears near the coast.

Tundra may occur not only in the far north but also in the higher elevations of mountains. Thus, alpine tundra can occur in the tropics at the very highest mountaintops where nightly temperatures drop to below freezing. In these situations, of course, daylight varies little from the 12 hours per day throughout the year. So instead of an intense period of productivity, vegetation in the tropical alpine tundra exhibits slow but steady rates of photosynthesis and growth all year.

Of course, not all communities fit neatly into these six major biome types. As with most things ecological, there exist characteristic regions where one biome type grades into another. For example, coniferous forests also occur in some temperate lowlands. A temperate rain forest extends along the coast all the way from Alaska into northern California. Again, most of the trees are conifers. That forest has some of the world's tallest trees—sitka spruce to the north and sequoia redwoods to the south. In the eastern United States, most of New Jersey's coastal plain, which is sandy, nutrient-poor soil, is dominated by pine barrens. This is a type of scrub forest with grasses and low shrubs growing among the open stands of pine and oak trees. Open stands of pine trees also occur in the coastal plains of North Carolina, South Carolina, Georgia, and Florida, and these forested regions are maintained by frequent fires. Tropical seasonal forests may be apparent where rainfall is heavy (between 125 and 250 cm a year) but occurs in a distinct wet season, as in India or Vietnam. In such monsoon forests, leaves may be shed in the dry season. Another distinct biome type is chaparral, a Mediterranean scrub habitat adapted for fire, which is common along the coastlines of southern Europe, California, South Africa, and southwest Australia.

## 13.7  MOUNTAIN COMMUNITIES

Mountain ranges must be treated still differently. Biome type relies predominantly on climate, and on mountains temperature decreases with increasing altitude. This decrease is a result of a process known as *adiabatic cooling*. Increasing elevation means decrease in air pressure. When wind is blown across the Earth's surface and up over mountains, it expands because of the reduced pressure; as it expands it cools, at a rate of about 10°C for every 1,000 m, as long as no water vapor or cloud formation occurs. (Adiabatic cooling is also the principle behind the function of a refrigerator—freon gas cools as it expands coming out of the compressor.) Higher elevations are also cooler because the less dense air allows a higher rate of heat loss by radiation back through the atmosphere. A vertical ascent of 600 m is roughly equivalent to a trek north of 1,000 km. Precipitation changes with altitude, too, generally increasing in desert elevations but decreasing on the leeward side of slopes, which are in a rain shadow. Approaching clouds have usually dumped all their moisture on the windward side. Thus, biome type may change from temperate forest through taiga and into tundra on an elevation gradient in the Rocky Mountains, and even from tropical forest to tundra on the highest peaks of the Andes in tropical South America.

## 13.8  MARINE COMMUNITIES

Within aquatic environments, biome types can also be recognized, such as rivers, freshwater lakes, and, within saltwater oceans, the intertidal rocky shore, sandy shores, the neritic zone (encompassing shallow waters over continental shelves), coral reefs, seagrass beds, and the pelagic zone or open ocean. In each of these, the physical "climate" is different, varying in such parameters as salinity, oxygen content, current strength, and availability of light.

Marine environments are among the most extensive and uniform on Earth. Marine ecosystems are found over nearly three-quarters of the Earth's surface. Like those in freshwater areas, marine communities are affected by the depth at which they occur. The shallow zone where the land meets the water is called the *intertidal zone*. Beyond the interdial zone is the *neritic zone*, the shallow regions over the continental shelves. Past the continental shelf is the open ocean or *oceanic zone*, which may reach very great depth. This is also often referred to as the *pelagic zone*, at the bottom of which is the sea floor or benthic zone. Just as in freshwater systems, we may also recognize the area of water near the surface where light penetrates as the photic zone, and, below, the dark aphotic zone. Phytoplankton, zooplankton, and most fish species occur in the photic zone. In the *aphotic zone*, production by plants is virtually zero and only a few invertebrates and luminescent fish live.

The area where a freshwater stream or river merges with the ocean is called an *estuary*. It is often bordered by extensive intertidal mudflats or salt marshes. The main plants of estuaries are, in temperate areas, including the United States, *Spartina* salt-marsh grasses. In tropical areas of the world, mangroves replace them along the mudflats. Most of the plant material is not eaten by herbivores but dies and rots on the mudflats. It is decomposed by bacteria, and fungi and pieces of plant material are food for nematodes, snails, crabs, and some fish. Clams, oysters, barnacles, and other filter feeders may live in the estuaries, too. The area

*intertidal — meritic — pelagic — abyssal*

is important as nursery ground for shrimp and fish. Salt marshes are so productive that they are used as feeding grounds for ducks, geese, and other water fowl or migratory birds.

The intertidal zone, where the land meets the sea, is alternately submerged and exposed by the daily cycle of tides (Photo 13.5). The resident organisms are subject to huge daily variation in the availability of seawater and in temperature. They are also battered by waves, especially during storms. In temperate areas, they may be subject to freezing in the winter or very hot temperatures in the summer. At low tides, they may be subject to predation by a variety of animals, including birds and mammals. High tides bring the predatory fishes. There is commonly a vertical zonation. This is most evident on rocky shores that may have three broad zones. The upper littoral is submerged only during the highest tides. The midlittoral is submerged during the highest regular tide and exposed during the lowest tide each day. Life here may be quite rich and consist of green algae, sea anemones, snails, hermit crabs, and small fishes living in tide pools. Competition for space on the rock face may be quite intense. The lower littoral is exposed only during the lowest tide, and diversity and richness of organisms is great. Along sandy and muddy shores, few large plants can grow because the sand or mud is constantly shifted around by the tide. Instead, the ecosystem contains burrowing marine worms, crabs, and small isopods.

Coral reefs exist in warm tropical waters in the neritic zone. (Photo 13.6) This is a conspicuous and distinct biome. Currents and waves constantly renew nutrient supplies, and sunlight penetrates to the ocean floor allowing photosynthesis. Coral reefs are made of organ-

Photo 13.5    *There is a distinct vertical zonation on rocky shores; upper littoral (submerged only during the highest tides), mid-littoral (submerged during normal tides), and lower littoral (exposed only in the lowest tides). Purple sea starts* (Pisaster ochraceus) *in the mid-littoral zone at the Olympic National Park.* (Zipp, Photo Researchers, Inc. 2H2948.)

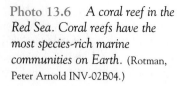

Photo 13.6    *A coral reef in the Red Sea. Coral reefs have the most species-rich marine communities on Earth.* (Rotman, Peter Arnold INV-02B04.)

isms that secrete hard external skeletons made of calcium carbonate. These skeletons vary in shape, forming a substrate on which other corals and algae grow. An immense variety of microorganisms, invertebrates, and fish live among the coral, making the reef one of the most interesting and rich biomes on Earth. Probably 30 to 40 percent of all fish species on Earth are found on coral reefs (Ehrlich 1975). Prominent herbivores include snails, sea urchins, and fish. These are in turn consumed by octopus, seastars, and carnivorous fish.

In the pelagic zone, nutrient concentrations are typically low, though the waters may be periodically enriched by upwellings of the ocean that carry mineral nutrients from the bottom waters to the surface. Pelagic waters are mostly cold, only being warm near the surface. This is where photosynthetic plankton grow and reproduce. Their activity counts for nearly half the photosynthetic activity on Earth. Many scientists have suggested that if phytoplankton productivity is increased, much of the increased carbon dioxide from fossil-fuel burning would be soaked up, and global warming would be slowed. One of the limiting factors seems to be the availability of iron, and huge experimental additions of iron to the Pacific have increased phytoplankton production (Van Scoy and Coale 1994).

Zooplankton, including worms, copepods, shrimplike creatures, jellyfish, and the small larvae of invertebrates and fish graze on the phytoplankton. The biome also includes free swimming animals called *nekton*, which can move against the currents to locate food. The phytoplankton and zooplankton move with the current. The nekton include large squids, fish, sea turtles, and marine mammals that feed on either plankton or each other. Only a few of these live at great depth. Here, the fish may have enlarged eyes, enabling them to see in the dim light. Others have luminescent organs that attract mates and prey. A number of marine animals are migratory, following seasonally available food sources or moving between summer breeding grounds and their winter feeding range.

Recently other communities have been discovered that exist near the openings of deep-ocean volcanic vents (black smokers) in midocean ridges (Ballard 1977). The primary producers there are giant worms, which are nourished by symbiotic chemosynthetic bacteria that produce ATP by oxidizing sulphides and reducing carbon dioxide to organic compounds. In other areas, animals that harbor chemoautotrophic bacteria seem to be the most common. Hessler, Lonsdale, and Hawkins (1988) discuss the common occurrence in the vents near the Philippines of a "hairy snail" that contains bacteria in its gills that oxidize sulphur to produce energy, and Smith (1985) discusses the dense beds of mussels, *Bathymodiolus thermophilus*, that can be found along the Galápagos rift. Water in these areas is often 20°C warmer than in surrounding areas.

## 13.9  FRESHWATER COMMUNITIES

Freshwater habitats are traditionally divided into standing-water **lentic** habitats (from the Latin *lenis*, calm—lakes, ponds, and swamps) and running-water **lotic** habitats (*lotus*, washed—rivers and streams). Natural lakes are most common in regions that have been subject to geological change within the past 20,000 years, such as the glaciated regions of northern Europe and North America. They are also common in regions of recent uplift from the sea, such as Florida, and in regions subject to volcanic activity. Volcanic lakes that are formed either in extinct craters or in valleys dammed by volcanic action are among the most beautiful in the world. Geographically ancient areas such as the Appalachian Mountains of the eastern United States contain few natural lakes.

The ecology of lentic habitats is largely governed by the unusual properties of water. First, water is at its least dense when frozen; ice floats. From a fish's point of view, this property is advantageous because a frozen surface insulates the rest of the lake from freezing. If ice sank, all temperate lakes would freeze solid in winter, and no fish would exist in lakes outside the tropics. Water is at its densest at 4°C. Thus, as long as no water in the lake is colder than 4°C, the warmest water is at the surface, and temperature declines with depth, though not in a linear fashion. Normally several layers are present (Fig. 13.5). There is an upper layer called the **epilimnion** that is warmed by the sun and mixed well by the wind. Below lies the **hypolimnion,** a cool layer too far below the surface to be warmed or mixed. The transition zone between the two is known as the **thermocline.** There are other divisions within a lake based on light availability. The upper layer, where light penetrates, is the **autotrophic** or **euphotic** zone. Below, in darkness, is the profundal zone, where **heterotrophs** live, depending on the rain of material from above for subsistence. The depth of the euphotic zone depends on light availability and water clarity. The level at which photosynthate production equals energy used up by respiration is the

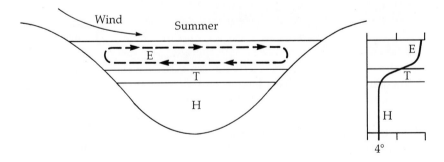

**Figure 13.5**  *Cross-section of lake stratification and profile of temperature with depth. E = epilimnion; T = thermocline; H = hypolimnion. Temperature scale on the right starts at 0 degrees C and increases to the right.*

lower limit of the euphotic zone and is known as the **compensation level** or compensation point. In the summer, in temperate lakes, the compensation level is usually above the thermocline. Oxygen-producing green plants cannot live in the hypolimnion, which becomes oxygen-depleted, a phenomenon known as **summer stagnation.**

The degree of summer stagnation in temperate lakes is partly determined by the degree of "productivity" of a lake. The least productive lakes are termed **oligotrophic.** Such lakes generally have low nutrient contents, largely as a result of their underlying substrate and young geologic age. Young lakes have not had a chance to accumulate as many dissolved nutrients as have older ones. Oligotrophic lakes are relatively clear, and their compensation levels may lie below the thermocline. In this situation, photosynthesis can take place in the hypolimnion, adding oxygen. Low nutrient concentrations keep the algae and rooted plants in the epilimnion sparse, and little debris rains down upon the inhabitants of the hypolimnion. As a result, oligotrophic lakes are clear and often contain desirable fish, such as trout. Even though few nutrients are present in oligotrophic lakes, eventually they do begin to accumulate; sediments are deposited and both algae and rooted vegetation begin to bloom. Organic matter accumulates on the lake bottom, respiration of bottom dwellers increases, the water becomes more turbid, and the oxygen levels of the water go down. Fish such as trout are excluded by bass and sunfish, which thrive in warm water and at low oxygen concentrations. This process of aging and degradation is natural and is termed **eutrophication;** its end result is a eutrophic lake. Eutrophication, however, can be greatly speeded up by human influences, which increase nutrient concentrations by the input of sewage and fertilizers from agricultural runoff. This is often termed *cultural eutrophication*. A measure of eutrophication is given by the dissolved oxygen concentration or **biochemical oxygen demand (BOD).** The BOD is the difference between the production of oxygen by plants and the amount of oxygen needed for the respiration of the organisms in the water. It is normally measured in the laboratory as the number of milligrams of oxygen consumed per liter of water in five days at 20°C. The stratified nature of temperate-zone lakes in summer does not last all year (Fig. 13.6). In the fall, the upper layers cool and sink, carrying oxygen to the bottom

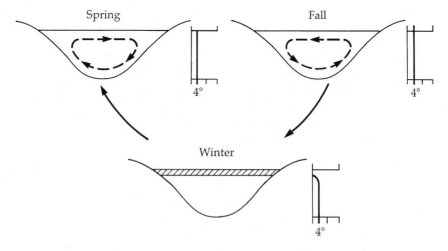

Figure 13.6 *Annual cycle of a lake. See Figure 13.5 for further details.*

of the lake. The lake is thoroughly mixed by this action and by frontal storms, and the thermocline disappears. In winter, the surface usually freezes, and no turnover of water occurs; once again, a gradient is set up. Then, as spring returns, the ice melts and sinks, producing another mixing, the spring overturn. In contrast to temperate lakes, tropical lakes are often **isothermal** (that is, all at one temperature) or at most exhibit only a weak temperature gradient from top to bottom. Little mixing occurs, and deep lakes are generally unproductive, with oxygen-poor, fishless lower depths, as the builders of tropical dams learn (to their chagrin). Worse still, most water from dams is drawn off from the base, meaning that the streams below dams are much less oxygen-rich than those above them. Shallow lakes do not show thermal stratification in any region.

Abiotic phenomena other than nutrient content and dissolved oxygen content may be important to the lake communities also. Water **pH** is particularly relevant, as fish enthusiasts know. For example, some *Poecilia* species, such as live-bearing mollies, breed only in alkaline (high pH) waters, whereas *Hyphessobrycon*, neon tetras, breed only in low pH. Changes in the pH of lake water have been frequent over the past 50 years because of the impact of **acid rain** (Chapter 20). Fish exterminations by acid rain have been recorded in over 300 lakes in the Adirondack Mountains of the northeastern United States.

Some ecologists also differentiate lakes into other zones within them called *littoral, limnetic,* and *profundal,* which show the greatest differences in terms of organisms present. The littoral zone is a shallow well-lit zone extending all around the shore to the depth where rooted aquatic plants stop growing. Usually, the greatest variety of organisms are found in the littoral zone, which is home to plants, snails, frogs, and many fish. The limnetic zone includes the open, sunlit waters down to the depth where photosynthesis can no longer occur. Here we find plankton, communities of floating weakly swimming organisms, mostly microscopic. The phytoplankton include tiny plantlike organisms such as diatoms, green algae, and cyanobacteria. The zooplankton, or tiny animals, include rotifers and copepods. The final zone of lakes, the profundal zone, is the deep, open water below the depth of effective light penetration. Shallow lakes may not have a profundal zone. Detritus, or rotting material, from the limnetic zone sinks to the profundal zone and into the bottom sediments that contain communities of bacterial decomposers. The decomposers release nutrients to the water through their activities.

Lotic or running-water habitats (Photo 13.7) generally have a fauna and flora completely different from those of lentic waters. Plants and animals are adapted so as to remain in place despite an often strong current. Nutrient inputs and phytoplankton blooms do not occur because each would be quickly washed away. Current also mixes water thoroughly, providing a well-aerated regime. Animals of lotic systems are therefore not well adapted for low-oxygen environments and are particularly susceptible to high-BOD (oxygen-reducing) pollutants. Fish such as trout may be present in rivers with cool temperatures, high oxygen, and clear water. In warmer, murkier waters, catfish and carp may be abundant.

When the freshwater environment is considered as a whole, the algae are the most important producers, and the aquatic spermatophytes rank second. Among the animal consumers, the bulk of the biomass is due to aquatic insects, Crustacea, and fish. Other orders rank much lower in importance, though in specific instances, any one may loom larger in the "economy" of the system.

Photo 13.7   *North Fork of Payette River, central Idaho. Running-water habitats contain animals adapted to remain in place despite a strong current.* (Frazier, Photo Researchers, Inc. 2B9590.)

# Summary

1. The common association of the same species in environments with similar physical characteristics led some early ecologists to accept the existence of well-defined communities that contained these species. Other ecologists argued that communities are no more than the sum of their parts, with no special emergent properties.

2. Large-scale communities are referred to as *biomes* and include the tropical forest biome, temperate forest, desert, grassland, taiga, tundra, salt marsh, coral reef, open ocean, freshwater lakes, rivers and streams, and many other minor biome types that grade into these major biome types.

**Discussion Question:** Which do you think is the most meaningful scale on which to examine communities? For example, are the insects and other invertebrates that inhabit rotting logs a more tightly knit and therefore more biologically meaningful type of community than the temperate grassland community?

# Species Diversity
# and Community
# Stability

Wallace (1878) was perhaps the first to point out that animal life was generally more varied and abundant on an area-for-area basis in the tropics than in more temperate latitudes. The same is true for plant life. A naturalist may have an intuitive feel for diversity when observing and comparing two different habitats, but to express this feeling on paper, various indices of species diversity are often used. The simplest measure of diversity is to count the number of species (Photo 14.1); the result is termed the **species richness.** (McIntosh 1967). Diversity of communities can be compared this way, but it is often difficult to identify a true resident species of the community, as opposed to a transient. It may be hard, for example, to distinguish resident and migratory birds. Furthermore, in a sample of 100 individuals from a habitat, the species richness of a community of two species, each of population size 50, would equal that of a community in which the population sizes of the same two species were 1 and 99. In actuality, the first community must be considered more diverse; one would be much more likely to encounter both species there than in the second community.

## 14.1 RAREFACTION

Counts of species numbers are very dependent on sample size: the larger the sample, the greater the expected number of species.

One method of avoiding incompatibility of measurements resulting from samples of different sizes is called *rarefaction*. This involves calculating the number of species expected from each sample if all the samples were reduced to a standard size (such as 1,000 individuals). The

**Photo 14.1**   *Diversity—a mixed herd at a water hole in Mkuze, Zululand. Most measures of diversity include number of individuals and number of species. There are 24 individuals in this photo, and 5 species—warthog, zebra, nyala, impala, baboon.* (Dennis, Photo Researchers, Inc. 7V7646.)

correct formula for doing this was derived independently by Hurlbert (1971) and Simberloff (1972):

$$E(S) = \sum_{i=1}^{s} \left\{ 1 - \left[ \binom{N-N_i}{n} \bigg/ \binom{N}{n} \right] \right\}$$

where $E(S)$ is the expected number of species in the rarefied sample, $n$ is the standardized sample size, $N$ is the total number of individuals in the sample to be rarefied, and $N_i$ is the number of individuals in the ith species in the sample to be rarefied.

The term $\binom{N}{n}$ is a "combination" which is calculated as,

$$\binom{N}{n} = \frac{N!}{n!\,(N-n)!}$$

where $N!$ is a factorial; for example, $5! = 5 \times 4 \times 3 \times 2 \times 1 = 120$

A worked example (following Magurran 1988) is given below. Imagine that two moth traps that have been operated for different lengths of time yield the following data:

| | Number of Individuals | |
|---|---|---|
| Species | Trap A | Trap B |
| 1 | 9 | 1 |
| 2 | 3 | 0 |
| 3 | 0 | 1 |
| 4 | 4 | 0 |
| 5 | 2 | 0 |
| 6 | 1 | 0 |
| 7 | 1 | 1 |
| 8 | 0 | 2 |
| 9 | 1 | 0 |
| 10 | 0 | 5 |
| 11 | 1 | 3 |
| 12 | 1 | 0 |

| | | |
|---|---|---|
| Number of Species ($S$) | 9 | 6 |
| Number of individuals ($N$) | 23 | 13 |

How many species would we have expected in Trap A if it too contained 13 individuals? First, take each species from Trap A, and insert it into the formula. For species 1 in trap A,

$$N = 23$$
$$n = 13$$
$$N_i = 9$$
$$N - N_i = 14$$
$$\binom{N}{n} = \frac{23!}{13!\,(23-13)!}$$
$$\binom{N-N_i}{n} = \frac{14!}{13!\,(14-13)!}$$

therefore:

$$\left\{ 1 - \left[ \left( \frac{14!}{13! \times 1!} \right) \Big/ \left( \frac{23!}{13! \times 10!} \right) \right] \right\} = \{1 - [14/1144066]\} = 1 - 0.00 = 1.00$$

The results for each species are then summed:

| $N_i$ | Expected |
|-------|----------|
| 9 | 1.00 |
| 3 | 0.93 |
| 4 | 0.98 |
| 2 | 0.82 |
| 1 | 0.57 |
| 1 | 0.57 |
| 1 | 0.57 |
| 1 | 0.57 |
| 1 | 0.57 |
| E(S) | 6.58 |

Thus, if Trap A contained 13 individuals, we would expect it to contain 6.58 species—about the same as Trap B. In addition to the rarefaction technique for measuring richness, several richness indices have been proposed:

1. Margalef (1969) $\qquad\qquad\qquad\quad R_1 = (S-1)/\ln N$
2. Menhinick (1964) $\qquad\qquad\qquad R_2 = S/\sqrt{N}$
3. Odum, Cantlon, and Kornicher (1960) $R_3 = S/\log N$

Using these indices, the richness values of the two moth traps listed above becomes:

|  | Trap A | Trap B | Difference in Index values |
|---|--------|--------|----------------------------|
| Margalef | 2.55 | 1.95 | 30.8% |
| Menhinick | 1.88 | 1.66 | 13.2% |
| Odum | 6.61 | 5.39 | 22.2% |

It is clear that the size of the difference in richness between traps A and B depends on the richness index used.

Unfortunately, species counts do not always provide a useful method for making inferences about the underlying community. Two communities can have very different relationships of species importances and yet have the same numbers of species in samples of a particular size. Consider sampling from two communities with 1,000 individuals each, the first with three species divided 33%: 33%: 33% and the second with 11 species divided 90%: 1%: 1%: 1%: 1%: 1%: 1%: 1%: 1%: 1%: 1% (Fig. 14.1). The second community is obviously richer in species, but the first community will appear richer for sample sizes less than 23 individuals. Clearly, there must be some way to measure diversity, taking into account species richness and abundance of individuals of each species. Such measures are commonly called *diversity indices* or *heterogeneity indices*.

**Figure 14.1**   *The relationship between the expected number of species and the sample size for two communities of 1000 individuals.* (After Peet 1974.)

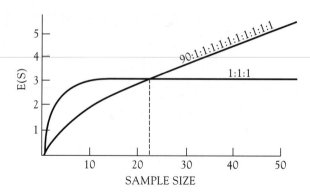

## 14.2  DIVERSITY INDICES

There are two types of diversity measures, dominance indices and information statistic indices.

### Dominance Indices

Dominance indices, so named because they give more weight to common or dominant species, were first introduced into the ecological literature by Simpson (1949) who had become aware of an approach by Yule (1949) to characterize vocabulary used by different authors. Simpson's index gives the probability of any two individuals drawn at random from an infinitely large community belonging to different species as:

$$D = \sum_{i=1}^{S} p_i^2$$

where $p_i$ is the proportion of the individuals in species $i$. For finite or real communities:

$$D = \sum \left( \frac{n_i (n_i - 1)}{N(N-1)} \right)$$

where $n_i$ is the number of individuals in the $i$th species, and $N$ is the total number of individuals. As $D$ increases, diversity actually decreases. To avoid this seeming contradiction, Simpson's index is often expressed as 1–D or 1/D so that increasing values mean increasing diversity. A worked example follows:

| Tree species | Number of individuals |
|:---:|:---:|
| 1 | 100 |
| 2 | 50 |
| 3 | 30 |
| 4 | 20 |
| 5 | 1 |

For the above hypothetical data set on tree abundance and richness in a community, the calculation of $D$ is $[ (100 \times 99) / (201 \times 200) + (50 \times 49) / (201 \times 200) +..... + (1 \times 0) / (201 \times 200) = 0.338$.

Then $1/D = 1/0.338 = 2.96$.

The disadvantage of Simpson's index is that it is heavily weighted toward the most abundant species. Thus, the addition of a few rare species of trees with one individual will fail to change the index. This is obvious by examining the contribution of tree species #5 (above) to the overall value of the index—it is zero. Two other indices do not suffer from this problem. McIntosh (1967) proposed the index

$$D = \frac{N - U}{N - \sqrt{N}} \quad \text{where } U = \sqrt{(\Sigma p_i^2)}$$

where $p_i$ = the proportional abundance of the $i$th species.

Berger and Parker (1970) proposed the index

$$d = N_{max}/N$$

where $N_{max}$ = the number of individuals in the most abundant species.

As with the Simpson index, the reciprocal form is usually adopted. Although the indices of McIntosh and Berger and Parker are not biased toward common species, both are influenced by sample size. There is clearly no one "perfect" index. However, May (1975) has concluded that the Berger–Parker index is one of the most satisfactory diversity measures available.

## Information Statistic Indices

These indices are based on the rationale that diversity in a natural system can be measured in a similar way to the information contained in a code or message.

## Shannon Index

The Shannon index, $H'$ (Shannon and Weaver 1949), assumes all species are represented in the sample and are randomly sampled

$$H' = -\Sigma p_i \ln p_i$$

where $p_i$ is the proportion of individuals found in the $i$th species. A worked example is given below:

| Species | Abundance | $p_i$ | $p_i \ln p_i$ |
|---------|-----------|-------|---------------|
| 1 | 50 | 0.5 | .347 |
| 2 | 30 | 0.3 | .361 |
| 3 | 10 | 0.1 | .230 |
| 4 | 9 | 0.09 | .214 |
| 5 | 1 | 0.01 | .046 |
| Total    5 | 100 | 1.00 | 1.201 |

Remember that the Shannon index has a minus sign in the calculation, so the index actually becomes 1.201, not –1.201.

The most important source of error comes from the failure to include all species from the community in a sample, but this error decreases as the proportion of species represented in the sample increases. Values of the Shannon diversity index for real communities are often found to fall between 1.0 and 6.0. The maximum diversity of a sample is $H_{max}$ when all species are equally abundant. This is equivalent to ln S. We can see how the actual diversity value compares to the maximum possible diversity by using a measure called *evenness*. The evenness of the sample is obtained from the formula

$$\text{Evenness} = H' / H_{max} = H' / \ln S$$

E is constrained between 0 and 1.0. As with $H'$, this evenness measure assumes all species in the community are present in the sample.

## Brillouin Index

When the randomness of a sample cannot be guaranteed, as for example in light trapping where different species of insect are differentially attracted to light, the Brillouin index $H_B$ can be used:

$$H_B = \frac{\ln N! - \sum \ln n_i!}{N}$$

A worked example follows:

| Species | Number of individuals | ln $n_i$! |
|---------|----------------------|-----------|
| 1 | 5 | 4.79 |
| 2 | 5 | 4.79 |
| 3 | 5 | 4.79 |
| 4 | 5 | 4.79 |
| 5 | 5 | 4.79 |
| (S)=5 | N=25 | $\sum (\ln n_i!) = 23.95$ |

$$H_B = \frac{\ln 25! - \ln 23.95}{25} = \frac{58.00 - 3.17}{25} = 2.193$$

Again, evenness is estimated from $E = \dfrac{H_B}{H_{Bmax}}$

Generally, the Brillouin index gives a lower value for the same data than the Shannon index. One major difference is that the Shannon index does not change providing the number of species and their proportional abundance remain constant, while the Brillouin index does. Whether or not this is desirable is open to debate. For example:

|  | | Individuals in | |
|---|---|---|---|
| Species | Sample 1 | Sample 2 |
| 1 | 3 | 5 |
| 2 | 3 | 5 |
| 3 | 3 | 5 |
| 4 | 3 | 5 |
| 5 | 3 | 5 |
| Shannon H′ | 1.609 | 1.609 |
| Brillouin $H_B$ | 1.263 | 2.193 |

A second major difference in the two indices is that the use of factorials in the equations quickly produces huge numbers that are unwieldy. The Shannon index is often chosen for its computational simplicity. Values of diversity are often correlated (Magurran 1988) so that the choice of the "correct" diversity measure may not be as critical as first feared. However, there is no doubt that in certain circumstances, different indices can give different results (Hurlbert 1971).

A comparison of the performance and characteristics of most important indices are outlined in the table (Table 14.1). Some indices are more applicable in some situations than others. For conservation of biodiversity, indices sensitive to rare species might be useful; for comparing communities from very different sample sizes, the Simpson and Berger–Parker indices would not be used.

## Beyond Standard Diversity Indices: Ordinal Indices

Clearly, certain indices of diversity are better suited to some situations than others. In conservation, the importance of rare species needs to be emphasized. However, it is also clear that it would be good to incorporate much more biological information in a measure of a

TABLE 14.1 **A comparison of the effectiveness of different diversity indices. (After Magurran 1988.)**

| Index | Discriminant ability | Sensitivity to sample size | Biased towards richness (R) or dominance of a few species (D) | Calculation | Widely used |
|---|---|---|---|---|---|
| S (species richness) | Good | High | R | Simple | Yes |
| Margalef | Good | High | R | Simple | No |
| Shannon | Moderate | Moderate | R | Intermediate | Yes |
| Brillouin | Moderate | Moderate | R | Complex | No |
| MacIntosh | Good | Moderate | R | Intermediate | No |
| Simpson | Moderate | Low | D | Intermediate | Yes |
| Berger–Parker | Poor | Low | D | Simple | No |

community—especially when trying to use these indices as measures of which types of community to conserve. For instance, should we give more weight to a taxonomically isolated species in a community than to a species that is not much different from its sister species? Should we given more weight to a more "weighty" species—that is, a large, rare predator more than a small, rare nematode? Is this just a question of human preference, or would the larger, rare predator be more important in the community, using up more energy or acting as a keystone species? Indices that attempt to rank species in importance can be thought of as *ordinal indices*. Indices that treat all species as equals are *cardinal indices*. All the indices discussed so far have been cardinal.

Counting the number of animal or plant species (S) in a weight class and the numbers of individuals (N) in a weight class and using ordinal indices within weight classes may be a useful next step up from pure cardinal indices and has been done for insects (Morse, Stork, and Lawton 1988). This allows us to at least tell at which size class most diversity is located. Some species, of course, could be counted in more than one weight class because they grow in size as they age. This adds further complications to the calculation of indices.

Vane-Wright, Humphries, and Williams (1991) have explored how to weight "taxonomically rare" species in diversity indices (Fig.14.2).

Species counts to be used in a diversity index are multiplied by a weighting factor. In a cladogram, the species that have the most branches between the stem and the tip are set to 1; then the sister group is given a score (W) equal to the sum of all the other branch values—in this case, 8. However, this may overweight the value of the taxonomically distinct species. For example, for reptiles, May (1990) pointed out that the two living species of tuatara lizards (Photo 14.2) would be weighted equally to the sum of all 6,000 other species of snakes and lizards. A second approach is based on an information index (l), based on the number of branchings in the tree that include the species whose characters are being measured. The sum of the l values is then divided by the value for the individual species itself. This contribution is then expressed as a percentage. Vane-Wright, Humphries, and Williams used this technique on the *Bombus subrircus* group of bumblebees, distributed worldwide. If a simple

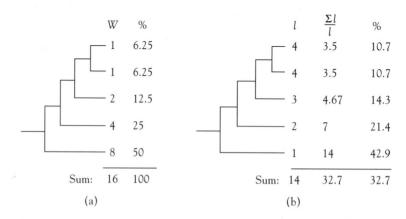

**Figure 14.2**   *Two ways of measuring taxonomic distinctiveness, (a) via common ancestry, where w represents a weighted score, and (b) measures of genetic distance apart, where l represents an information index.*

**Photo 14.2**   *Sphenodon punctuatus, the tuatara, from New Zealand, the most taxonomically isolated lizard in the world. How can we weight rare or unusual species in diversity indices?* (Concalosi, Peter Arnold 95776.)

species count were used, then maximum diversity occurred in Ecuador (10 species = 23 percent of world total) but if taxonomic distinctness was accounted for, maximum diversity occurred in the Gansu region of China (23 percent of world total).

## 14.3  RANK ABUNDANCE DIAGRAMS

Description of whole communities by one statistic, either richness or diversity, runs the risk of losing much valuable information. A more complete picture of the distribution of species abundances in a community is to plot the proportional abundance (usually on a log scale) against rank of abundance. The abundance of the most abundant species appears on the extreme left and the rarest species on the extreme right (Fig. 14.3). A rank-abundance diagram can be drawn for the number of individuals, biomass, ground area covered (for plants), and other variables. There are several different forms of rank-abundance diagrams: geometric, log series, log-normal, and broken stick. Species diversity can be calculated for each of the rank-abundance relationships, but the calculations are too complex for this level and the reader is referred to Magurran (1988). A biological explanation for the geometric series is that the first

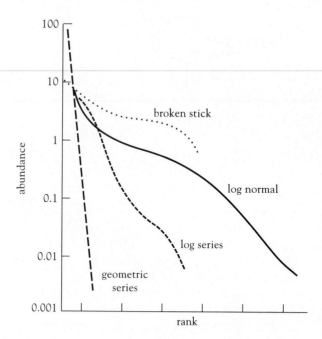

**Figure 14.3**   *Rank abundance plots illustrating the typical shape of four species abundance models: geometric series, log series, log normal, and broken stick. In these graphs the abundance of each species is plotted on a logarithmic scale against the species' rank, in order from the most abundant to least abundant species. Species abundances may in some instances be expressed as percentages to provide a more direct comparison between communities with different numbers of species.*

species colonizing an area appropriates a fraction of the resource and by competitive interaction preempts that fraction. The second colonizing species preempts a similar fraction of the remaining resource, and so on with further colonists. Fits to this model have been found for plants from a subalpine fir-forest community (Whittaker 1975) and for benthos in a polluted fjord (Gray 1981). A more equitable distribution is represented by the "broken-stick" model of MacArthur (1957). Here the resources (the stick) are divided at the same time and at random into segments over the whole stick length (implying instantaneous colonization by all species or more-intense competition). The segments are ranked into decreasing length order. The abundance of each species is assumed to be proportional to the length of the stick segment. Birds, fish, ophiuroid worms, and predatory gastropods were found to fit the broken-stick model well (MacArthur 1960; King 1964). An ecological explanation for the log-series and log-normal distribution refers to the broken-stick model, but the stick is broken sequentially not instantaneously. The stick is broken at random into two parts; then one part is chosen at random and broken again giving three parts. One of the three is chosen at random and broken again, and so on. After a large number of breaks, the product is a log-normal distribution of lengths.

The log-normal distribution owes its place in ecology to Frank Preston (1948), who obtained a normal "bell"-shaped curve when plotting number of species and log-species

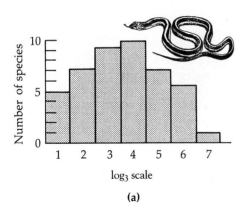

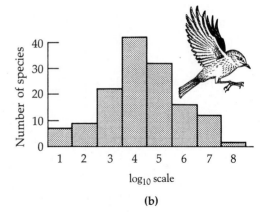

log$_3$ scale

**(a)**

log$_{10}$ scale

**(b)**

**Figure 14.4** *The log-normal distribution. The "normal," symmetrical bell-shaped curve is achieved by logging the species abundance on the x-axis. A variety of log bases can be used. (**a**) Successive classes refer to treblings of numbers of individuals. Thus in this example showing the diversity of snakes in Panama, the upper bounds of the classes are 1, 4, 13, 40, 121, 364, and 1093 individuals. Although widely used by Williams (1964), log$_3$ is rarely employed today. (**b**) log$_{10}$. Classes in log$_{10}$ represent increases in order of magnitude 1; 10; 100; 1000; 10,000; 100,000. The choice of log base is most appropriate for very large data sets, as for example in this case, the diversity of birds in Britain. In all cases the y-axis shows the number of species per class.* (Modified from Williams 1964.)

abundance on the x-axis (Fig. 14.4) Preston also recognized the existence of the truncated log-normal distribution in which plots of individuals per species on a logarithmic scale, against the numbers of species followed a normal distribution but were truncated to the left of the mode (Fig. 14.5).

The truncation was explained as being due to species that were represented in the habitat but not in the sample: if larger samples were taken, more species would be obtained, and the mode would move to the right. Many authors had termed this truncated log-normal a log-series distribution, but Preston's theory showed how more-intense sampling would often reveal the true distribution—a log-normal distribution.

Other authors have suggested that the **lognormal distribution** arises in communities where the total numbers of species are large, and the relative abundance of these species is thought to be determined by many factors operating independently (May 1975). It is therefore the expected statistical distribution for many biological communities. By contrast, the geometric series arises with communities of relatively few species and is thought to occur where a single environmental factor is of dominating importance. That the lognormal distribution is due to many factors operating independently has also been argued mathematically (Ehrlich and Roughgarden 1987). If a random-number generator (or a phone book) is used to generate numbers from 1 to 1,000, the frequency distribution would be a flat line (a uniform distribution): each number would have an equal probability of occurrence. Now, if pairs of these numbers are taken, and their averages, x, are taken, the distribution of x̄s will be more normally distributed than the original numbers, the x̄s themselves. In fact, the distribution of

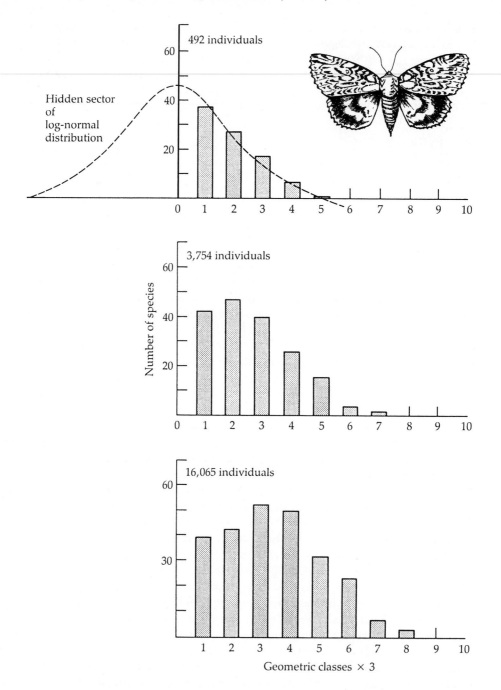

**Figure 14.5** *Distribution of abundance in moths captured by light traps in England. In the top figure, the true distribution of abundance is hidden behind the veil of the y axis. As sampling becomes more intense, the distribution pattern moves to the right to reveal the distribution of rarer species.* (Modified from Williams 1964.)

$\bar{x}$s will look like a triangle with a peak at 500. If $\bar{x}$s are generated from triplets of numbers, the distribution of $\bar{x}$ would appear even more normally distributed. As $n$, the number of $x$s averaged to form $\bar{x}$, increases toward infinity, the distribution of $x$ tends toward an exact normal distribution. For practical purposes, the distribution is statistically indistinguishable from the normal distribution where $n \geq 10$. This phenomenon is described by the **central limit theorem** of statistics, which states that a normal distribution results for a quantity whose values are determined by the product of many small random effects; thus, log-normal series tend to indicate randomness in nature. Finally, Gray (1987) has suggested that early successional stages in communities fit the geometric series; then, as succession proceeds, the log-normal series fits the data well, and finally, at climax, the community may return to a geometric series.

## 14.4   DIVERSITY GRADIENTS

Questions of diversity indices notwithstanding, it has long been known that the diversity of many plant and animal taxa is higher in the tropics than in more temperate regions. For example, the number of ant species in Alaska is 7, in Iowa 73, in Cuba 101, in Trinidad 134, and in Brazil 222. There are 293 species of snakes in Mexico, 126 in the United States, and only 22 in Canada. More than 1,000 species of fish have been found in the Amazon, whereas Central America only has 456, and the Great Lakes of North America have 142. Species diversity of North American mammals increases from Arctic Canada to the Mexican border (Fig. 14.6a), and so does the diversity of the birds (Fig. 14.6b). It should also be noticed that there are more birds than mammals in any given region of the United States. Bird-species richness increases 12-fold in the 60° of latitude shown, whereas mammal diversity only increases eight times. Further examples of latitudinal gradients in species diversity are given in Table 14.2. Sometimes, a reverse pattern is found, such as those for the sandpipers, family Scoloparidae, whose diversity increases toward the Arctic regions (Cook 1969); for aphids, whose diversity is highest in temperate realms; and for helminth parasites of whales and sea lions (Rohde 1982). Diversity of trees in North America is also not well linked to latitudinal gradients (Currie and Paquin 1987) (Fig. 14.7). Trees do not grow well in deserts of the U.S. Southwest, despite decreases in latitude. Many theories for such temperate-to-tropical progressions have been advanced, and although it is difficult to "prove" any one of them, some of the more feasible are listed below:

### Biotic Explanations

*Spatial-heterogeneity theory:* Generally, there are more plant species in the tropics, which in turn support higher numbers of herbivorous animal species and hence carnivores. Diversity of vegetation increases numbers of herbivores in two ways: by increasing the numbers of monophagous herbivores directly, and by creating a more diverse architectural complexity, providing more niches to occupy. MacArthur and MacArthur (1961) related bird-species diversity to both plant-species diversity and foliage-height diversity. (Birds usually visit many different plants for food, so their diversity is often linked to that of their plants. In contrast, insects often spend their entire lives on one type of plant.) Of course, the

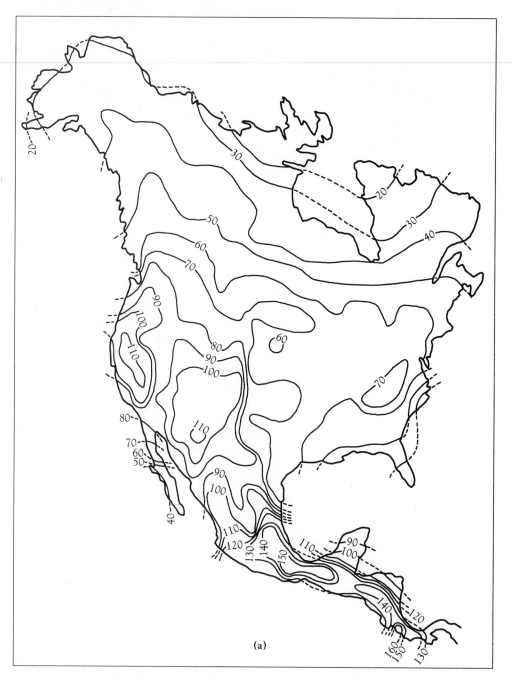

(a)

**Figure 14.6a**  *Geographic variation in species diversity of North America. (**a**) Mammals. (**b**) Land birds. Note the pronounced latitudinal gradients in both groups and the high diversity in the southwestern United States and northern Mexico, a region of great topographic relief and habitat diversity.* ([a] Redrawn from Simpson 1964; [b] redrawn from Cook 1969.)

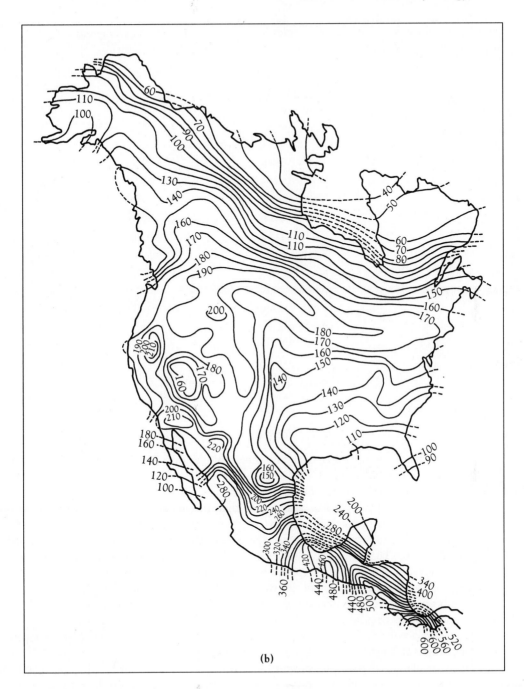

**(b)**

Figure 14.6b

TABLE 14.2 *Latitudinal gradients in species diversity in various taxonomic groups. (From Brown and Gibson 1983.)*

| Taxon | Region | Latitudinal range | Range in species richness |
|---|---|---|---|
| Land mammals | North America | 8° – 66°N | 160 – 20 |
| Bats (Chiroptera) | North America | 8° – 66°N | 80 – 1 |
| Quadrupedal land mammals (all orders except Chiroptera) | North America | 8° – 66°N | 80 – 20 |
| Breeding land birds | North America | 8° – 66°N | 600 – 50 |
| Reptiles | United States | 30° – 45°N | 60 – 10 |
| Amphibians | United States | 30° – 45°N | 40 – 10 |
| Marine fishes | California coast | 32° – 42°N | 229 – 119 |
| Ants | South America | 20° – 55°S | 220 – 2 |
| Calanid Crustacea | North Pacific | 0° – 80°N | 80 – 10 |
| Gastropod mollusks | Atlantic coast of North America | 25° – 50°N | 300 – 35 |
| Bivalve mollusks | Atlantic coast of North America | 25° – 50°N | 200 – 30 |
| Planktonic Foraminifera | World oceans | 0° – 70° | 16 – 2 |

spatial-heterogeneity theory does not address the reason for the higher numbers of plant species themselves. Presumably it is due to microhabitat or topographical diversity, although this hypothesis remains untested and circular.

*Competition theory:* It has been argued that, in temperate climates, natural selection is controlled mainly by harsh physical extremes and that species are generally *r* selected. In the more-constant tropical temperatures, species are thought to become more *K* selected, to compete more keenly, and to interact more. This keen competition would reduce species-niche breadth, allowing more species to pack along the resource axes (Dobzhansky 1950). Again, no critical evaluation of this theory has been performed, and niche parameters have not been measured for a sufficient variety of species groups to determine how niche breadths are affected by tropical-to-polar gradients.

*Predation theory:* This theory, proposed by Paine (1966), runs contrary to the competition hypothesis and argues that there are more predators and parasites in the tropics and that these hold populations of their prey down to such low levels that more resources remain and competition is reduced, allowing more species to coexist. The increased diversity, in turn, promotes more predators. Paine provided evidence to support this theory from studies on the intertidal communities of the U.S. Northwest coast at Milkkaw Bay and Tatoosh Island in Washington State, where the food web was fairly constant and the starfish *Pisaster* was the top predator:

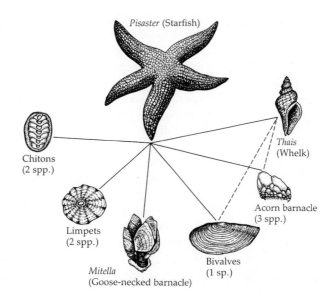

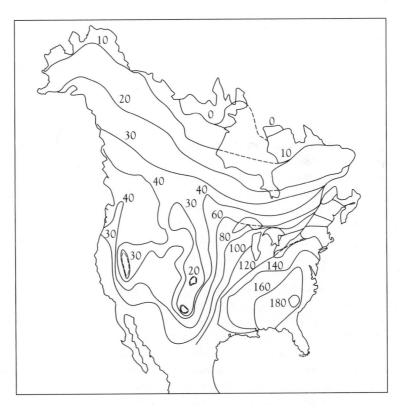

**Fig. 14.7** *Latitudinal gradient and tree species richness in North America.* (After Currie and Paquin 1987.)

*Thais* preyed on bivalves and acorn barnacles. *Pisaster* preyed on those groups, on *Thais* and on chitons, limpets, and *Mitella* as well. After removal of *Pisaster* from a section of the shore, diversity decreased from 15 to 8 species. A bivalve, *Mytilus,* increased, crowding out other species. In unmanipulated sections of shore, *Pisaster* tended to remove *Mytilus* and other species, preventing any one species from monopolizing space. Removal of any other single species from the system would not affect species diversity so drastically as *Pisaster* removal. For this reason, species like *Pisaster* are often termed **keystone species,** by analogy with the keystone that holds all the other stones in an arch in place. Top predators commonly specialize on the most abundant prey, developing a **search image** for it. Such a phenomenon, wherein predation allows the **coexistence** of more prey species, was noted even by Darwin (1859), who observed more grass species coexisting in areas grazed by sheep or rabbits than in ungrazed areas. Otters act as keystone species in the kelp beds of the Pacific coast (Estes, Jameson, and Rhode 1982), and elephants may function as keystone species on the African savannas. Keystone species are not always predators. In the southeastern United States, gopher tortoises have been regarded as keystone species because their burrows provide homes for an array of mice, possums, frogs, snakes, and insects. Without tortoises' burrows, many of these creatures would be unable to survive in the sandhill areas where they are found (Sunquist 1988).

Of course, for such a system to explain tropical diversity, the predation would have to be intense on the majority of species at all **trophic levels,** and few data are available as yet to test such an idea. Indeed, whether or not *Pisaster* acts as a keystone species even on most of the shores of the Pacific coast has recently been brought into question. On some other Pacific rocky intertidal sites, other processes may instead be operating (Foster 1991, but see reply by Paine 1991).

*Animal-pollinators theory:* In the tropics and other humid parts of the world, winds are less frequent and of lower intensity than in temperate regions. This effect is accentuated by dense vegetative cover. Therefore, most plants are pollinated by animals: insects, birds, and bats. Even some grasses that are typically wind-pollinated throughout most of the world are probably pollinated by insects in the tropics. Usually, and particularly in the case of bee pollinators, associations build up between plants and specific pollinators, increasing the reproductive isolation between plant populations with a consequent increase in speciation rates. Coevolution of plants and pollinators then ensures high animal-pollinator speciation as well.

## Abiotic Explanations

*Ecological time theory:* Proponents of this theory argue that communities diversify with time and that temperate regions have younger communities than tropical ones because of recent glaciations and severe climatic disruption. According to this view, species that could possibly live in temperate regions have not migrated back from the unglaciated areas into which the Ice Ages drove them, or resident species have not yet evolved new forms to exploit vacant **niches.** Lake Baikal in the Soviet Union is an ancient unglaciated temperate lake and contains a very diverse fauna; for example, there are 580 species of benthic invertebrates (Kozhov 1963). A comparable lake in glaciated north Canada, Great Slave Lake, contains only four species in the same zone (Sanders 1968). In a useful analogy, tropical and temperate habitats can be compared to equal-sized libraries. The numbers of species, books in each library, is dependent on different things. In the tropics, the library is full, and size of books and available shelf space dictate the number of volumes held. In a temperate situation, there is plenty of available shelf space, and the number of books depends on their rate of purchase by the library and the length of time since the library opened.

Another test of the time hypothesis has been provided by areas recently colonized by plants and animals, where species richness can be assessed over time. Since the end of the last Ice Age, trees have recolonized Britain, and in the last 2,000 years, humans have introduced trees as well. Tree-species richness is much lower in glaciated Europe than in North America, much of which was unglaciated. Southwood (1961) was the first to examine the number of insect colonists associated with each tree, and he found good correlations of insect diversity with length of tree tenure in Britain. Strong (1974ab), however, showed that insect abundance was better correlated with the area over which a tree species could be found (see the area hypothesis, below). Furthermore, Strong, McCoy, and Rey (1977) provided more detailed information on another system, sugarcane, its pest loads, and the dates of introduction into at least 75 regions of the world during the last 3,000 years. They found no support for the time hypothesis but good support for the area hypothesis, though there have been several criticisms of this argument, usually based on the poor quality of data available on areas of sugarcane plantings (for example, see Kuris, Blaustein, and Alio 1980). Others (Birks 1980) have argued for the time hypothesis using improved estimates of time of colonization of British trees from radiocarbon dating of pollen profiles, but the time hypothesis still looks weak. In particular, there seems no reason why marine organisms couldn't easily shift their distribution patterns in times of glaciations; yet, the diversity gradient still exists in marine habitats.

*Climatic stability:* Temperate organisms are, of necessity, well adapted to harsh physical conditions and to climatic change. The tropics are climatically much more stable, and organisms there are less well adapted to cope with small temperature and moisture fluctuations. Because in fact there are more minor fluctuations in the tropics than major ones in temperate regions, the climatic-stability hypothesis argues, more species will specialize and adapt to each of the many small resultant niches in the tropics. In support of this argument are data which show more species of benthic invertebrates in deep-sea environments, where conditions are relatively stable, than in shallow waters, where physical factors are much more variable (Sanders 1968). However, there are often extreme variations in temperature, salinity, and currents in tropical shallow waters, such as coral reefs, and these have extremely high diversity.

*Productivity theory:* This theory has been termed the **species-energy hypothesis** by Wright (1983), who first proposed its use. The theory proposes that greater production results in greater diversity; that is, a broader base to the energy pyramid permits more species in that pyramid. A common modification is that there is "room" for obligate fruit-eating birds (such as parrots) or raptorial reptile eaters in the tropics, but not in temperate regions (Orians 1969). Fruits appear year-round in the tropics, but a parrot would starve in a temperate winter. It is also argued that a longer growing season not only increases productivity, but also allows component species to partition the environment temporally as well as spatially, thereby permitting the coexistence of more species. Thus for species with annual life cycles, such as insects, some species could feed on leaves early in the year and others later. Currie and Paquin (1987) showed that species richness of trees in North America is best predicted by evapotranspiration rate (see Fig. 5.17). Realized annual evapotranspiration is correlated with primary production and is therefore a measure of available energy. Currie (1991) later expanded his arguments to show how richness in North American birds, reptiles, amphibians, and mammals was also linked to energy. Turner, Gratehouse, and Carey (1987) demonstrated a correlation between the diversity of British butterflies, exo-thermal species, and sunshine and temperature during the months they were on the wing, again suggesting a relationship between energy and species diversity. A simple prediction from this theory is that the number of resident species in seasonal habitats should change according to the seasons. Turner, Lennon, and Lawrenson (1988) have shown that this is true for British birds. The number of birds present in Britain in the winter is less than that in summer, and this pattern is consistent with the amounts of energy present. However, some tropical seas have low productivity but high diversity. Also, eutrophic lakes have high productivity but low diversity. Finally, coastal salt marshes have high productivity but low diversity.

*Area theory:* This idea is based on the notion that in larger areas the chances of isolation between populations increase, with corresponding increases in the chances of speciation (Terborgh 1973). It has also frequently been shown that larger areas support more species. Thus, large areas of climatic similarity will have greater species diversity. On a worldwide scale, there is a symmetry of climates between polar regions and temperate areas, but only in the tropics do we see the symmetrically opposite climates adjacent, creating one large area. In a slight variation of this idea, Darlington (1959) argued that most dominant species evolved in the largest areas (which he also argued were the equatorial zones) and diffused out, creating a species-diversity gradient. Neither theory, however, seems able to explain why, if diversity is linked to area, there should not be more species in the vast contiguous land mass of Asia. North America has a larger area than Central America but has many fewer species of birds and mammals. Neither the surface nor the shelf areas of the northern Pacific are greater than those of the northern Atlantic; nevertheless, the northern Pacific has a three-times-larger relative species diversity of parasitic Monogenea.

Of course, these eight theories are not exhaustive or mutually exclusive and can be combined in many permutations. Nevertheless, there is a strong tendency among ecologists to search for a common cause. Unfortunately, as pointed out by Rohde (1992), most "biotic" explanations are circular; that is, explanations that invoke increased competition or predation or disease are secondary explanations (Table 14.3). A primary explanation is still needed to invoke why these mechanisms might themselves be more or less important in certain areas. As for the "abiotic" explanations, there may be good correlations of various abiotic variables, like evapotranspiration rates, with species richness but there is no reason why increased productivity promotes diversity and not simply higher population densities of just a few species. Furthermore, there often are numerous exceptions to these patterns. Rohde concluded that the best explanation of the temperate-tropical diversity gradient lies in terms of "evolutionary speed," which creates more species in the tropics. Higher energy levels in the tropics promote:

   a. shorter generation times,
   b. higher mutation rates,
   c. acceleration of selection leading to fixation of favorable mutants.

TABLE 14.3 *Explanations of latitudinal gradients in species diversity and their flaws. (After Rohde 1992.)*

| Type of Flaw | Reason |
|---|---|
| Circular | Competition |
| | Predation |
| | Disease |
| | Spatial heterogeneity |
| Insufficient evidence | Climatic stability |
| | Productivity |
| | Area |
| | Ecological time |
| None as yet | Evolutionary speed |

As yet, there is insufficient evidence to support or refute this theory (but see Jablonski 1993).

Why the large concern over species diversity? There are at least two major reasons. First, species diversity is the basis for many fields of scientific research and education (U.S. Congress Office of Technology Assessment, 1987). A variety of species provides a variety of raw materials for biomedical research and drug synthesis. Desert pupfishes, found only in the U.S. Southwest, tolerate salinity twice that of sea water and are valuable models for research on human-kidney disease. The armadillo is one of only two nonhuman species known to contract leprosy. These animals now serve as research models with which to find cures for human diseases (see Myers 1983; Chapter 18 in this book). Second, most ecological research is based on trying to find order in seeming chaos, to discern patterns and create holistic theories from a mass of biological data. One prominent idea in ecology is that diversity begets stability. More diverse communities may be more stable and less subject to change. Stable communities are perhaps easier for humans to manage. Addressing this problem requires examination of the concept of **stability.**

## 14.5 STABILITY

A community is often seen as stable when no change can be detected in the identities or population sizes of member species. The frame of reference for detecting change may encompass a study of a few years or, preferably, a few decades. Long-term data on a few communities shows a "constancy" over time. Bird watching is a popular pastime, and long-term data exist for certain areas that shows a stability over time (Fig. 14.8). However, such long-term data does obscure some potential changes in the community—such as the fact that some species go extinct or emigrate and others immigrate. A definition of *stability* usually infers a stable equilibrium for populations or communities. A community in equilibrium for a long time is likely to be stable. On the other hand, a stable community need never be at equilibrium. For example, a cycle of predator and prey may exhibit a stable periodicity over years—yet equilibrium in numbers is never reached (Williamson 1987).

Despite the apparent simplicity of the stability concept, stability is actually difficult to define. There are in fact many different ways of thinking about stability (Orians 1975):

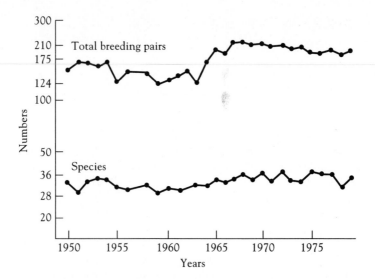

Figure 14.8    *Time plots of the total number of breeding pairs, and the number of species breeding, on the same logarithmic scale, for the birds of Eastern Wood, Brookham Common. The relative lack of change is taken to represent stability in the community.*  (After Williamson 1987.)

a.  Resistance to change.
b.  Resilience—refers to the return to equilibrium after a perturbation:
    (i)   elasticity—how quickly a community can return
    (ii)  amplitude—how much a disturbance it can return from.

These independent concepts are sometimes correlated and sometimes not. Lakes are often only weakly resistant, and they concentrate pollutants from a variety of sources. Deserts on the other hand are very resistant. Rivers are resilient because the fast-flowing water often cleanses them quickly.

## Evidence for Stability

To compare stabilities of different communities it is necessary to:

1.  a.  Determine the stable point of a community  ⎫
    b.  Apply a force                             ⎬  This permits a
    c.  See if the community changes              ⎭  measure of resistance

or

    d.  Determine the stable equilibrium point    ⎫
    e.  Apply a disturbance                       ⎬  This permits a
    f.  Measure the time the community takes to    ⎭  measure of resiliance
        return to "normal"

Both of these procedures are difficult to do because:

1. It is hard to ensure the community is stable when measured.
2. There is often insufficient time between natural disturbances (for example, storms or fires) for communities to return to stable points.

Simberloff and Wilson (1969) defaunated mangrove islands in the Florida Keys and examined recolonization. Heatwole and Levins (1972) argued that the trophic structure of the community was remarkably constant following recolonization, despite the fact that species identity and taxonomic composition varied widely on recolonized islands. They took this as strong evidence of a stable community. However, Simberloff (1976a) himself noted that the recolonization patterns observed could not be shown to be different from random. Second, the concept of stability refers to the "bouncing back" of perturbed communities, not their recolonization patterns following extinction, when completely different rules (often called assembly rules) may apply.

On the other hand, data from counts of wild game in East Africa did indicate the existence of stable communities (Prins and Douglas-Hamilton 1990). Data from counts of 13 species of large herbivores from 1959 through 1984, with counts every three or four years, showed little variability. Only buffalo numbers change significantly. Five functional groups could be distinguished:

1. Buffalo
2. Elephants as grazers
3. Elephants as browsers
4. Other grazers
5. Other browsers.

Under this scheme, there was no change in numbers in groups during the 25 years of the study, except for buffalo, suggesting an overall constancy of biomass in each of these herbivore groupings or guilds. The only exception was a gradual increase in buffalo numbers following an outbreak of rinderpest disease in 1959. Prins and Douglas-Hamilton implicated competition as a main mechanism contributing to stability.

## Multiple Stable States

Must a community return exactly to how it was before following a disturbance for it to be labeled *stable*? If, following an oil spill, populations of 90 of 100 intertidal organisms declined significantly and only 85 of these populations recovered, would the community have recovered, or would it be a different community? Also if a "recovered" community had different abundances of species of those of the original pre–oil-spill community, would we consider the community to have changed? In other words, can communities exist in multiple stable states? Connell and Sousa (1983) wrestled with these ideas and concluded that there are few multiple stable states in nature. Their three main arguments were:

(1) Assemblages occupying an area before and after a disturbance are interpreted as alternative stable states when, in fact, the physical variables at the site have been changed, for example, increased nutrient levels, agricultural run-off, water temperatures, sediment loads, and hunting or fishing effort.

(2) One state is artificially maintained by the constant addition or removal of predators or competitors by humans. This is true of some freshwater-fish communities in England and the Great Lakes.

(3) Data are not of taken for more than a generation or a year so as to establish long-term stability.

## Diversity and Stability

Does diversity promote stability? The idea here is that the effects of severe factors (catastrophes) will be cushioned by large numbers of interacting species and will not produce as drastic effects on species in the community as they would an individual species.

There is evidence for a link between diversity and stability:

1. Laboratory experiments by Gause (1934) confirmed the difficulty of achieving numerical stability in simple systems.
2. Small, faunistically simple islands are much more vulnerable to invading species than are continents. Most natural species on remote oceanic islands have been selected for high dispersal ability, not competitive dominance.
3. Outbreaks of pests are often found on cultivated land or on land disturbed by humans, both of which are areas with few naturally occurring species.
4. Tropical rain forests do not have insect outbreaks like those common in temperate forests.
5. Pesticides have caused pest outbreaks by the elimination of predators and parasites from the insect community of crop plants.
6. In a review of 40 food webs, the complexity of food webs in stable communities have been found to be greater than the complexity of food webs in fluctuating environments (Briand 1983).

There is also evidence against a link between diversity and stability:

1. The fluctuations of microtine rodents (lemmings, voles, and so on) are as violent in relatively complex temperate ecosystems as they are in simple Arctic environments.
2. Goodman (1975) has argued that the stability of tropical ecosystems is a myth and that there are reports of cases in which insects nearly completely defoliated Brazil-nut trees and monkeys succumbed in large numbers to epidemics.
3. Rain forests seem particularly susceptible to human-made perturbations (May 1979).
4. Agricultural systems may suffer from outbreaks, not because of their simple nature, but because their individual components are often bioengineered and have no coevolutionary history whatever, in complete contrast to the long associations evident in forest biomes (Murdoch 1975).

5. May (1973) argued that increasing complexity actually reduces stability in hypothetical models. When trophic links are assembled at random, the more diverse communities are more unstable than the simple communities. May cautioned ecologists that, if diversity causes stability in nature, it does not so as a direct consequence of the mathematics of the situation.

What is needed to sort out the link between stability and diversity is a rigorous experimental approach. The closest study to do this so far has been the laboratory study of Hairston et al. (1968) with microorganisms. *Paramecium*, when cultured with bacterial prey, showed less tendency to go extinct with only one species of prey present (extinction rates = 32 percent of cultures) than with two or three prey species available (extinction rates = 61 percent and 70 percent respectively). When the number of *Paramecium* species was increased to three, the results depended on which particular species was added to which other two. It was evident that species were not simply interchangeable numerical units. Finally, when a third trophic level was added, in the form of the predator protozoa *Didinium* and *Woodruffia*, there was a further decrease in stability because *Paramecium* were usually forced to extinction regardless of how may species of *Paramecium* were present or whether one or two predator species were present. In these simple systems, diversity did not automatically lead to stability. Most important, species cannot always be regarded as simply interchangeable. This point casts doubt on mathematical models of stability, all of which treat species as simple equivalents.

What then is the conclusion about how communities are organized around such concepts as stability and diversity? The older, more conventional view can be termed the **equilibrium** hypothesis, in which local population sizes fluctuate little from equilibrium values, which are thought to be determined by predation, parasitism, and competition. In this view, communities are stable and disturbances are damped out; species diversity is determined by the diversity of available niches. In the more modern **nonequilibrium** hypothesis, species composition is viewed as constantly changing and never in balance. Stability is elusive, and persistence (the length of time for which a community exists) and resilience are the most apt measures of community behavior. The most probable mechanism for the nonequilibrium hypothesis is the intermediate-disturbance hypothesis of Connell (1978) (Fig. 14.9), in which highest diversity is maintained at intermediate levels of disturbance. The rationale is that, at high levels of disturbance (for example, where environmental variables are extreme—low temperature, high temperature, low rainfall, high wind, high frequency of fire, and the like), only good colonists, *r*-selected species, will survive, giving rise to low diversity. This theory also predicts that, at low rates of disturbance, competitively dominant species will outcompete all other species and only a few *K*-selected species will persist, again giving low diversity. The most diverse communities lie somewhere in between. Natural communities seem to fit into this model fairly well. Tropical rain forests and coral reefs are both examples of communities with high species diversity, and both were used as evidence for the old equilibrium hypothesis. However, Connell (1978) has pointed out that coral reefs maintain their highest diversity in areas disturbed by hurricanes. He has also argued that the richest tropical forests occur where disturbance by storms or people has been documented. Reice (1994) argues that disturbance operates in habitats of all types—naturally disturbed areas are nearly always more diverse than undisturbed areas.

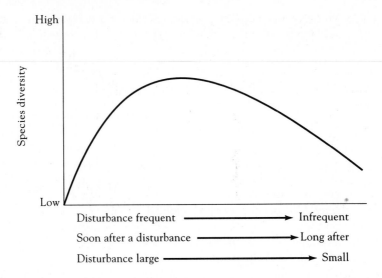

Figure 14.9    The intermediate-disturbance hypothesis of community organization. *The species composition of a community is seldom in a state of equilibrium, and high species diversity can be maintained only at intermediate levels of disturbances such as fires or windstorms.*    (Redrawn from Connell 1978.)

Sousa (1979) has also provided an elegant experimental verification of the intermediate-disturbance hypothesis in a marine intertidal situation. He found small boulders, easily disturbed by waves, to carry a mean of 1.7 species of sessile plant and animal species; large boulders, rarely moved by waves, had a mean of 2.5 species; and intermediate-sized boulders, 3.7 species. Sousa then cemented small boulders to the ocean floor to show that these results were a result of rock stability, not rock size. Furthermore, diversity not only changed in response to disturbance but often did so in a predictable way, in a sequence of events known as **succession.**

# Summary

1. Biodiversity may be expressed in many ways: as species richness (numbers of species) or by various indices which take numbers of species and numbers of individuals of each species into account. Dominance indices include Simpson's index, McIntosh's index, and the Berger–Park index. Information statistic indices include the Shannon index and the Brillouin index.

Each of these indices may give slightly different values from other indices, and each is based on certain assumptions. Choosing the appropriate index for the appropriate situation depends on what sort of question is being asked.

2. Most diversity indices can be referred to as cardinal indices: they treat every species as equal. However, every species is not always equal. For example, in striving to preserve biodiversity, rare and unusual species might be viewed as more important than common or pedestrian species. Some ecologists give more weight to rare species in calculating diversity. Indices that attempt to weight rare species, or any other type of species, are known as *ordinal indices*. The development of ordinal indices is undergoing much development at the present time.

3. Diversity indices attempt to describe whole communities with just one statistic. A more complete description of a community can be obtained by plotting the proportional abundance of every species against its rank of abundance. The result is a rank-abundance diagram. The form of the rank-abundance diagram can be one of at least four slightly different shapes that are referred to as geometric, log series, log-normal, and broken-stick.

4. Whatever measure of diversity is used, there is little argument that diversity increases from the poles to the equator. Thus, tropical areas have more species than temperate areas which, in turn, have more species than arctic areas. There have been a variety of explanations for this phenomena. These explanations can be grouped under two headings: biotic explanations and abiotic explanations. However, as pointed out by Rohde (1992), most of these explanations seem circular or incomplete. Perhaps the best explanation is that evolutionary speed is higher in the tropics and this creates more species.

5. Diversity has often been linked to stability. Stability can be thought of in different ways—as resistance to change or as resilience, which refers to return to equilibrium after perturbation. Resilience in turn can be divided into two—elasticity, which measures how quickly a community can return, and amplitude, which measures how big of a disturbance it can return from. The evidence for and against links of diversity with stability are many and varied, and no firm conclusions have, as yet, been made.

*Discussion Question:* In developing ordinal indices of species diversity, what factors would you use in weighting your indices: rarity (how would you measure it?), size, color, presence of fur or feathers, or any other features you can think of? How might we set about establishing a link between diversity and stability in nature?

# Community Change

D isturbance in most communities is provided by catastrophic events such as fire, grazing, or erosion. Following such events, bare ground or "light gaps" (patches of clear sky in the canopy) often result in forests, and clear substrate may result in intertidal situations. Species diversity then gradually changes as the community returns to "normal" (Photo 15.1) This change can be predictable and orderly and is termed succession. Primary succession occurs when plants invade an area in which no plants have grown before, such as bare rocks or new lakes created by the retreat of glaciers. In primary succession on land, the plants must often build the soil; so a long time is required for the process—hundreds to thousands of years. Only a tiny proportion of the Earth's surface is currently undergoing primary succession, though some argue that because our climate is not stable, primary succession never ends. Primary succession has practical use in the reclamation of spoiled lands. Secondary succession can be considered as a modification of the longer-lasting primary succession: Clearing a natural forest and farming the land for several years is an example of severe forest disturbance that may lead to a distinct secondary succession when farming stops. The ploughing and lack of plant cover cause substantial changes in the soil, particularly the loss of organic matter and nutrients. If the field is then abandoned, the succession that follows can be quite different from one that develops after a natural disturbance.

## 15.1 SUCCESSION

The concept of succession was first developed by botanists who monitored floristic changes along sharp environmental gradients, for example, from seashore and sand-dune halophytes back to scrub and finally mature woodland. Clements is often viewed of the father of

(a)

(b)

(c)

Photo 15.1   *The appearance and disappearance of plants following a disturbance often occurs in a predictable way, termed succession. An example of this type of vegetation change is shown after a September fire in California chaparral near Los Angeles. Photo (**a**) taken September 12, 1979; (**b**) December 24, 1979; and (**c**) April 20, 1980.*   (McHugh, Photo Researchers, Inc. 5D7573, 5D7571, 5D7569.)

successional theory. His early work (Clements 1916) emphasized succession as a deterministic phenomenon with a community proceeding to some distinct end-point or climax community. A key assumption was that each invading species made the environment a little different (say, a little less salty or a little more shady) so that it then became suitable for more K-selected species, which invaded and outcompeted the earlier residents (Clements 1936). This process, known as *facilitation*, supposedly continued until the most K-selected species had invaded, when the community was said to be at climax. Retrogression in this sequence was not possible unless another disturbance intervened. The **climax community** for any given region was thought to be determined by climate and soil conditions. This early view was challenged by Gleason (1926) and others who interpreted communities as random assemblages of species with change inevitable in the long run.

Succession following the retreat of Alaskan glaciers was argued to fit the Clementsian pattern well (Cooper 1923, Crocker and Major 1955). Over the past 200 years, the glaciers in the Northern Hemisphere have undergone dramatic retreats, up to 100 km in some cases (Fig. 15.1), sure evidence of a climatic change. As glaciers retreat, they leave moraines, deposits of soil and stones brought forward by the ice, whose age has often been determined by direct observation. The basic bare soil has a pH of 7.0 and has low nitrogen content with little organic matter. In the early stages, it is first colonized by fireweed (*Epilobium*) and *Dryas*. In the intermediate stages, alder trees, *Alnus sinuata*, invade the area and, in the mature stage, these are followed by spruce trees. At first, ecologists suggested that each species facilitated the entry into the community of the next species. However, Chapin et al. (1994) have shown that facilitation occurs only during part of the process. The taller alders shade out the shorter early colonizing plants. Because *Dryas* and alders can fix atmospheric nitrogen, the nitrogen content of the soil increases dramatically. This facilitates the invasion of spuce trees. After about 50 years, the dense stands of alder (*Alnus* sp.) begin to be shaded out by Sitka spruce, which after another 120 years may form a dense forest. The alders are shaded out and die. Spruce trees cannot fix nitrogen directly; they take it from the soil, lessening the available nitrogen and stabilizing the pH at about 5.0. Facilitation, originally thought to fuel the entire sequence of succession, was important in the establishment of spruce, but competition is also important in driving the whole chain of events. The entire succession sequence from bare substrate to climax is called a **sere.**

The decomposition of plant material, such as logs, also incorporates elements of facilitation. A now-classic experiment by Edwards and Heath (1963) demonstrated this phenomenon. They put oak and beech leaves in nylon bags in the soil and examined decomposition rates. By varying the mesh size of the bags, they could vary the sizes of the decomposers entering them. They obtained the following results:

| Bag Mesh Size (mm) | Fauna that could Enter Bags | Percent Oak Leaves Gone in Nine Months |
|---|---|---|
| 7.0 | All | 93 |
| 0.5 | Small invertebrates, microorganisms | 38 |
| 0.003 | Microorganisms only | 0 |

Although micro-organisms are very important in decay, they cannot begin their work until particle size is reduced by larger organisms. In the soil, earthworms are most important in the initial decay process. The ultimate form of facilitation is enablement, where species B cannot invade unless species A has already invaded first. If this type of phenomenon occurs, the community is often thought to obey certain "assembly rules" (Drake 1990) and is analogous to the pieces of a jigsaw only fitting together in a certain order (Fig. 15.2). Drake et al. (1993) and Fox and Brown (1993) discuss assembly rules theory in more detail.

Other types of succession do not show elements of facilitation at all. After the retreat of the glaciers from the Great Lakes area, lake levels began to drop, and with each drop a raised beach was left. These raised beaches, still evident today parallel to the shores of Lake Michigan (Photo 15.2), are about 7, 12, and 17 m above the present water level. They are identical in substrate and differ only in age, presenting a good chance to study succession. Cowles (1899) first worked on the system and documented a succession from bare sand to

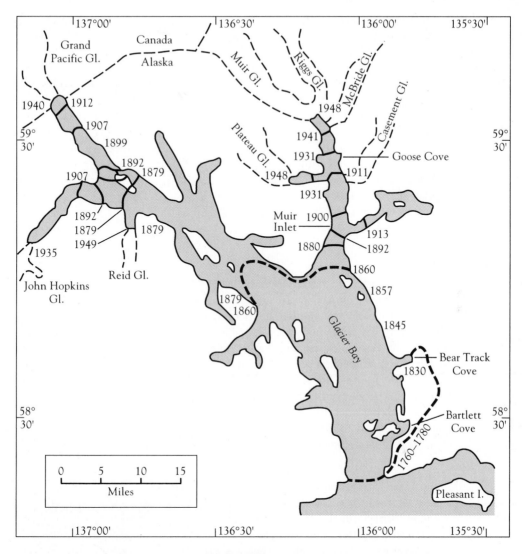

**Figure 15.1**   *Glacier Bay fjord complex of southeastern Alaska showing the rate of ice recession since 1760.*   (Redrawn from Crocker and Major 1955.)

marram grass (*Ammophila Breviligulata*) in six years; sand reed grass (*Calamovilfa longifolia*), little bluestem (*Andropogon scoparius*), sand cherry trees (*Prunus* sp.), and willows (*Salix* sp.) after 20 years; jack pines and white pines after 50–100 years; and black-oaks at about 150 years. Cowles predicted the climax community would eventually be beech-maple forest. However, Olson (1958) studied very old sand dunes in this area (about 12,000 years old) and could see no further evidence of succession beyond the black-oak stage. Olson also noted that, though soil pH changed from 7.6 at the start of succession to 4.0 after the first 1,000 years, from then on it stabilized, and conditions for the establishment of climax beech-maple forest seemed to diminish. Beech and maple require neutral pH and more water.

Figure 15.2     *Depiction of the "puzzle" analogy for community assembly. If species A colonizes first, then either species B or species C can colonize—competition is assumed. Depending on whether B or C is successful, species D is either a potential colonist or finds the community resistant to invasion. However, a real jigsaw puzzle has but one possible complete picture; a biological community may exist in alternative state.* (After Drake 1990.)

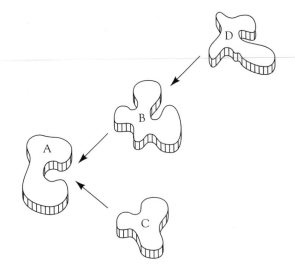

Beech-maple "climax" forest is generally found on moister soils with higher nutrient contents. Black-oak communities are ineffective at nutrient cycling through leaf litter, so on dunes, this type of community appears stable. This type of effect is known as inhibition. It is also probably true that the actual "target" climax community is likely to keep changing. In New England, old-field successions might take 100–300 years to reach climax, but frequency of major disturbance—fire or hurricane—is every 70 years or so. In some communities inhibition occurs because of the effects of leaf litter. Facelli and Facelli (1993) removed the litter of *Setaria faberii*, an early successional species in New Jersey oil fields, and noted that the removal increased the biomass of a later species, *Erigeron annuis*. Release of phytotoxic compounds from decomposing litter or physical obstruction may be part of the reason.

Photo 15.2     *Raised beaches, like these at Warren Dunes, Lake Michigan, Berrien County, provide a good opportunity to study succession.* (Lemker, Earth Sciences 485343.)

Such differences in the patterns of succession in nature prompted Connell and Slatyer (1977) to outline three different models governing succession. The classical view of succession, proposed by Clements (1936), which supposes an orderly progression to a predictable climax community, was termed by Connell and Slatyer the **facilitation model** as each species makes the environment more suitable for the next. For the type of succession found on very old Michigan sand dunes, they formulated the **inhibition model** because some colonists tend to prevent subsequent colonization by other species. Thus, black oaks appear to lessen the chance that beeches and maples will invade. In this model, succession depends on chance events (essentially who arrives first). Succession proceeds as colonists die, but it is not orderly. Connell and Slatyer then proposed a third model, which they termed the **tolerance model,** in essence intermediate between the other two. In this model, any species can start the succession, but the eventual climax community is reached in a somewhat orderly fashion. The best evidence for the tolerance model actually came from Egler's (1954) work on floral succession. Egler showed that in many flower communities, there is a tendency in succession for most species to be present at the outset, as buried seeds or roots. Whichever species germinated first, or grew from roots, would start the succession sequence. Egler had termed this idea of succession "initial floristic composition sequence." All three models predict that the most likely earlier colonists will be $r$-selected species, weeds in many cases. The key distinction between the models is in how succession proceeds. In the classical facilitation model, species replacement is facilitated by previous colonists; in the inhibition model it is inhibited by the action of previous colonists; in the tolerance model, it is unaffected (Fig. 15.3).

Ten years after the models were proposed and many data later, Connell, Noble, and Slatyer (1987) suggested that these models represent not strict alternatives but rather the ends of a continuum of effects of earlier or later species. In the real world, components from each may be prevalent in any one given study system. Thus, for the succession of alder to spruce on Alaskan floodplains, "stochastic, life history, facilitative, competitive, and herbivory processes all affect the interaction between alder and spruce during succession and no single successional process or model adequately describes successional change" (Walker and Chapin 1987, p. 131; see also Chapin et al. 1994). Walker and Chapin (1987) have discussed the relative importance of each of these major factors on communities of a different successional state and have presented their results in a figure, which outlines how major processes determine change in plant-species composition (Fig. 15.4). Their conclusion is that succession is a complex process driven by many processes acting simultaneously in any given situation. Thus, the effects of such factors as competition, seed-arrival time, insect and mammalian herbivory, and stochastic events vary in importance according to the stage of succession—early, middle, or late. For example, competition is not very important to colonizing species in the early stages of succession—there are very few competitors present. However, in mature communities, competition can be an important force. Similarly, stochastic events such as fire can devastate early successional communities but may have less effect on species change in mature communities where such species as large trees are able to withstand periodic burns. With a little patience you can think your way through each of the graphs in Walker and Chapin's figure.

Connell and Slatyer's original models are also further discussed by Horn (1976), who has provided a simple model of succession that assumes succession is a Markovian replacement process. First, Horn gathered together information on a community, in this case the tree species present in a New Jersey forest. Next, he mapped the seedlings underneath the

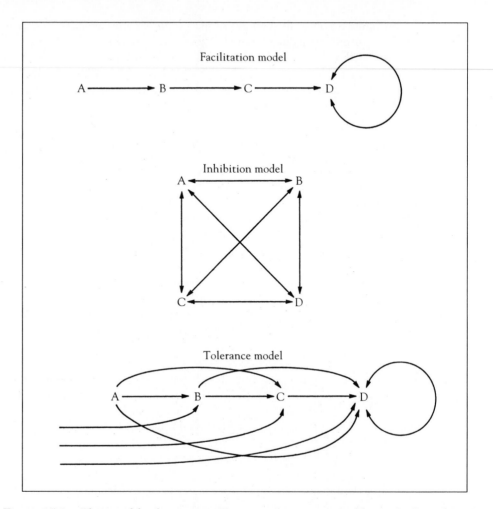

Figure 15.3   *Three models of succession. Four species are represented by A, B, C, and D. An arrow indicates "is replaced by." The facilitation model is the classic model of succession. In the inhibition model, all replacements are possible, and much depends on who gets there first. The tolerance model is represented by a competitive hierarchy in which later species can outcompete earlier species but can also invade in their absence.*   (Redrawn from Horn 1976.)

crowns of the forest trees; these saplings were assumed to replace the mature trees in time. Horn assumed that the higher the number of seedlings of a particular species under a tree, the higher the likelihood that this species would replace the mature tree. Horn thus constructed a series of replacement probabilities like those shown in Table 15.1. For example, there were a total of 837 saplings under large grey birch trees (Horn 1975). Among these, there were no grey birch saplings. Janzen [1970] has argued that this phenomenon, at least in tropical situations, results because seed eaters are often host-specific and congregate under a parent tree, devouring seeds there. Only when seeds happen to fall under different species are they likely to survive. Thus, each tree casts a "seed shadow" in which survival of

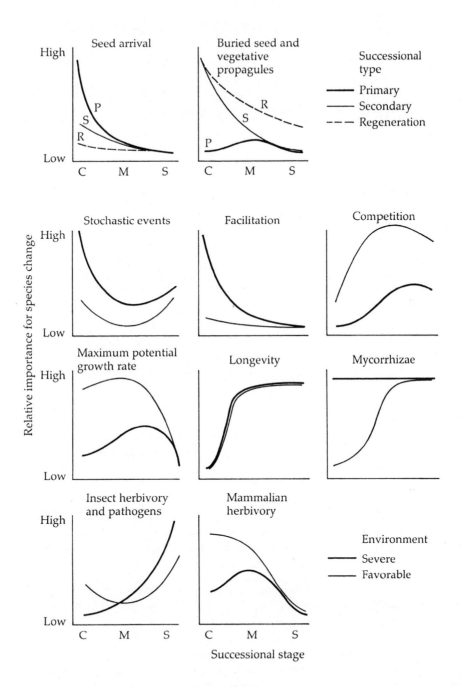

**Figure 15.4**  *Influence of succession and environmental severity on major successional processes that determine change in species composition during colonization (C), maturation (M), or senescence (S) stages of succession.*  (Redrawn from Walker and Chapin 1987.)

**TABLE 15.1** *A 50-year tree-by-tree transition matrix showing the probability of replacement of one individual by another of the same or different species 50 years hence in a New Jersey forest. The diagonal is the percent of trees replaced in 50 years by another of their own kind plus 5% of the original trees left standing (assuming a death rate of 95% for Grey Birch in 50 years, for Blackgum and Red Maple in 150 years, and for Beech in 300 years). Off-diagonal terms are percent of trees replaced by another species in 50 years. (From Horn 1976.)*

| Present occupant | Occupant 50 years hence | | | |
|---|---|---|---|---|
| | Grey birch | Blackgum | Red maple | Beech |
| Grey birch | 0+0.05 | 0.36 | 0.50 | 0.09 |
| Blackgum | 0.01 | 0.20 + 0.37 | 0.25 | 0.17 |
| Red maple | 0.0 | 0.14 | 0.18 + 0.37 | 0.31 |
| Beech | 0.0 | 0.01 | 0.03 | 0.35 + 0.61 |

its own kind is reduced—hence, the high value of dispersive seeds. The replacement probability for grey birch under grey birch is then $0/837 = 0$. Replacement values can be summed to give the total predicted abundance of a species. Beginning with an observed distribution of the canopy species in a stand in New Jersey known to be 25 years old, Horn modeled the change in species composition over several centuries (Table 15.2). Theoretically, red maple is predicted to assume dominance quickly while grey birch disappears. Beech should slowly increase to predominate later. All these predictions agree with what happens in nature. Thus, Horn's model predicts a deterministic successional outcome, independent of the initial forest composition—a prediction very similar to Connell and Slatyer's (1977) tolerance model.

Noble (1981) has provided a nice diagrammatic summary of succession as viewed by Clements, Egler, Horn and Connell and Slatyer (Fig. 15.5). Included in this figure is direct succession. This is the process when, in simple communities like lakes, deserts, or tundra the vegetation simply replaces itself. Noble also notes that real case histories often show a mix of these succession types.

Finally, it is worth remembering that there are some cases in which communities change not in a directional fashion but to a new state. This type of event, more severe and with longer-lasting effects, is often produced by the introduction of a new species into the com-

**TABLE 15.2** *The predicted percentage composition of a New Jersey forest consisting initially of 100% grey birch. (From Horn 1976.)*

| Age of forest (years): | 0 | 50 | 100 | 150 | 200 | ∞ | Data from old forest |
|---|---|---|---|---|---|---|---|
| Grey birch | 100 | 5 | 1 | 0 | 0 | 0 | 0 |
| Blackgum | 0 | 36 | 29 | 23 | 18 | 5 | 3 |
| Red maple | 0 | 50 | 39 | 30 | 24 | 9 | 4 |
| Beech | 0 | 9 | 31 | 47 | 58 | 86 | 93 |

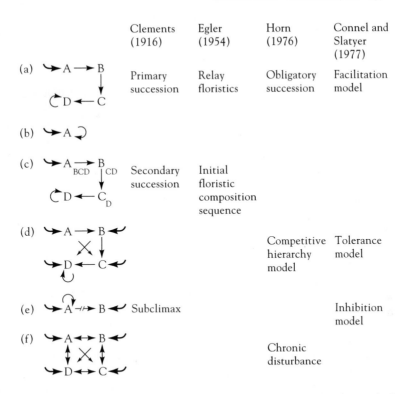

Figure 15.5   *Replacement sequences in succession proposed by different authors. The letters A–D in (a)–(f) represent hypothetical vegetation types or dominant species; subscript letters in (c) indicate that species are present as minor components or as propagules—for simplicity, they have been omitted from (d)–(f); thin arrows represent species or vegetation sequences in time; bold arrows represent alternative starting points for succession after disturbance. (b) is usually referred to as direct succession.* (After Noble 1981.)

munity. Thus, *Endothia parasitica*, the chestnut blight, has forever altered the structure of temperate forests in eastern North America by killing trees, and Dutch elm disease has had the same effect in Europe and the United States. The introduction of weed plants can have similar effects; for example, the arrival of prickly pear cactus in Australia has forever changed the landscape there, though the effects are not so violent now that the cactus has been controlled somewhat by a natural enemy that eats it, the moth *Cactoblastis* (see Chapter 10).

## 15.2  ISLAND BIOGEOGRAPHY

Experimental studies on succession were stimulated years ago by observations of succession on offshore islands that had been totally defaunated by volcanic eruptions, such as the one that rocked Krakatau in 1883. The 1980 eruption of Mount St. Helens in Washington state provided another good opportunity to study natural recolonization. In the six years following this eruption, Wood and del Moral (1987) monitored vascular-plant invasion of the barren substrates. They found that dispersal was limited in many species and that nurse plants played a key role in trapping seeds and promoting seedling establishment. They concluded that the path of early succession depended on spatial position of an area and dispersal abilities of species in the seed pool nearby. Chapin et al. (1994) also concluded that seed dispersal affected primary succession on Alaskan glaciers. Ultimately, who colonized what depended as much on the distance to the nearest floral or faunal source as on succession processes.

It was MacArthur and Wilson (1963, 1967) who first formally developed a comprehensive theory to explain the influence of a variety of variables on succession on islands in their "equilibrium theory of insular zoogeography." They suggested:

1. The number of species on an island, S, tended toward an equilibrium number, $_S$. Initially, the colonization curve of an empty area will start steeply and flatten out toward an asymptote, but many processes that start from empty and go to full show this type of curve (Williamson 1989).

2. $\hat{S}$ is the result of a balance between the rate of immigration and the rate of extinction. Thus species may come and then go extinct, but $\hat{S}$ remains the same (see Fig. 15.6).

3. $\hat{S}$ is determined only by the island's area and position, which influence the rates of migration. Extinction rates were argued to be greater on smaller islands, and colonization rates greater on islands near source pools (Fig. 15.7). The particular shape of the relationship was not predicted, merely that it was monotonic (Williamson 1989).

4. Equilibrium is dynamic, and following actual colonization we should see the number of species stay constant through time, but their identities will change. Thus we will see turnover of species.

These predictions relied on population-level phenomena and ignored any competition between species or other community effects. Extinction was more likely on small islands purely by stochastic processes, by chance alone, because population sizes would be smaller there.

Since its original exposition, MacArthur and Wilson's theory of island biogeography has undergone some modifications:

1. The rate of colonization depends on an island's size, as well as on its distance from a source pool of potential colonists because larger islands present larger "targets" for colonists than do smaller islands (Whitehead and Jones 1969).

2. The "rescue effect" (Brown and Kodric-Brown 1977). Distance from an island to a source pool of potential colonists affects the rate of extinction as well as the rate of colonization. This is so because immigration of individuals of taxa already resident on the island slows the rate of extinction of those taxa by keeping population sizes higher than they would be in the absence of immigration.

3. Patches of particular habitat on continents were viewed as "islands" in a sea of other, unsuitable, habitats (Kilburn 1966; Vuilleumier 1970). Janzen (1968, 1973) extended this concept by proposing host-plant species are islands in "evolutionary time" to their associated herbivore fauna, and individual plants are islands in "contemporary time" for individual herbivores, usually insects.

**Figure 15.6**   *MacArthur and Wilson's equilibrium model of a biota of a single island. The equilibrial species number (Ŝ) is reached at the intersection point between the curve of rate of immigration of new species not already on the island and the curve of extinction of species from the island.*   (Redrawn from MacArthur and Wilson 1967.)

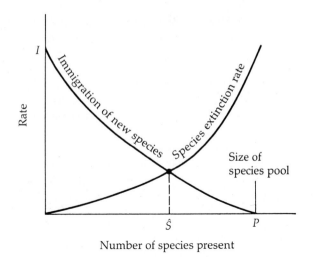

The McArthur–Wilson theory generated some falsifiable predictions:

1. Turnover of species should be considerable.
2. Log Ŝ increases more rapidly with log A on distant islands than on islands near a source of potential immigrants.
3. Log Ŝ increases more rapidly with the reciprocal of distance on smaller islands than large.
4. The rate of turnover, *T*, can be estimated from the time taken for an initially uninhabited island to reach 90 percent of its equilibrial number ($t_{90}$) using the equation

$$T = 1.16\hat{S}/t_{90}$$

**Figure 15.7**   *Equilibrium models of biotas of islands with varying distances from the principal source area and of varying size. An increase in distance (near to far) lowers the immigration curve; an increase in island area (small to large) lowers the extinction curve.*   (Redrawn from MacArthur and Wilson 1967.)

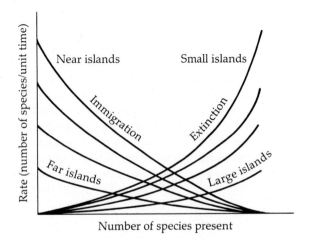

## Evidence in Support of the Theory of Island Biogeography

Gilbert (1980) suggested that, in order to support the MacArthur–Wilson model, there should be data to suggest that:

1. A close relationship exists between the area of different islands and the number of species they contain.
2. The number of species on an island remains constant.
3. There is turnover of species in the system.

### 1.   Species-area relationships

**(a) Oceanic islands**

Preston (1962) provided many examples which showed a species-area relationship: amphibians, reptiles and beetles on the West Indies (Fig. 15.8), ants in Melanesia, vertebrates on islands in Lake Michigan, and land plants in the Galápagos. At least 19 data sets from islands were provided in a review by Quinn and Harrison (1988). However, there are many data sets that show no relationship between area and species richness (Gilbert 1980) and many more where the richness of island fauna is better predicted by variables other than area, for example elevation or soil type. Gilbert (1980) suggested that MacArthur and Wilson were aware of this variation in the data and carefully selected the examples they used in their book to support their hypothesis.

Among those studies for whom the species-area relationship was viewed as proved, one of the early topics of discussion was the significance of variations in the value of $z$, the slope of the regression line in the equation.

$$S = cA^z$$

or, in logarithmic form,

$$\log S = \log c + z \log A$$

where $S$ = number of species; $c$ = a constant measuring the number of species per unit area, that is, per unit of forest, grassland, or desert, and $A$ = area.

Preston (1962) calculated that "theoretically" the value of $z$ should be 0.263. However, values of 0.3 were often found in island studies. Such values were suggested to be permissible in "nonequilibrial" and "isolated situations" (Schoener 1976), but Gilbert (1980) suggested that they were glaring examples of untestable and hoc hypotheses erected to explain conveniently what amounts to pure speculation. There was rarely argument to whether $z$ values of 0.2 or 0.3 verified Preston's arguments or not.

**(b) Habitat "islands"**

Data from many more than 100 studies of islands or analogous habitat isolates are available for study (Quinn and Harrison 1988). Two examples are shown in Figs. 15.9 and 15.10 for

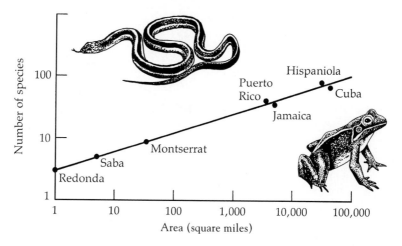

**Figure 15.8** *Species-area curve for the amphibians and reptiles of the West Indies.* (Modified from MacArthur and Wilson 1967.)

North American birds and flowering plants in England. Values of $z$ obtained from these continental areas were often found to be lower than those from truly insular situations, being 0.15 to 0.25. This means that as larger areas are sampled, fewer new species are added in continents as on islands. The rationale here is that each area in the continental studies, except the largest, contains some transient species from adjacent habitats, so the slope of the

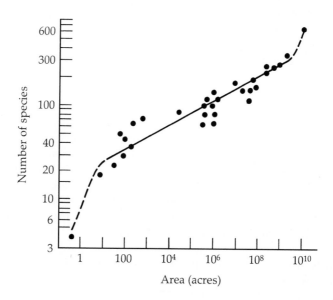

**Figure 15.9** *Species-area curve for North American birds. The points range from a 0.5 acre (0.2 hectare) plot with three species in Pennsylvania to the whole United States and Canada (4.6 billion acres, 1.86 billion hectares) with 625 species.* (Redrawn from Preston 1960.)

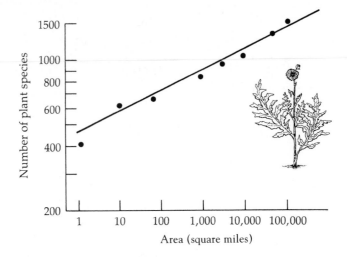

**Figure 15.10**   *Species-area relationship for flowering plants in England. The smallest plot is 1 mi.²
in Surrey, and the largest plot is the whole of England, 87,417 mi.² (227,284 km²).* (Modified from
Williams 1964.)

species-area curve is shallow. Islands are actual isolates with reduced migration rates, so the
number of transients in an area is minimal, thus steepening the slope of the curve.

Values of $z$ can also be affected by dispersal abilities of animals. Brown (1978) studied
the distribution of boreal mammals and birds in the isolated ranges of the Great Basin in the
United States. The mountain ranges were essentially isolated from one another, and the
mammalian fauna was a relict community of a bygone age when rainfall was higher and this
type of boreal habitat was contiguous. Each mountaintop was essentially a forest island in a
sea of desert (Fig.15.11a). The species-area relationship for birds (Fig.15.11b) had a slope of
0.165, and that for mammals, 0.326 (Fig.15.11c). The slope of the line for mammals was
more like that found on islands.

The reason is that there is no mammalian migration between mountaintops. In this situation, mountaintops behave as true islands. In contrast, birds disperse much more, and the
$z$ value in their case is much more in line with a mainland type of relationship. It is also
worth noting that, because birds disperse so well, there is also no effect of distance of a mountaintop from a possible source pool. This situation is unusual because it goes against the usual
belief that colonization of an area is dependent on distance from a source pool. For most
organisms, in most situations, distance is an important parameter.

Should too much importance be attached to $z$ values? Connor and McCoy (1979)
argued that $z$ values of between 0.2 and 0.4 can be expected by chance alone given the variation in values of species richness and area. But this has been disputed by Sugihara (1981)
(see reply by Connor et al. 1983). Gilbert (1980) has provided many examples where values
of $z$ vary from 0.15 to 0.43.

The evolutionary aspect of Janzen's ideas were elaborated first by Donald Strong
(1974a,b) who found a species-area relationship between the geographical area of distribution
of species and the number of insect herbivores (Fig. 15.12). However, Van Valen (1975)

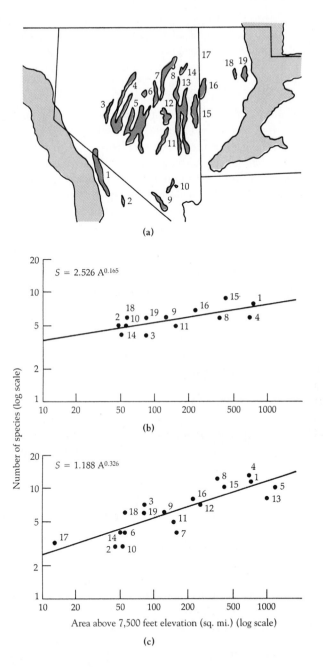

**Figure 15.11**   *Island biogeography applied to mountaintops.* (**a**) *Map of the Great Basin region of the western United States showing the isolated mountain ranges between the Rocky Mountains on the east (right) and the Sierra Nevada on the west (left).* (**b**) *Species-area relationship for the resident boreal birds of the mountaintops in the Great Basin.* (**c**) *Species-area relationship for the boreal mammal species. Numbers refer to sample areas on the map.* (Redrawn from Brown 1978.)

pointed out that the slope of his regression line was not significantly different from zero or from one. Other critics (for example, Claridge and Wilson 1978; Birks, 1980; Kuris, Blaustein, and Alio 1980) pointed to inadequacies in Strong's data. As is often usual in ecological debates, Strong and his coworkers rebutted many of these arguments (Rey, McCoy, and Strong 1981). It is often very difficult to decide in these cases who is right and who is wrong. Readers are encouraged to read original papers for themselves. However, it is worth noting that since Strong's original work, many other workers have documented species-area relationships for insect herbivores. For example, Paul Opler (1974) found a similar regression for California oaks and their leaf miners. The species-area controversy is not just an academic issue. It is important because island-biogeographic theory has come to be accepted uncritically by conservationists and park planners and is regularly incorporated into refuge design (Boecklen and Simberloff 1986). Island-**biogeography** theory was thought to be particularly useful in the continuing debate over the design of wildlife preserves, particularly in the question of whether planners should design many small preserves or a few large ones (given the unlikely prospect of choice). The International Union for Conservation of Nature and Natural Resources (IUCN), a major worldwide conservation group, has stated that refuge design criteria and management practice should be in accord with the equilibrium theory of island biogeography (IUCN 1980). As will be made clear later (Chapter 19), this idea is on shaky ground at best, and more consideration should really be given to the **autoecology** of target species.

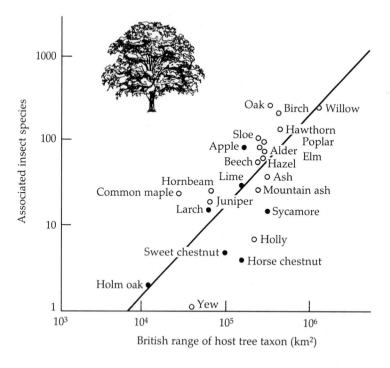

Figure 15.12   *The species-area relationship of the insects of British trees (r = 0.78, P < 0.001). Closed dots indicate introduced taxa; open dots indicate British natives.* (Modified from Strong 1974a.)

The existence of species-area relationships does not necessarily validate the theory of island biogeography. Indeed species-area relationships greatly antedate island-biogeography theory (Rey, McCoy, and Strong 1981). It would have to be assumed that all islands contain identical habitats and are at the same distance from a source pool of potential colonists for us to accept species-area relationships as validating the theory of island biogeography. Also, rates of colonization would have to be shown to be identical for each island, and rates of extinction would have to be inversely related to the size of the islands. These assumptions are clearly not often valid. This short-cut method of proving the operation of island-biogeographic processes is clearly invalid. Also Gilbert (1980) has documented cases that failed to find evidence of a species-area relationship, though there has yet to be a detailed accounting procedure of how many studies support the species-area relationship and how many do not. One more recent example of failure to support the species-area relationship was provided by Loman and von Schantz (1991), who showed that there were often more birds in small habitat islands in Finland than on large ones. Possible reasons could include the fact that edges of habitats are very productive areas for birds, so small habitat areas have relatively more edge and thus more birds per unit area. This result has important implications for the conservation of bird habitat.

Another short-cut method often implied to verify the island-biogeography theory is to compare lists of taxa resident on an island at different times. If lists are available at regular time intervals for different islands/areas of different sizes, then inferences can be drawn. Such inferences often include the existence of turnover and support for equilibrium theory. However, such lists are often compiled only at two different time periods and by different people using different methods for different amounts of effort, so that the results derived from such comparisons should be viewed with caution (Schrader–Frechette and McCoy 1993).

Finally, species-area relationships can be caused by many phenomena, not just increased extinction rates on small islands. Among the most important of these alternative reasons is probably that larger areas contain a greater diversity of habitats, and this probably has the greatest effect on diversity of organisms.

## 2. Colonization and extinction

Often, artificial substrates have been put out in the field to act as potential sites of colonization (Schoener 1974; Schoener and Spiller 1978). For example, tiles of various sizes might be placed in the marine intertidal zone to be colonized by barnacles and other fouling organisms. Colonization is often shown, but rates of extinction and colonization usually cannot be observed directly, mostly because of the length of time between observations and also because the source of colonists cannot be determined, particularly in aquatic studies.

## 3. Establishing turnover

Gilbert (1980) managed to find 25 investigations carried out in order to demonstrate turnover. He dismissed virtually all of them as flawed in method, statistics, or quality of data. Only one, that of Simberloff and Wilson (1969, 1970) is seen as being of merit. Simberloff and Wilson censused small (11–25 m in diameter) red mangrove (*Rhizophora Mangle*) islands in the Florida Keys for all terrestrial arthropods. They then fumigated some of them (experimental islands) with methyl biomide to kill all animals. (Photo 15.3) Periodically thereafter, they censused all islands for several years. After 250 days, most islands had similar numbers

**Photo 15.3**   *Experimental defaunation of a mangrove islet in the Florida Keys.* (*a*) *Construction of a scaffold frame.* (*b*) *Installation of a large tent into which insecticide was introduced, killing all life. Commercial pest control operators from Miami were hired to perform the fumigation. Tent and scaffold were removed after defaunation, and recolonization was monitored.*   (Photos by Daniel Simberloff, Florida State University.)

of arthropods (Fig.15.13). Simberloff and Wilson observed colonization and extinction, but the periodic nature of the sampling meant that it was difficult to determine the exact rates. The data did indicate that colonization rates during the first 150 days were higher on nearer islands than on farther islands. Calculated rates of turnover were found to be very low—1.5 extinctions per year. However, again, the length of time between censuses was so long that pseudoturnover could have occurred—colonization and subsequent extinction all occurring between censuses—masking true rates of turnover. Simberloff (1974) initially seemed to view his mangrove experiments as good evidence in support of the MacArthur–Wilson theory, which he embraced, but later he seems to have had a change of heart (Simberloff 1976b), suggesting that the data are weak support for the MacArthur–Wilson theory and that very few other studies seemed to support it either. Part of this about-face has to deal with the reinterpretation of his own data, which showed initial random colonizations of mangroves through a set of well-defined communities into a final assortative equilibrium with low turnover. This implies the existence of biological processes such as competition between colonists or predation and other biotic interactions that shaped final community structure. Under this scenario, extinctions and colonizations could not be regarded as having certain probabilities on islands of different sizes or distances from the source pool without knowing the existing species present.

It had become clear that the theory of island biogeography treated the dynamics of different colonizing species as essentially the same, with community properties essentially unimportant. As Simberloff (1978b) and Schoener and Spiller (1987) have concluded, turnover probably only involves a subset of transient or unimportant species with the more important species being permanent after colonization.

Predefaunation surveys

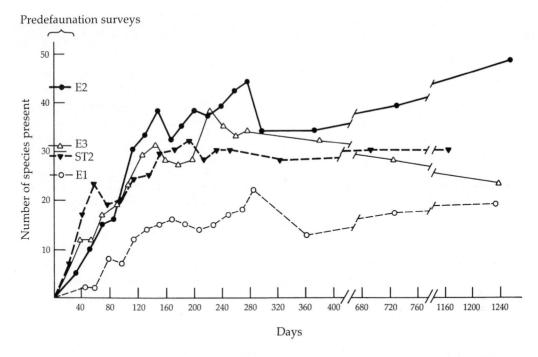

**Figure 15.13** *Colonization curves for arthropods on mangrove islands near Sugarloaf Key, south Florida, after defaunation by fumigation. Predefaunation species numbers are on the ordinate.* (Redrawn from Simberloff 1978.)

Finally, it is interesting to note that the equilibrium theory of island biogeography, first credited to MacArthur and Wilson (1963), was actually presented 15 years earlier by a British lepidopterist named Munroe in his doctorial thesis, which dealt specifically with the distribution of butterflies in the West Indies (Munroe, 1948). Munroe noted that the bigger an island, the larger its butterfly fauna, but he never actively pursued this aspect further, instead sticking to taxonomic interests. In rediscovering this work, Brown and Lomolino (1989, p. 1957) note:

Important as new ideas are in the progress of science, they are often not the unique inspirations of genius that are portrayed in the textbooks. On the one hand, scientific revolutions usually do depend on major conceptual innovations. On the other hand, in order for those insights to have impact, they must be promoted cogently at a receptive stage in the development of a discipline. It is not sufficient to have a good idea; it is even more important to develop and publicize it.

# Summary

1. Primary succession occurs when species invade an area in which no organisms have grown before, such as land unearthed after retreating glaciers. Secondary succession is the change of species composition following a change in land usage, such as the return of forests to old fields after agriculture has stopped.

2. Early views of succession viewed the entire process as facilitative, where each species makes the environment more suitable for the next. However, later work revealed the existence of inhibition, where some colonists actually prevented colonization by other species. In 1977, Connell and Slatyer recognized the existence of a third type of succession, which they termed *tolerance*—in essence, intermediate between the other two. In this model, any species can start the succession, but the eventual climax community is reached in a somewhat orderly fashion. It has since been recognized that these three models represent not strict alternatives but the ends of a continuum of effects of earlier of later species. Herbivory, disease, and other factors can all affect how succession proceeds.

3. Succession may also be affected by the distance of an area undergoing succession from source areas of colonists and the size of that area. The influence of these variables on succession is often termed *island biogeography*.

4. Island-biogeography theory predicts that the number of species in an area, $S$, is determined by a balance between immigration and emigration (extinction). There should be turnover of species. There is much data to suggest that species richness increases with island size, but there is little data to back up other predictions of the theory of island biogeography.

---

*Discussion Question:*  If agriculture on once-virgin tropical-forest areas were to stop, how could we speed up the process of secondary succession so that tropical forests returned?

# Ecosystems Ecology

Three links in the food chain: seven-spot ladybird larva eating black bean aphids feeding on a plant leaf. How much energy is passed from one link to another? How many links are there in food chains. Ecosystems ecology focuses on the flow of energy and nutrients through food chains and attempts to answer questions like this. (Cattlin, Photo Researchers, Inc. 2N0578.)

The term **ecosystem** was coined by the British plant ecologist Tansley (1935) to include not only the **community** of organisms in an environment, but also the whole complex of physical factors around them. Ecosystem ecology involves organisms and their **abiotic** environment and concerns the movement of energy and materials through communities. The concept can be applied at any scale: a drop of water inhabited by protozoa is an ecosystem, and a lake and its biota constitute another. Lovelock (1979) took this idea to its extreme and regarded the whole Earth as one totally interlocked ecosystem, which he named Gaia, after the ancient Greek Earth goddess. In this viewpoint, Gaia was reminiscent of one superorganism, forever regulating temperature, oxygen, and moisture levels to ensure the continuation of life. Lovelock pointed out that levels of such things have not changed appreciably in hundreds of millions of years, whereas on the basis of physical changes alone, such alterations would have been expected.

Most ecosystems can never really be regarded as having definite boundaries. Reiners (1986) has argued that, for this reason and others, the ecosystem remains the least coherent of the organization levels of ecology. He suggests it lacks a logical system of interconnected principles and a well-understood and widely accepted focus. Even in a clearly defined pond ecosystem (Fig. 16.1), waterfowl may be moving in and out. The big advantage of ecosystems ecology, however, is the common currency of energy or nutrients, which allows the biology of communities and populations to be compared between and within trophic levels, something no other ecological discipline can boast.

In attempts to determine the importance of individual units within the scheme of an ecosystem, at least three major constituents can be measured. The first is **biomass,** the standing crop of an organism. Attaching too much importance to biomass, however, may lead to erroneous conclusions in some cases. In the applied world of timber technology, a small standing crop may indicate small potential harvest or yield, but this is not necessarily true if the tree species in question has a high growth rate—new biomass will be produced rapidly and a higher rate of harvest could be sustained (Fig. 16.2). In this situation, energy flow may be more critical. Then the community is regarded as an energy transformer; energy flow is the second constituent. Third, the ecosystem may be most limited by the availability of a rare chemical or mineral. In this case, the flow of limiting chemicals through ecosystems becomes the most important factor in understanding how systems work.

CHAPTER 16

# Trophic Structure

*(handwritten notes:)*
corn    1° pro
cattle    1° con
MAN    2° con
BAC    3° con

## 16.1 TROPHIC LEVELS

There are two major ways in which organisms derive energy; **autotrophs** pick up energy from the sun and nonliving sources, whereas **heterotrophs** eat living matter or, in the case of saprophytes and decomposers, dead material originally derived from living autotrophs. The transfer of food energy from the source, in plants, through herbivores to carnivores occurs through the **food chain.** Two examples of food chains are

|  Example 1 | Example 2 |
| --- | --- |
| Sun | Sun |
| <u>Producer</u><br>corn | <u>Producer</u><br>pine tree |
| <u>Primary</u> <u>consumer</u><br>corn earworm/caterpillar | <u>Herbivore</u><br>aphid |
| <u>Secondary</u> <u>consumer</u><br>ichneumonid wasp/parasite | <u>Carnivore</u><br>spider |
| <u>Tertiary</u> <u>consumer</u><br>spider | <u>Secondary</u> <u>carnivore</u><br>chickadee |

Every step in the food chain is termed a **trophic level,** and **herbivores** and **carnivores** feed at different trophic levels. **Omnivores** feed on at least two trophic levels. Each pair of trophic

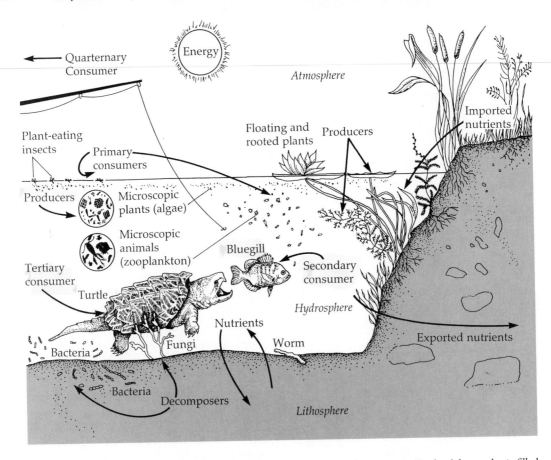

**Figure 16.1** *Producer, consumer, and decomposer in a pond ecosystem. Each of these roles is filled by a number of different organisms. For example, additional secondary or tertiary consumers might be water snakes, snapping turtles, and various birds of prey. Although ecosystems are often thought of as closed systems, none really is. Typically, both living and nonliving things are imported and exported.*

levels is connected by a chain known as a *link*. Thus, the trophic level of a species is one more than the chain length. Charles Elton (1927) was one of the first to explain the importance of food chains to ecology. He pointed out that most chains are actually fairly short and contain no more than five or six links. Some of the most important links in the food chain are those involving detritivores. It is generally underappreciated that much primary production is not consumed by herbivores but dies and rots on the ground to be consumed by detritivores (see Fig. 16.3). Bacteria and fungi are thus common constituents of many food chains. Often the rate at which mineral resources such as phosphorus and nitrogen, locked up in dead material, are released and again made available for growing plants is controlled by the rate of action of the decomposers.

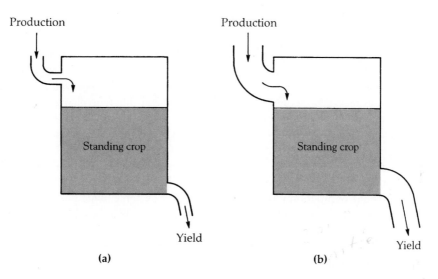

Figure 16.2 *Hypothetical illustration of two equilibrium communities (where input equals output). (a) Low input, low output, slow turnover. (b) High input, high output, rapid turnover. Standing crop is not related to production or yield because turnover time for all systems is not a constant.*

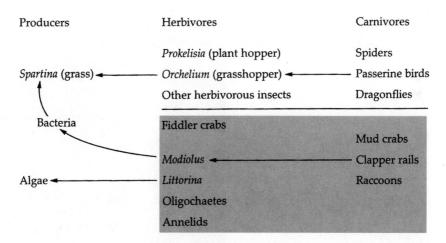

Figure 16.3 *Food web of a Georgia salt marsh with groups listed in their approximate order of importance. Often Spartina grass is the only plant in this food web. Detritivores and the species that feed on them (under the dotted line) constitute a large fraction of the total species. Note that although newer studies have revealed the presence of additional insect herbivores (Stiling, Brodbeck, and Strong 1982; Stiling and Strong 1983), the general pattern remains unaltered. (Redrawn from Teal 1962.)*

## 16.2 FOOD WEBS

Few ecosystems are so simple that they are characterized by a single, unbranched food chain. Several types of primary consumers usually feed on the same plant species; for example, one may find several insect types feeding on one tree and some vertebrate grazers. Also, any species of primary consumer usually eats several different plants. Such branching of food chains occurs at other trophic levels as well. For instance, frogs eat several different types of insect species, which also may be eaten by different types of birds. Owls may eat primary consumers such as field mice and also prey on higher level organisms like snakes. It is more correct, then, to draw relationships between these plants and animals, not as a simple chain but as a more elaborate interwoven food web. The classification of organisms by trophic levels is one of function rather than species. For example, male horseflies are herbivores, feeding on nectar and plant juices, whereas females are blood-sucking ectoparasites.

Most complete food webs are extremely complicated, even when a simplified subset is considered, as for example the species inhabiting pitcher plants in West Malaysia (Photo 16.1, Fig. 16.4).

**Photo 16.1**    *Pitcher plant,* Nepenthes macfarlanei, *in Montane forest at Bintang, Malaysia. Pitcher plants like this have their own self-contained ecosystems which lend themselves well to food web analysis.*
(Fletcher and Baylis, Photo Researchers, Inc. 7Y8275.)

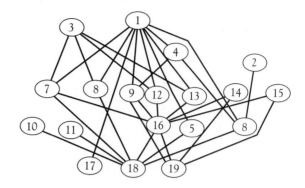

**Figure 16.4**  *A food web of the insects in the pitcher plant* Nepenthes albomarginata *in West Malaysia. Each line represents a trophic linkage; predators are higher in the figure than their prey. Key:    (1)    Misumenops    nepenthicola;    (2)    Encyrtid    (near    Trachinaephagus); (3)* Toxorhynchites klossi; *(4)* Lestodiplosis syringopais; *(5)* Megaselia *sp.* (nepenthina); *(6)* Endonepenthia schuitemakeri; *(7)* Triperoides tenax; *(8)* T. bambusa; *(9)* Dasyhelea nepenthicola; *(10)* Nepenthosyrphus *sp.; (11)* Pierretia urceola; *(12)* Culex curtipalpis; *(13)* C. lucaris; *(14)* Anotidae *sp. 1; (15)* Anotidae *sp. 2; (16) bacteria and protozoa; (17) live insects; (18) recently drowned insects; (19) older organic debris.*
*Number of top predators is 7: species* **1, 2, 3, 10, 11, 14, 15.**
*Number of basal species is 3: species* **17, 18, 19.**
*Number of intermediate species is 9: the rest.*
*Number of linkages: intermediate to top 14; basal to top 4; intermediate to intermediate 8; basal to intermediate 7.* (After Beaver 1985.)

Most food webs are imperfectly known because they are so vast. Ecologists have recognized three kinds of food web:

a.  Source webs—one or more kinds of organisms and the organisms that eat them, their predators, and so on
b.  Sink webs—one or more kinds of organism, the organisms they eat, their other prey, and so on
c.  Community webs—a group of species within a defined area or habitat (like the pitcher plant). Most food-web theory employs these webs. More than 200 such webs have been reported in the literature.

Food webs can form a useful starting point for the analysis of ecosystem organization (Cohen 1978; Pimm 1982). The relative complexity of a food web can be denoted by a measure known as **connectance,** where

$$\text{connectance} = \frac{\text{actual number of interspecific interactions}}{\text{potential number of interspecific interactions}}$$

so that for $n$ species

$$\text{the number of potential interspecific interactions} = \frac{n(n-1)}{2}$$

The number of links per species is called linkage density, $d$. In Fig. 16.4, there are 19 "species," 33 actual interactions, $(18 \times 18)/2 = 171$ potential interactions, and a connectance of 0.19. Linkage density = $33/19 = 1.74$.

A number of generalizations have been made about food webs and some of these are summarized by Pimm, Lawton, and Cohen (1991):

1. Cycles are rare. That is species A eats species B, species B eats species C, and species C eats species A. Cannibalism is a cycle in which one species feeds on itself.
2. The average proportion of top predators (nothing preys on them), intermediate species (with both predators above and prey below), and basal species (autotrophs) remains constant in webs regardless of the number of species.
3. The proportion of trophic links between top predators and intermediates, intermediates and intermediates, basal species and intermediates, and basal to top remains constant.
4. Linkage density is often constant, except for webs with large numbers of species. Links per species is scale invariant and is roughly equal to two. (Cohen, Briand, and Newman 1990).
5. Connectance remains constant as the number of species in the food web increases (Martinez 1992). This is the opposite of Cohen's (1990) scale invariant law (#4). Imagine an insectivorous bird that feeds in two communities, A and B. A has twice the number of insects than B does. Martinez (1992) argues that a bird would eat twice as many species of insect in community A as in community B. Because this would likely apply to all species in the community, so connectance changes as species richness increases.
6. Omnivory is rare.
7. Contrary to intuition, food-chain lengths do not differ greatly among ecosystems with different primary productivities. This was elegantly shown by Pimm and Kitching (1987), who fertilized certain habitats, increasing primary productivity, but noted no increase in numbers of trophic links. However, Schoener (1989) presented evidence that food-chain lengths are linked to the amount of productive space, a quantity that combined productivity with area (or volume) occupied by the food web.
8. There are only slight differences in chain lengths where consumers are vertebrates, compared with where they are all invertebrates.
9. Chain length is smaller on smaller islands.
10. Chains are shorter in areas with frequent natural or experimental disturbances.
11. Chains are shorter in two-dimensional habitats such as grasslands than three-dimensional habitats such as forests or reefs.
12. The ratio of the number of prey to the number of predators is about one to one (Briand and Cohen 1984; Cohen and Briand 1984).

## Problems with Food-web Theory

While food-web theory has undoubtedly lead to an increase in the understanding of biological systems, numerous authors have pointed out numerous problems with the theory and the data:

1. Predation on "minor" species is often omitted; linkages are often informal and idiosyncratic. Often, authors are not sufficiently interested in organisms other than in their area of speciality so that some links are never drawn. For example, Polis (1991) noted that food webs in the real world are much more complex than reported in the food-web literature. In a relatively simple desert system in the Coachella Valley, California, Polis noted 174 species of plants, 138 vertebrates, 55 spiders and scorpions, an estimated 2,000–3,000 species of insects, and an unknown number of microorganisms and nematodes in a single food web. Polis compared his web to catalogues of published webs by authors such as Joel Cohen and Keith Schoenly (Table 16.1). Chains were longer, omnivory and loops were common, and connectivity was greater.

TABLE 16.1 *Comparison of statistics from the Coachella Valley food web with means of the statistics from cataloged webs. (After Polis 1991.)*

|  | Cohen et al. and Briand | Schoenly et al. | Coachella |
|---|---|---|---|
| Total number of "kinds" or species (S) | 16.7 | 24.3 | 30 |
| Total number of links per web (L) | 31 | 43.1 | 289 |
| Number of links per species (L/S) | 1.99 | 2.2 | 9.6 |
| Number of prey per predator | 2.5 | 2.35 | 10.7 |
| Number of predators per prey | 3.2 | 2.88 | 9.6 |
| Total prey taxa/total predator taxa | .88 | .64 | 1.11 |
| Minimum chain length | 2.22 | 1 | 3 |
| Maximum chain length | 5.19 | 7 | 12 |
| Mean chain length | 2.71 | 2.89 | 7.34 |
| Connectance |  | .25 | .49 |
| Basal species (%) | 19 | 16 | 10 |
| Intermediate species (%) | 52.5 | 38 | 90 |
| Top predators (%) | 28.5 | 46.5 | 0 |
| Primarily or secondarily herbivorous (%) |  | 14.6 | 60 |
| Primarily or secondarily saprovorous (%) | 21 | 35.5 | 37 |
| Omnivorous (%) | 27 | 22 | 78 |
| Consumers with "self-loop" (%) | <1.0 | <1.0 | 74 |
| Consumers with mutual predation loop (%) | <1.0 | <1.0 | 53 |

2. Data on quantities of food consumed, that is, thicknesses of connecting links, were usually absent. In the first example where interaction strengths have been calculated in a food web (Paine 1992), most connections were found to be weak so that the food web was essentially very simple. This is opposite to Polis's argument that food webs are really more complicated than originally thought.

3. There is little data on the importance of chemical nutrients. One apparently feeble link may be very important if it supplies a limiting chemical.

4. There is little data on temporal variation. Some species may constitute a large proportion of the web at one point in the year but not at other times.

5. Where does detritus fit? Most authors put it at the bottom of the food chain. The correct procedure is to place material according to the number of acts of assimilation it has undergone since being part of a green plant. Thus, dung, because it is unassimilated, is equivalent to green plants, but carcasses should be placed at the same level as the bodies when they were alive. Treating carcasses in this way gives many food chains a higher number of trophic levels (Cousins 1987).

6. In food-web theory, species are often aggregated into "trophic species," for example, whales, insect larvae, plankton. For example, "insects" are likely to be lumped together because of the difficulties of identifying them all. This disguises much important biology. Aggregation is rife in many published webs. The rejoinder to this criticism is that if aggregation has introduced any errors, patterns from aggregated webs should differ from nonaggregated webs where all species are known. Pimm, Lawton, and Cohen (1991) suggest that most patterns appear across all types of webs. A test is provided by taking completely known webs, aggregating them, and comparing patterns. For instance, in Fig. 16.4, the three pairs of biological species (10 and 11; 12 and 13; 14 and 15) that have the same sets of predators and prey could be joined into three trophic species. The web could be further aggregated by joining 7, 8, 12, and 13, which share similar species of predators and prey. Aggregation of webs, in this case using trophic criteria, affects webs' properties only slightly; in contrast, progressive aggregation by taxonomic affinities alters webs' properties more rapidly. However, Paine (1988) showed how connectance values of between 0.31 and 0.61 occur for the same rocky intertidal community in New England, depending on which scientists studied the system and which methods they used for aggregation. The disconcerting point about this is that these different values for connectance in the same intertidal system span about half the range of 0.05 to 0.60 that is given in 40 published webs presented in Briand (1983).

7. It is hard to define web boundaries. Some highly mobile species, such as predatory gulls, may be very important predators in food webs, but they are almost always underrepresented in webs, as are fish that prey on intertidal organisms during periods of high tide.

8. Many species, such as starfish or fish, exhibit size- or age-related changes in diet and are not easy to assign a single position in a web.

9. Many individuals do not fit into discrete trophic levels. For example, a hawk feeding on a small bird that fed on an insect that fed on a plant would be feeding at trophic level 4. If we insert two extra trophic levels—for instance, the insect is attacked by a parasite that in turn is eaten by a spider before the small bird attacks it—then the hawk would be feeding at trophic level 6.

10. Constancy of the 1:1 predator-prey ratio may be an arithmetical artifact. Many taxa can be recorded as both predator and prey and can hence be double counted. Closs,

Watterson, and Donnelly (1993) argue that in many webs the proportion of species that are double counted is large, so the observed ratio of predator species to prey will tend toward one.

## 16.3  GUILDS

Sometimes, trophic levels can be split up into functional units called guilds (Root 1967, 1973). Herbivores, for example, may comprise of the leaf-chewing guild, the sap-sucking guild, the stem-boring guild, and the leaf-mining guild. The stem-boring guild of salt-marsh cord grass, *Spartina alterniflora*, a common salt-marsh plant, is shown in Fig. 16.5. Species in a

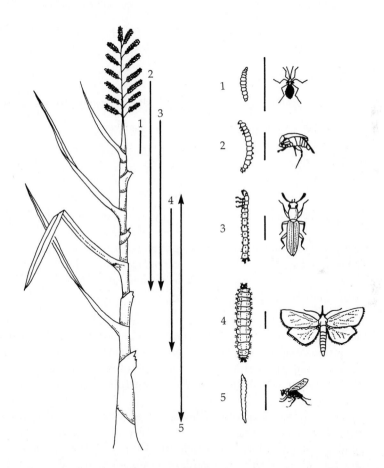

Figure 16.5   *The stem-boring guild associated with salt-marsh cord grass,* Spartina alterniflora, *on the Gulf Coast of North America.* (*1*) Calamomyia alterniflorae (Diptera). (*2*) Mordellistena splendens (Coleoptera). (*3*) Languria taedata (Coleoptera). (*4*) Chilo plejadellus (Lepidoptera). (*5*) Thrypticus violaceus (Diptera). *Arrows indicate where in the stem the larva of each species is found and the direction it bores. Adults, shown at extreme right, are free-living. The length of each scale line between the larva and adult of each species represents 0.5 cm in the drawing.*

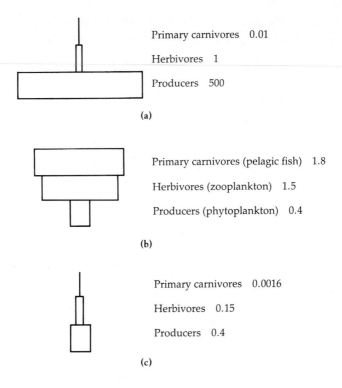

Primary carnivores   0.01

Herbivores   1

Producers   500

**(a)**

Primary carnivores (pelagic fish)   1.8

Herbivores (zooplankton)   1.5

Producers (phytoplankton)   0.4

**(b)**

Primary carnivores   0.0016

Herbivores   0.15

Producers   0.4

**(c)**

**Figure 16.6** *Biomass pyramids. (**a**) A pyramid of biomass seen in an old field in Georgia (g dry wt/m²) (Odum 1957). (**b**) An inverted pyramid of biomass in the English Channel (g dry wt/m²). Phytoplankton has a lower standing crop than the zooplankton, and the zooplankton biomass is less than that for the fish. (**c**) Production rates of organisms in the English Channel (g dry wt/m²/day). The inversion in (**b**) is made possible by the high rate of production by producers in that system.* (Redrawn from Price 1975.)

guild may be expected to exhibit certain properties, among the most important of which is that severe competition may occur between members because they are feeding in the same manner. Competition studies, therefore, often focus on guilds. Of course, according to this line of thinking, competition between conspecific individuals should always be even more intense.

Guilds must be correctly defined, not an arbitrary subset of species from a community. Errors of omission are critical in that they may change conclusions as to the intensity, or lack thereof, of competition (Simberloff and Dyan 1991). For example, does the desert-seed-feeding guild feed on all seeds, seeds of only a few plants, seeds of particular sizes, or seeds in particular habitats? Any of these might be appropriate depending on an investigator's frame of reference. A large raptor may overlap in prey items taken with a medium-sized raptor but not with a small raptor. Are all three raptors part of the same guild?

How much overlap in diet does there have to be for species to constitute the same guild? In his original definition, Root (1967) omitted some species of birds from the foliage-gleaning guild because they took prey from this resource habitat only occasionally. Jaksic

(1981), and Jaksic and Medel (1990) suggested a 50 percent overlap in diet as the minimum value for guild association. However, is the overlap of 50 percent by prey types or 50 percent by prey weight. For example, a red fox is likely to eat many insects, which would comprise 90 percent of its prey types, but the other 10 percent would be mammals, especially rabbits, which constitute 80 percent of the prey by weight. So is the fox an insectivore or not?

One widely accepted phenomenon evident from a study of food chains is that top predators tend to be rather large and sparsely distributed, whereas herbivores are smaller and more common. This generalization is often termed **the pyramid of numbers.** In a small pond, the numbers of protozoa may run into millions, and those of *Daphnia* and *Cyclops* (their predators) into hundreds of thousands, whereas there will be fewer beetle larvae and even fewer fish. One can think of several exceptions to this pyramid. The elm tree, one producer, supports many herbivorous beetles, caterpillars, and so on, which in turn support even more predators and parasites. The best way to reconcile this apparent exception is to weigh the organisms in each trophic level. The elm tree weights 27 metric tons, the herbivores 50 g, and the predators, say, 5 g. It is clear that the elm tree is not a real exception. Inverted pyramids can still occur, even when biomass is used as the measure (Fig. 16.6). In the English Channel, the biomass of phytoplankton supports a higher biomass of zooplankton. This is possible because the production rate of phytoplankton is much higher than that of zooplankton, and the small phytoplankton standing crop (biomass at any one point in time) processes large amounts of energy. The most realistic pyramid is thus the energy pyramid, which never becomes inverted.

# Summary

 1. An ecosystem includes not only a community of organisms in an environment, but also the flow of energy and nutrients through that community.

2. There are two major ways in which organisms derive energy: autotrophs derive energy from the sun and are usually plants; heterotrophs eat living or dead matter originally made by autotrophs.

3. Every step in the food chain is termed a trophic level, and herbivores and carnivores feed at different trophic levels. Omnivores feed on at least two trophic levels. Most food chains are aggregated into food webs.

4. Most food webs are imperfectly known because they are so vast. At least three different types of web are recognized: source webs, sink webs, and community webs.

5. At least 12 generalizations about properties of food webs have been made in the literature. Most of these are based on numbers of species feeding at each trophic level and the number of connecting links between them.

6. There are numerous problems associated with food-web theory based on the imperfect knowledge of food webs. Many of these focus on the lack of knowledge about strength of connecting links between species and taxonomic lumping of species into certain groups or guilds.

7. Some ecologists split up trophic levels into functional units called *guilds*. For example, among herbivores the leaf-chewing guild may be recognized as may the stem-boring guild and the root-feeding guild. However, the precise limits of guilds are not easy to define: Where would one place a species that feeds or leaves *and* stems?

*Discussion Question:* Would there be any meaningful way to delimit food webs more precisely—such as including only prey items that make up more than 5 percent of the diet (by some measure)?

# 17

# *Energy and Nutrient Flow*

Viewing species and individuals as energy transformers in a community has a great strength, which is that the calorie is used as a lowest common denominator and species and individuals can be reduced to caloric equivalents. In the search for common theories in communities, an ecosystems approach may provide great insights into how biological systems function by closely examining energy flow.

The first law of thermodynamics states that energy can be neither created nor destroyed. The second law states that, in every energy transformation, potential energy is reduced because heat energy is lost from the system in the process. Thus, as food passes from one organism to another, the potential energy contained in the system is dissipated as heat. Therefore, there is a unidirectional flow of energy through the system, with little possibility of recycling (Fig. 17.1). The process of energy transfer is inefficient: as a rule of thumb, about 10 percent is transferred from one level to the next, and 90 percent is lost. This may be why some food chains remain short. In contrast, chemicals are not dissipated and remain in the ecosystem indefinitely unless erosion occurs. Chemicals constantly circulate or recycle in the system, becoming more concentrated in higher organisms (Fig. 17.2), often leading to disastrous results in the case of chemical pollutants. This is especially true for the insecticide DDT.

DDT, dichlorodiphenyltrichloroethane, was first synthesized in 1874. In 1939, its insecticide quality was recognized by the scientist Paul H. Müller in Switzerland, who won a Nobel prize in 1948 for that discovery and subsequent research on the uses of DDT. The first important application of DDT was in human health programs during and after World War II, and at that time its use in agriculture also began. The global production of DDT peaked in 1970 when 175 million kg were manufactured. The peak of production in the United States was 90 million kg in 1964. Most industrialized countries banned the use of DDT after the early 1970s, and only a few less-developed countries still use it today.

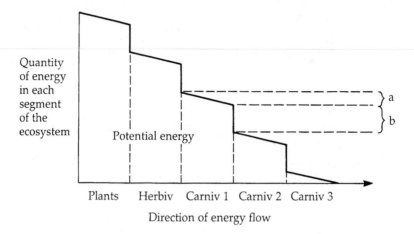

**Figure 17.1**   *Energy flow through the ecosystem. At each trophic level maintenance energy is lost as heat (a); energy is also lost as heat in each transformation from one trophic level to the next (b).*

DDT has several chemical and physical properties that profoundly influence the nature of its ecological impact. First, DDT is persistent in the environment. It is not easily degraded to other, less toxic chemicals by microorganisms or by physical agents such as sun and heat. The typical persistence in soil of DDT is about 10 years. This is two to three times as much as other organochlorine insecticides. The good news is that because of the outlawing of DDT in the United States, amounts in the soils are by now negligible. Another important characteristic of DDT is its low solubility in water (less than 0.1 ppm) and its high solubility in fats or lipids, a characteristic that is shared with other chlorinated hydrocarbons. In the environment, most lipids are present in living tissue. Therefore, because of its high lipid solubility, DDT has a great affinity for organisms, and it tends to bioconcentrate by a factor of several orders of magnitude. Furthermore, because organisms at the top of a food web are

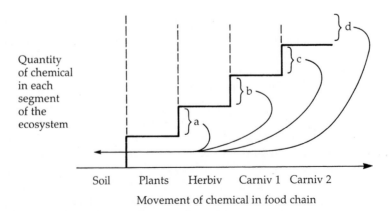

**Figure 17.2**   *The cycling of a chemical in the ecosystem assuming no erosion. Chemicals in organisms not consumed by the higher trophic level (a) to (d) slowly return to the soil as they are released through decay.*

effective at accumulating DDT from their food, they tend to have an especially large concentration of DDT in their lipids. A typical pattern of food-web accumulation of DDT is illustrated in Fig. 17.3 which summarizes a pattern of DDT residues in a Lake Michigan food chain. The prime reason for the introduction of DDT was to control mosquitoes. The largest concentration of DDT was found in gulls, an opportunistic species that often fed on small fish. Large residues were also present in some game fish that became unfit for human consumption.

A relatively complete energy-flow diagram for a Georgia salt marsh (essentially an energetic version of the food web outlined in Fig. 16.3) is shown in Fig. 17.4. As with many ecosystems, most **primary production** goes to the decomposers.

In the salt marsh, the **producers** are also the most important **consumers;** in other words, most primary production is used in plant respiration. The bacteria are next in importance; as **decomposers,** they degrade about one-seventh of the energy that plants use. Animal consumers are a poor third in importance, degrading only about one-seventh of the energy the bacteria use, though it must be stressed that, since this study was performed, many additional species of insects have been found to feed on *Spartina* in the U.S. Southeast (Rey 1981; McCoy and Rey 1987, and references therein), including leaf miners (Stiling, Brodbeck, and Strong 1982) and stem borers (Stiling and Strong 1983) that were not originally noticed in Teal's pioneer studies. The relative importance of these new herbivores, however, is probably less than that of the more abundant plant hoppers and grasshoppers that Teal (1962) recognized.

## 17.1  PRIMARY PRODUCTION

The process of **photosynthesis** is the cornerstone of all life and the starting point for studies of community metabolism. The bulk of the Earth's living mantle is green plants (99.9 percent by

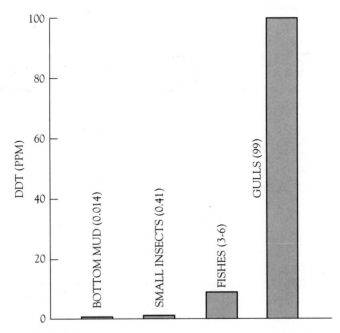

**Figure 17.3**  *DDT concentration in a Lake Michigan food chain. DDT load in gulls was about 240 times that of small insects.*

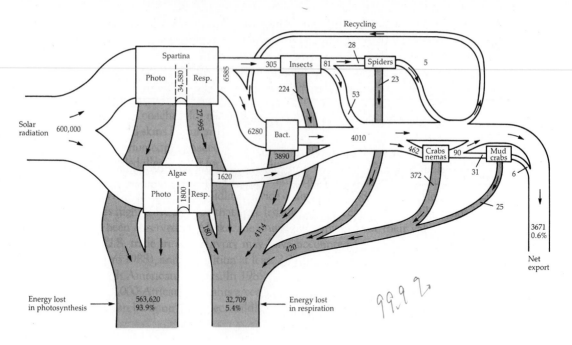

**Figure 17.4** *Energy-flow diagram for a Georgia salt marsh. Nemas = nematodes. Units are kilocalories per square meter per year.* (Redrawn from Teal 1962.)

weight) (Whittaker 1975); only a small fraction of life is animal. **Gross primary production** is equivalent to the energy fixed in photosynthesis, and **net primary production** is gross primary production minus energy lost by plant respiration. The simplest method of measuring net primary production is the harvest method. The amount of plant material produced per unit of time, $\Delta B$, is equivalent to the biomass change in the community between time 1 and time 2. Two possible losses must be recognized: $L$ = biomass lost by death of plants, and $G$ = loss due to consumer organisms. Knowing these two elements, one can estimate net primary production as

$$\text{net primary production} = \Delta B + L + G$$

The energy equivalent of the production in biomass can then be obtained if one burns the biomass in a bomb calorimeter, but it must be remembered that this method cannot effectively gauge new root growth. In herbaceous communites, below-ground root biomass can be 40 percent of total biomass. Also, Wallace and O'Hop (1985) showed that, in some cases where grazing by animals is high, net productivity can be grossly underestimated when based solely on estimates of standing-crop biomass. In their study on beetle herbivory on water lilies, larval production of beetles alone—without adjustments for egestion, respiration, or adult feeding—surpassed plant biomass availability. Therefore, rapid macrophyte turnover was going on to support the beetle population.

A good estimate of global primary production is 110–120 × 10⁹ metric tons dry weight yr⁻¹ on land and 50–60 × 10⁹ metric tons in the seas (Lieth 1975; Whittaker 1975). How does primary production vary over the different types of vegetation on the Earth? In general, primary production is highest in the tropical rain forest and decreases progressively toward the poles (McNaughton et al. 1991) (Fig. 17.5, Table 17.1). Productivity of the open ocean is

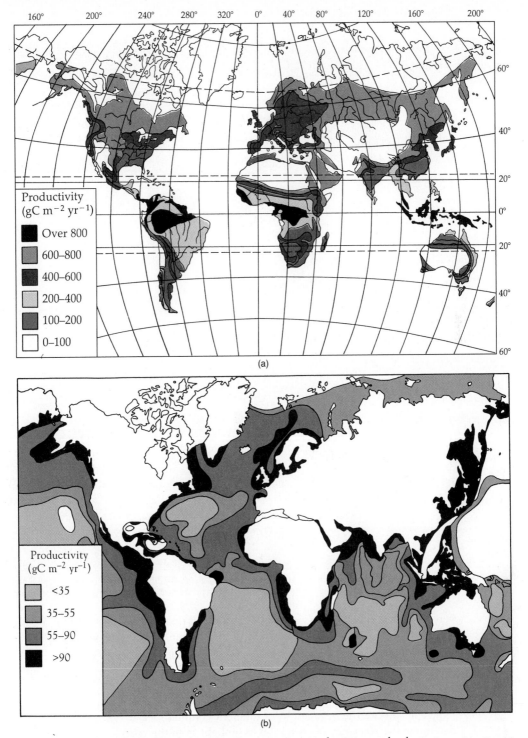

**Figure 17.5** (*a*) *Worldwide pattern of net primary productivity on land.* (Redrawn from Reichle 1970.) (*b*) *Worldwide pattern of net primary productivity in the oceans.* (Redrawn from Koblentz-Mishke, Volkounsky, and Kabanova 1970.)

TABLE 17.1 *Primary production and plant biomass for the Earth. (From Whittaker and Likens, cited by Whittaker 1975.)*

| Ecosystem Type | Area ($10^6$ km²) | Mean Net Primary Productivity (g m$^{-2}$ yr$^{-1}$) | World Net Primary Production ($10^9$ tons yr$^{-1}$) | Mean Biomass (kg m$^{-2}$) | World Biomass ($10^9$ tons) |
|---|---|---|---|---|---|
| Tropical rain forest | 17.0 | 2,200 | 37.4 | 45 | 765 |
| Tropical seasonal forest | 7.5 | 1,600 | 12.0 | 35 | 260 |
| Temperate evergreen forest | 5.0 | 1,300 | 6.5 | 35 | 175 |
| Temperate deciduous forest | 7.0 | 1,200 | 8.4 | 30 | 210 |
| Boreal forest | 12.0 | 800 | 9.6 | 20 | 240 |
| Woodlands and shrubland | 8.5 | 700 | 6.0 | 6 | 50 |
| Savanna | 15.0 | 900 | 13.5 | 4 | 60 |
| Temperate grassland | 9.0 | 600 | 5.4 | 1.6 | 14 |
| Tundra and alpine | 8.0 | 140 | 1.1 | 0.6 | 5 |
| Desert and semidesert scrub | 18.0 | 90 | 1.6 | 0.7 | 13 |
| Extreme desert, rock, sand, ice | 24.0 | 3 | 0.07 | 0.02 | 0.5 |
| Cultivated land | 14.0 | 650 | 9.1 | 1 | 14 |
| Swamp and marsh | 2.0 | 2,000 | 4.0 | 15 | 30 |
| Lake and stream | 2.0 | 250 | 0.5 | 0.02 | 0.05 |
| Total continental | 149 | 773 | 115 | 12.3 | 1,837 |
| Open ocean | 332.0 | 125 | 41.5 | 0.003 | 1.0 |
| Upwelling zones | 0.4 | 500 | 0.2 | 0.02 | 0.008 |
| Continental shelf | 26.6 | 360 | 9.6 | 0.01 | 0.27 |
| Algal beds and reefs | 0.6 | 2,500 | 1.6 | 2 | 1.2 |
| Estuaries | 1.4 | 1,500 | 2.1 | 1 | 1.4 |
| Total Marine | 361 | 152 | 55.0 | 0.01 | 3.9 |
| Grand Total | 510 | 333 | 170 | 3.6 | 1,841 |

very low, approximately the same as that of the Arctic tundra. In agricultural situations, primary production of crops may fall short of that of natural ecosystems, possibly because of a short growing season, or it may be increased by the application of fertilizers.

## Efficiency Measures of Primary Production

How efficient are the vegetations of different communities as energy converters? One can determine the efficiency of utilization of sunlight from the ratio

$$\text{efficiency of gross primary production} = \frac{\text{energy fixed by gross primary production}}{\text{energy in incident sunlight}}$$

Phytoplankton communities have very low efficiencies of usually less than 0.5 percent, herbaceous communities 1–2 percent, and crops generally less than 1.5 percent; the highest values occur in forests—2–3.5 percent (Cooper 1975). Perhaps 50–70 percent of the energy fixed by photosynthesis is lost in **respiration,** so usually less than 1 percent of the sun's energy is actually converted into net primary production. In the temperate zone, this percentage is equivalent to 300–600 calories of primary production $cm^{-2}$.

## The Limits to Primary Production

What limits primary production? In terrestrial systems, water is a major determinant, and production shows an almost linear increase with annual precipitation, at least in arid regions (Fig. 17.6). However, the temperature conditions are also important, as there is a much greater range of temperatures over land than over water. Rosenzweig (1968) showed that actual evapotranspiration rate could predict the above-ground production with good accuracy in North America. Lieth (1975) showed that, in addition to evapotranspiration rate, length of growing season was well correlated with net primary production of forests, at least in North America (Fig. 17.7). It also explains why tropical wet forests, with a very long growing season, are so productive. It also might help explain why conifers are so predominant in northern realms: they effectively extend the short growing season at high latitudes by retaining their leaves (needles) for long durations.

Nutrient deficiency, particularly of nitrogen and phosphorus, can limit primary productivity, too, as agricultural practitioners know only too well. Fertilizers are commonly used to boost productivity of annual crops. What is less-commonly appreciated is that harvesting timber also removes large amounts of nutrients from the forest ecosystem, and such losses

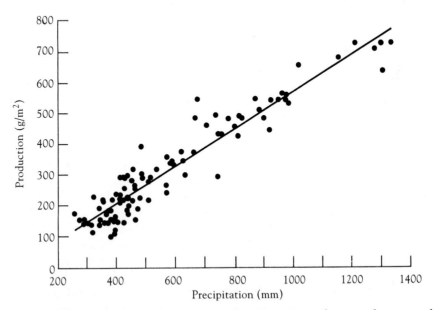

Figure 17.6   *Relationship between mean annual precipitation and mean above-ground net primary production for 100 major land resource areas across the central grassland region, Great Plains, of the United States.* (Redrawn from Sala et al. 1988.)

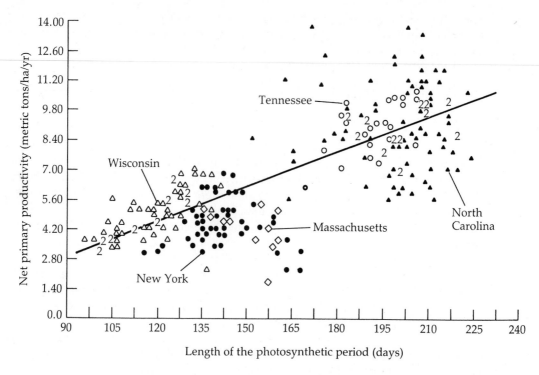

**Figure 17.7** *Relationship between net primary production and the length of the growing season for stations in the eastern deciduous forest region of North America.* (Redrawn from Lieth 1975.)

should also be made good by nutrient supplements. Rennie (1957) showed that pines removed fewer nutrients from soil than hardwoods. Some foresters have provided good evidence to show that fertilization increases timber yields. Among the most impressive studies are those involving fertilization of pines (*Pinus radiata* ) with phosphate after a fire. Fertilized trees grew to more than twice the size of unfertilized ones after 15 years (Gentle, Humphreys, and Lambert 1965). The basal area of pines on unfertilized plots was only 12 m²/ha, compared to 27 m²/ha in fertilized areas.

Similar results are obtained in natural ecosystems, for example, fertilized *Metrosideros polymorpha* trees on Hawaii showed an annual rate of growth up to double that of unfertilized controls (Gerrish, Mueller-Dombois, and Bridges 1988). In Britain, unfertilized grassland swards yield about 2.5 tons of dry matter ha⁻¹ yr⁻¹, and grass/legume swards about 6 tons ha⁻¹ yr⁻¹. Adding 400 kg ha⁻¹ nitrogen increases the yield of both types to about 10 tons ha⁻¹ yr⁻¹. Peak grass yields of 12 tons ha⁻¹ yr⁻¹ require 625 kg ha⁻¹ nitrogen, but peak profitability occurs at an application rate of 400 ha⁻¹ (Jackson and Williams 1979).

If nutrient availability is so critical to plants, there should be an obvious relationship between soil nutrient content and plant production. Surprisingly, such relationships have not commonly been found (Gessel 1962), which tends to contradict results from fertilization studies. The probable explanation is that in many cases, nutrients are locked up in a form unavailable to plants. An analogous problem exists in estimating what fraction of plant biomass is available to animals. In many cases, nitrogen availability is critical to animals, but measures of total percent nitrogen in plant tissues may not give a good indication of what is available to herbivores because much nitrogen is locked up in indigestible forms (Bernays 1983) or is unavailable because of the presence of digestibility-reducing substances (Mattson 1980; Wint 1983; Brodbeck and Strong 1987; see also Chapter 10).

## Aquatic Systems

Of the factors limiting primary production in aquatic ecosystems, among the most important are available light and available nutrients. Light is particularly likely to be in short supply because water absorbs light very readily. Even in "clear" water, only about 5–10 percent of the radiation may be present at depths of only 20 m. Too much light can also inhibit the growth of green plants by overheating them. Such a phenomenon can be found in tropical and subtropical surface waters throughout the year, where maximum primary production occurs several meters beneath the surface of the sea.

The most important nutrients affecting primary productivity in aquatic systems again are nitrogen and phosphorus. Important only locally in terrestrial systems, both often limit production in the oceans, where they occur in low concentrations. Few nutrients are tied up in the standing crop, in contrast to terrestrial systems, especially forests, where large amounts of nutrients occur in plants themselves. Rich, fertile soil contains about 5 percent organic matter and up to 0.5 percent nitrogen. One square meter of soil surface can support 50 kg dry weight of plant matter. In the ocean, the richest water only contains 0.00005 percent nitrogen, and 1 $m^2$ could support no more than 5 g dry weight of phytoplankton (Ryther 1963). Enrichment of the sea by addition of nitrogen and phosphorus can result in substantial algal blooms. Such enrichment occurs naturally in areas of upwellings, such as the Antarctic or the coasts of Peru and California, where cold, nutrient-rich, deepwater is brought to the surface by strong currents, resulting in very productive ecosystems.

Phosphorus is particularly important in limiting productivity in freshwater lakes. During the late 1960s, some lakes in North America became polluted by runoff from the land of rainwater enriched with phosphorus from fertilizer application or from sewage. This new input into the lakes caused huge blooms of blue-green algae, which clogged the lake and increased turbidity, a process termed **eutrophication.** Eventually, the problem was traced to increased phosphorus input (see also Chapter 20). In a dramatic series of experiments conducted in Canada, that was begun in the late 1960s, lakes were treated with a whole range of chemicals, including phosphorus, and other lakes were left unmanipulated as controls. Only where phosphorus levels were greatly increased was the character of the lake drastically altered by algal blooms and all the other attendant problems of eutrophication (Schindler 1974, 1977). Carefully designed experiments had clearly implicated phosphorus as the main limiting factor for primary production.

For some systems, aquatic and terrestrial, one particular nutrient may be limiting at first, but if the level of this particular factor is raised, then another factor becomes limiting. This type of sequential progression of limiting factors is illustrated in Fig. 17.8. Such a phenomenon was demonstrated in nature by Menzel and Ryther (1961), who studied primary productivity of the Sargasso Sea. They found that iron levels were the most obvious limiting factor, but that nitrogen and phosphorus were likely to be limiting in the presence of sufficient iron. Increasing iron levels in the world's oceans may permit increased phytoplankton

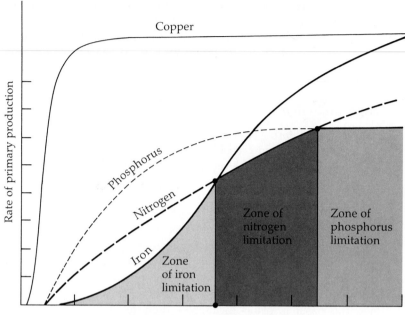

**Figure 17.8**  *Hypothetical sequence of nutrient factors that limit primary productivity. The rate of primary production follows the heavy line and is limited first by iron, then (as more iron becomes available) by nitrogen, and finally by phosphorus. Some nutrients, such as copper, may always be present in superabundant amounts.* (Redrawn from Krebs 1985a.)

production, which will use up more carbon dioxide in the atmosphere and possibly reduce global warming (see Chapter 20).

## 17.2  SECONDARY PRODUCTION

The biomass of plants that accumulates in a community as a result of photosynthesis can eventually go in one of two directions: to herbivores or to detritus feeders. Most of the biomass dies in place and is available to detritivores. This is known as dead organic matter (DOM). Heal and Maclean (1975) calculated that detritivores carried out over 80 percent of the consumption of matter, often "working it over" on a number of occasions to extract the most energy from it. However, it is the herbivores that feed on the living-plant biomass and, as a result, constitute the greatest selective pressure on living plants. Energy flow in such species will therefore be considered in more detail.

The energy for an individual animal can be seen as a series of dichotomies:

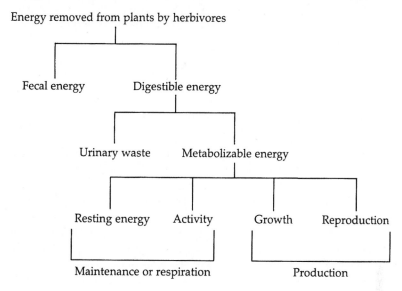

Thus energy can end up in waste products, it can be used in movement, or it can be allocated to growth and reproduction.

## Efficiency Measures of Secondary Production

How are the components of **secondary production** in animal communities measured? Metabolizable or assimilation energy can be measured very simply in the laboratory, where food intake (gross energy intake) and fecal and urine production can easily be recorded. In the field, such measurements are extremely difficult, and the usual approach is to measure assimilation energy by means of the relation

$$\text{assimilation energy} = \text{respiration} + \text{net production}$$

Respiration is measured while an animal is confined in a cage; oxygen consumption, carbon-dioxide output, and heat production are monitored directly. Net production can be measured by the growth of individuals and reproduction, the birth of new animals. Production as biomass can again be converted to energy measures by determination of the caloric value of a unit weight of the species. In warm-blooded animals, the resting energy or minimum rate of metabolism, often referred to as the **basal metabolic rate** is a simple function of body size:

$$\text{basal metabolism (kcal/day)} = 70 \times (\text{weight})^{3/4}$$

The basal metabolism is measured under resting conditions, with no food in the stomach and at a temperature at which no energy for cooling or heat production is required. Such a situation is rarely achieved in the field, and Brody (1945) estimated metabolism in the field for maintenance to be approximately twice the basal rate.

In viewing an herbivore as an energy transformer, three main types of efficiency can be measured in the animal world and used to compare species from different communities:

$$\text{growth efficiency or production efficiency} = \frac{\text{net productivity}}{\text{assimilation}}$$

Growth efficiency in vertebrates is generally less than that in invertebrates because much more energy is used in respiration. Even within vertebrates, much variation occurs. In mammals, more than 95 percent of assimilation energy may be used in respiration, whereas in cold-blooded animals, this figure is less, reflecting the energy cost of homeothermy. Thus, growth efficiencies of around 10 percent are common for ectotherms, and 1–2 percent for endotherms. One consequence is that sparsely vegetated deserts can support healthy populations of snakes and lizards, where mammals might easily starve. The large monitor lizard known as the Komodo dragon eats the equivalent of about a pig a month, its own weight every two months, whereas a cheetah consumes something like four times its own weight in the same period (Kruuk and Turner 1967). It is also interesting to note that growth efficiencies are higher in young animals than in old (35 percent, as opposed to 3–5 percent)—hence, the practice of raising young chickens and calves for meat. Further, smaller species often have higher growth rates than larger species. For a given amount of hay, rabbits produce the same quantity of meat as beef cattle, but do so four times as quickly.

Efficiency between trophic levels can be measured by two other indices:

$$\text{Lindeman's efficiency} = \frac{\text{assimilation at trophic level } n}{\text{assimilation at trophic level } n-1}$$

named after ecologist R.J. Lindeman (1942) in recognition of his classic work on ecological energetics, and

$$\text{consumption efficiency} = \frac{\text{intake at trophic level } n}{\text{net productivity at trophic level } n-1}$$

Lindeman's efficiency often appears to be around 10 percent for each set of trophic levels, although some data on marine food chains shows it can exceed 30 percent (Steele 1974).

Consumption efficiency measures the relative pressure of one trophic level on the one beneath it and seems generally to fall in the range of 0–10 percent for terrestrial herbivores, meaning that 90–100 percent of the net terrestrial plant production goes into the decomposer chain. In aquatic systems, zooplankton may be more efficient grazers and values of 10–40 percent have been reported. For carnivores, however, consumption efficiencies may reach 50–100 percent, showing that animals are much better equipped to handle meat than they are dead material or chemically protected plants.

Whittaker (1975) concluded that the percentages of net primary production going to animal consumption in tropical rain forests and in temperate rain forests and grasslands are, respectively, only 7.5 percent and 10 percent. In open oceans, however, the figure is 40 percent, and in zones of upwelling 35 percent, indicating that zooplankton are better croppers of phytoplankton than terrestrial herbivores are of their food plants. In the general scheme of things, birds and mammals, the animals most keenly observed by humans, contribute almost nothing to secondary production and very little to consumption. Insects are a little more important, but soil animals, including nematodes, remain the major factor. Thus, ecosystems ecology, unlike population ecology, does not stress the importance of higher trophic levels because they are relatively unimportant in energy flow.

## The Limits to Secondary Production

What controls secondary production is a complex question, but it is generally thought to be limited to a high degree by available primary production. McNaughton et al. (1989) have documented a tight correlation between primary productivity in a variety of ecosystems and the biomass of and consumption by herbivores. Existence of this correlation is not as obvious as one might think because it means that secondary chemicals that make plants poisonous or distasteful to herbivores are much less important influences on consumption at the ecosystem level than they are at the plant and population levels (see Chapter 10). When host quality and quantity are increased experimentally by an input of fertilizer, herbivore biomass often increases, too. After reviewing 18 laboratory and field nutrition-interaction studies, Onuf (1978) concluded that a general correlation could be found between nitrogen levels and the susceptibility of plants to insect attack. Leaf-feeding insects removed four times as much foliage from mangrove, *Rhizophora mangle*, on a high-fertility site enriched by the droppings from a large colony of egrets and pelicans as from a nearby low-fertility site (Onuf, Teal, and Valiela 1977). Nitrogen concentration in the phloem sap of grasses increases under fertilization, and sucking species such as aphids and plant hoppers become much more successful and prolific (Prestidge and McNeill 1983). Vince, Valiela, and Teal (1981) found a dramatic increase in insect biomass on fertilized salt marshes of the sort where Teal (1962) had done his pioneering energy-flow studies. But fertilization does not always increase the growth rates of herbivores. Stark (1964), in a review of 15 tree-insect interactions involving nitrogen fertilizers, showed that insect survival was reduced. Auerbach and Strong (1981) could find no increase in growth rates or nitrogen-accumulation rates of two species of tropical hispine beetles feeding on *Heliconia* spp. in Costa Rica. In such cases, natural enemies are often more commonly involved in determining the numbers of herbivores in the field, but this type of argument belongs more in the realm of control of population numbers than in an ecosystem chapter. However, Strauss (1987) has shown that fertilization can increase levels of generalist predators on certain insects, which indirectly decreases the abundance of other herbivores, even on fertilized areas. Fertilized *Artemisia ludoviciana* plants in Minnesota supported greater numbers of phloem- and seed-feeding insects and also had a concurrent increase in patrolling by aphid-tending ants. Such high levels of aggressive predators reduced the levels of other foliage feeders such as chrysomelid beetles, even though such beetles readily feed on fertilized foliage.

Finally, Odum (1969) suggested that the energy relations of communities can also be used as a measure of their maturity (Table 17.2). According to Odum, gross and net primary productivity are very low in mature communities, high in developing ones. Energy in a complex, highly mature system is shunted into maintenance of order, and less is used for production of new materials. Food chains are shorter in the developing communities; those

**TABLE 17.2** *Model of ecosystem development: general trends in 24 variables during ecological succession. (From Odum 1969.)*

| Ecosystem Attributes | Developmental Stages | Mature Stages |
|---|---|---|
| **Community Energetics** | | |
| 1.  Gross production/community respiration (P/R ratio) | Greater or less than 1 | Approaches 1 |
| 2.  Gross production/standing crop biomass(P/B ratio) | High | Low |
| 3.  Biomass supported/unit energy flow (B/E ratio) | Low | High |
| 4.  Net community production (yield) | High | Low |
| 5.  Food chains | Linear, predominantly grazing | Weblike, predominantly detritus |
| **Community Structure** | | |
| 6.  Total organic matter | Small | Large |
| 7.  Inorganic nutrients | Extrabiotic | Intrabiotic |
| 8.  Species diversity—variety component | Low | High |
| 9.  Species diversity—equitability component | Low | High |
| 10. Biochemical diversity | Low | High |
| 11. Stratification and spatial heterogeneity (pattern diversity) | Poorly organized | Well organized |
| **Life History** | | |
| 12. Niche specialization | Broad | Narrow |
| 13. Size of organism | Small | Large |
| 14. Life cycles | Short, simple | Long, complex |
| **Nutrient Cycling** | | |
| 15. Mineral cycles | Open | Closed |
| 16. Nutrient exchange rate, between organisms and environment | Rapid | Slow |
| 17. Role of detritus in nutrient regeneration | Unimportant | Important |
| **Selection Pressure** | | |
| 18. Growth form | For rapid growth | For feedback control |
| 19. Production | (r selection) | (K selection) |
| **Overall Homeostasis** | | |
| 20. Internal symbiosis | Undeveloped | Developed |
| 21. Nutrient conservation | Poor | Good |
| 22. Stability (resistance to external perturbations) | Poor | Good |
| 23. Entropy | High | Low |
| 24. Information | Low | High |

in mature communities are more complex. Odum also viewed the ratio of production to biomass (*P/B*) as the single most important measure of maturity in an ecosystem; *P/B* ratios become lower as maturity increases. Of much interest to applied ecologists is that high *P/B* ratios mean high crop yields, so agricultural practices should be aimed specifically at exploiting immature ecosystems rather than mature ones.

## 17.3  NUTRIENT CYCLES

As previously discussed, nutrients such as nitrogen and phosphorus often limit primary or secondary production. For example, McNaughton (1988) reports that the mineral content of foods is an important determinant of the spatial distribution of animals within the Serengeti National Park, Tanzania. Areas of grassland containing higher concentrations of magnesium, sodium, and phosphorus support higher densities of large herbivores than areas of low concentrations of these minerals. It is often argued that ecosystems can best be understood not from the path of energy through them but by the paths of nutrients, the **biogeochemical cycles.** Because chemicals are not dissipated but remain in the ecosystem indefinitely, they tend to accumulate in individuals or species, which then act as "pools" of nutrients. The rate of nutrient movement between pools is called the **flux rate** and is measured as the quantity of nutrient passing from one pool to another per unit of time. Nutrients cycle between pools through meteorological, geological, or biological transport mechanisms. Meteorological inputs include dissolved matter in rain and snow, atmospheric gases, and dust blown by the wind. This is how sulfur alters the pH of lakes. It is emitted through smokestacks and becomes incorporated in the rain, which then falls as "acid rain." Geological inputs include elements transported by surface and subsurface drainage, and biological inputs result from the movements of animals or animal parts between ecosystems. The turnover times of nutrients in the various compartments of ecosystems seem to be longer with increasing latitude. Jordan and Kline (1972) quote 10.5 years for the cycling time of nutrients in a tropical rain forest and 42.7 years for the taiga in the Soviet Union.

Nutrient cycles have often been studied by means of the introduction of radioactive tracers into ecosystems. For example, Whittaker (1961) followed the introduction of $^{32}$P-labeled phosphoric acid into an aquarium. He found a definite sequence of nutrient uptake (Fig. 17.9):

1. $^{32}$P was rapidly taken up by phytoplankton and subsequently discharged.
2. Filamentous algae on the sides and bottom slowly picked up $^{32}$P.
3. Crustaceans grazing on algae picked up $^{32}$P even more slowly.
4. $^{32}$P began to accumulate in bottom sediment and was tied up in less-active forms.

This type of cycle is broadly similar to that occurring in natural lakes; phosphorus and other nutrients accumulate in the bottom sediment, largely unavailable for use, which is why phosphorus is so often limiting in lake ecosystems.

Nutrient cycles can be divided into two broad types:

*Local cycles*, such as the phosphorus cycle just described, which involve elements with no mechanism for long-distance transfer.

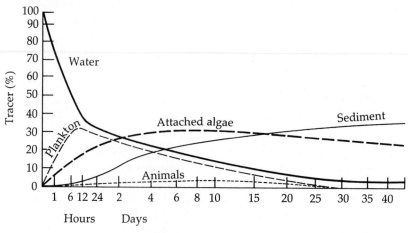

Figure 17.9    *Movement of radiophosphorus in an aquarium microcosm. Percentage of the tracer present at a given time (after correction for radioactive decay) is on the vertical axis; time after tracer introduction (on a square-root scale) is on the horizontal.* (Redrawn from Whittaker 1961.)

*Global cycles,* which involve interchange between the atmosphere and the ecosystem and are particularly applicable to elements such as nitrogen, carbon, oxygen, and water. Global nutrient cycles unite all of the world's living organisms into one giant ecosystem called the **biosphere,** the whole Earth system.

## The Phosphorus Cycle

The phosphorus cycle is a relatively simple cycle. This is because phosphorus does not have an atmospheric component; that is, phosphorus is not moved around by the wind or the rain. Phosphorus tends to cycle only locally over short time periods. Exact rates of movement of phosphorus between different components of ecosystems vary, but the general patterns are outlined in Fig. 17.10 . Any long distance transfers of phosphorus involve movement from land, to sediments in the sea, and then back to land. The Earth's crust is the main storehouse for this particular mineral.

Generally, small losses from terrestrial systems caused by leaching through the action of rain are balanced in gains from the weathering of rocks. The geologic components from the phosphorus cycle take millions of years to release significant amounts of phosphorus from rocks into the living ecosystem. Compared to this, the ecosystem phase of the phosphorus cycle is much more rapid. All living organisms require phosphorus, which becomes incorporated into substances and, which helps give plants and animals their energy. Plants have the metabolic means to absorb dissolved ionized forms of phosphorus. The most important form of phosphorus occurs as phosphate. Plants can take up phosphate rapidly and efficiently. In fact, they can do this so quickly that they often reduce soil concentrations of phosphorus to extremely low levels. Herbivores obtain their phosphorus only from eating plants, and carnivores obtain it by eating herbivores. Herbivores and carnivores excrete phosphorus as a

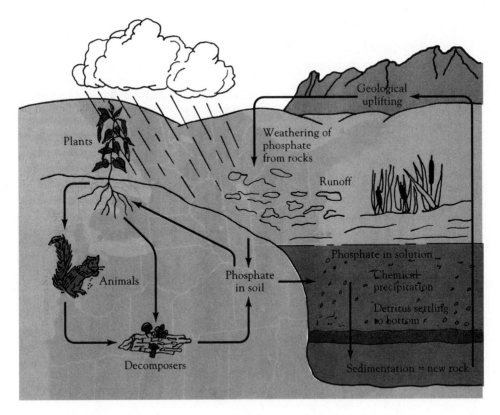

**Figure 17.10** *The phosphorus cycle. Phosphorus, which does not have an atmospheric component, tends to cycle locally. Small losses from terrestrial systems caused by leaching are balanced by gains from the weathering of rocks. In aquatic systems, some phosphorus is lost from the ecosystem because of chemical processes that cause precipitation or through settling of detritus to the bottom, where sedimentation may lock away some of the nutrient before biotic processes can claim it. This may be why phosphorus is more often a limiting nutrient in aquatic systems. This general pattern of local cycling applies to many other nutrients, including trace elements.*

waste product in urine and feces. Phosphorus is, of course, also released to the soil when plant or animal matter decomposes.

In aquatic systems, plants can take up phosphate even quicker than in terrestrial systems. In most aquatic systems, phosphorus is the limiting element. In other words, the more phosphorus that is added, the more aquatic productivity increases. Of course, an overabundance of phosphate can be damaging and can cause blooms of algae that clog up lakes. This process is known as *eutrophication*, a phenomenon more fully discussed in Chapter 20. Cultural eutrophication results when humans induce eutrophication in a lake by releasing phosphorus rich effluent into it. The phosphorus cycle can be modified by humans in a variety of ways, for example, by mining phosphate deposits for fertilizers and by polluting underground and surface waters with phosphorus-rich effluent from agricultural runoff or sewage discharge.

## The Carbon Cycle

A simplified picture of the carbon cycle is shown in Fig. 17.11. Carbon dioxide is present in the atmosphere at the relatively low concentration of about 0.03 percent. Autotrophs (primarily plants) acquire carbon dioxide from the atmosphere and incorporate it into the organic matter of their own biomass via photosynthesis. Plant respiration returns some $CO_2$ to the atmosphere. Each year, plants remove approximately one-seventh of the $CO_2$ in the atmosphere. Decomposition of plants eventually recycles most of this carbon back into the atmosphere as $CO_2$, although fires can oxidize organic material to carbon dioxide much faster. Herbivores can also return some carbon dioxide to the atmosphere, eating plants and breathing out $CO_2$, but the amount flowing through this part of the cycle is probably minimal.

The amount of carbon dioxide in the atmosphere can increase considerably when volcanoes erupt. It can also vary with the seasons in temperate environments. Concentrations of carbon dioxide are lowest during the Northern Hemisphere's summer and highest during the winter (see Fig. 20.1). This is so because there is more land in the Northern Hemisphere than in the Southern Hemisphere and, therefore, more vegetation. The vegetation has a maximum photosynthetic activity during the summer, reducing the global amount of carbon dioxide. During the winter, the vegetation respires more carbon dioxide than it uses for photosynthesis, causing a global increase in the gas. The combustion of fossil fuels is thought to have caused a far greater amount of $CO_2$ to enter the atmosphere than normal (see Chapter 20).

## The Nitrogen Cycle

The nitrogen cycle (Fig. 17.12 ) is a good example of a global cycle. There are five basic steps in the cycle:

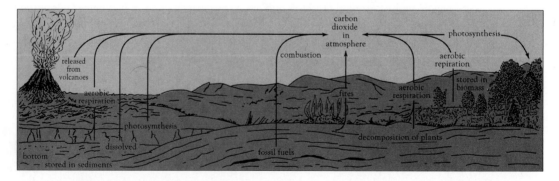

Figure 17.11     *Simplified diagram of the carbon cycle. Carbon dioxide in the atmosphere is absorbed by plants during photosynthesis, when compounds containing carbon and oxygen are assembled into living biomass; carbon dioxide is released by respiring plants and animals. Carbon is converted to other compounds during decomposition. Carbon can also become locked up in peat, coal, oil, and gas and then subsequently released during the combustion of fossil fuels. Some carbon becomes locked up in the bottom sediments of oceans. Additional carbon enters the atmosphere during volcanic eruptions.*

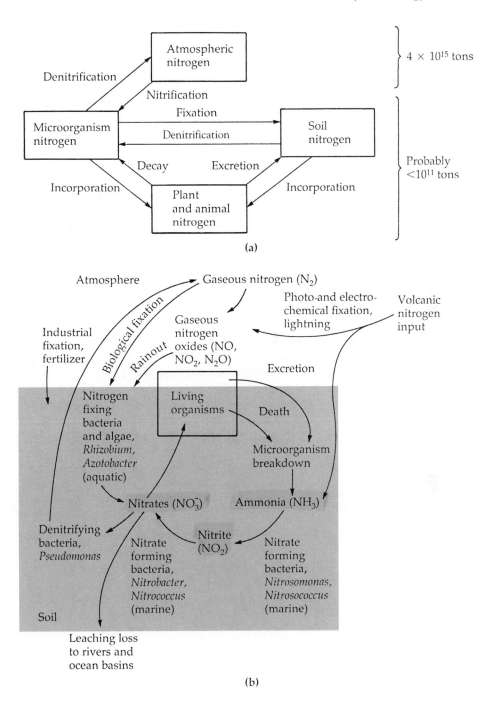

**Figure 17.12**   *The nitrogen cycle.* (**a**) *Major relationships between the very large atmosphere pool of gaseous nitrogen and the biosphere.* (**b**) *The complex interrelationships of the soil-based portion of the cycle.*

1. Nitrogen fixation
2. Nitrification
3. Assimilation
4. Ammonification
5. Denitrification

## 1. Nitrogen fixation

The quantity of nitrogen tied up in living organisms is very small compared with the total amount of nitrogen in the atmosphere. Despite the atmosphere being almost 80 percent nitrogen, plants cannot assimilate this form of nitrogen. Only certain bacteria can fix nitrogen—that is, reduce atmospheric nitrogen to ammonia, which can be used to synthesize other biological compounds. In fact, almost all nitrogen available for plants comes from nitrogen-fixing bacteria, or in aquatic environments, cyanobacteria, and there is severe competition between plants for this available nitrogen. This is why nitrogen is often a limiting factor for plant growth and why fertilizers contain so much nitrogen and have such a pronounced effect on plant growth. It is usually true that limiting nutrients such as nitrogen, and indeed phosphorus, are usually bound up tightly in the ecosystems in living components, whereas nonlimiting elements such as sodium do not accumulate in the food chain. Organisms that fix nitrogen, of course, are fulfilling their own metabolic requirements, but they release excess ammonia, and this is the nitrogen that becomes available to other organisms. The most important of these bacteria in the soil are called *Rhizobium*, which live in the special swellings called *nodules* in plants called *legumes*.

In terms of the global nitrogen budget, industrial fixation of nitrogen for the production of fertilizer makes a significant contribution to the pool of nitrogen containing materials in the soils and waters of agricultural regions. Industrial fixation of nitrogen may occur in other processes, too.

## 2. Nitrification

For plants, the most useful form of nitrogen in the soil is not ammonia but nitrates. Nitrates are formed from ammonia by more bacteria such as *Nitrosomonas, Nitrococcus, and Nitrobacter*.

## 3. Assimilation

Plant roots assimilate nitrogen mainly in the form of nitrates, and animals assimilate their nitrogen by eating plants.

## 4. Ammonification

Ammonia can also be formed in the soil through the decomposition of plants and animals and the release of animal waste. This process is again carried out by many bacteria and fungi. Again, the ammonia in the soil is generally not used directly by plants.

## 5. Denitrification

Detrification is the reduction of nitrates to gaseous nitrogen, and denitrifying bacteria perform almost the reverse of their nitrogen-fixing counterparts.

The most important components of the nitrogen cycle are the nitrogen compounds in the soil and the water, not in the atmosphere. Although nitrogen fixation from the air has been important in the gradual build-up of a pool of available nitrogen, in most systems it contributes only a fraction of the nitrogen assimilated by total vegetation. However, locally, direct nitrogen fixation by bacteria in the root nodules of legumes can be important. Legumes can grow in poor soils, devoid of nitrogen, and thus avoid competition with many other plants. Nitrogen-fixing plants can be very important in agriculture for this reason and because they replenish soil nitrogen. In agricultural systems, because so much nitrogen is removed when the plants are harvested, large quantities of nitrogen have to be re-added into the soil in the form of fertilizers.

Nitrogen availability can be critical in limiting individual species and population cycles. One of the most striking examples was often thought to be that of the brown lemming, which lives in the tundra areas of North America and Eurasia. The ecosystem of the tundra is simple compared with more temperate or tropical ecosystems, and the lemming is often the major herbivore. Every three to four years, numbers of these small rodents build up, only to crash again in a never-ending cycle (Fig. 17.13). This ecosystem has been studied in some detail on the Arctic coastal-plain tundra near Point Barrow, Alaska. The traditional story is as follows. As lemming numbers increase in winter, their feces and urine stimulate plant growth,

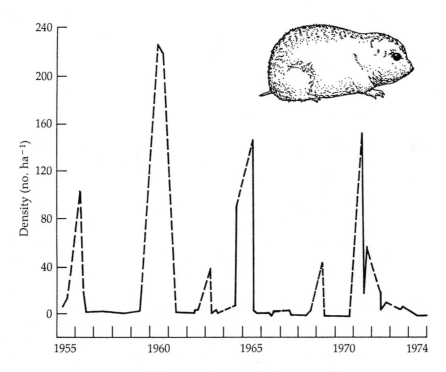

**Figure 17.13** *Estimated lemming densities in the coastal tundra at Barrow, Alaska, for a 20-year period.* (Redrawn from Batzli et al. 1980.)

which in turn increases lemming production. Eventually, the lemming numbers become so great that the vascular plant cover is thinned out. More sunlight reaches the ground, thawing out the soil overlying the **permafrost,** and plant roots can penetrate a little further. Permafrost lower down prevents water movement, so nutrients are not **leached** out but accumulate in peat, which is not available to the plants. The intense grazing and lack of root nutrients soon reduce forage quality. This trend, coupled with high predation rates, social strife, and dramatic emigration, reduces lemming numbers drastically. Low nutrient quality also prevents breeding the following year. Gradually, during the next two to three years, the vegetation slowly recovers, nutrients increase, and the cycle begins again. This whole description of the process that controls population variation was termed the *nutrient-recovery hypothesis* by Pitelka (1964) and Schultz (1964), and although it is an attractive idea, it must be stressed that not all the elements of the cycle have been proven beyond question. Much work remains to be done, and Batzli et al. (1980) have pointed out some inconsistencies between data and theory.

First, food supplementation in times of high densities should prevent population decline, but early experiments failed to produce this result. It is possible, however, that the failure resulted because predators were attracted from nearby unprofitable areas and killed higher numbers of rodents. When Ford and Pitelka (1984) provided supplementary food *and* prevented predation in California, they found that experimental populations declined modestly compared to the natural crash of control populations that summer. It is also worth noting that "intrinsic" changes in the behavior of the rodents themselves may be causing variations in population numbers (see also Chapter 12). As densities of animals increase, aggressiveness between individuals goes up, affecting hormonal levels (Krebs 1985b). This process, in turn, promotes a huge dispersal of animals away from centers of high population density (Stenseth 1983), so the population cycles of lemmings may essentially be caused and regulated by the behavior of the animals themselves.

Similar processes probably occur in other ecosystems as well but are harder to unravel. The ecosystem of the Arctic is simple: There are about 100 species of plant, but only about 10 are important, comprising 90 percent of the plant biomass. The lemming is the only major herbivore, and it has two main predators and six minor avian and mammalian enemies. Other ecosystems are much more complex. In addition, much carbon, nitrogen, and phosphorus is held in the soil in the Arctic; only about 2 percent is held in living material. This proportion is in sharp contrast to those in temperate and tropical forests, where between 20 and 70 percent of these nutrients are found in living plants (Fig. 17.14). Finally, it must be emphasized that Arctic systems are commonly not in a state of equilibrium; much organic matter is slowly accumulating as peat. This is not the case in many other ecosystems. Thus, although some of the peculiarities of other ecosystems could probably be explained by studies of nutrient cycling, the answers are likely to be different from those of the Arctic.

Nutrient availability on a large scale can have a severe impact on the type of vegetation that is supported. In the Northern Hemisphere, where many areas have been glaciated, soils are derived from pulverized bedrock (till) and are very fertile. Those in the Southern Hemisphere, however, have not commonly undergone such a phenomenon. These soils are old, weathered, and infertile, derived from Gondwanaland. Such areas include Australia,

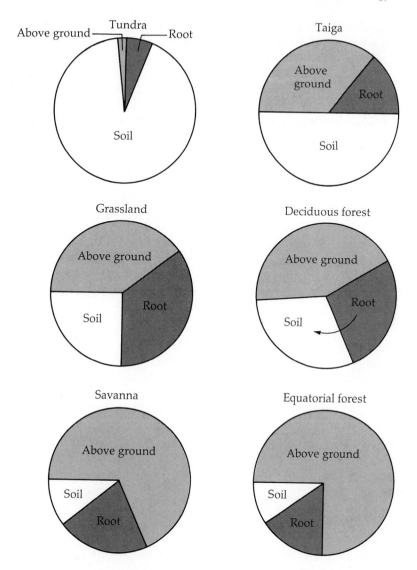

**Figure 17.14** *Deposition of nitrogen in the three organic-matter compartments (above ground, root, and soil) for each of six biome types. In arctic ecosystems most nitrogen is held in the soil. Soil nutrient availability may thus be critical in the population processes of arctic life. However, in most other ecosystems nitrogen is held more in living matter and biotic interactions may be more important in ecosystem processes.*   (Redrawn from Swift, Heal, and Anderson 1979.)

South America, and India. Among the most severely limiting nutrients is phosphorus, and it is interesting that *Eucalyptus* species are well adapted to grow on soils with a low phosphorus content. In general, the vegetational communities that grow on such impoverished soils can be termed **oligotrophic** in much the same fashion as crystal-clear lakes. Nutrient-rich temperate soils are then **eutrophic.** However, because vegetational types have evolved specifically to cope with local conditions, productivity of natural vegetation differs little between oligotrophic and eutrophic systems. When disturbed, however, for agriculture, the nutrient-poor systems quickly lose their productive potential, whereas the nutrient-rich ones are not so easily perturbed. This factor must be considered in the search for reasons for the often poor success rates of tropical agriculture.

# Summary

1. In ecosystems, energy is lost with each transfer up the food chain. In contrast, chemicals are not dissipated; they remain in the ecosystem and often concentrate at higher trophic levels.

2. In most ecosystems, plant material goes not to herbivores but to decomposers after the plant dies. Most primary production is used in plant respiration, so plants are the most important consumers. Bacteria as decomposers are next in importance and degrade about one-seventh of the energy that plants use. Herbivores degrade only about one-seventh of the energy the bacteria use.

3. In general, primary production is highest in tropical forests and decreases progressively toward the poles. There are exceptions to this trend. Temperate salt marshes are as productive as many tropical areas. Tropical deserts, because of a lack of moisture, are not as productive as temperate grasslands. Thus, temperature and rainfall both limit primary productivity.

4. Nutrient deficiency, particularly of nitrogen and phosphorus, can limit primary productivity, too. In aquatic systems, light availability can also be important. In freshwater lakes, excess phosphorus can cause huge algal blooms and turbid water, a process known as *eutrophication*.

5. The limit to secondary productivity is available primary productivity. This means that, at a large scale, plant defenses do not effectively reduce consumption by herbivores.

6. Nutrients may cycle locally, like phosphorus in a lake, or globally, like carbon and nitrogen.

7. The phosphorous cycle is relatively simple because it does not have a large atmospheric component. The Earth's crust is the main storehouse for phosphorus.

8. In the carbon cycle, carbon dioxide is present in low concentrations in the atmosphere (0.03 percent) and at higher levels in plants. Each year, plants remove about one-seventh of the $CO_2$ in the atmosphere. Decomposition eventually returns most of this $CO_2$ back to the air. Forest fires can accelerate this process.

9. The nitrogen cycle is, like the carbon cycle, a global cycle. There are five basic steps to the nitrogen cycle:

a. Nitrogen fixation of atmospheric $N_2$ to ammonia by soil bacteria.

b. Nitrification—the formation of nitrates from ammonia by soil bacteria.

c. Assimilation of nitrates by plant roots.

d. Ammonification—the production of ammonia through plant decomposition and the production of animal wastes.

e. Denitrification—the reduction of nitrates to gaseous nitrogen. The limiting step is generally the first one, fixation of atmospheric nitrogen by bacteria. As in so many biogeochemical cycles, humans can radically change chemical fluxes. In this case, industrial fixation of nitrogen for the production of fertilizer can change local nitrogen budgets immensely.

---

*Discussion Question:* What advantages and disadvantages does the study of whole ecosystems, through the analysis of nutrient or energy flows, have over population or community ecology, which emphasizes biotic interactions?

# Applied Ecology

The influence of humans is felt by many plants and animals, as this porcupine is finding out. Applied ecology tries to use ecological theory to solve problems relating to real-world issues such as the conservation of animals, the effects of pollution, and how to sustainably harvest a population. (Rue, Photo Researchers, Inc. 6F 0905.)

Applied ecology is the study of people's influence on ecological systems and, concurrently, the management of those systems and resources for society's benefit. Many problems instantly arise when one faces the prospect of ecological management—what constitutes an environmental problem depends partly on public perception. An organism or habitat that has been incorporated into art, poetry, or national symbolism becomes an object of special concern—hence, the overriding importance of the bald eagle to the United States. However, some agreement on the important issues in applied ecology have been reached by U.S. national committees (see table on following page). These include the prevention of pollution, the conservation of natural areas, and the preservation of the Earth's genetic diversity.

It is very difficult to provide a blueprint for success in solving ecological problems. Each case must be dealt with largely on its own merits. There are so many unusual and unique ecological variables associated with each case study that broad generalizations are of little use. Useful predictions are more often based on local field experience than on formal theory. Strong (1980) has argued that predictability in biology will never be as advanced as it is in atomistic sciences such as chemistry and physics. He suggests that we cannot compare organisms to atoms; whereas there are only a relatively small number of atom types (106 at last count), there are many millions of organisms (see Chapter 2). Thus, although scientists can discover certain laws linking similar atoms, the profusion of different and independent organisms suggests that there will be few universal solutions to ecological problems. Although ecological problem solving cannot be reduced to a formula, the Committee on the Applications of Ecological Theory to Environmental Problems (National Research Council 1986) recommended that nine general guidelines be followed:

*Involve scientists from the beginning:* Scientists can help to identify nonobvious goals, can translate environmental goals into scientific objectives, can show what information is needed to answer major questions, and are usually important in determining the values attached by society to ecosystem components.

*Treat projects as experiments:* Pilot-scale experiments were conducted to study the potential effects of the Alaska pipeline on caribou movements (Cameron et al. 1979). In the process, a great deal more was learned about caribou movements than would have been possible if the pipeline had not been constructed or if it had been constructed without prior experimentation. Frequently, a major project itself can function as a large-scale experiment if careful baseline studies are done beforehand.

*Publish information so that others can learn from any mistakes:* Larkin (1984) believes that literature review can provide 50 percent of the information needed in most initial impact

**Important issues in applied ecology. (From National Research Council 1986.)**

Legal requirements

    Air and water quality standards

    Public health

    Rare, threatened, and endangered species

    Protected areas or habitats

Aesthetic values

    Landscape appeal

    Attractive communities

    Appealing species (e.g., large ungulates, colorful birds, cacti)

    Species at higher trophic levels (e.g., eagles and tigers)

    Clean air and water

Economic concerns

    Species or habitats of recreational or commercial interest

    Ecosystem components

Environmental values and concerns

    Ecosystem rarity or uniqueness

    Sensitivity of species or ecosystems to stress

    Ecosystem "naturalness"

    Genetic resources

    Ecosystem services

    Recovery potential of ecosystems

    "Keystone" species

assessments and as much as 75 percent when coupled with brief reconnaissance surveys. Looking for analogous studies is extremely useful; even if some are not perfect, reading them can be useful in "scoping out a project."

*Set proper boundaries on projects:* Well-intentioned efforts to conserve animals in reserves can be undermined if they are planned for too short a term or too small an area (Soulé and Wilcox 1980).

*Use natural-history information:* Nonrandom harvests of animals, in which unequal numbers of the sexes are taken (for example, the preferential hunting of male deer [Law 1979] or the unwitting selection of [larger] male fish through size selectivity of fishing gear [Moav, Brody, and Hulata 1978; Richer 1981]) can lead to uneven and catastrophic sex ratios. The carefully designed Garki malaria-control project in Africa and its attendant mathematical predictions were thrown into turmoil because incorrect assumptions were made about the biology of the mosquito vectors, which did not land inside buildings (and thus were not exposed to contact insecticides) as often as predicted (Molineaux and Gramiccia 1980; May 1986).

*Be aware of interactions:* When **DDT** was first introduced, no one imagined it would eventually show up in marine fish (Buckley 1986). Also, in northern Australia, when water buffalo were introduced, they brought with them their own blood-sucking fly, a species that bred in cattle dung and transmitted an organism sometimes fatal to cattle. Australia's native dung beetles were accustomed only to the small pellets of grazing marsupials and could not tackle the large pats of the buffalo. The blood-sucking flies proliferated. Eventually, African dung beetles, used to large pats, had to be brought in to compete with the flies for dung, which they successfully did, reducing fly populations (Moon 1980).

*Be alert for possible cumulative effects:* Recent awareness of such serious cumulative effects as **acid rain,** loss of tropical forests, and the threatened extinction of many species has brought increasing political and scientific attention to these effects. Increasing the intensity of harvesting can lead to population collapse if a critical threshold is passed, especially in conjunction with other environmental changes (Beverton 1983).

*Plan for heterogeneity in space and time:* Plankton communities undergo seasonal turnover, and any population changes must be viewed in the light of natural variations in population densities.

*Prepare for uncertainty and think probabilistically:* Attempts to achieve maximum sustainable yields, MSYs, in many fisheries can result in more variable yields, greater population fluctuations, and a greater likelihood of population collapse (May 1980).

Sometimes, it is simply too hard for ecologists to move from pure science to applied science. As Wigglesworth (1955) explained, pure scientists are used to attempting to solve problem A but, if that proves too difficult, to switching to problem B and even to coming across a solution to problem C. In applied cases, however, there is no escape from problem A.

Most ecological problems in today's world are linked to the adverse effects people have on the environment. There is, therefore, much interest in knowing how human population levels will change in the future.

# CHAPTER 18

# *Human Population Growth and the Direct Exploitation of Wildlife*

## 18.1 HUMAN POPULATION GROWTH

Though the data are scanty before A.D. 1650, they consistently show that populations have a very slow rate of growth. High infant-mortality rates and low longevity meant that increases were small; in Roman times, the average life expectancy for men was about 30 years. Even then, humans severely affected the environment. In Bronze Age Britain, for example, lime trees (*Tilia* sp.) may have been selectively removed because the sod underneath them was agriculturally rich (Turner 1962). Forty percent of the mammal population that lived with humans in the Olduvai gorge in Tanzania are now extinct (Ager 1991).

Up until the beginning of agriculture and domestication of animals, about 10,000 B.C., the average rate of population growth has been very roughly estimated at about 0.0001 percent per annum. After the establishment of agriculture, the world's population grew to about three hundred million by A.D. 1, and 800 million by the year 1750, but still the average growth rate was well below 0.1 percent per annum. The modern period of rapid population growth may be regarded as starting at about 1750. Average annual growth rates climbed to about 0.5 percent between 1750 and 1900, 0.8 percent in the first half of this century, and 1.7 percent in the second half. Advances in medicine and nutrition are certainly responsible for a large part of the growth in reproductive rate and longevity of humans. In this relatively tiny period of human history (equivalent to the last 1 minute on a 24-hour clock), the world's human population increased from 0.8 billion in 1750 to 5.3 billion in 1990. Thus, in less than 0.1 percent of human history has occurred more than 80 percent of the increase in human numbers. This rapid increase in human growth rates in recent times is illustrated in Figure 18.1.

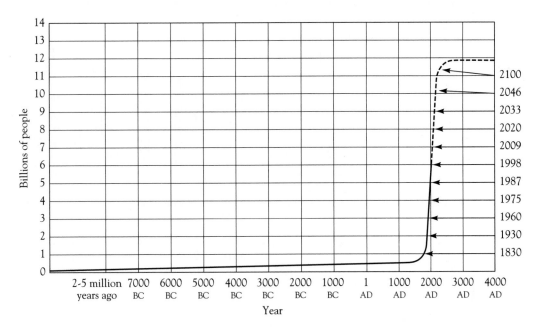

**Figure 18.1**   *The world population explosion. For most of human history, the population grew very slowly, but in modern times it has suddenly "exploded." Where and when it will level off is the subject of much debate.*

It was estimated that in 1990 the world's population was increasing at the rate of three people every second. In 1984, United Nations' projections estimated that the most likely course of world population was one of growth towards an eventual total stabilizing at around 10 billion people by the year 2100. However, human growth rates have not been reduced as quickly as was anticipated. In 1994, United Nations estimates pointed to a world population stabilizing at around 11 billion toward the year 2100.

In the developed countries, the average annual rate of population increase has slowed considerably, from 1.19 percent in 1960–1965 to a projected 0.48 percent in 1990–1995. In the developing world, however, average annual growth rates rose from 2.35 percent in 1960–1965 to 2.38 percent in 1970–1975 before dropping to 2.1 percent in 1980–1985. The results of these growth rates are that in developed nations, the population has seemed to stabilize at around one billion people, but the number of people in developing countries is still increasing (Fig. 18.2). The result is that the world population growth is still rising and is expected to rise to more than 6 billion people by the year 2000.

Estimates of birth rates and death rates can be broken down into geographical regions (Fig. 18.3.) African countries clearly have the highest birth rates. Overall, the global birthrate has decreased from an average of 37 per thousand during 1950–1955 to 27 per thousand for the period 1980–1985. Thus, although the world's population is still increasing, the rate of increase has slowed.

Global population growth can also be expressed as fertility rates, the average number of live births that would typically be borne by a woman of childbearing age (Table 18.1). Again,

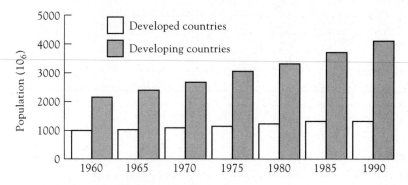

Figure 18.2   *Population growth in developed and developing world regions, 1960–1990.*

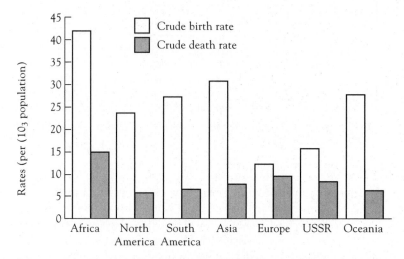

Figure 18.3   *Crude birth rates (number of live births per $10^3$ population) and crude deaths (number of deaths per $10^3$ population) by world region, 1990.*

the fertility rate differs considerably between countries. In Africa, the total fertility rate of 6.0 has scarcely declined since the 1950s when it was around 6.6 children per woman. However, in Latin America and Southeastern Asia, fertility rates have declined considerably from the 1970s and are now at around 4.0. In 1960–1965, the global average number of children born per woman was almost five. In 1990, the average was 3.3. In more developed regions, the fertility rate has declined steadily and is now expected to remain fairly stable at around 1.9, for the next few decades. Part of the reason for the differences among countries has been linked to contraceptive use. For example, a 1988 study indicated that in developed countries around 70 percent of couples, where the woman was of reproductive age, were using some form of contraceptive. In developing countries, this level was estimated to be around 45 percent . In Africa, only an estimated 14 percent of couples use contraceptives,

TABLE 18.1 *Total fertility rate, the average number of children per woman in her reproductive lifetime, 1990 data.*

| Region | Fertility rate |
| --- | --- |
| Africa | 6.0 |
| South America | 4.0 |
| Asia | 4.0 |
| Oceania | 3.0 |
| North America | 3.0 |
| Former USSR | 2.3 |
| Europe | 2.0 |
| United States | 1.9 |
| World | 3.0 |

whereas the level of use is around 50 percent in Asia and about 56 percent in Latin America. Some of the most startling reductions in fertility have been experienced in China. In the 1950s and 1960s, the total fertility rate in China was six children per woman. During the 1970s, family-planning services were offered by the government and incentives were also offered to couples who agreed to limit themselves to one child. In 1990, an estimated 75 percent of the population used birth control, and the fertility rate had dropped to 2.2.

Government-supported family-planning programs have been implemented in an increasing number of countries during the last two decades. In 1976, only 97 governments provided direct support for family planning, compared with 125 in 1988. In contrast, the number of governments limiting access to family planning fell from 15 in 1976 to 7 in 1988. However, as of 1989, there were still 31 countries in the developing world where couples have virtually no access to modern family-planning methods. One could also argue that reductions in birth rates have not occurred in some developing countries because women wish to have large families, but there are many surveys which show that women in developing countries would like to have fewer children than they already have.

Even though Africa has a high rate of population increase, its total population, at about 650 million people, is far below the 3 billion that inhabit Asia (Table 18.2). Thus, in terms of the global population level, even a small decrease in the rate of population increase in Asia would reduce the absolute numbers of humans by more than a substantial decrease in the rate of population increase in Africa. The density of people in Africa, at 53 persons per square kilometer, is much less than the 330 persons per square kilometer in Asia.

It is interesting that despite the concerns in many parts of the world about increase in human population growth rates, in some countries the concern is that the growth rates are not enough. This attitude has been prevalent in some western European countries. Since about 1965 the "baby-boom" in these countries, together with other more-developed countries such as Australia, Canada, New Zealand, and the United States, has turned into a "baby-bust," and total fertility rates have fallen below the national replacement level of 2.1 children per woman (Fig. 18.4). This rate of replacement, 2.1 children per woman, allows for

TABLE 18.2  *Trends in world population size, rate of increase, and density of people.*

| Region | Population | | | | | Rate of increase (%) | | | | Density of people per Km² |
|---|---|---|---|---|---|---|---|---|---|---|
| | 1960 | 1970 | 1980 | 1985 | 1990 | 1960–65 | 1970–75 | 1980–85 | 1990–95 | |
| WORLD | 3,019 | 3,698 | 4,450 | 4,854 | 5,292 | 1.99 | 1.96 | 1.74 | 1.71 | 115 |
| Developed regions | 945 | 1,049 | 1,136 | 1,174 | 1,205 | 1.19 | 0.86 | 0.65 | 0.48 | 77 |
| Developing regions | 2,075 | 2,649 | 3,314 | 3,680 | 4,087 | 2.35 | 2.38 | 2.10 | 2.06 | 302 |
| Africa | 281 | 363 | 481 | 557 | 648 | 2.48 | 2.69 | 2.95 | 3.01 | 53 |
| Latin America | 218 | 285 | 362 | 404 | 448 | 2.80 | 2.48 | 2.19 | 1.94 | 16 |
| Northern America | 199 | 226 | 252 | 265 | 276 | 1.49 | 1.06 | 1.00 | 0.71 | 164 |
| Asia | 1,677 | 2,101 | 2,583 | 2,834 | 3,108 | 2.19 | 2.27 | 1.86 | 1.82 | 330 |
| Europe | 425 | 460 | 484 | 492 | 498 | 0.91 | 0.58 | 0.32 | 0.22 | 160 |
| USSR | 214 | 243 | 266 | 277 | 288 | 1.49 | 0.96 | 0.84 | 0.68 | 13 |
| Oceania | 16 | 19 | 23 | 25 | 26 | 2.09 | 1.78 | 1.55 | 1.34 | 16 |

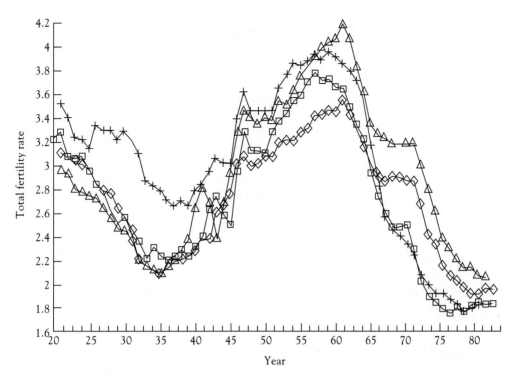

**Figure 18.4**    *Total fertility rates, 1920–1983, for Australia (◊), Canada (+), New Zealand (Δ), and the United States (□).*

natural mortality and, in the end, produces a natural increase of about 2.0, ensuring that the population neither decreases nor increases. Some politicians fear that a negative growth rate could affect the political and economic structure of some nations.

This fear of population decline is not new. It last flourished during the Great Depression in the years between World War I and II. Jacques-Bené Chirac, the conservative former prime minister of France, described the demographic situation of France in the 1980s as terrifying and of Europe as "vanishing." Aleksandr Solzhenitsyn lamented the demise of Russian culture and commented that with Russia's low birthrate, within a century the Russian people would be diminished by one-half, and their culture would almost vanish from the face of the Earth.

There are three main ecological concerns with a burgeoning human population. The first is the greater chance for a plant or animal species to be pushed to extinction through direct exploitation, hunting, fishing, or harvesting. The second is the replacement of natural habitats by human altered habitats, plantation forests or agricultural fields that will have strong indirect effects on wildlife, threatening many more species with extinction. The third concern is pollution. Pollutants in rivers, streams, even in the air may affect the distribution patterns of plants and animals. There are also more chances for people to introduce foreign species into parts of the globe where they will out-compete native species—a sort of biological pollution. It is with these three concepts in mind—direct exploitation, habitat destruction, and pollution—that our discussion of applied ecology will proceed.

## 18.2  DIRECT EXPLOITATION

In many areas, animals and plants are exploited in their own right. In developing countries, wild game is a principal source of meat, providing 75 percent of the animal protein in Ghana and Zaire and 20 percent along Brazil's Trans-Amazon highway. A large proportion is also taken by hunters and poachers for markets in Europe, North America, or Japan. Wildlife may be traded as hides and skins for fur and leather companies, as exotic meat for exotic dishes, or as products for the perfume, pharmaceutical, or "aphrodisiac" industries. Photo 18.1 shows an example, a giraffe killed illegally just for its tail. Alive, wild animals are sold to the pet trade, zoos, and biomedical research. In 1979 the estimated turnover for smuggled Australian wildlife was $30 million. The U.S. Fish and Wildlife Service estimated in 1984 that the illegal traffic in birds was at least as high as the 223,000 that were legally imported each year. Small boatloads of walrus heads have been observed in Alaska, the animals decapitated for their valuable tusks (Allen 1980). The U.S. trade in carved ivory may have accounted for some 32,000 elephants in 1986 (Di Silvestro 1988), and in addition at least 2,000 more may have been legally killed to supply hides to North American markets. In 1989, the international ivory trade consumed tusks from more than 80,000 African elephants a year, taken from a total population of approximately 625,000. What can be done to protect wildlife? Should we ban trade in wildlife products?

Photo 18.1   *A giraffe poached in Kenya for its tail. The rest of the carcass has been left to rot.*   (Photo by Thane Riney, Food and Agriculture Organization.)

Let's return to the case of the elephant. In southern Africa, numbers of elephants were actually increasing in 1989 in South Africa, Zimbabwe, and Botswana (Armstrong and Bridgland 1989). Careful management and culling of herds allowed numbers to increase and to support a valuable ivory trade. These countries were loath to see this source of revenue denied them, and they claimed that similar success stories in other African countries are possible with good management. Besides, an official ban on ivory, they argued, would simply drive the whole business underground, which in turn would cause prices to skyrocket, further encouraging heavy poaching. In Zambia, availability of local hunting licenses drastically reduced poaching because local poachers were generally willing to apply for licenses. Professional poaching by outsiders continues and is heavily combatted.

Wildlife imports are surprising in their extent. The imports of wildlife into Britain during just six months in the 1970s were staggering (Table 18.3) and included species that supposedly should not even have been traded. Taken to its limit, hunting can and does exter-

---

**TABLE 18.3  Selected imports of wildlife into Britain (January–July 1976).  An asterisk indicates species that should not be legally traded.  (From Burton 1976.)**

| | |
|---|---|
| *Leopard (Panthera pardus) | 661 skins |
| *Jaguar (Panthera onca) | 279 skins |
| Margay (Felis wiedii) | 9,277 skins |
| Geoffroy's cat (Felis geoffroyi) | 14,169 skins |
| Ocelot (Felis pardalis) | 16,619 skins |
| Tiger cat (Felis tigrina) | 9,000 skins |
| *Polar bear (Thalarctos maritimus) | 101 skins |
| Malayan bear (Ursus malayanus) | 100 live |
| Elephant (Loxondonta africana) | 42,871 kg ivory |
| Estuarine crocodile (Crocodilus porosus) | 100 live |
| Estuarine crocodile (Crocodilus porosus) | 523 skins |
| Caiman (Caiman crocodilus) | 310 live |
| Caiman (Caiman crocodilus) | 5,021 skins |
| Tortoises (Testudo graeca) | 245,026 live |
| (Testudo horsefieldi) | 25,000 live |
| *Hawksbill turtle (Eretmochelys imbricata) | 322 kg tortoise shells |
| Monitor lizards *(Varanus flavescens) | 646,389 skins |
| *(Varanus bengalensis) | 462 live |
| *(Varanus griseus) | 245,254 skins |
| (Varanus exanthematicus) | 25,000 skins |
| (Varanus niloticus) | 25,002 skins |
| (Varanus salvator) | 22,008 skins |
| Iguana (Iguana spp.) | 25,000 skins |
| Tegu (Tupinambis spp.) | 40,013 skins |

minate species in the wild. The Arabian oryx is a case in point. The last known wild Arabian oryx was reportedly killed in 1972. Luckily, three were taken captive in 1962 in the Rub 'al Khali, the great southern desert of Saudi Arabia, just as a hunting party was preparing to shoot them for fun. These animals, together with a small number of others donated by members of the royal families of Saudi Arabia and Kuwait, were bred in captivity, and a herd of 400 was formed in zoos in the southwestern United States. The story completed its full circle when captively bred animals were released in Oman in 1980 (see Chapter 19).

## The Optimal Yield Problem

How many deer can be shot before a marked effect on population numbers occurs? How can people harvest a large part of a population without causing long-term changes in its equilibrium numbers? Removing too many young individuals may seriously impair population-age structure and hence population size of future generations, and removal of a large fraction of the dominant reproductive individuals would also seriously affect reproductive output. Unfortunately, this scenario is at loggerheads with an economic viewpoint. Current profits (which can be invested at a favorable rate of interest) are more valuable than future profits. Economically, it is best to overexploit a population now.

## 1. Maximum Sustainable Yield

In theory, if its environment is constant, every population has a maximum sustainable yield (MSY), the largest number of individuals that can be removed from the population without causing long-term changes. In practice, the MSY is often difficult to determine, let alone achieve. The MSY is a result of both the reproductive rate and the number of individuals reproducing. Both yields and population sizes fluctuate more as the MSY is approached, the effect being most pronounced in large mammals such as whales (May 1980).

In debates over the effects of harvesting or hunting animal populations, two schools of thought have emerged. In one, hunting is seen as an extra strain on the population in addition to those already acting on it. This view is referred to as the *totally additive mortality hypothesis*. In the other, hunting is seen as a source of mortality that merely replaces other sources of mortality; in other words, shooting game merely removes individuals that would be killed in some other way. This view is known as the *compensatory mortality hypothesis* (Anderson and Burnham 1976). Surprisingly, the data seem to support the compensatory hypothesis rather than the totally additive mortality hypothesis (Nichols et al. 1984) at least for populations of mallards, *Anas platyrhynchos*. Hunting waterfowl is probably the single most important hunting activity in North America (Martin, Pospahala, and Nichols 1979), so the data on this issue are among the best. Nearly 700 species of migratory birds occur in North America north of Mexico, and nearly 100 of these are shot as game. It has been estimated that the fall population of ducks is more than 100 million, and hunting them provides more than 17 million days of hunting recreation for 2.4 million hunters annually.

Of course, the acid test for comparing the totally additive mortality and the compensatory mortality hypotheses would be to ban hunting over a large experimental area and then to examine subsequent mortality rates. This experiment is unlikely to be politically or socially acceptable. Data like these are necessary, however, to answer the pressing question of how much game it is acceptable to take; already hunting is responsible for between 35 and

66 percent of known mortality in U.S. black ducks (*Anas rubripes*) and mallards (Krementz et al. 1988).

Additional studies on other organisms seem to indicate that mortality of populations due to shooting is "made up" by lessened mortality in other areas. This was true for grey squirrel populations in the United States (Mosby 1969) (often the game animal most hunted by hunters) and for wood pigeons in the United Kingdom, shot both for the pot and as an agricultural pest (Murton, Westwood, and Isaacson 1974). This is not to say that harvesting cannot seriously affect population numbers. If the harvesting technique is more thorough— for example, trawling for fish—then overexploitation can cause population stocks to crash.

## 2. Calculating Yields

Out of all applied fields, it is in fisheries that the most work on the optimal yield problem has been done. Commercial fisheries generate a vast amount of money, and the problem of overfishing has been addressed since at least the 1920s when commercial stocks of many species began to dwindle.

For a harvested population, the important measurement is the *yield*, expressed in terms of either weight or numbers over a particular time period to give a catch per unit effort. Catch per unit effort can then be compared year after year to determine how well a particular managed resource is doing. Remember that the maximum yield of a population is related to the maximum population increase. Greatest population increase, $dN/dt$, occurs according to the equation

$$\frac{dN}{dt} = rN\left(\frac{K-N}{K}\right)$$

at the midpoint of the logistic curve, Fig. 6.5. Thus, maximum yield is obtained from populations at less than maximum density, when they are constantly trying to expand their own population densities into unutilized resource areas. Adjusted for fishing losses, $dN/dt$ becomes

$$\frac{dN}{dt} = rN\left(\frac{K-N}{K}\right) - qXN$$

where $q$ is a constant and $X$ equals the amount of fishing effort; so $qX$ equals fishing mortality rate. As fishing and hence fishing losses increase, $qXN$ can begin to affect the total catch (yield) severely. This relationship is clearly shown in Fig. 18.5, which details the relationship between fishing effort and total catch for the Peruvian anchovy. The Peruvian anchovy fishery was the largest fishery in the world until 1972, when it collapsed. In 1972, 12.3 million metric tons were harvested, and this one species alone comprised 18 percent of the world's total harvest of fish. Note that from 1964 to 1971, the catch was close to the supposed maximum. People may not have been the only cause of population collapse. In 1972, the phenomenon known as El Niño, a major climatic change, a warming of the oceans' volcanic activity, occurred, permitting warm tropical water to move into the normally cool, nutrient-rich upwellings near the Peruvian coast. The anchovies failed to spawn, and adult fish moved south to cooler waters. Whether or not the fishery might have been saved if fishing had

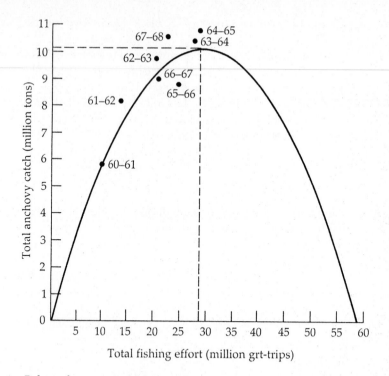

**Figure 18.5**    *Relation between total fishing effort and total catch for the Peruvian anchovy fishery, 1960–1968. The effects of humans and seabirds are combined in these data. The parabola represents the logistic model fitted to these data. Arrows indicate maximum-sustained-yield-appropriate fishing effort.*    (Redrawn from Boersma and Gulland 1973.)

ceased or if its intensity had been reduced to allow recovery is a matter for speculation. From the logistic equation, MSY can be estimated as

$$MSY = Nr$$

where $r$ is the rate of natural increase and $N$ is the average number of animals present throughout the year. However, this approach assumes a constant rate of cropping throughout the year. If there is one "season," then MSY is best estimated by the equation

$$MSY = N\,(1 - e^{-r})$$

which is likely to be a lower value. If $r = 0.14$ and $N = 10,000$, then

$$MSY = 10,000\,(1 - e^{-0.14}) = 1,306$$

But if the harvest is spread out over the year, then $N$ is likely to be reduced by natural deaths. $N$ might be reduced by 10 percent so that $N = 9,000$. Then

$$MSY = 9,000 \times 0.14 = 1,260$$

In practice, cropping is often limited to seasons if only to give animals some respite during breeding. Several techniques are available for estimating MSY of a population (Eltringham 1984).

In a previously uncropped population, one proceeds largely by trial and error, by guessing the MSY and taking a constant number of animals, year after year. It is necessary to count the population before and after harvest (Caughley 1977). For each year, compute

$$log_e \frac{N_{t+1}}{N_{t-C}}$$

where $N_t$ is the number at time $t$ and $C$ is the crop. Plot this value against $N_t$, and the y intercept is the rate of natural increase $r$. Now,

$$MSY = \frac{r}{2} \times \frac{K}{2}$$

where $K$ is the original size of the population.

If a population has been harvested for some time, then other methods can be applied to estimate MSY. For example, $r$ can be estimated from the equation

$$r = \frac{KH}{(K-H)}$$

where $H$ is the harvest in past years and $N$ is the size of the population from which it has been taken. MSY is now calculated as before.

Finally, Caughley (1976) has suggested

$$MSY = \frac{CK^4}{4N^2 (K2-N^2)}$$

where, as before, $N$ is the standing crop (size of cropped population), $C$ is the annual crop, and $K$ is the original, unharvested population size.

## 3. Economic Yield

The ultimate yield of any harvest is, of course, not in biomass but in dollars. Surprisingly, it is often poor business to operate a fishery or any other type of harvest at maximum yield (Gordon 1954). Consider Fig. 18.6, a simple economic model for a fishery. Total costs are assumed to be proportional to fishing effort, and revenue is directly proportional to yield, so that the yield curve of Fig. 18.5 is identical to the revenue curve of Fig. 18.6. The important point is that maximum economic revenue occurs at a lower fishing density than does maximum yield. It might seem that sound economic practice would also ensure a safe biological management strategy. Of course, if one country's total costs rose at a lower rate

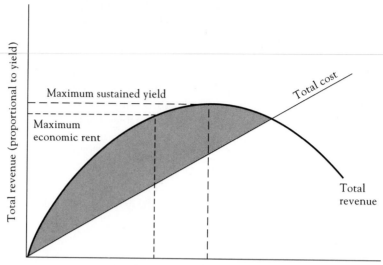

Figure 18.6   *A simple economic model of a fishery in which costs are directly related to fishing effort and revenue is directly related to yield. The shaded zone is the area in which revenue exceeds costs. Maximum economic return is achieved when the revenue-cost difference is maximal, and this is below maximum sustained yield in this model.* (Redrawn from Krebs 1985a.)

than another's, perhaps because of cheaper labor or more-efficient gear, then the maximum yield and maximum economic returns for the two would differ. This is a common source of conflict in international waters; what is sound economic practice for one group is not good management for the other.

Some basic principles still emerge from a consideration of MSYs. First, the most difficult (yet most critical) aspect of population management is undoubtedly accurate population censuses. This still remains one of the most difficult tasks in wildlife biology. Accuracies of greater than 90 percent are very unusual, and accuracies are often as low as 20 to 30 percent (Eltringham 1984). Second, in the absence of other information, a good rule of thumb is not to harvest more than 10 percent of the population at carrying capacity.

Despite the existence of good scientific methods to identify sustainable yields and prevent overharvesting, there is case upon case of mismanagement of wildlife populations, forest trees and fish stocks. It is the fisheries industry that provides some of the best data on overharvesting.

## 4. Overfishing

Overfishing is the bane of modern fisheries technology. The annual world catch is 15–20 million tons (20–24 percent) lower than it might be if the resource were not overfished. Particularly favored fish species have shown dramatic declines, included among which are the Asian sardine, the California sardine, the northwest Pacific salmon, the Atlantic Scandinavian herring and, in 1993, the Canadian cod. Photo 18.2 shows the vast fish-meal stocks in Peru in the 1960s before collapse of the Peruvian anchovy populations, as a result of overfishing, in the 1970s. Others have shown considerable signs of strain, such as North Sea herring and yellowfin tuna. Worse still is the accidental capture and killing of nontarget animals. In shrimp fishing in the Gulf of Mexico, the ratio of fish discards to shrimp caught ranges from 3:1 to 20:1. The problem of *incidental intake*, as it is known, is particularly acute in marine environments (Allen 1980) and can lead to controversy. An example is the debate in Florida in the late 1980s and early 1990s over the use of turtle exclusion devices, designed to allow sea turtles to escape from shrimp nets. They save the lives of turtles but reduce the efficiency of the nets in catching shrimp.

In the United States, management of marine fisheries has not been distinguished by many great successes. Striped bass (*Roccus saxatilis*), haddock (*Melanogrammus aeglefinus*), and Atlantic halibut (*Hippoglossus hippoglossus*) were once abundant on the East Coast, as were Pacific sardines (*Sardinops sagax*) and chub mackerel (*Scomber japonicus*) in California.

International regulatory measures intended to reduce catches, such as net size and catch limitations, are difficult to enforce, and often once a species has been overfished, it may not be possible for it to regain its place in the energy pathways of the ecosystem. The Pacific sardine of the California Current system was overfished in the 1930s and replaced by a competitor, the anchovy *Engraulis mordox* (Fig. 18.7). At the end of the 1950s, the latter's bio-

Photo 18.2   *Fish-meal stocks in Peru. This photo conveys the enormous scale of the Peruvian fishery, which collapsed a few years later.*  (Photo by R. Coral, Food and Agriculture Organization.)

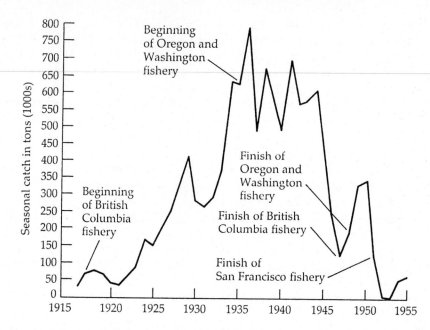

**Figure 18.7** *History of the Pacific Coast sardine fishery, at one time the first-ranking fishery in North America in weight of fish landed and third-ranking in value (after tuna and salmon fisheries). The decline is attributed to overfishing, and there has been no recovery between 1955 and 1970.* (Redrawn from Dasmann 1976.)

mass was similar to that of the sardine 30 years before, and it has clearly ousted the former species (Ehrlich and Ehrlich 1970). Sometimes, overfishing alone cannot be given the blame for species decline. For example, it is not yet clear whether changes in ocean conditions due to climatic events such as El Niño are equally important as overfishing (Hilborn and Walters 1992). Climatic changes can also occur over a wide time span, such as hundreds of years, with concurrent gradual effects on wildlife (Bryson and Murray 1977).

A definite example of overexploitation involves the whaling industry (or rather involved it) (Photo 18.3, Fig. 18.8). In 1982 the International Whaling Commission (IWC) resolved to discontinue commercial whaling within three years. In 1988, the proposal actually took effect, and the number of whales caught dropped from 15,000 in 1980 to several hundred in 1988. However, a biologically depletive program of cropping, mainly by Japan and the former Soviet Union, continued under the transparent excuse of scientific whaling. Baker and Palumbi (1994) used molecular genetic approaches to show that whale products available on the Japanese retail market included species that were imported illegally and others that had been hunted or processed illegally. However, it is encouraging that, with the advent of such technologies, such illegalities can be brought to light. If managed properly, whale populations could make an important contribution to marine harvests, amounting to about 10 percent of the total yield of marine products on a sustainable basis. At the moment, whales are being mined rather than harvested.

Photo 18.3   *For centuries humans have "mined" whales from the sea, as this painting of a Dutch whaling fleet in the Artic Circle around 1720 shows. Only since 1988 has there been a total ban on commerical whaling, and since then whale numbers have increased. But how long will this ban be enforced?* (Bernard, Animlas Animals 609026M.)

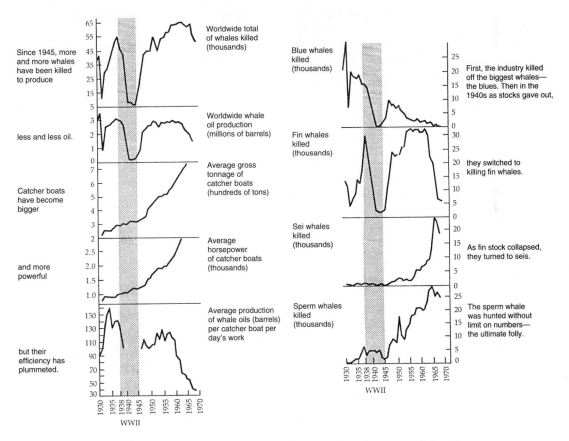

Figure 18.8    *The decline in whale numbers, related to the inputs of the industry.* (Redrawn from Payne 1973.)

# Summary

1. Most modern environmental problems can be linked to an ever-increasing human population. The human population is one of the few that seems to be increasing exponentially, having increased from 0.8 billion in 1750 to 5.3 billion in 1990.

2. There are three main ecological concerns with a burgeoning human population: natural habitat destruction, direct exploitation of wildlife (for example, overfishing), and pollution. All increase the likelihood of species extinction.

3. In theory there is a maximum sustainable yield, or MSY, which is the largest number of individuals that can be removed from a population without causing long-term changes. In practice, the MSY is difficult to achieve—populations are constantly being overharvested.

4. Hunting can be viewed as an additional source of mortality on a population or a compensatory source of mortality—removing individuals that would have been killed in some other way. Surprisingly, the data seem to suggest that most hunting acts as a compensatory mortality rather than an additive mortality.

*Discussion Question:* If good models exist to predict maximum sustainable yields, why are so many populations (like marine fish species and forest trees) overharvested to the point of extinction?

# Patterns of Resource Use

For convenience, habitats can be grouped according to degree of degradation. Each group then serves as a jumping-off point for further studies. Such a listing might be, from least degraded to most degraded,

*Wilderness areas:* These are lands virtually untouched by humans.

*Lands used for outdoor recreation and nature conservation:* They may include "unnaturally" protected areas, where, for example, forest fires that would normally occur are stopped or herds of game that would occur naturally at reduced densities as a result of lowered carrying capacity are held at higher densities. National parks, where nest boxes and winter feed may be provided, are a good example.

*Forests:* The effects of forestry are harder to quantify because the extent of forestry differs from area to area. In Europe, forestry involves planting forests and harvesting them, whereas in other areas, especially the tropics, forestry involves harvesting natural forests and can cause devastating changes, including the leaching of nutrients from soil.

*Agriculture:* Farm lands lie toward the highly manipulated end of the scale. Crop monocultures require huge energy inputs in the form of fertilizers, pesticides, and herbicides. Furthermore, the systems are often prone to epidemics of pests or diseases.

*Urban environments and derelict land:* The ultimate form of land degradation

## 19.1 WILDERNESS AREAS

Designated **wilderness** areas may be thought of as large, unmanaged regions containing whole sets of **biotas** (as distinct from parks and reserves, which are usually smaller and are preserved with particular landscapes, recreations, or communities in mind). One rationale for wilderness preservation is that it is everyone's spiritual right to enjoy such areas. Another is that they maintain a gene pool of wild organisms (for possible future economic gain) and provide natural communities for scientific research into behavior and ecology. Third, many people have argued that species have a right to exist in their native habitat. Of course, designating an area as a wilderness is often likely to attract some people, whose numbers make management necessary.

Fifty-two percent of the Earth's land area is wilderness, but most of that is in Antarctica, Greenand or deserts (Hannah et al. 1994) . When this total is adjusted for rock, ice, and barren land, 27 percent remains undisturbed. Only 4 percent of the U.S. land area is protected as wilderness, with most of it in Alaska. Only 1.8 percent of land in the lower 48 states is wilderness. East of the Mississippi, half of the wilderness is in two places: Everglades National Park in Florida and Minnesota's Boundary Waters Canoe Area.

Even the largest wilderness area in the world, Antarctica, contains point settlements. The Antarctic Treaty of 1959 among "user" nations ensured that only such settlements would be permitted and only for scientific research. Theoretically, in a point settlement, everything—all foods and supplies—are brought in and *everything*—all waste materials and byproducts—are shipped out. No resources are taken from the environment, which should remain unaltered. However, the discovery of reserves of coal, oil, and natural gas in Antarctica and the possibility of harvesting krill could cause severe economic discord among user nations, upsetting the status quo of the agreement.

Some wilderness areas also contain small groups of humans who exist as hunter gatherers. The hunter-gatherer way of life was once the dominant mode of existence on Earth, but today, hunters and gatherers form perhaps 0.001 percent of the global population. They include such tribes as the Bushmen of southern and southwestern Africa, the Pygmies of central Africa, the Australian Aborigines, and the Indians of "interior" Brazil and Venezuela. Some North American Native American and Inuit groups also live largely on hunted food, although contact with western culture is high. Manipulation of the environment in these groups is minimal.

## 19.2  PROTECTED ECOSYSTEMS: NATIONAL PARKS

About 5 percent of the world's land surface has been set aside for protection of biological resources. There is much variation from country to country in the amount of land so set aside. For example, New Zealand, with a population of 3.3 million people, has a varied array of nature reserves amounting to nearly 19 percent of its area. Great Britain, with a population of 55 million and a population density maybe 20 times that of New Zealand, has designated a similar area for various kinds of nature reserves, though these contain more people. The United States has 10.5 percent of its land designated as protected.

The national parks system of the United States is composed of various different categories of preserved land, such as national parks, national monuments, national preserves, national seashores and lakeshores, and national recreation areas. These are illustrated in Figure 19.1. Besides the national parks system, each state in the United States generally has state parks and within the states, and counties may have their own county parks, so there is considerably more preserved land than that illustrated in Figure 19.1. Recreational visits to national park service areas in the United States showed a steady increase from the 1970s

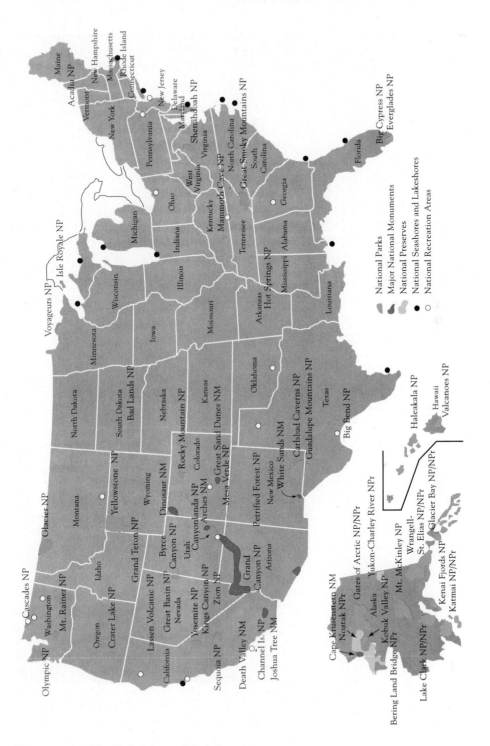

Figure 19.1 *The U.S. National Park System.*

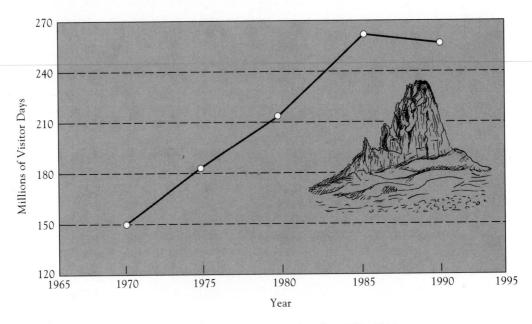

Figure 19.2    *Recreation visits in U.S. National Parks, 1970–1990.*

through the 1980s (Fig. 19.2), underscoring the importance of acquiring new land for new national parks so that the old ones are not overused by visitors. Overuse by the public tends to devalue the very resources that draw people to the park in the first place—solitude, lack of noise, views unimpeded by people, and so on. It is encouraging that in the United States, as in the world as a whole, the cumulative growth of protected areas is increasing both in number of sites and in area protected (Fig. 19.3). The creation of Greenland National Park

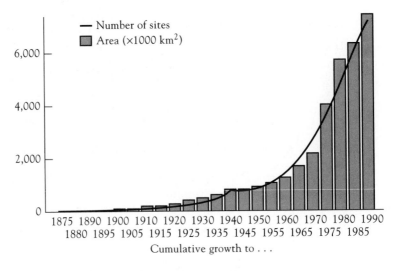

Figure 19.3 *Cumulative growth of the world's protected areas.*

in 1974, which covers some 97 million ha, and the Great Barrier Reef Marine Park in Australia in the 1980s, which covers some 34 million ha, gave a dramatic boost to the area under protection in the world. As of 1992, there were 8,491 sites covering 7,734,900 km², or some 5.19 percent of the Earth's land area.

Table 19.1 shows the extent to which major biomes are covered by protected areas. Temperate grasslands and lake systems are poorly represented in the protected area network, and this indicates an area requiring attention. It is probably true that temperate grasslands constitute some of the richest agricultural land in the world, and this is why the number of preserves in such biomes are low. Tropical forests and grasslands appear quite well represented.

## The Theory of Reserve Design

An essential requirement of conservation is to ascertain how many individuals of a species should be conserved in order to ensure its survival in a particular area. There are two approaches to estimating such a Minimum Viable Population (MVP) size, the demographic and the genetic. The process of applying either demographic or genetic MVP models to particular species or populations and proposing the management interventions necessary is known as Population Viability Analysis (PVA) (Menges 1990; Shaffer 1990).

TABLE 19.1 *Distribution and coverage of protected areas by biome type. (World Conservation Monitoring Centre 1992.)*

| Biome Type | Protected Areas Number | Protected Areas Area (km²) | Biome Area (km²) | % of Total Area |
|---|---|---|---|---|
| Subtropical/temperate rainforests/woodlands | 935 | 366,100 | 3,928,000 | 9.32 |
| Mixed mountain systems | 1,265 | 819,600 | 10,633,000 | 7.71 |
| Mixed island systems | 501 | 246,300 | 3,244,000 | 7.59 |
| Tundra communities | 81 | 1,643,400 | 22,017,000 | 7.46 |
| Tropical humid forests | 501 | 522,000 | 10,513,000 | 4.96 |
| Tropical dry forests/woodlands | 807 | 818,300 | 17,313,000 | 4.73 |
| Evergreen sclerophyllous forests | 786 | 177,400 | 3,757,000 | 4.72 |
| Tropical grasslands/savannas | 56 | 198,200 | 4,265,000 | 4.65 |
| Warm deserts/semideserts | 296 | 957,700 | 24,280,000 | 3.94 |
| Cold-winter deserts | 139 | 364,700 | 9,250,000 | 3.94 |
| Temperate broad-leaf forests | 1,509 | 357,000 | 11,249,000 | 3.17 |
| Temperate needle-leaf forests/woodlands | 440 | 487,000 | 17,026,000 | 2.86 |
| Lake systems | 18 | 6,600 | 518,000 | 1.28 |
| Temperate grasslands | 196 | 70,000 | 8,977,000 | 0.78 |
| Classification unknown | 961 | 700,800 | 0 | NA |
| TOTAL | 8,491 | 7,734,900 | 146,968,000 | 5.26 |

Normally, a MVP is taken to be the size of population that has a 95 percent probability of persistence for $x$ number of years, where for consistency $x$ is usually taken as either 100 or 1,000. The results from applying MVP analysis to different species shows that over a range of body masses from shrews to rhinoceroses, MVP sizes range from hundreds to millions (Belovsky 1987). The resultant Minimum Area Required (MAR) to support these populations ranges from tens to millions of square kilometers. Generally, as body mass increases, MVP size decreases. This is an interesting result. However, although larger mammals may require fewer numbers in viable populations, they still require proportionately larger ranges. MARs are larger for carnivores than for herbivores.

MVPs can also be examined from a genetic perspective, in which not only the number of individuals surviving but their genetic variation, sometimes called *heterozygosity*, is considered important. In the long term, this genetic variation is a necessary adaptation to potential future changes in the environment. Such changes may include moderations to habitats because of global warming, global cooling, or acidification to lakes. In the short term, heterozygosity is positively correlated with features associated with survival including disease, resistance, and growth.

In many outbreeding species (species that do not breed with close relatives) genetic variation can quickly be lost through breeding with closely related individuals (inbreeding) that leads to low levels of heterozygosity and lowered offspring numbers and survival, a phenomenon known as *inbreeding depression* (Chapter 2). It has been proposed that, an effective population size of 50 is the MVP required to guard against the negative effects of inbreeding over the short term, with no immigration or introduction of unrelated stock. Even with 50 individuals in a population, individuals will eventually inbreed, and populations will generally become inbred over the long term. Thus, in the long term, an effective population size of 500 has been suggested as a more suitable genetic MVP because in a population of this size, rates of mutation will renew genetic variation as quickly as it is lost by inbreeding and genetic drift.

These figures, 50 and 500, have been thought of as "magic numbers" in refuge design and have been raised to the status of a rule in management circles. However, genetically based MVP sizes are likely to vary between species. For example, a species that exhibits a boom-and-crash population cycle will require a larger MVP than one which inhabits a stable environment and whose population is relatively stable. Also, there are many species that have effective population sizes of less than 50 even with 50 individuals present. Prominent among these are species with harem mating structures, where only a few males mate with a majority of females. Here the effective population size is much smaller than 50 because all the males don't breed in a population. In some of these species, the effects of inbreeding do not always have severe effects. Similarly, many self-fertilizing plants have little genetic variation.

In practical terms, even if 500 is accepted as a minimum viable population size, in some animals 500 individuals cover a lot of territory. For example, to maintain a red-cockaded-woodpecker population of about 500 individuals requires an area of at least 25,000 ha. Few, if any, large mammal species would be adequately conserved with the current scale of protection, as most protected populations are too small to constitute MVPs.

There has been much discussion about the shape, design and management of nature reserves since the early 1970s. The theories of park design are summarized in Figure 19.4. The debate on reserve design has centered on island bioeographic theory (see Chapter 15), which suggests that large parks should hold more species than small parks. It is widely agreed that the ideal strategy would be to have many large refuges. Of course, large refuges cost large amounts of money, so many countries have to settle for preserving small areas of land rather than large ones. The debate then becomes: Should single large areas be preserved, or several small ones?—(the so-called SLOSS debate, Simberloff 1986). Single large preserves may buffer populations against catastrophe and possible extinctions, but many studies suggest that a few, dispersed, small sites often contain at least as many species as does a single site of equal area. Usually multiple small sites contain more species. This is because a series of small sites is likely to have a broader variety of habitats contained within it than one large site. Hence, in practical terms, it may actually be better to have several small sites.

Quinn and Harrison (1988) have reviewed data from more than 30 censuses for 15 island groups where all data were reported, island by island, species identities of individuals were confirmed, and at least six islands were included in each survey. The resultant species-area relationships for vertebrates, land plants, and insects show that collections of small islands generally harbored more species than do comparable areas composed of one or a few large islands. National park faunas were shown to be richer in collections of small parks than in the larger parks. This finding has an important bearing on future land purchases for conservation purposes. Furthermore, in 29 out of the 30 cases, the small-islands curve saturated with species more rapidly than did the large-island curve, and in 18 out of the 30 cases, the two curves never intersected, indicating that smaller, more numerous islands always had more species than fewer large ones.

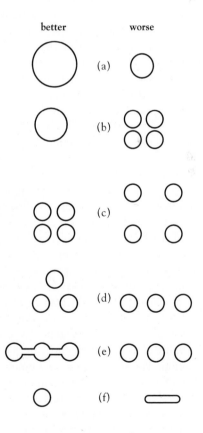

**Figure 19.4**  *The theoretical design of nature reserves based on the tenets of island biogeography.* (**a**) *A larger reserve will hold more species and have lower extinction rates.* (**b**) *Given a certain area available, it should be fragmented into as few pieces as possible because species incapable of dispersing readily among fragments may be confined to areas too small to avoid extinction.* (**c**) *If an area must be fragmented, the pieces should be as close as possible to minimize inhibition of dispersal.* (**d**) *A cluster of pieces in contrast to a linear arrangement will enhance dispersal among the fragments.* (**e**) *Maintaining or creating corridors of natural habitat between fragments may enhance dispersal.* (**f**) *A circular-shaped area will minimize the amount of "edge" habitat adjacent to nonreserve areas.* (After Wilson and Willis 1975.)

Of course, at some point, a potential refuge is simply too small to conserve for long the species for which it is designed, and that minimum critical area varies for different species. Detailed ecological study is probably the best means of telling just how large minimum critical areas are and where to locate them for given species. In this way, the habitat requirements for individual animals are documented and the minimum area required for one individual can be multiplied by the number of individuals desired in the population to give a minimum area requirement.

Another area of contention in the design of nature reserves is how close to situate small reserves if these are chosen over large ones. Thus, should we have three or four small reserves close to each other, or should they be further apart? In practice, it is probably true that having small sites far apart would preserve more species than having them close together, for no other reason than distant sites are likely to incorporate slightly different habitats.

A variation of the argument as to whether small reserves should be situated close together or far apart is the suggestion that they should be linked together by biotic corridors and networks, thin strips of land that may permit migration of species between small patches. This argument is based on the fact that corridors may facilitate movements of organisms with poor powers of dispersal between habitat fragments. If a disaster befalls a population in one small reserve, it can be recolonized by immigrants from neighboring populations. This obviates the need for humans to physically move new plants or animals into an area. Many types of habitat have been considered for corridors, like hedgerows in Europe, riparian habitat (along rivers) in the United States, and strips of tropical forest. Reserves with occasional interrefuge migrations may appear to be optimal for genetic mixing of populations. However, there are many disadvantages associated with corridors (Simberloff et al. 1993; Bonner 1994). Corridors also facilitate the spread of disease, invasive species, and fire between small reserves. Having isolated populations greatly lessens the chances that all of them will succumb to the same catastrophe. For example, there are many small oceanic islands that have had their ground nesting birds wiped out by introduced predators such as goats, rats and even lighthouse keepers' cats. It is not widely appreciated that even smaller oceanic islands have remained bastions for wildlife because they were not invaded by exotic species. Had these small oceanic islands been connected, the spread of exotic species would have been greatly facilitated, and there would have been many more human induced extinctions than at present.

It has been said that the application of island biogeography theory to conservation was a worthwhile experiment, but experience and further deliberation have shown that it is not very helpful. Habitat considerations of individual species are ususally much more important. In reality, there is rarely any choice as to the size, shape, and location of nature reserves, and this constitutes a strong reason for being critical of adopting theories from biogeography for establishment of nature reserves. Management practicalities, costs of acquisition, and management, and also politics, usually override ecological considerations, especially in developing countries, where management costs for large reserves may be relatively high. Economic and ecological considerations are both justified in choosing which areas to preserve. Typically, many countries protect areas in those regions that are the least economically valuable rather than choosing areas to ensure a balanced representation of the country's biota. For example, in the United States, most National Parks have been chosen for their scenic beauty, not the fact that they preserve the richest habitat for wildlife.

While existing protected areas may be able to conserve populations of plants and animals for a very long period of time, it is unlikely that extinctions will be prevented forever. This is because nearly all national parks, even the largest, are not large enough to support the process of speciation within them. Thus, as habitats change or climates modify, species will not have enough genetic variation to adapt to such changes and populations may well become extinct (Soulé 1980) (Fig. 19.5). The basis for this conclusion was a survey of vertebrate taxa on islands, looking for the minimum-sized island on which there was evidence for in-situ speciation for a particular group. The results suggested that small mammals such as rodents have enough room for speciation on islands such as Cuba and Luzon (about 110,000 km$^2$), but birds or large mammals as large or larger than jackals or vervet monkeys require an area the size of Madagascar (nearly 600,000 km$^2$). Higher plants appear to fall somewhere in between these extremes. For large organisms, natural speciation requires too much space, and probably too much time, for humans to be able to construct necessary nature reserves to allow such a process to continue. Perhaps the best we can hope for is that existing species will continue to survive for relatively long periods of time.

## The Practice of Preserving the Environment

It has been argued that rather than concern ourselves with island biogeography theory when designing nature reserves, we should better consider how to finance their protection—this will dictate how big reserves can be. The amount of money spent to protect nature reserves may better determine species extinction rates than reserve area. According to island-biogeography theory, large protected areas minimize the risk of extinctions because they contain sizable populations. In Africa, several parks such as Serengeti, Tsavo, Selous, and Luangwa in Zambia are large enough to fulfill theoretical ideals. However, in the 1980s, populations of black rhinos and elephants declined dramatically within these areas because of poaching, showing that there was a wide gap between theory and reality. In reality, poaching is a far more serious threat than the threat of genetic isolation.

In reality, the rates of decline of rhinos and elephants, due largely to poaching, are related directly to conservation effort and spending (Milner-Gulland and Leader-Williams

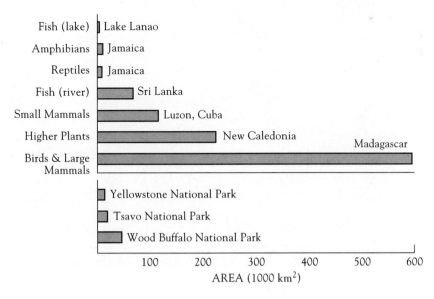

Figure 19.5    *Smallest "islands" within which speciation of vertebrate taxa has occurred, compared with three large national parks.* (From Soule 1980.)

**Figure 19.6** *Relationship between change in black rhino numbers between 1980 and 1984, and conservation spending in various African countries.* (Leader-Williams and Alban 1988.)

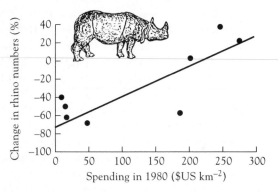

1992; Leader-Williams, Dalban, and Berry 1990) (see Fig. 19.6). The conclusion is that nature reserves must be adequately funded if local extinctions are to be avoided. Instead of conservation biology concentrating on theoretical implications of the size of protected areas, they should perhaps be more concerned with the direct level of funding for protection of species. Indeed, the successes of conservation in Africa have resulted from situations where resources were concentrated in particular areas to preserve certain species. For example, the few remaining rhinos in Kenya (Photo 19.1), the lowland gorillas and pygmy chimpanzees in

**Photo 19.1** *Debate over the best size and shape of a nature reserve seems out of place when in reality, protection of many species like this white rhino in Meru National Park, Kenya, is better linked to adequate protection by park rangers from poachers.* (Moore, Animals, Animals 616427M.)

Africa, and the vicuna in South America have all shown the greatest stabilization of numbers in areas that have been heavily patrolled and where resources have been concentrated. If parks are too large, they may not be adequately patrolled.

Also, it is not always remembered in theoretical debates that species are not simply numbers in an equation and cannot be treated as such. Some plants and animals are of more interest to conservationists than others, and island biogeographic theory fails because it does not draw any distinctions between desirable and undesirable species. Some areas such as oligotrophic lakes are species-poor, yet are preserved because of their pristine appearance. Other areas, such as agriculturally improved grassland or even old railway sidings, can be very rich in plant species; yet we often do not preserve these areas. Richness seems to be important only when it suits us; otherwise we ignore it.

Finally, Mabey (1980) and others have argued that conservation has catered principally to an elite minority and that government agencies and trusts select preserves on pseudoscientific grounds for ecological specialists. Can everyone afford to drive to national parks? What does the public really want? Goldsmith (1983) suggests that their principal interest is near to home, not at some distant reserve. They would like the "feeling" of nature in their immediate neighborhoods. They want to see pretty flowers, birds, and butterflies in the countryside. Swallowtail butterflies and bluebell flowers may do just as well as panthers and bobcats, which are very secretive anyway. At the moment, much conservation seems to specialize on rare species and untouched habitat. Local conservation projects, closer to urban centers, may also be worthwhile. Photo 19.2 of Central Park in New York City illustrates the importance even of relatively small urban parks in conserving undeveloped land for recreational purposes. Such parks provide the only access some urban dwellers have to wooded or open areas. It might be argued that the importance of different taxonomic groups could be assessed from the membership of the organizations that study them. For example, in Britain the Royal Society for Protection of Birds has about 250,000 members, and the Botanical Society of the British Isles has 2,300. But of course membership is affected by the marketing efficiency of the organization as well as the size, color, and attractiveness of the organisms concerned, and without sound habitat management for plants, the suitability of an area for birds would certainly deteriorate.

Photo 19.2    *Central Park, New York City, 1905. Before the advent of the automobile, city parks were even more heavily used than today because of the lack of access to other open areas.*
(Photo from Library of Congress, no. LC-D401-9285.)

## 19.3 FORESTS

Forests provide the basic habitat for a large proportion of the world's wildlife. Trees are important as a source of food and shelter. Forests also have stabilizing effects on stream flow that provides habitats for fish. Fallen leaves, branches, and trunks provide crucial habitats and resources for many animals. The relationship between forest and wildlife is so intertwined and complex that little can be done to a forest that doesn't have impact on some form of wildlife. An obvious impact is the effect of clear-cutting or burning a forest (Photo 19.3). The habitat of some animals is lost, cover and nesting or denning sites are removed, and seed-producing or fruit-producing trees are destroyed. But clear-cutting can create essential habitats for other species that prefer gaps in the forest as long as a mosaic of for-

Photo 19.3    *Mount St. Helens, Washington, 1970. Clear-cutting a forest has a severe impact on wildlife because whole areas are denuded of trees.* (Harrison, Grant Heilman CP4-150A.)

est and clear-cut is left. For example, young trees are especially important food for deer, elk, and moose.

There is also an impact on wildlife populations when a forest is allowed to proceed toward climax through succession. The larger herbivorous species dependent on low growing plant forms diminish and are replaced by a greater diversity of small carnivorous animals feeding mostly on other animals. Thus, the concept of what is "good" or "bad" forest management for wildlife may vary according to the outlook or preferences of the individual viewing the situation. The forest management scheme favorable for ruffed grouse, woodcock, and white-tailed deer in the Great Lake states would be detrimental to ducks, bald eagles, and Kirtland's warblers.

Use of a forest by an animal may be obligatory (meaning that the animal cannot exist in its wild form without trees) or may be facultative (that is, the animal may use trees if they are available, but can survive without them). Among North American terrestrial wildlife, about half of the birds and more than 10 percent of the mammals have an obligatory relationship to trees or forest cover. Of some 567 species of birds on the continent, at least 272 can be considered to have an obligatory relationship with trees. Similarly, of the 348 species of mammals in North America, north of Mexico, about 49 have an obligatory relationship to trees (Table 19.2).

Trees provide the needs of many birds and a considerable number of mammals. For example, most woodpeckers nest in holes excavated in trees, and their food usually consists of insects collected on or in trees. Chickadees and nuthatches also nest in tree cavities, and their food consists largely of the adult and immature stages of insects taken from the bark, branches, and foliage of trees. Many species of squirrels depend on trees for their nests as well as for most of their food. Raccoons commonly use tree cavities for their dens but do most of their foraging on the ground. Several animals utilize trees when it is convenient but can survive quite well in the absence of a forest. These are considered to be facultative forest users.

TABLE 19.2 *Numbers of species of North American warm-blooded vertebrates dependent on trees.*

| Normal Habitat | Total Birds | Birds Considered Tree Obligates | | Total Mammals | Mammals Considered Tree Obligates | |
|---|---|---|---|---|---|---|
| | | Food | Nest | | Food | Nest |
| Aquatic | 82 | 0 | 13 | 4 | 1 | 0 |
| Shoreline | 65 | 0 | 11 | 5 | 0 | 0 |
| Marshes | 19 | 0 | 0 | 37 | 0 | 0 |
| Grasslands and tundra | 83 | 0 | 27 | 184 | 0 | 0 |
| Brush | 91 | 0 | 14 | 42 | 0 | 0 |
| Trees | | | | | | |
|   Single or groups | 0 | 0 | 41 | 6 | 6 | 0 |
|   Forests | 178 | 159 | 160 | 34 | 27 | 23 |
| Aerial foragers | 49 | 0 | 29 | 39 | 1 | 6 |

Probably few trees are not used as food by some form of wildlife at one time or another. Among 224 North American herbivores (birds and mammals), resources provided by trees constitute an important part of the diet of 89 species or 40 percent of the total. Among the most valuable trees in North America, in terms of wildlife use, are oaks whose acorns are used most often and occur in the diet of at least 96 species of wildlife. Other trees used as food by more than 25 species of animals include pine, cherries, hackberries, junipers, mulberries, maples, blackgum, beech, and aspen. The buds, fruit, or seeds of these trees are the parts usually used as food by wildlife. Collectively, fruits and seeds are called *mast*, and for many species of wildlife, the annual mast crop is a major determinant of abundance from year to year. Reduction in the mast crop can result in a decline in the numbers of squirrels, and in some regions, deer may suffer. Where snow is an important part of the animal's life cycle, herbivorous species such as hares, deer, and moose depend on shrubby plants or the young growth of trees for their food resource, and they take these throughout the winter. Beaver and porcupines live mostly on the bark and cambium layers of various trees.

About one-third of the world's land surface is covered with forests, whose continental distribution is outlined in Table 19.3. Data are usually scarce, often out of date and sometimes inaccurate, but it is generally thought that the forest area is stable or increasing in the developed world where some balance between agriculture and forestry has been struck. For the world as a whole, in the 1980s, the rate of deforestation for closed forests was about 0.39 percent per year, for open forests or woodlands 0.30 percent, and for all forests averaged together 0.38 percent or 15,517 thousand ha per year. The rate of reforestation for the world as a whole during the same period of time was 14,713,000 ha per year, but most of this was probably plantation forest whereas the deforestation includes many natural forests. Tropical forests are certainly declining. In Africa, 0.6 percent of the total tropical forest is lost each year, the main cause being conversion to cropland, not cutting for timber. In South America, 0.7 percent is lost per year, and in Asia, 0.9 percent. There are many claims at their present rate of decline, for example, that the moist tropical forest will be gone in approximately 50

TABLE 19.3 *Forest areas of the world (land area in millions of km².)*

| Region | Total Land Area | Forest Area Coniferous | Forest Area Broad-Leaved | Total | Percentage of Total Forest |
|---|---|---|---|---|---|
| North America | 18 | 4.0 | 2.6 | 6.6 | 18.2 |
| Latin America | 20 | 0.3 | 7.4 | 7.7 | 21.3 |
| Europe (excl. former Soviet Union) | 5 | 0.9 | 0.6 | 1.5 | 4.1 |
| Africa | 30 | 0.1 | 6.9 | 7.0 | 19.1 |
| Asia (excl. former Soviet Union and Japan) | 27 | 0.7 | 4.1 | 4.8 | 13.3 |
| Japan | 1 | 0.1 | 0.1 | 0.2 | 0.6 |
| Former Soviet Union | 21 | 6.0 | 1.7 | 7.7 | 21.3 |
| Pacific area | 8 | 0.1 | 0.8 | 0.9 | 2.2 |
| World (excl. Antarctica) | 130 | 12.2 | 24.2 | 36.4 | 100.0 |

years. Some transformation of forests is necessary to generate badly needed capital and agricultural land. To deny countries the opportunity to do this would be to force them to keep living in the nineteenth century. On the other hand, the methods used should ensure maintenance of long-term productivity of the soil and the plants growing on it and should minimize the effects on native wildlife.

Deforestation not only directly removes wildlife habitat; it may also affect the distribution patterns of life via indirect effects. For example, increased soil erosion causes stream and river siltation and affects aquatic life. Acceleration of nutrients may also cause eutrophication. Loss of trees for firewood in marginal areas may promote desertification.

## 19.4  AGRICULTURAL LAND

The source of much of the food consumed by people is terrestrial agriculture, and agricultural land represents the most manipulated of all the nonurban ecosystems. The scouring of the land to plant agricultural crops creates soil erosion, increased flooding, declining soil fertility, silting of the rivers, deforestation, and desertification. Even though acid rain, pollution, nuclear waste, eutrophication, and the preservation of land for parks and reserves are pressing issues, they seem to pale in comparison with the prospect of huge amounts of land being used for agriculture. Not only is the agricultural system human-made, but the plant and animal components have also been genetically engineered. The area of land under cultivation in different parts of the globe is illustrated in Table 19.4. Europe gives over by far the greatest percentage of its land to crops and pasturelands, and Africa the least, which helps to explain why Europe often has food surpluses and Africa often suffers famine. Energetically, the most efficient agricultural systems include wet-rice culture

TABLE 19.4 *Agricultural and grazing lands, 1985, in thousands of hectares. Permanent crops includes tree crops such as rubber, citrus, and cocoa. (From FAO Production Yearbook 1986.)*

| | Continental area excluding inland water | Arable land and permanent crops | % | Permanent pastures and meadows | % |
|---|---|---|---|---|---|
| World | 1,3078,873 | 1,476,483 | 11.3 | 3,170,822 | 24.2 |
| Africa | 2,964,595 | 184,869 | 6.2 | 788,841 | 26.6 |
| North and Central America | 2,242,075 | 274,626 | 12.2 | 367,062 | 16.4 |
| South America | 1,781,851 | 140,638 | 7.9 | 458,364 | 25.7 |
| Asia | 2,757,252 | 454,253 | 16.5 | 644,669 | 23.4 |
| Europe | 487,067 | 139,625 | 28.7 | 84,260 | 17.3 |
| Oceania | 850,967 | 50,285 | 5.9 | 453,026 | 53.2 |
| Former U.S.S.R. | 2,240,220 | 232,187 | 10.4 | 374,600 | 16.7 |

and shifting agriculture (Fig. 19.7), hence their popular and, in the case of shifting agriculture, unfortunate appeal.

The two basic agricultural systems today are *shifting agriculture*—the total manipulation of natural systems, but only for one to five years—and *sedentary agriculture*—permanent replacement of natural systems.

Shifting cultivation—often called **slash-and-burn cultivation**—is today largely confined to tropical forests and savannas, where the burning of natural vegetation provides minerals for crop uptake. Crops are planted so as to provide as complete cover as possible, reducing the **leaching** of the soil. Plots are normally abandoned when nutrient drain reduces

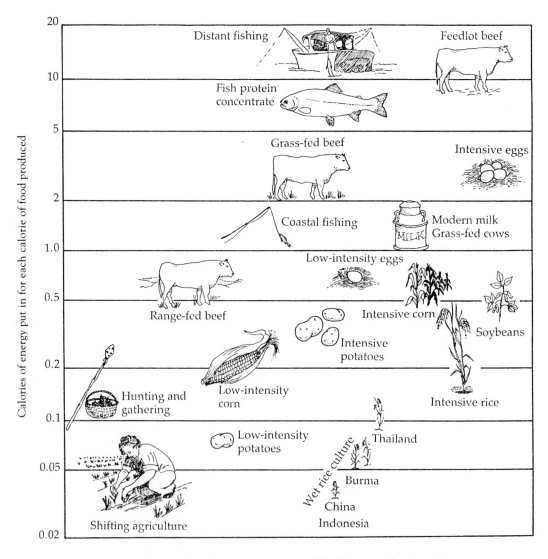

**Figure 19.7** *Comparative energy intensiveness of food crops in the United States.*

fertility or when weeds reach an unacceptable level. Breakdown of this system occurs when plots are recultivated too soon because mineral nutrient cycles never build up again, and fertility is not renewed. Thus, the system cannot cope with intensive crop production in populated areas but suffices in forested land where regeneration eventually occurs.

Sedentary cultivation represents the permanent manipulation of an ecosystem; the natural biota is removed and replaced with domesticated plants and animals. The soil is extremely important in this case, for it is the long-term reservoir of all nutrients, although it can be supplemented by chemical fertilizers or organic excreta (the latter is preferable because it helps maintain the crumb structure of the soil). The paddy culture of rice is an exception—it is essentially an aquarium system with earth boundaries instead of glass. The soil is largely a rooting medium, and water supplies the essential nutrients. Thus, paddy rice can be grown on poor soils.

Paramount among the world's crops are rice, wheat, and maize, which comprise ⅚ of the world grain crop. Meat is a much scarcer product, being inefficient to produce. It has the advantage, however, of being nutritionally superior to many plants. Furthermore, animals are often adaptable to climate, and the proteins are not locked away behind a cell wall of cellulose not easily broken down by human stomachs.

## The Ecological Consequences of Agriculture

The extension and intensification of food production is not carried out without environmental side effects that can affect the distribution of wildlife:

1. Pollution of water bodies. Runoff water often contains high levels of nitrogen and phosphorus from fertilizers, and residual pesticides, which contaminate streams. These can all affect communities of aquatic plants and animals.

2. Loss of topsoil. Some 4 billion tons of topsoil are washed into U.S. waterways each year. Topsoil can also be whipped away by wind. The most spectacular and well-known example of soil erosion in the United States was the creation of a huge dust bowl from Kansas across southwest Oklahoma to western Texas in the 1930s as a result of an extensive monoculture of wheat. Photo 19.4 shows a portion of the aftermath. Clearly, whole plant communities and their associated herbivores can become extinct in the face of severe soil erosion. In many cases, however, people may not be entirely to blame. Early in the nineteenth century, the Great Plains were known as the Great American Desert—not a strictly accurate designation but correct in that patterns of rainfall could be very erratic and almost desertlike at times. The term *Great Plains* came into general use after the 1860s as a result of propaganda by the railway companies, who wished to see the plains settled, creating business for the trains. In good years of relatively high rainfall, farms thrived, but in bad years—not uncommon—the plains were much better suited to grazing than to farming. A series of drought years brought on wind-blown erosion of topsoil and economic disorder (Eckholm 1976). The intertwined roots of hardy native grasses that normally shielded the soil from these droughts and winds were absent, and in the 1930s, snowlike drifts of sand began to smother fields.

3. Continued irrigation for agriculture can cause salting of the land, as was being discovered in California in the 1970s. Even "fresh" water contains minute quantities of dissolved salts, and as water evaporates or is used by plants, the salt is left behind and accu-

Photo 19.4    *The result of severe wind erosion in Cimarron County, Oklahoma, 1936.* (Photo from H. Hull, Food and Agriculture Organization.)

mulates in the soil. Mesopotamia, probably the world's oldest irrigated area, on the plains of the Tigris and the Euphrates, had a fertility legendary throughout the Old World. The area now contains some of the world's lowest crop yields. In the 1970s, some areas of southern Iraq glistened like fields of freshly fallen snow (Eckholm 1976).

4. Severe grazing by livestock can induce desertification. Cutting herd numbers is the only answer, and, paradoxically, a reduction of 50 percent could double meat output. The reason lies in the physiological nature of growth. Of all the food consumed by a grazing animal, roughly the first half is used for maintenance, the next quarter for reproduction, and the final quarter for milk production, growth, and fat storage. Any cutback in feeding as a result of overgrazing results in a reduction of these final functions rather than the others.

5. Finally, agriculture can result in a severe loss of wildlife through habitat removal due to deforestation, drainage of swamps, and hedgerow removal.

What about the prospect of using more wild game as food? Wild game is often more catholic in its use of forage than domestic varieties and may eat plants that currently go un-harvested. A case in point is a large South American rodent, the capybara, which feeds on aquatic weeds. Attempts to validate the idea that game ranching should be more sound economically than cattle raising have been made since the 1960s, especially in Africa, but have usually proved unsuccessful (Caughley 1976). More recently, some contrary data have appeared. For example, Freudenberger (1982) has suggested that game farming in South Africa is economic and that the net return from springbok is $0.25 per ha greater than that for sheep, mainly because of the farmer's lower production costs (see also Lewis 1977, 1978; Hopcraft 1982). Unfortunately, the economics of the situation are not directly related to natural **carrying capacity.** For one thing, ecological carrying capacity is not the same as economic carrying capacity, the density of animals that will allow maximal sustained harvesting, and later reports showed that economic carrying capacity does not necessarily differ between wild ecosystems and domestic ones (Walker 1976). Second, the quality and desirability of meat may well be greater from cattle than from wild game. By analogy, British or American people would not buy the meat of horses or dogs for human consumption, even though people in other countries may consume them with a passion, at least in the case of horsemeat.

## 19.5  DERELICT LAND

Apart from affecting the environment through agriculture and by damming rivers, people alter the landscape or geomorphology in many other ways (Table 19.5). Without question, strip mining devastates the land more severely than any other form of physical excavation

**TABLE 19.5  *Some anthropogenous landforms. (From Goudie 1986.)***

| Feature | Cause |
| --- | --- |
| Pits, ponds, and lakes | Mining |
| Spoil heaps (mine tailings) | Mining |
| Terracing | Agriculture |
| Cuttings | Transport |
| Embankments, impoundments | Transport, river and coast management |
| Dikes | River and coast management |
| Mounds | Defense, memorials |
| Craters | War, bombing |
| Canals | Transport, irrigation |
| Reservoirs | Water management |
| Subsidence depressions, sink holes | Mineral and water extraction |
| Moats | Defense |

(Photo 19.5). The world's largest excavation is the Brigham Canyon copper mine in Utah, which has involved the removal of 3,355 million tons of earth over an area of 7.21 km² to a depth of 774 m, seven times the amount of material moved to build the Panama Canal. In Britain as early as 1922, it was realized that "man is many times more powerful as an agent of denudation than all the atmospheric denuding forces combined" (Sherlock 1922, p. 333).

## 19.6 RESTORATION ECOLOGY

Despite the best efforts of conservationists, some species or communities can go extinct in a particular area following habitat loss. If there are other populations of the species and other remnants of the habitat in other geographical areas, it may be possible to take seeds of plants and young animals from these areas and use them to restore the damaged area. How best to do this is the province of restoration ecology.

One might define *restoration ecology* as the full or partial replacement of biological populations and/or their habitats that have been extinguished or diminished, and the substitution by the same or similar species that have social, economic, or ecological value. Restoration ecology can involve simply trying to return species to the wild, for example, returning peregrine falcons to suitable habitat in the eastern United States, returning Arabian oryx to Saudi Arabia, or returning the California condor to southern California. Or it

Photo 19.5    *Strip mining coal, Black Mesa, Arizona. Strip mining results in severe degradation to the land.*  (Brooks, Photo Researchers, Inc. OV 7187.)

can focus on restoring complete habitats; for example, following open-cast mining, huge tracts of disturbed land have to have a large part of their original quota of biological species such as grasses, shrubs, trees, and animals returned in order for the ecosystem to be anything like "normal" (Photo 19.6).

Sometimes, developers of land located on rare habitats are required to reconstruct the same amount of habitat, or usually more, in a different place. This process is known as *mitigation*. Usually, the developer contracts with an outside company, such as a consultancy company, to do the mitigation work. Once again, whole habitats are reconstructed. Along the coasts of Florida, hotels and condominiums built along the coastline often displace native coastal-plant communities, salt marshes, mangroves, and salt flats. New salt marshes and coastal ecosystems have to be built in different places. Often, the success rate is low. For example, in Florida, of 63 permits issued for wetland creation sites between 1985 and 1990, only four were in full compliance with the requriements of the permit. This suggests that the techniques of restoration ecology have still to be perfected.

Some ecologists and conservationists are fearful of the restoration process. Their concern is that admission of even partial effectiveness in restoration will be viewed as a license for further ecological destruction in the name of progress and growth. They fear that some of the wilderness area, national parks, and other ecosystems that now have exceptional protection may have this protection reduced as a consequence of the feasibility of repairing

Photo 19.6   *This old landfill in New York has been given a cover of green vegetation. In this case the original ecosystem has not been restored, but the land has certainly been upgraded.* (Degginger, Earth Sciences 650789.)

damage following exploitation of various resources in these systems. For example, there is much pressure to "open up" part of the Arctic Wildlife Refuge in Alaska because of the probability of huge oilfields in the area. However, large-scale restoration in such areas is likely to be prohibitively expensive. The costs of restoration are much greater than those of conservation because in many cases whole communities of plants have to be replanted, soil and water conditions have to be altered, and keystone herbivores brought into the area. This is generally much more expensive and demanding in terms of human time and money than simply conserving a habitat in the first place. Thus, only habitats and wildlife damaged as a result of small-scale oil spills could likely be restored using restoration ecology techniques. The whole process is analogous to the relationship between preventive medicine and surgery. Conservation can be viewed as preventive medicine, and restoration ecology as surgery. Surgery is almost always more expensive.

Restoration ecology is in its infancy, with many of the techniques yet to be well worked out and refined. Following the wreck of the oil tanker, *Torrey Canyon*, in 1967, some clean-up methods appear to have caused more damage to the indigenous biota than the oil itself. A natural equilibrium was restored more rapidly to areas where no intervention occurred than to those where clean-up processes were used. Careless use of suction devices, scrapers, oil dispersants, and the like may cause more stress to the ecosystem if improperly used than the material of the spill itself.

## The Sources and Uses of Damaged Land

If reclamation is actively carried out, three aims are possible. The first is restoration, in which an attempt is made to put back exactly what was there prior to the disturbance. It is unlikely that this state of grace can ever quite be achieved because the restored ecosystem will not be subject to exactly the same soil chemistry or microhabitat. However, there are many good successes. The second possibility is to aim for something that is similar to, but a little less than, full restoration, that we can call rehabilitation. It could, of course, be argued that all restoration is, in fact, rehabilitation, but we should consider aims here rather than achievements. In the case of rehabilitation, full restoration is not expected.

The third possibility is that no attempt is made to restore what was present originally. Instead, there is replacement of the original ecosystem by a different one. The replacement could be an ecosystem that is simpler but more productive, such as a deciduous forest being replaced after mining by a simple grassland to be used for public recreation. All of these possibilities could be appropriate in the reclamation process. There is a tendency among many environmentalists to feel that all reclamation should be restoration, and, indeed, federal and state laws in the United States seem to be tending toward this end point. In some cases, full restoration is appropriate and sensible, particularly in large-scale new operations, such as after surface mining for coal. However, there are many cases where restoration is so difficult or expensive as to be impractical, and we should aim for something less than was there originally. Replacement is particularly sensible for truly derelict land produced by an operation that occurred some years ago. The obvious example here is quarries, where entire topographies have been changed and it would difficult to recreate them.

## The Techniques of Restoration

The foundation of an economical and successful restoration program needs a clear understanding of the environment and the plants, animals, and people involved. A restoration program should proceed via:

1. **A knowledge of why a species or community went extinct in the first place.** This is obvious in the case of activities such as mining even though causes of extinction for individual species are not so clear. For example, the release of 1244 Hawaiian geese back into Hawaii over a 16-year period essentially failed in the 1970s. The goose, called the ne-ne, did not establish a self-sustaining population. No one knows why. Possibly predators were to blame. The ne-ne spends much time on the ground, and the adults molt when leading their young and so cannot fly, making both the adults and young vulnerable to predators.

2. **A natural history of similar ecosystems.** Information on the native vegetation, the soil characteristics, and the animals occupying comparable, undisturbed native ecosystems should be obtained, with as much information as possible on the interactions among plants, animals, and humans. When this information is available, a draft plan for restoration can be developed.

3. **Test plots.** Another step should be to establish test plots to evaluate the strategies for restoration that appear promising. During this time, a seed-collection program should be initiated, and seed nurseries should be established if needed to increase seed stocks.

4. **Soil preparation.** Soil preparation may include techniques such as deep ripping and discing to break up the soil surface and make it rough. Sometimes, the soil is sterilized by chemicals or by other techniques. Prominent among these is the technique of solarization that involves moistening the soil and covering it with transparent plastic, letting the sun heat the soil, thereby killing weed seeds and many pathogens.

5. **Revegetation**. The essential elements of a minimum-cost restoration effort are the introduction of appropriate seeds and related symbionts to sites that provide suitable soil and moisture conditions for rapid root growth and plant establishment. Weed control can help slow-growing native plants to compete. Controlled burning at the time that weed species are most vulnerable can also help control weeds.

6. **Advanced techniques.** Other restoration elements that may be of value include pest control (cages or fencing to protect plants from vertebrate herbivores, or chemical sprays to kill insects), irrigation programs, and fertilizer. Fertilizer should be used with care because it may increase shoot rather than root growth, increase weed competition, and make plants more palatable for pests. Mulching and composting can also provide many benefits.

7. **The reintroduction of animal components.** Several authors have recently discussed the problem of returning animals to the wild. Wilson and Stanley Price (1994) have examined 149 bird and mammal reintroductions, from both captive-bred and wild caught animals, involving 121 species. Most of these projects are ongoing, and it is, as yet, not possible to quantify them as successes or failures. However, for captive-born animals alone, reintroduction is not very successful—only 16 of 145 reintroductions, or 11 percent were classified as successes (Beck et al. 1994). Almost half the bird reintroduction projects

Photo 19.7   *Red wolf reintroduction in North Carolina by the Nature Conservancy. The techniques of restoration ecology and reintroduction of animals are still in their infancy* (Krasemann, Peter Arnold 49114.)

involve birds of prey, and 26 percent of mammal projects used carnivores. There is great opposition by farmers to the introduction of these types of animals. Some lynx *(Felix lynx)* reintroductions in Europe failed because of shootings by farmers. Most reintroduction projects have been undertaken when the principal cause of extirpation was hunting, and suitable habitat is still available for the reintroduced population (Photo 19.7). The reintroduction of animals is discussed in more detail below.

## Restoring Animals to Ecosystems

Griffith et al. (1989) have reviewed intentional releases of native birds and mammals to the wild in Australia, Canada, Hawaii, New Zealand, and the United States and tried to identify factors associated with restoration success. Nearly 700 translocations, on which the report was based, were conducted from 1973 to 1986. Several factors were associated with success of translocations (Table 19.6). Native game species constituted 90 percent of translocations and were more successful (86 percent) than were translocations of threatened, endangered, or sensitive species (46 percent).

Also associated with greater success was increased habitat quality. Likewise, translocations into the core of species' historical ranges were more successful than were those on the periphery or outside historical ranges. Herbivores were more likely to be translocated successfully than either carnivores or omnivores. Translocations into areas with potential competitors of similar life form were less successful than translocations into areas without competitors. Early breeders with large clutches were slightly more likely to be translocated successfully than were species that bred late and had small clutches. Wild caught animals

**TABLE 19.6** *Percentage success of intentional introductions or reintroductions (translocations) of native birds and mammals to the wild in Australia, Canada, Hawaii, New Zealand, and the United States between 1973 and 1986. (From Griffith et al. 1989.)*

| Variable | Translocations (n) | Success (%) |
|---|---|---|
| Threatened, endangered, or sensitive species | 80 | 44 |
| Native game | 118 | 86 |
| Release area habitat | | |
|     Excellent | 63 | 84 |
|     Good | 98 | 69 |
|     Fair or poor | 32 | 38 |
| Location of release | | |
|     Core of historic range | 133 | 76 |
|     Periphery or outside | 54 | 48 |
| Wild-caught | 163 | 75 |
| Captive-reared | 34 | 38 |
| Adult food habit | | |
|     Carnivore | 40 | 48 |
|     Herbivore | 145 | 77 |
|     Omnivore | 13 | 38 |
| Early breeder, large clutch | 102 | 75 |
| Late breeder, small clutch | 96 | 62 |
| Potential competitors | | |
|     Congeneric | 39 | 72 |
|     Similar | 48 | 52 |
|     Neither | 105 | 75 |

were more likely to be successful when translocated than those of exclusively captive-reared animals. Success was also higher when individuals came from big populations in the wild as compared with small populations and if populations were increasing rather than decreasing. There appears to be no consistent association of translocation success with number of releases, habitat improvement, whether the release was "hard" (that is, where no food or shelter were provided) or "soft," or where the release was immediate or delayed at the release site after a period of time spent in fenced areas.

The results of Griffith's survey support the following conclusions:

1. Larger founder populations are more successful.
2. Habitat suitability is important.
3. Increased number and size of clutches enhances successful invasion.
4. Herbivores are more successful invaders than carnivores.
5. Competing species in an area may prevent successful invasion.

It is also true that successful translocation is a function of the number of animals released, as shown in Figure 19.8. However, the increases in success associated with releasing of large numbers of organisms become asymptotic; that is, they tend to level off so that the efforts involved in introducing very large numbers of animals are not warranted. For birds, generally, releases larger than 80 to 120 animals do little to increase the chances that a translocation will be successful. Similar results are obtained for other taxa. For example, for large native game mammals, the asymptote was reached at releases of 20 to 40 animals. These data are consistent with the idea that threshold population sizes exist below which extinction is likely, possibly due to chance events affecting birth and death of individuals or possi-

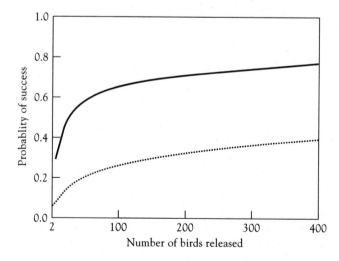

Figure 19.8   *Predicted probability of successful translocation as a function of the number of birds released in either excellent (solid line) or good (dashed line) habitat quality.* (From Griffith et al 1989.)

bly because below certain population densities, social interactions and mating success are disrupted.

Such data can be useful in planning release strategies. For example, we could predict that if 500 endangered birds were available for a translocation program, it would be wise to split them up and release them at three separate areas where each has an equal amount of birds. At each site, sufficient numbers of birds would be present to create a viable population. This would be in preference to, say, releasing all 500 birds at one site.

# Summary

 1. Habitats can be grouped according to degree of degradation. Such a listing might be, from least degraded to most degraded: wilderness areas, conservation areas, forests, agricultural land, urban environments.

2. Wilderness areas are large, unmanaged regions and constitute 52 percent of the Earth's surface, mostly in Antarctica, Greenland, or deserts. Only 27 percent of wilderness is not rock, ice, or barren land. The United States has only 4 percent of its area designated as wilderness.

3. Some national parks are designated to preserve specific plants or animals. The theory of park design has centered on two approaches. The first, Minimum Viable Population analysis, MVP, focuses on determining either the size of a population that has a 95-percent probability of survival for at least 100 years or the size of a population that will not undergo inbreeding or genetic drift (taken as 50 and 500 individuals, respectively). The second approach focuses on area and caters around the SLOSS debate—a single large or several small reserves. Most ecologists agree that this debate is ended because, in practice, a series of small reserves would generally encompass a wider variety of habitats and, thus, a richer array of species. In reality, park design may depend more on the ability of authorities to police a reserve adequately and prevent poaching than on theoretical constraints.

4. The ecological consequences of forestry and agriculture include the uses of chemical fertilizers and pesticides that affect nontarget organisms and pollute water bodies. Loss of topsoil can also severely degrade habitats. Finally, habitat alteration for both agriculture and silviculture can reduce the area of wildlife habitats.

5. Restoration ecology focuses on use of techniques to restore degraded land to either pristine wildlife habitat or to other habitat types. Restoration ecology requires a knowledge of local habitats and biota, soil preparation, and techniques to ensure successful reintroduction of plants and animals.

---

*Discussion Question:* Should we be concerned about maintaining small amounts of "green space" in cities and other urban habitats, or would money be better spent in preserving large tracts of relatively undisturbed land in pristine areas?

# 20

# Waste and Pollution

J ust about everything made or harvested by people turns to waste sooner or later, and an incredible amount is generated. Waste or pollutants may either be emitted into the air, like sulfur dioxide or carbon monoxide, find their way into water, like fertilizers, or be sprayed on land like pesticides. Nearly all forms of waste or pollutants have the potential to affect the distribution and abundance of life.

## 20.1 AIR POLLUTANTS

### Carbon Dioxide

In the greenhouse effect, gases in the atmosphere act somewhat like the glass panes of a greenhouse: they allow light, infrared radiation and some ultraviolet radiation, to pass through the troposphere. The earth absorbs much of this solar energy and degrades it to heat. Some of this heat escapes back into the atmosphere and is absorbed by molecules of greenhouse gases. The greenhouse effect is made possible by a group of gases, including water vapor, which together make up less than 1 percent of the total volume of the atmosphere. There are about 20 of these greenhouse gases. The four most important are carbon dioxide, nitrogen oxide, chlorofluorocarbons, and methane. Carbon dioxide is by far the most abundant. There is no doubt that $CO_2$ concentrations have been increasing globally (Fig. 20.1).

One of the main reasons for the increase in atmospheric $CO_2$ is the burning of fossil fuels. Global emissions of carbon dioxide from fossil-fuel combustion and cement manufacture have increased steadily during the past 40 years (Fig. 20.2).

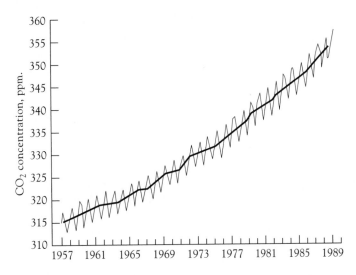

**Figure 20.1**   *Monthly mean concentrations of atmospheric $CO_2$ at Mauna Loa, Hawaii, 1958–1989. Similar trends are evident in Alaska, American Samoa, and the South Pole.*

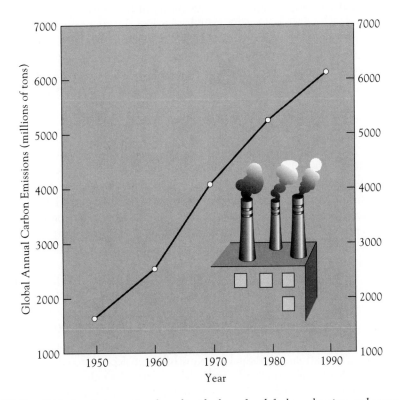

**Figure 20.2**   *Global emissions of carbon dioxide from fossil-fuel combustion and cement manufacture, 1950–1990.*

Seventy-five percent of increased global carbon emissions are thought to be from fuel-fossil burning. Augmenting that is the destruction of natural vegetation, which reduces the amount of $CO_2$ absorbed during photosynthesis, causing the level of atmospheric $CO_2$ to increase. This contributes only 25 percent to increased global $CO_2$ levels, but in some areas, such as South America and Asia the effect is particularly pronounced. The role of vegetation in controlling $CO_2$ through photosynthesis is clearly indicated in Figure 20.1 by variations in the levels of the gas during the year: in the Northern Hemisphere where most of the land mass is, levels go up in the winter and down in the summer growing season.

Computer models of global warming have yet to include accurately all the necessary mechanisms that may influence the degree of warming. For instance, higher temperatures associated with an intensified greenhouse effect would bring about more evaporation from the Earth's surface. This could lead to increased cloudiness as the rising water vapor condensed. The clouds, in turn, would reduce the amount of solar radiation reaching the surface and therefore cause a temperature reduction that might moderate the increase caused by the greenhouse effect. This is a negative feedback mechanism that could prevent escalation in global warming. However, there are also positive feedback mechanisms that augment the change. The colder northern waters of the world's oceans, for example, act as an important sink for $CO_2$, but their ability to absorb gas decreases as temperatures rise. Global warming would reduce the ability of the oceans to act as a sink. Instead of being absorbed by the oceans, $CO_2$ would remain in the atmosphere, thereby adding to the greenhouse effect. Finally, as permafrost melts with increased global warming, more methane would be produced from the resultant bogs and swamps—another positive feedback mechanism. Neither positive or negative feedback mechanisms have adequately been dealt with in current models of global warming. Thus, there is great uncertainty about the future extent of global warming that may occur. A 1990 survey of 18 global-surface warming simulations from seven different modeling groups had a range of 1.9° to 5.2°C if global atmospheric $CO_2$ doubled.

What are the long-term data on global warming and $CO_2$ levels? Ice cores have been drilled at a number of locations, the best known being Vostock in the former Soviet Union. It is possible to extract from these cores records of $CO_2$ concentration and temperature, based on measurements of gases trapped in air bubbles going back 160,000 years (Fig. 20.3). These records show much variation in both global temperature and $CO_2$ concentration. They also show a clear correlation between $CO_2$ and temperature. A 1993 article in the journal *Nature* on cores drilled in Greenland ice also showed changes of up to 10°C in the past (Raynaud et al. 1993).

What will be the effects of global warming on the distribution of plants and animals? About 18,000 years ago was the height of the last ice age. Global temperatures were about 5–6°C lower than present. Major portions of northern North America and northern Europe were under thick ice sheets. This ice age terminated 11,000 years ago, and the present (Holocene or Recent) geologic epoch started. It is an interglacial epoch, an era between so-called glaciations. A gradual warming has been accompanied by the melting of inland glaciers. This changed the landscape in higher latitudes radically. Tundra turned into forest, and previously glaciated areas became tundra. In many localities, a good picture of the changes in the flora can be obtained by pollen counts in lake sediments. These can be dated from

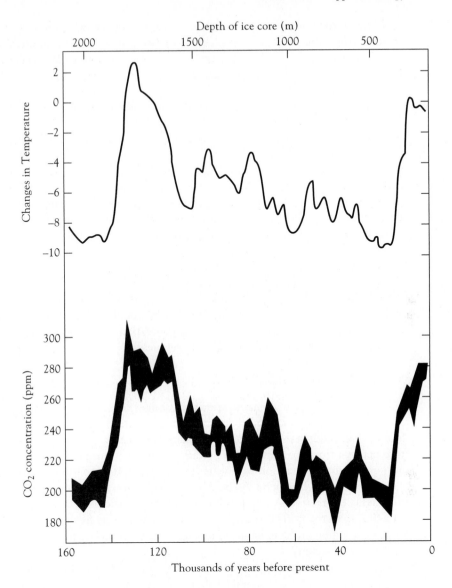

**Figure 20.3**  *Carbon dioxide and temperature during the past 160,000 years as estimated from the Vostok ice core.*  (After Barnola et al. 1987.)

various layers in a core taken from the lake bottoms by a process called *carbon 14* ($C^{14}$) *analysis*. One can see, for example, the succession of pollen from subarctic species to warmth-loving plants and trees, from birches and firs to oaks and maples. Animal remains show the retreat of musk ox and mammoth and the replacement by less woolly species. Sea-level rises caused by the melting ices can be documented. All this culminated between 8,000 and 5,000 years before the present warm period. Global temperatures were about 1.5°C warmer than presently. Since that time, some cooling has occurred again. The northern tree line has receded 300 kilometers. Old tree trunks can be still found in Canadian tundra. The northern glaciers have advanced, but since about 2,000 years before the present, the global climate has become relatively stabilized.

Because of a slight cooling in recent ages, some geologists have dubbed our era the "little ice age." Clearly, changes of only a few degrees Celsius can have a profound affect on distribution patterns of plants and animals (see Chapter 5). Recently, Baur and Baur (1993) noted the extinction of 16 of 29 populations of the land snail, *Arianta arbustorum*, in the Basel area of Switzerland. Habitat destruction due to urban development was the reason for the extinction of many of the populations, but suitable habitat was present at four other sites. However, urban encroachment had raised local temperatures, and Baur and Baur showed that egg-hatching success decreased substantially with only a slight alteration in temperature—this was the reason for the extinction of the other four populations of snails.

Other negative effects of global warming include alteration of the reproductive capacities of some vertebrates. The sex ratio of lizards and alligators would change, because sex ratio is determined by nest temperature. Warmer temperatures would perhaps mean more males of these species. Turtles, on the other hand, would produce mainly females. For insects, warmer temperatures might allow an extra generation to fit into a longer summer. In agriculture, more insect pests might lead to lower yields.

## Sulfur Dioxide

Sulfur dioxide, $SO_2$, comes mainly from the combustion of coal and oil to produce electricity. Sulfuric acid plants and the processing of metal ores containing sulfur also produce some $SO_2$. Urban-industrial areas emit large quantities of $SO_2$—New York City emits 2 million tons per year from coal, and Great Britain 5.9 million tons per year. The downwind transport of $SO_2$ and its reaction with water in the atmosphere to form sulfuric acid ($H_2SO_4$) is a major ecological headache. Although nitrogen oxides from car exhaust can also influence the acidity of precipitation, it is sulfur dioxide that is the primary cause of acid rain. Industrial $SO_2$ from the northeastern United States can acidify freshwater lakes in Canada and kill fish; a similar problem has been identified in Scandinavia from acid rain originating in Britain and Germany.

A relative measure of the acidity of rain is the number of hydrogen ions in a liter of water. In pure water, there would be about 1 hydrogen ion for every $10^7$ molecules of water; in rainwater, about 1 per $10^6$ water molecules because rainwater comes into contact with carbon dioxide in the atmosphere and carbonic acid, a very weak acid, is produced. Some rain falling in New England, on the other hand, has 1 hydrogen ion for every $10^4$ water molecules—that is 1 in 10,000, a 100-fold increase above normal. The reason is that the area is downwind from the industrial centers of the Northeast, where fossil fuels are burned in huge quantities. Acidity is usually measured, however, by **pH**, the negative logarithm of the hydrogen-ion concentration. Thus, for pure water, $-\log_{10}(0.0000001) = 7$. Many of the lakes of the Adirondack Mountains of the northeastern United States, have a pH value of less than 5, and some of these have few fish. Fish in acid lakes may also suc-cumb to toxic concentrations of metals such as aluminum, mercury, magnese, zinc, and lead. These are leached from the surrounding rock by acids. Thus, acid rain has direct and indi-rect effects. Aluminum may be deposited on the gills, where it inhibits breathing and even-tually leads to suffocation.

In southern Sweden, the gradual disappearance of fish from lakes has been noted for many years (Fleischer et al. 1993). First roach disappeared; then, 20 years later, pike; then, 10 years later again, eel and perch. In all cases, it is usually the young fish that are most sus-ceptible. Acid rain is clearly a serious problem, and the **air pollution** that crosses interna-tional borders to cause it can be a subject of intense political debate.

Besides acidifying lakes, there is also some evidence that terrestrial ecosystems show adverse affects from acid rain. Reduction in forest growth in Sweden, physical damage to trees in West Germany, and the death of sugar maples in Quebec and Vermont have all been linked to the increased acidity of the precipitation in these areas, together with increased ozone levels in the atmosphere. In addition, terrestrial wildlife can be affected by acid precipitation. Acid deposition reduces snail abundance on certain soils in Europe. Reduced snail abundance can lead to passerine birds producing eggs with thin and porous shells (Graveland et al. 1994). This, in turn, leads to clutch desertion and empty nests. In Russia, densities of voles is affected by high sulfur-dioxide concentrations from a copper-nickel smelter. The sulfur dioxide decreases the abundance of food plants in the area. Den-sity of voles increases as distance from the smelter increases (Kataev, Suomela, and Palokangas 1994).

One encouraging development is that during the last 10 years, emissions of sulfur diox-ide from European countries and in North America have dropped by about 20 percent. If countries conform to international agreements, another 20 percent reduction will be made in the 1990s. However, even more reduction is needed to eliminate further acidification in many regions.

## Chlorofluorocarbons

Chlorofluorocarbons (CFCs) are used in refrigerator cooling systems as aerosol propellants. They are also used in the production of polyurethane and polystyrene foam for coffee cups, fast-food containers, and foam insulation. In the stratosphere, CFCs react with ozone and deplete the ozone layer. The depletion of stratospheric ozone allows more UV light to pen-etrate to the Earth and results in an increase in skin cancer in humans. Other effects include reduced growth rates in plants such as tomatoes, lettuce, and peas. It is clear that reduced growth rates in native vegetation and increased mutation rates in animals also probably results—but these have not been studied. Alterations in the ozone layer may also affect global climate.

CFC production peaked in 1973, though production is now declining. The good news here is that by 1990, the European community and the United States banned all production of chlorofluorocarbons by 2000. However, the slow rate of decay of CFCs means that they will remain in the atmosphere for up to 40 years.

In summary, the effects of gaseous waste are seen at three scales:

**At the point source:** An example is the release of the volatile liquid methyl isocyanate from the Union Carbide plant at Bhophal, India, which caused more than 2,000 deaths in 1984. Methyl isocyanate, made by combining methylamine and phosgene, had been used there since 1980 in the production of the pesticide carbaryl for the domestic market.

**Regionally:** The local climate may be affected and wildlife may be affected. Acid rain is a good example.

**Globally:** The world's climate can be changed. The greenhouse effect is of paramount importance here.

## 20.2  PESTICIDES

The political controversy surrounding pesticides stems from their dual nature; they are valuable because they are lethal. Without the use of pesticides, preharvest crop losses in the United States would be about 18 percent from insects, 15 percent from diseases, and 9 percent from weeds, and the benefits from pesticide use are valued at many billions of dollars. With the use of pesticides, preharvest crop losses are estimated at 13 percent from insects, 12 percent from diseases, and 8 percent from weeds. Therefore, the use of pesticides causes a reduction in attack of 5 percent for insects, 3 percent for diseases, and 1 percent from weeds. This means that roughly 10 percent additional harvest is reaped with the use of pesticides. Of course, these are overall figures, and, in specific cases, much higher yields are possible because of pesticides. Pesticide use in agriculture can increase not only crop production but livestock production, too, and give greater longevity of food products on the shelf. In nonagricultural uses, pesticides have been important in reducing the incidence of malaria and other diseases, saving untold thousands of human lives.

A wide variety of pesticides are used in modern industrialized agriculture. More than 45,000 formulations (mixtures) and 600 active ingredients are registered pesticides in the United States. Only about 40 active ingredients, however, account for 75 percent of the use.

Despite the value of pesticides several factors give rise to environmental concern:

1. Some pesticides have a long life in the environment and can accumulate in food chains so that pesticide poisoning becomes serious in "higher" organisms. Consider the case of the peregrine falcon. The breeding population of British peregrines had been reasonably stable until about 1955, probably for some centuries, but by 1962, 92 percent of known prewar territories in southern England were deserted. Moreover, there were successful nestings in only 26 percent of the occupied territories. What was causing this rapid decline in numbers and lessened breeding success?

There were a variety of possible causes, including changes in climate, food, and diseases and also habitat loss. There seemed to be no good evidence for change in any of these factors, and pollutants were thought to provide the most likely explanation. Two compounds were implicated, dieldrin and DDT. DDT and its derivatives can inhibit the enzymes in the eggshell gland, which causes a reduced eggshell thickness and strength (Ratcliffe 1980) (Photo 20.1). It was not feasible to do laboratory tests on peregrines to demonstrate directly the toxic dose of pesticides for this species—it is not an easy species to keep in captivity and was moreover by then an endangered species. But there was evidence to suggest that consumption of two to three heavily contaminated pigeons could be sufficient to kill a peregrine. There were many incidences of pigeons found dead and dying in the 1950s from acute toxicity after eating seed treated with pesticides.

Because egg collectors had collected peregrine eggs over a long period of time, there was ample opportunity to go back and test the strength and thickness of these shells and compare them to modern shells (Fig. 20.4). The thinning of the eggshell started in 1947, shortly after the first recorded use of DDT in 1946, although widespread use did not occur until the 1950s. With the banning of many of the most toxic organochlorine pesticides such as DDT in the 1970s, a good opportunity arose to see if eggshell thickness increased again—which it did. Also the population numbers of the peregrine showed a rise, though this could also have

**Photo 20.1** *Duck's nest with clutch of eggs illustrating effects of DDT. DDT accumulates in food chains and causes thin egg shells, which break under the weight of the parent birds.* (Photo from U.S. Department of Agriculture.)

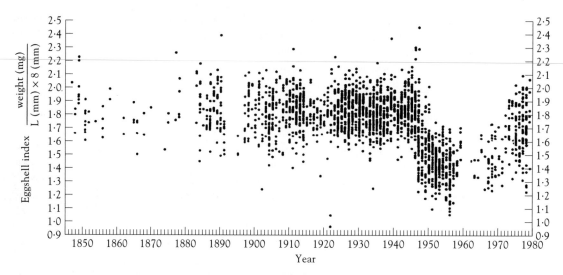

**Figure 20.4**    *Changes from 1845 to 1979 in Ratcliffe's eggshell index, (shell wt, mg/length, mm × breadth, mm), for British peregrine falcons* (Falco peregrinus). *This is the most extensive published set of data for eggshell thickness in birds and indicates the abrupt thinning that started in 1947.* (From Ratcliffe 1980.)

been affected by releases of captively bred peregrines, to improvement in habitat of the bird, and to increased numbers of its prey items. However, it is likely that cessation of pesticide use is a prime cause because numbers of in another British bird of prey, the sparrowhawk, increased dramatically over the period 1980–1989 following the withdrawal of the pesticides aldrin and dieldrin from their major use—on cereal seeds (Newton and Wyllie 1992). At the same time, pesticide residues in dead birds decreased, and eggshell thickness increased and breakage decreased.

2. Pesticides generally have a broad spectrum of toxicity and have a lethal effect on a wide variety of organisms, not limited to just insects but also including fish, birds, and mammals.

3. There are manufacturing accidents and accidents when applying pesticides to crops in the field.

4. Resistance can develop to the insecticide so that the pesticide essentially becomes ineffective.

5. Pesticides may kill the pest's natural enemies, allowing a resurgence of pests in the absence of their natural enemies. In both cases 4 and 5, different pesticides are often used to "get the job done."

6. People are exposed to pesticide by consuming residues in food and water. As long ago as the early 1970s, about 50 percent of food sampled by the U.S. FDA contained detectable levels of pesticide. This led to the so-called Delaney paradox—a ruling by Congress (the Delaney clause) that allowed absolutely no traces of pesticide residue in certain foods. However, with such sensitive detection equipment as is now in use, it is virtually impossible not to detect a minuscule fraction of pesticides. To ensure no traces of pesticides, many people demand that their produce be "organically" grown without the use of chemicals.

Clearly, pesticide use can drastically alter the distribution and abundance of all forms of life, via direct effects on, for example, honeybees to indirect effects through food webs to higher predators.

## 20.3  WATER POLLUTION

### Oil

In marine systems, a common contaminant is crude oil. Oil can kill directly through coating and asphyxiation, especially acute for intertidal life, or by poisoning by contact or ingestion, as in plants and preening birds, respectively. Water-soluble fractions can be lethal to fish and invertebrates and may disrupt the body insulation of birds, resulting in their death from hypothermia. The jackass penguin, endemic to South Africa, has suffered severe population declines from oil lost by tankers rounding the Cape of Good Hope.

Though natural seepage of oil into the oceans, mainly at junctions of tectonic plates, occurs at the rate of 600,000 tons per year, one large tanker can carry 200,000 tons of oil (Goldberg and Bertine 1975). The *Amoco Cadiz,* wrecked off France in 1978, was a 230,000 ton vessel. Oyster beds along the Brittany coast were totally ruined, fishing and resort industries were affected, and in tidal flats and salt marshes, effects lasted for seven years. Estimates of the total damages at the time were on the order of $30 million. It is important to remember that clean-up operations with detergents can prove more devastating to marine life than the oil itself, which is what happened in Britain after the *Torrey Canyon* disaster (Walsh 1973). Sometimes, it is best merely to encourage the natural breakdown of oil. Bioremediation following recent oil spills suggest that fertilizer applications significantly increase the rate of biodegradation of oil by existing microorganisms (Bragg et al. 1994).

A particularly severe oil spill resulted from the wreck of the *Exxon Valdez* in Alaska at 12:04 A.M. on Good Friday, March 24, 1989 (see Photo 20.2). The rocks at Bligh Island tore five huge gashes in the hull of the ship, one 6 ft high by 200 ft long. The result was one of the worst oil spills in U.S. waters; 10.7 million gal spilled, most of it in the first 12 hours. The accident took place in fine weather, clear visibility, and no traffic and was clearly the result of human error. A week after the spill, the resultant slick covered nearly 900 sq mi. Hundreds of miles of shoreline were covered with oil, in places as much as 6 in. deep. Officially, 27,000 birds, 872 sea otters, and untold numbers of fishes died, although the true numbers are probably higher because many dead birds and otters probably sank and were not recovered. These deaths resulted because birds and otters depend on the insulation provided by their feathers and fur to help them maintain proper body temperature. A coating of oil destroys that insulating property, and the animals literally die of exposure.

The effects also carried over into the terrestrial ecosystem when bears, otters, and bald eagles feasted on the oily carrion washed up on the beach. Sitka black-tailed deer ate kelp on the beaches. Few of these animals were expected to be found dead on the beaches because they generally return to their normal habitats before the effects become apparent. Still, the Fish and Wildlife Service found more than 100 dead eagles, and most pairs in the area failed to produce young that year. Much money was spent in the rescue and rehabilitation of oiled wildlife, including sea otters and birds. It is valuable to examine the success of these types of procedures, particularly as it seems that money might be better spent in other protective measures (Estes

Photo 20.2
(a)  *Aftermath of a massive oil spill. The oil tanker Exxon Valdez, run aground off the Alaska coast, 1989, showing remaining oil being offloaded to another tanker and a boom placed in an attempt to minimize the amount of oil washed up on the shore.*

(b)   *The adjacent shoreline, covered with oil despite containment efforts. Cleaning efforts were slow and laborious. Steam cleaning was sometimes employed. Although the hot water removed the oil more effectively than cold, it killed the animals that lived on the shore.* (Photos by U.S. Coast Guard.)

1991). In total, 357 sea otters were captured and delivered to rehabilitation facilities. Of these, 123 died in captivity. Thirty-seven of the 234 survivors were judged unsuitable for return to the wild and were transferred to aquaria and other permanent holding facilities; 25 of these animals were still alive 10 months later. The remaining 197 survivors were released by August 1989, 45 of them with surgically implanted radios. Twenty-two of the instrumented animals were dead (11) or missing (11) the following spring, thus indicating relatively low post-release survival of the captured and treated animals. At best, 222 sea otters (the 197 released and 25 in captivity) were captured and rehabilitated. A total of 878 dead sea otters were found, and this probably represents only 20 percent of the total killed by oil.

Capture and rehabilitation costs for sea otters alone was $18.2 million. Assuming that 222 otters were saved (maximum possible), costs exceeded $80,000 per animal. Post–oil-spill capture and rehabilitation probably cannot be used to substantially reduce the otter losses. Perhaps a more effective strategy to protect populations of sea otters is to spend the equivalent amount of money to reduce the risks and to enhance populations of threatened species in anticipation of potential catastrophic loss. Money could perhaps be better spent in properly documenting population densities and studying their population biologies. Such documentation is necessary if we are to know the impact of any impending catastrophe.

## Eutrophication

Eutrophication is the enrichment of waters with nutrients, primarily phosphorous and nitrogen. This usually leads to enhanced plant growth. These changes may occur as a result of anthropogenic changes (cultural eutrophication) or as a result of succession, the natural aging of a lake (natural eutrophication). As outlined in Chapter 13, bodies of water that are not rich in nutrients are called *oligotrophic*, and those rich in nutrients are called *eutrophic*. The most common attributes of a eutrophic lake are blooms of algae that make the water more turbid, more unattractive to swimmers and less suitable to certain kind of fish.

What is the state of eutrophication of lakes around the world? The data indicate that there is great disparity between countries and regions. Of the three-quarters of a million lakes in Canada, the great majority, 75 percent, are still oligotrophic, although the lakes of the southern, more densely populated regions are predominantly eutrophic, which reflects the increase in eutrophication associated with anthropogenic effects. Among the smaller U.S. lakes, there are many that are surrounded by farmland. Their condition, therefore, reflects eutrophication from agriculture and animal husbandry. The result is that up to 70 percent of U.S. lakes may be eutrophic. In Europe, too, many lakes seem to be eutrophic.

However, the great majority of lakes and rivers in the eutrophic category are relatively small in both surface area and volume, averaging only 2.2 km$^3$ in volume compared to 67.6 km$^3$ for oligotrophic lakes. Smaller lakes are perhaps more susceptible to cultural eutrophication than bigger lakes. Because there are many small lakes and few big ones, the percentage of lakes that are eutrophic is high. However, the volume of freshwater in a eutrophic condition is small, only 12 percent of the volume compared to 52 percent classified as oligotrophic. The other 36 percent of lake water is in intermediate condition. It is clearly difficult to reach a consensus on the state of eutrophication of U.S. lakes. On the other hand, many of the smaller water bodies are important either for drinking-water supply or for recreational purposes and should be cleaned up. There is clearly room for improvement.

The data for rivers is less definitive, but eutrophication effects are generally less acute than for standing waters because nutrient inputs are often quickly washed away. On the other hand, the data for human-constructed reservoirs shows much higher rates of eutrophication than for natural lakes, although the volume of these waters is small in comparison to the area and volume of natural lakes.

How can we control cultural eutrophication? First, we have to know what the anthropogenic causes are. The degree of eutrophication for the Great Lakes shows a striking resemblance to the maps of human population density along the shores. Similarly, areas mostly affected by eutrophication in the Mediterranean coincide with densely populated lands, most of which are either areas of intensive agriculture or high industrial development. High population densities lead to cultural eutrophication via three pathways:

1. A strong tendency for urban waste to increase and be discharged directly into waterways. An added factor to this since the 1940s is the use of detergents containing polyphosphates.
2. Rapid industrialization with a corresponding increase in industrial wastes of all kinds.
3. Intensification of agriculture and the increased use of chemical fertilizers, especially those containing phosphorous, concentration of livestock breeding, and direct discharge of agricultural wastes, rich in nitrogen, into waterways.

The measures to control eutrophication fall into two main headings: preventive measures and corrective measures. Preventive measures include:

1. Treatment of waste waters (removal of phosphorous and nitrogen).
2. Diversion of waste waters from lakes or rivers.
3. Primary sedimentation basins in waste-water streams to let phosphorus- and nitrogen-rich material settle to the bottom where it can be removed.
4. Watershed protection (reforestation, restriction of livestock, controlled fertilization/irrigation).
5. Substitution of phosphate detergents by other detergents not rich in phosphorus.

Corrective measures to try to bring eutrophic lakes back to an oligotrophic condition include:

1. Physical manipulation (withdrawal of water; aeration to increase levels of oxygen).
2. Chemical manipulation (application of herbicides to kill blooms of algae or large plants [macrophytes]).
3. Biological manipulation (mechanical harvesting of algae, macrophytes; direct manipulation of the food chain by adding exotic fish).

Treatment of waste waters is already underway in many areas. The reduction of polyphosphates in detergent has been imposed by law in Canada and some U.S. states bordering on the Great Lakes. In Sweden, 80 percent of treatment plants include a third stage for the elimination of phosphorous, and only 20 percent of waters discharged into the waterways receive no treatment. Strategies for control of nitrogen outputs, especially from agricultural activities, are in less well-developed stages, but there are some dramatic success stories in lake restoration.

It would be unfair to create the impression that water pollution is a totally insoluble problem or that levels of pollution must inevitably rise. The complete case study of the eutrophication of Lake Washington and its reversal to an unpolluted condition has been well documented by Lehman (1986). Lake Washington at Seattle is a moderately deep (65 m) warm basin that discharges into Puget Sound via a system of locks and canals built in 1916. Seattle began discharging raw sewage into Lake Washington at the beginning of the twentieth century, but in 1926 this trend began to change as sewage was diverted into treatment plants and thence into Puget Sound. By 1941, the last sewer outfall into the lake was removed, but thereafter the suburbs of Seattle began to expand, and by 1953, ten new treatment plants had sprung up around the lake and were discharging 80 million liters into it daily. Alternative options were not as readily available to the suburbs as they had been to Seattle itself. By this time, both the scientists at the University of Washington and the lay public were aware that algae had increased in the lake, and a species indicative of classical lake deterioration, *Oscillatoria rubescens*, was found in the lake for the first time. This alga had been connected with the early stages of decline in water quality in the classical examples from Europe, particularly Lake Zurich, and in Madison, Wisconsin (Hasler 1947). Scientists monitoring the Lake Washington situation were able to predict with great certainty the demise of the lake (Edmondson 1979), and the newspapers were quick to pick up these predictions. Such attention by the media paid off because by 1958, local politicians voted money to clean up the lake. Ground-breaking ceremonies for the clean-up campaign were held in 1961, not a moment too soon as this body of water had already been christened "Lake Stinko" by the local press. Visibility in lake water had declined from 4 m in 1950 to less 1 m in 1962. One by one, the waste-treatment plants around the lake had their effluent diverted. The trend of lake deterioration stopped in 1964, and by 1965 algal abundance was decreasing and water transparency increasing. Surprisingly, *Oscillatoria* persisted into the 1970s, but by 1975, it was gone and phosphorus concentrations had leveled off. The lake was clean again; visibility was as high as 12 m at times! The whole process was later repeated in Canada, this time in a controlled-environmental setup, by limnologists in a large-scale series of experimental lakes in Ontario. The results showed beyond a shadow of doubt that phosphorus was the master controlling agent of eutrophication (Schindler 1977) (Photo 20.3).

## 20.4  RADIOACTIVE WASTES

The two major concerns about atomic energy is that it generates radioactive waste and that the possibility of an accident in a nuclear-power plant could spew radiation over a wide area.

From radioactive wastes comes ionizing radiation—radiation with sufficient energy for its interactions with matter to produce an ejected electron and a positively charged ion. The biological problem is that such interactions in the cells of living organisms can cause genetic and physiological damage and even death. Most phases of the fuel cycle of a nuclear reactors produce

Photo 20.3   *Lake 226 in the Experimental Lakes area of northwestern Ontario, showing the role of phosphorus in eutrophication. The far basin, fertilized with phosphorus, nitrogen, and carbon, is covered with a bloom of the blue-green alga* Anabaena spiroides. *The near basin, fertilized with nitrogen and carbon, showed no changes in algal abundance. Photo taken September 4, 1973.* (Photo by D.W. Schindler.)

radioactive substances. Some are very short lived and can be released in carefully controlled quantities to the atmosphere or bodies of water, but others can have long half-lives and must be kept away from the environment and from people for a very long time. Strontium-90 has a half-life of 28 years, which means that it would take 500 years to drop to one-millionth of its original activity and the recommended level of safety. Plutonium has a half-life of 24,400 years. Fortunately, these wastes are not produced in large quantities: one ton of spent nuclear fuel when reprocessed gives 500 liters of waste. The disposal of these wastes, however, causes many problems because they must be isolated from the biosphere for 250,000 years, a high societal commitment. Suggestions for disposal have included sinking them into the Antarctic ice cap, firing them into the sun, or placing them on interplate subductions to be carried into the Earth's core. The most popular, however, is geological disposal in rock formations without any groundwater circulation; salt deposits (especially abandoned salt mines) are favored.

Studies of the effects of ionizing radiation on ecosystems have been carried out in oak-pine forests (Woodwell and Rebuck 1967), in tropical rain forests (Odum et al. 1970), in hardwood forests in the northern (Murphy, Sharitz, and Murphy 1977) and southern (Cotter and McGinnis 1965) United States, and in a shortgrass prairie (Fraley and Whicker 1971). Photo 20.4 shows an example of one such study in North America. One useful generalization to emerge is that the sensitivity of plants depends on the ratio of photosynthetic tissue to total tissue (Woodwell 1967, 1970). Thus, the most sensitive plants are trees, which have a relatively low ratio of photosynthetic mass (leaves) to nonphotosynthetic mass (stem and root), then herbs, grasses, and finally algae, which are usually quite resistant. Among trees, pines, which are evergreen, are more sensitive than deciduous hardwoods, in which the investment in leaves is smaller. A generalization that applies to both plants and animals is that sensitivity to radiation is correlated with size, the largest species being the most sensitive (Woodwell 1967), as they are to stress in general (Woodwell 1970).

Problems resulting from nuclear-power plants are not simply going to go away; 1 kg of uranium-235 produces as much commercial energy as 2,500 tons of coal, and nuclear power is thus a very attractive energy source. In addition, use of nuclear power would lessen the production of $SO_2$ and $CO_2$ from the burning of coal. Acid rain and the threat of global warming would both diminish.

## Nuclear Accidents

The greatest risk of nuclear power may lie not in waste production but in accidents at the plant. No full **meltdown** has yet occurred in the United States, but there have been a large

Photo 20.4   *Effects of ionizing radiation on a forest ecosystem. This forest was irradiated deliberately as part of an experiment.* (Photo by G.M. Woodwell, Woods Hole Marine Biological Laboratory.)

number of minor accidents, the most famous of which was the Three Mile Island incident on March 28, 1979, in Hamsburg, Pennsylvania. A faulty water pump caused overheating, which was followed by a whole series of human errors, which in turn allowed a small amount of radioactive water to be released into a nearby river. No deaths or serious injuries were reported, but the clean-up cost at least $1.5 billion and lasted more than 10 years. Other locations have not been so lucky; the Chernobyl disaster in Russia occurred on April 26, 1986. Two explosions inside one of the four water-cooled reactors blew the roof off the reactor building and set the graphite core on fire. As usual, human error was to blame; engineers had turned the automatic safety systems off for an experimental test. At least 31 plant workers died from exposure to radiation, and 200 others suffered acute radiation sickness. Winds carried radioactive material over much of Europe (Fig. 20.5). In the nearest village, trees in an area about 6 km² were killed within a few days of the accident (Medvedev 1994). By 1991, the area of dying trees had reached 38 km². In this area, rodents showed strong population declines, and even insect populations, normally highly radio-resistant, were depressed.

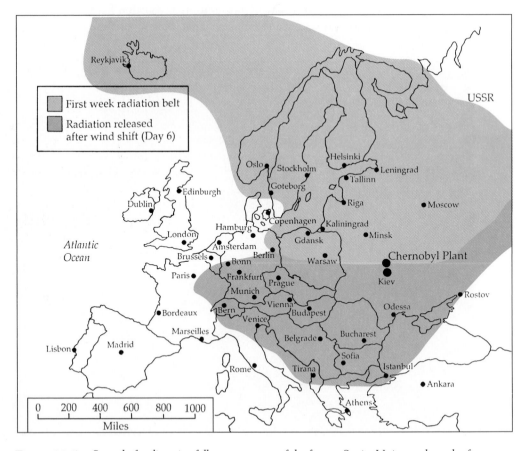

Figure 20.5    *Spread of radioactive fallout over parts of the former Soviet Union and much of eastern and western Europe after the Chernobyl accident.* (Redrawn from Miller 1988.)

# *Summary*

 1. Pollutants have the capability of changing the distribution patterns of plants and animals. Some of the main pollutant categories are air pollutants, pesticides, water pollutants, and radioactive waste.

2. The main air pollutants are carbon dioxide, sulfur dioxide, and chlorofluorocarbons. Carbon dioxide, released mainly from the combustion of fossil fuels, is the main cause of the greenhouse effect whereby the earth warms. If global warming occurs, the distribution patterns of plants and animals may change, with the possibility of local extinctions. Sulfur dioxide can react with rainwater to produce acid rain. Acid rain can damage forests, but its effects are most severe in bodies of freshwater where the resultant acidity is damaging to aquatic life. Chlorofluorocarbons were used primarily in refrigerator cooling systems and as aerosol propellants. In the stratosphere, CFCs, react with ozone and deplete the ozone layer. This allows more ultraviolet light to reach the Earth's surface and damage plants and animals.

3. Pesticides are powerful broad-spectrum poisons. They can affect a variety of nontarget organisms, not just pests. Pesticides often accumulate in food chains so that poisoning becomes frequent in top predators.

4. Eutrophication, an increase in algal production and water turbidity, results from the enrichment of water with nutrients, particularly phosphorus and nitrogen. Causes include fertilizer runoff from agricultural fields and the discharge of urban waste into waterways. Treatment of waste waters can greatly reduce eutrophication.

5. Two major concerns about atomic energy are fear of accidents and concern over the generation of radioactive waste. The 1986 Chernobyl explosion underscored the great ecological damage that can result from radioactive accidents.

*Discussion Question:* Which types of pollutants should we be most concerned about in terms of their effects on the distribution and abundance of plants and animals? With the same question in mind, which is the best form of energy for society to use and why? (Bear in mind that solar power and wind power cannot yet fully supply society's needs)

# CHAPTER 21

# *Exotic Species*

Not only can people radically alter the environment around them by introducing chemical pollutants to the air, land, and water, they can also be important agents in the deliberate or accidental spread of exotic plants and animals around the globe. Some of the most important introductions have involved domesticated plants and animals. Such deliberate movements have had many effects, some good, some bad. Accidental introductions nearly always have bad effects. To some people, the introduction of exotic species into new environments qualifies as biological pollution.

## 21.1 HOW INTRODUCTIONS OCCUR

### *Deliberate Introductions*

Many plants have been introduced deliberately, for example, as new fodder for sheep to bolster the wool trade in England or as ground cover to prevent erosion in the United States (Photo 21.1). The Texas longhorn, pride of the Texas cattle industry, was introduced. The Sitka spruce (*Picea sitchensis*), present mainstay of the British Forestry Commission's plantations, was introduced. Most organisms introduced in the eighteenth and nineteenth centuries had some sort of economic merit, but later introductions were often motivated by curiosity or decorative value. In Britain, in the early 1900s, whole rooms full of exotic birds were released in suburban gardens in the hope that they might adapt to local conditions and provide exotic bird song. Some nineteenth-century New Yorkers wanted all the birds in the works of Shakespeare, and in Hawaii a group called the Hui Manu collected money from schoolchildren to produce a lowland avifauna that is totally exotic.

**Photo 21.1**   *Kudzu, Pueraria lobata, a Japanese vine that was deliberately introduced to combat erosion but now chokes native vegetation in the southeastern United States.*   (Photo by Peter Stiling.)

Some introduced plants will always remain dependent on people for their survival-requiring watering or for competing weeds to be removed. Other introductions, however, do quite well on their own in an exotic environment. The potato, native to South America, grows unaided in the mountains of Lesotho, and the peach in New Zealand, the guava in the Philippines, and coffee in Haiti are all perennials that have established themselves as wild-growing populations. In Paraguay, oranges (which originated in southeast Asia and the East Indies) do well in the native vegetation. Many species of mammals and birds have been introduced as game animals and have done extremely well on their own. These include the brown trout which was brought from Europe to many areas of the world, Thar mountain sheep from the Himalayas to New Zealand, pheasants from Asia to Europe and North America, Barbary sheep from North Africa to New Mexico, red deer from England to New Zealand, and many other examples.

Worldwide, humans have succeeded in deliberately establishing at least 330 species of nonnative birds and mammals in more than 1500 separate cases. Twenty percent of the introductions have taken place on continents, 20 percent on islands near continents (shelf islands), and the majority (60 percent) on oceanic islands. Most introductions have been made to North America, Europe, and Australia (Fig. 21.1).

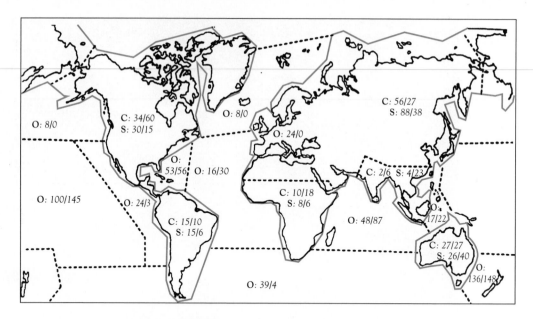

**Figure 21.1**    *Number of introductions on continents (C), continental shelf islands (S), and oceanic islands (O) (mammals/birds).*    (After Ebenhard 1988.)

## Accidental Introductions

Many introductions are not deliberate but inadvertent; they can nevertheless have drastic results. Human visitors to islands (called *ceys*) in Australia's Great Barrier Reef transport many alien plants and animals. The number of alien species in the flora on each of ten ceys is dependent not on island size (as might be predicted by island-biogeography theory) but on the amount of human-visitor traffic to that cey (Chaloupka and Domm 1986; Fig. 21.2).

The actual mechanism is proposed to be external attachment to clothing and footwear or in soil adhering to footwear. This overall pattern is one that tends to hold up in a variety of situations. This general relationship, increased species with increased visitor load, has been found to hold for 11 out of 13 nature reserves in North America and South Africa (MacDonald et al. 1989). This is something of a dilemma in terms of management—too few visitors cuts down on income and political support; too many and the very nature of the desired preserve is altered.

Plants have also been dispersed accidentally by moving objects such as vehicles, by crop-seed movements, and among soils (such as ships' ballast or road material). More than 50 percent of the 210 major U.S. weed species are invaders from outside the United States (Pimentel 1986). The percentage of introduced plant species in the faunas of different countries is staggering (Table 21.1). In California, the establishment of alien species has skyrocketed in recent years (Fig. 21.3), but it is worth remembering that number of species is not always a good measure of the extent of invasion. Numerous examples are known of single species that, on their own, have a huge effect, such as *Casuarina litorea* in the Bahamas, which creates such a dense

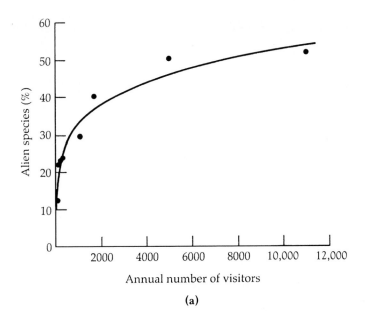

**(a)**

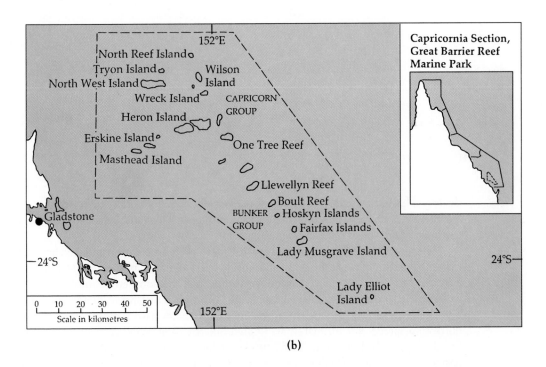

**(b)**

Figure 21.2  *Humans as agents in the spread of exotic species. (**a**) Percentage of alien plants recorded from each of 10 Australian ceys in the Great Barrier Reef as a function of the annual numbers of human visitors. (**b**) Location of study area.* (Redrawn from Chaloupka and Domm 1986.)

TABLE 21.1 *Percentages of introduced plant species in selected floras. (From data of Moore 1983 and Heywood 1989.)*

| Island | Number of native species | Number of alien species | Percent of alien species in flora |
|---|---|---|---|
| New Zealand | 1790 | 1570 | 47 |
| Campbell Island | 128 | 81 | 39.0 |
| South Georgia | 26 | 54 | 67.5 |
| Kerguelen | 29 | 33 | 53.2 |
| Tristan da Cunha | 70 | 97 | 58.6 |
| Falklands | 160 | 89 | 35.7 |
| Tierra del Fuego | 430 | 128 | 23.0 |
| Australia/Barbuda | 900 | 180 | 10 |
| Guadeloupe | 1668 | 149 | 9 |
| Hawaii | 12–1300 | 228 | 17.5–19 |
| Java | 4598 | 313 | 7 |
| Australia | 15–20,000 | 1500–2000 | 10 |
| Sydney | 1500 | 4–500 | 26–33 |
| Victoria | 2750 | 850 | 27.5 |
| Austria | 3000 | 300 | 10 |
| Canada | 3160 | 881 | 28 |
| Ecuador (Rio Palenque) | 1100 | 175 | 15 |
| Finland | 1250 | 120 | 10 |
| France | 4400 | 500 | 11 |
| Spain | 4900 | 750 | 15 |

shade and has such a toxic effect that few native plants can grow under it. In less than 100 years, *Andropogon pertusus* has become the commonest grass in lowland Jamaica, and bracken fern, *Pteridium aquilinum*, is rampant almost globally.

The most frequently introduced species are those most easily introduced accidentally—insects, small and inconspicuous and easily overlooked at borders and at international customs. Fire ants, introduced from Argentina into the southern United States in ships' ballast, first landed at Mobile, Alabama, in 1918 and have spread rapidly throughout the Southeast, drastically altering the appearance of old fields with their prominent nests. Their northward spread may ultimately be limited only by their inability to overwinter in colder environments. At last count, 1,554 species of insects had been introduced into the continental United States (Sailer 1983), representing 1.7 percent of the continental fauna. In Hawaii, the situation is even worse. There, 1,476 insect species have been introduced, which represent 29 percent of the islands' total. On the remote island of Tristan da Cunha in the South Atlantic, 38 percent of the insects are not native; 32 species are introduced and 84 are

**Figure 21.3**  *Estimation of the establishment of alien plant species in California since 1750.*
(Redrawn from Frenkel 1970.)

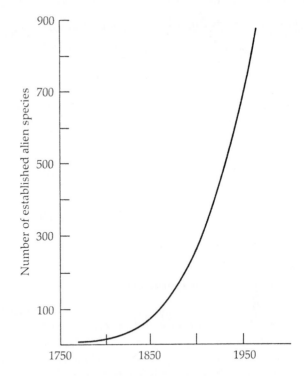

native. About 40 percent of the pests species of crops in the United States have been introduced, against which many control agents, biological and chemical, have been released. The economic damaged caused by introduced insects in the United States was assessed at $70 billion annually (Pimentel 1986).

Water carried by ships as ballast can be instrumental in the spread of aquatic organisms. A sample of Japanese ballast water released in Oregon contained 367 taxa (Carlton and Geller 1993). Transport of entire coastal planktonic communities makes coastal areas among the most threatened in the world. In 1988, the zebra mussel, *Dreissena polymorpha,* a native of the Caspian Sea, appeared in the Great Lakes. It now clogs water intake pipes, sinks buoys, and fouls boats.

## 21.2  THE EFFECTS OF INTRODUCTIONS

Introductions can have drastic influences on the native flora and fauna, sometimes outcompeting it or eating it (Table 21.2).

Reviews of introductions around the world have shown that in 20 percent of the cases of mammal introductions, a plant or habitat was affected because of herbivory (Ebenhard 1988). Another 16 percent of the mammalian species, mainly strictly carnivorous predators, caused ecological effects through predation. Introduced birds are relatively unimportant both as herbivores and predators. Negative effects on a native species because of competition are more common on islands. Similarly, island faunas and floras are more vulnerable to introduced predators and herbivores.

TABLE 21.2 *The effects of exotic species on native biota.*

| Type of exotic animal | Effect |
|---|---|
| Introduced plants | 1. Reduction of native plants through competition |
|  | 2. Increase of native herbivores |
| Introduced herbivores | 1. Reduction of native plants |
|  | 2. Reduction of native herbivores through indirect competition for plants |
|  | 3. Increase of native predators because of new availability of prey |
| Introduced predators | 1. Predation of native herbivores |
|  | 2. Reduction of native predators through indirect competition over herbivores |
| All of the above | 1. The spreading of parasites and diseases from exotic species to native species |
|  | 2. Genetic changes of native species through hybridization with exotic species |

## Effects of Herbivores

Sometimes, plant species can be eliminated by introduced herbivores. On Three Kings Islands, just north of New Zealand, 143 plant species were recorded in 1888. In 1934, after the introduction of goats, only 70 species remained. Goats had the same effect on the South Atlantic island of St. Helena, where 10 of 30 endemic plants have gone extinct and a further 15 are endangered. Introduced herbivores can also have a severe impact on forests. In New Zealand, introduced red deer, feral pigs, and sika deer have greatly reduced forest regeneration. On Hawaii, goats and other ungulates have disrupted the reproduction of the tree, *Acacia koa,* through grazing on shoots. Up until now, regeneration has not seriously been threatened because the seeds of native plants can lay dormant in the soil for at least 50 years. After this time, the seed supply will be exhausted, and a natural recovery of the plant will be impeded.

When forest canopy is opened up by browsing, more light reaches the ground, and more shade-tolerant species may be replaced by other species. Also, in heavily grazed habitats, annual plant species replace perennials, as observed in goat-derived vegetations on Hawaii and in the Netherlands due to rabbit grazing. In all, 160 of 510 total introductions of mammal herbivores have been deemed to cause a habitat change; that is, 24 percent of these introductions have caused severe changes in native plant composition.

## Effects of Predators

Introduced predators can have effects ranging from moderate to severe. For example, feral cats on the Crozet Islands and Kerguelen have completely exterminated smaller birds from the main islands, but on smaller cat-free islets, sea birds still occur. In the most extreme example known, the lighthouse-keeper's cat on Stephen Island arrived in 1894, and within a year, this one animal had completely exterminated the population of the Stephen Island wren. It has been estimated that 61 bird taxa have been exterminated by introduced predators, the most important predators being cats (the cause of 33 extinctions), rats (14 extinctions), and mongooses (nine extinctions).

A total of 49 mammalian species with more-or-less predatory habits have been introduced worldwide, and 20 of them have reportedly caused one or more indigenous populations to shift their abundance or go extinct. A further eight bird species have also caused such ecological effects (Ebenhard 1988). Only one predatory marsupial has been introduced, the opossum, to a number of West Indian islands. It is reported to threaten the bridled quail dove on Dominica. Seven species of monkey, which have omnivorous diets, have also been thought to have an effect on birds through their egg and nest predation. For example, the crab-eating macaque on Mauritius threatens both the Mauritius parakeet and the Mauritius kestrel. Four of five introduced dogs and foxes have had a negative effect on prey populations, and all six species of weasels and martin that have been introduced have had negative effects on prey populations. But probably the most dangerous predator ever introduced by humans has been the domestic cat, with 38 known or probable cases of abundance-shifting predation; that is, it has caused a decrease in prey population. The cat has been brought to all continents (except Antarctica) and to many islands, often intended as a means of rat control.

In all, 135 out of 425 total cases of predator introductions into separate areas of the world have resulted in decreases in abundance of native prey species. This amounts to 32 percent, a considerable proportion. This calculation includes more omnivorous species, such as the domestic pig. If only strict predators are considered, the proportion rises to 52 percent. It is clear that introduced predators can have a severe effect on native fauna.

What are the main vertebrate species affected by introduced predators? Table 21.3 summarizes the data. Reptiles are often vulnerable when newly hatched or as juveniles. Ground nesting birds seem particularly vulnerable especially to mammalian predators. Many of these breed in large colonies on the ground on oceanic islands. When humans introduce rats, cats, and mongooses, the result is often disastrous—for example, the great-winged petrel nests in colonies on Whale Island off New Zealand in the presence of brown rats. In dense colonies, the predation pressure on young is 10–35 percent, while in smaller colonies it may be 100 percent. On the islands known as the Kermadecs, cats and rats inflict 80-percent fatality on eggs and chicks of the sooty tern and threaten survival of the colony. Mammals seem less negatively affected through predation, though the data are less extensive for this group.

TABLE 21.3 *Number of predator-prey relationships involving abundance-shifts in the prey. Distribution of different prey categories with birds and mammals as predators. (After Ebenhard 1988.)*

| Prey | Avian predators | Percentage | Mammalian predators | Percentage |
|---|---|---|---|---|
| Reptiles | 0 | 0 | 53 | 21 |
| Ground-nesting birds | 8 | 36 | 103 | 41 |
| Tree-nesting birds | 14 | 64 | 65 | 26 |
| Mammals | 0 | 0 | 29 | 12 |
| Total | 22 | 100 | 250 | 100 |

## Effects of Competitors

Competition is notoriously difficult to demonstrate in the field, but lack of evidence is not proof that competition does not occur. In New Zealand, there is a large overlap in food choice among red deer, feral goats, bushy-tailed possums, and the native kokako bird. Species share resources, but this does not prove competition because it is not known whether or not any of the populations are affected. Similarly, in England, the introduced grey squirrel and the native red squirrel show more-or-less exclusive distributions that may be a result of competition, but it has recently been argued that the red squirrel was declining in certain areas and that the grey squirrel simply took over these vacant habitats.

Competition can be of two types, interference or exploitative. Interference competition involves aggressive behavior between two species. For example, on the Japanese Island of Oshima, interference competition has been observed between the introduced gray-bellied squirrel and the oriental white-eye bird. The squirrel chases the bird away from the flowers of camellia, which not only affects the bird but also the plant because it is pollinated by white-eyes. In exploitative competition, species compete for a common resource. For example, nesting boxes made for eastern bluebirds on Bermuda are used by introduced great kiskadees as perches, making it impossible for the bluebirds to nest in them. Table 21.4 summarizes some data regarding the extent of competition from introduced species. Competitive interactions from introduced birds seem more common (18 percent) than those due to introduced mammals (9 percent). Nevertheless, mammals have been found to be serious competitors in some instances. This is especially true in Australia, where one single order, the *Marsupialia*, has become highly diversified to fill the roles of insectivores, predators, grazers, fruit-eaters, and so on. The rabbit is a serious competitor for burrows with an animal called the boodie, the only burrowing kangaroo, and with the common rabbit bandicoot. Among the predators, the cat is probably competing with dunnarts, and the dingo (feral dog) has been thought to have contributed to the exclusion of the thylacine (native marsupial wolflike animal) from mainland Australia. Goats and sheep probably compete with a range of kangaroo species, especially the brush-tailed rock wallaby and larger species such as the red kangaroo and the western gray kangaroo. Sometimes, competition has been suggested to be more intense between introduced species closely related to native species than between distantly related species. For example, the American beaver has been introduced into Finland, where it excludes the European beaver, with the result that the two species now have separate geographic distributions.

TABLE 21.4 *Number of introductions of mammals and birds resulting in potential competition. Data indicate number of introductions with/without potential competition with percentage introductions with potential competition in parentheses. (After Ebenhard 1988.)*

| Category | Continents | Continental shelf islands | Oceanic islands | Total |
|---|---|---|---|---|
| Mammals | 24/60 (29%) | 6/108 (5%) | 15/291 (5%) | 45/109 (9%) |
| Herbivores | 15/34 (31%) | 3/54 (5%) | 13/119 (10%) | 31/207 (13%) |
| Birds | 11/29 (28%) | 1/40 (2%) | 22/88 (20%) | 34/157 (18%) |

TABLE 21.5 *Competition objects. Only cases involving an introduced bird or mammal species.* *(Data from Ebenhard 1988.)*

| Objects | Introduced mammals | | Introduced birds | |
|---|---|---|---|---|
| | No. of cases | % | No. of cases | % |
| Food | 31 | 63 | 8 | 18 |
| Nest-site | 3 | 6 | 29 | 64 |
| Space | 7 | 14 | 1 | 2 |
| Not known | 8 | 16 | 7 | 16 |
| Total | 49 | 100 | 45 | 100 |

What generally is being competed for? When mammals are introduced, competition is strongest over food, while for introduced birds, competition is strongest for nest sites (Table 21.5).

## Other Effects of Introduced Species

Other effects of introduced species include the introduction of diseases and parasites with exotic animals. For example, the extinction of large parts of the endemic Hawaiian avifauna has been blamed on introduced diseases brought by exotic birds (van Riper, Goff, and Laird 1986; Brewer 1988). Avian malaria was probably present in Hawaii long before humans arrived, brought by migrating wading birds. With the arrival of humans, domestic fowl, ducks, and gamebirds constituted an ever-present reservoir for the disease, but still no vector existed on Hawaii. In 1826, a party from Captain Cook's expedition went ashore there to fill the ship's water tanks. They emptied out the dregs of the old water, collected on Mexico's west coast and full of mosquito larvae.

It is often argued that indigenous species are more susceptible to introduced diseases than are the species bringing them in. This is ascribed to the fact that they have no natural resistance because they never had been exposed to the new antigen before. This could have happened on Christmas Island in the India Ocean, where the endemic bulldog rat vanished soon after the arrival of ship rats. Sick and dying bulldog rats were observed in large numbers. Indeed, rats are probably the worst introducers of diseases on a worldwide scale. It is not the rats themselves but rather their ectoparasites that bring the disease. Rat ticks are responsible for spread of scrub typhus, and the Indian rat flea brings the bubonic plague. Domestic stock have brought diseases, such as rinderpest and foot-and-mouth disease, around the world.

Another problem with introductions is hybridization between introduced and native species. Under natural conditions, this takes place only very infrequently as there are usually ecological, morphological, and behavioral barriers against hybridization. On the Seychelles Islands, the local subspecies of the Madagascar turtledove has been destroyed through hybridization with introduced doves. In Japan, the native subspecies of the Siberian white weasel has suffered the same fate, due to introduced Korean subspecies. In Britain, it is thought that few pure-bred polecats remain because of domestic ferret releases in the wild.

Other effects of introduced species include acceleration of soil-erosion rates. For example, in Channel Islands National Park, California, irreversible loss of topsoil through gully and sheet erosion is thought to have been initiated by destruction of vegetation cover by feral mammals, especially rabbits. In coastal areas of Britain, the opposite has happened where introduced salt-marsh plants have collected silt and debris, thus raising the level of the mudflats. In the Great Smoky Mountains, feral pigs rooting in the deciduous forest have accelerated leaching of calcium, phosphorous, zinc, copper, and magnesium from leaf litter in soil. In Hawaii, invasive introduced plants such as *Myrica faya* actively fix nitrogen in association with symbiotic bacteria, resulting in much elevated nitrogen levels on otherwise nitrogen-poor sites on young larva flows. Introduced plants may also affect hydrological cycles by utilizing more of the annual precipitation than was used by native vegetation that they replaced. Finally, fire regime may be altered by the invasion of introduced grasses. In most cases the changes result in more frequent fires.

## 21.3  WHAT MAKES A SUCCESSFUL INVADER?

What is the most frequent type of introduction? Knowing the answer to this question might help us find out what makes a successful invader. The emptying of ballasts from the holds of ships at east-coast and west-coast ports was undoubtedly responsible for the introduction of many weeds such as the white campion through Philadelphia and the Australian fireweed through San Francisco. Also, many alien plants arrived during major periods of farm establishments as contaminants in seed lots of wheat, clover, and alfalfa and were spread rapidly by a comprehensive railroad system. Often, domestic animals facilitate the intraregional spread of alien plants. There is, no doubt, a close connection between anthropogenic disturbance and invasions. Many of the invading plants in California had initial points of establishment at either or both Los Angeles and San Francisco. Thus, species that can live in close association with humans are certainly favored. For example, exotic birds are often ground foragers in grasses in suburban habitats. Luckily, they are at a disadvantage in native forests, and as such they may often have few effects on native birds.

Why is one organism a good invader and another is not? Table 21.6 indicates some possible distinctions between successful invaders and unsuccessful invaders. The ability to subsist on a relatively wide variety of foods is often a prerequisite of being a successful colonist.

TABLE 21.6  *Possible concomitants of invasion potential.*

| Successful Invaders | Unsuccessful Invaders |
| --- | --- |
| Abundant in original range | Rare in original range |
| Polyphagous | Monophagous or oligophagous |
| Short generation times | Long generation times |
| Much genetic variability | Little genetic variability |
| Fertilized female able to colonize alone | Fertilized female not able to colonize alone |
| Larger than most relatives | Smaller than most relatives |
| Associated with *H. sapiens* | Not associated with *H. sapiens* |
| Able to function in a wide range of physical conditions | Able to function only in a narrow range of physical conditions |

Herbivores that are monophagous (that is, they feed on one or a few species of plant) will normally be handicapped relative to polyphagous species. Successful colonists do tend to be plants and animals that are relatively abundant and widely distributed where they are endemic. Also, characteristics lending themselves to transport by humans, both purposely and accidentally, and to adaptation to habitats created by them have been important attributes of successful colonization. Most of the truly widespread terrestrial animals today are ones considered pests or commensals of human beings and are dependent on them for more than transport.

Some authors suggest it is nearly impossible to predict which invaders will be successful and which will not. Even the close relatives of some invaders are not prone to invade themselves. For example, the European cabbage butterfly, has in the last century invaded Bermuda, North America, Australia, Hawaii, and other Pacific Islands, while other closely related European species have not yet crossed the Atlantic Ocean. Only one of six species of serranid fishes (groupers, sea trout, and their relatives) introduced purposely to Hawaiian waters (where the native serranid fauna is depauperate) became successfully established. The house sparrow occupied the entire United States, with the help of additional releases by humans, in a little more than 50 years after it was first successfully introduced. The closely related tree sparrow, however, has been confined to its one point of introduction, St. Louis, where it was introduced in 1870. It has not spread much further afield, despite having a range in Eurasia almost as large as that as the house sparrow. Two extraordinarily successful invaders are the black rat and the Norway, or brown, rat. Both species are found throughout the world, generally in association with human beings. But some 43 other species of the subgenus *Rattus*, a few of which are commensal with humans in limited areas, have not become ubiquitous. Of course, *Homo sapiens* have been very successful invaders, while their closest living relatives, *Pan troglodytes* and *Gorilla gorilla* have been pushed nearly to the brink of extinction. Indeed, it is generally a futile exercise to predict which members of any taxon will be successful invaders (Simberloff 1986; Williams and Brown 1986). In Britain, it has been predicted that the probability that an established invader will become a pest is around 10 percent, and it is thought that at least half of the thousand-odd species of recorded invaders have become established.

### What Areas Are Most Easily Invaded?

Islands are often thought more vulnerable to invasions than continents (but see Simberloff 1995). First, a large number of mainly oceanic islands has been seriously affected by introduced species, but this could be because more exotics have been introduced into islands. Also, many islands are thought to be species impoverished. That is not to say that any empty niches necessarily exist, but it is a possibility. The absence of terrestrial predators has also been argued to be a key factor in insular ecosystems. This doesn't necessarily mean that continental areas are resistant to exotics—just that the faunas may have had preadaptations against competitors and predators.

## 21.4  COMBATING THE SPREAD OF EXOTICS

There are two major means of spread of exotics, referred to as Steady Advance and Satellite. In Satellite invasions, new founder populations become established at distinct distances away from the original population by means of road or rail transport. In Steady Advance, there is a front of invasion that steadily progresses through an area. The most rapid steady advance of an introduced mammal was recorded for the muskrat in Central Europe and Finland. The muskrats moved their frontier by 28 to 40 km per year. Ungulates generally are slower, about 0.8 to 2 km per year in New Zealand, but the chamois managed 8.5 km per year.

How do we combat the problem of introduced species? The problem has certainly been compounded by the efficiency of modern jet travel, which provides transport for invaders from one habitat to another in a matter of hours. Increased vigilance by detection and quarantine stations may be the key. The U.S. Department of Agriculture funds APHIS (the Animal and Plant Health Inspection Service), and individual states may have comparable local agencies. For example, the California Department of Food and Agriculture is unusually vigilant in its pursuit of invading species: In 1985, it intercepted 459 gypsy-moth individuals at borders stations. From October 1, 1978, through September 30, 1979, APHIS intercepted 18,644 plant pests (Anonymous 1981); of these, 14,002 (75 percent) were insects. What proportion of the invaders are actually caught? Capture of 10 percent of the organisms coming in is considered a success (Dahlsten 1986). Besides, the action of these agencies probably acts as a useful deterrent to individuals who might otherwise import millions more-illegal arthropod aliens.

Only rarely is it possible to eradicate an introduced species, even among vertebrates. The painstaking physical removal of individuals is the usual solution for plants (Photo 21.2). A mammal called the coypu (*Myocastor coypus*) was introduced in Britain in 1929 and probably escaped from a fur farm in 1937. It was confined to the marshy areas of two counties, Norfolk and Suffolk. Although coypu numbers rose as high as 200,000 in the 1960s, government control policies have now eradicated this animal; in January 1989, the government announced that the coypu-control organization had completed its task. Coypus had become pests, undercutting river

**Photo 21.2** *Signboard on Long Key in the Florida Keys. Combating introduced species sometimes involves tedious manual removal as was necessary here.* (Photo by Peter Stiling.)

banks and encouraging further erosion and floods, as well as feeding on an assortment of crops. The government had set itself the target of eradication of coypus within 10 years of the start of the campaign in April 1981 and succeeded before schedule! But trapping has its problems, not the least of which is the capture of native species. For example, the Scottish muskrat population originated from a mere nine animals that escaped from captivity in Perthshire in 1927. A campaign to eradicate the muskrat from Scotland started in 1932. The campaign involved traps that caught at least 26 other species of vertebrates, and there was a death rate of about seven other vertebrates for every muskrat killed. Most of the other mammals were rats that would probably be acceptable to the general public. However, a few otters and many species of birds were also caught, and such a side effect is probably unacceptable. In Australia, there has been experimentation with appropriate traps for the dingo. At least 20 species of protected wildlife were trapped at a rate of 2 to 3 individuals per dingo trap.

Finally, poisoning is also used in some cases. In New Zealand, poison was incorporated into chipped carrots and oats used to control the invasive brush-tailed possum. Native birds were also poisoned as well as several other introduced species. It is probably true that no method of control, traps or poisons, can be aimed solely at the target species.

Of the species that have been successfully controlled, low dispersal ability has been thought of as being of paramount importance. The successful trapping of coypu was perhaps facilitated because the animal did not disperse far. In contrast, trapping campaigns against mink have not been successful, and rabbits have also proved difficult to control in many countries. The mink has a far greater dispersal ability than the coypu, and rabbits have one of the largest mean annual dispersal distances. Invasive birds are notoriously difficult to control, and this may be due to their high mobility.

# Summary

1. The introduction of exotic species into new environments is seen by many as biological pollution. Such introduction can be either accidental or deliberate.

2. Many plants and animals have been deliberately introduced, worldwide, for agriculture, ranching forestry. At least 330 species of nonnative birds and mammals are known to have been introduced around the globe.

3. Accidental introductions have occurred commonly from ships ballast—either in the off-loading of soil or water. In this way, many pest-plants and invertebrates (such as fire ants) have been introduced. Fifty percent of U.S. weeds are exotic species.

4. The effects of exotics are numerous and include predation, competition, herbivory, parasitism, and hybridization with native species. Such effects can result in the extinction of native species.

5. Successful invaders include species with a broad diet and those that can live in close proximity to humans. Islands have often been thought to be more vulnerable to invasions than continents.

6. To combat the spread of exotic species, increased vigilance at quarantine stations is necessary. At present, capture of 10 percent of invading organisms is considered a success. Eradication of established exotics is only rarely possible and has only been successful for a handful of mammals.

*Discussion Question:* On balance, are exotic species a benefit or a detriment?

# Epilogue

Chandni Chowk, India, once a grand avenue for imperial processions in the 1600s, now a crowded bazaar. Curbing human population growth is the key to minimizing environmental degradation. (Chickering/ Porterfield, Photo Researchers Inc. 6W 9756.)

I t is at once obvious that the effects of human populations determine the distribution patterns of many species of plants and animals. Only rarely do humans have a positive impact on species welfare, as for example when urban populations feed birds, which, because of increased food intake, lay more eggs and fledge young at a higher rate (Powell, Powell, and Paul 1988). There are also the so-called *synanthropic species,* which actually profit from inadvertent human activity. Rats, mice, sparrows, pigeons, and gulls fall into this category and perhaps the blackbird (*Turdus merula*), too, once a shy forest bird and now a regular and bold inhabitant of European gardens. By and large, however, human activity correlates well with general extinction. It has been suggested (Martin and Klein 1984) that global patterns of extinctions of large land mammals were caused by rapid and substantial changes of climate at the termination of the last glacial period. Evidence is strong, however, that humans are more to blame (see graphs on following page). In 1600, there were approximately 4,226 living species of mammals; since then 36 have become extinct (0.75 percent), and a further 120 (2.84 percent) are believed to be in great danger of extinction. Martin (1982) has argued that the global pattern of extinction of large land mammals follows the footsteps of Paleolithic settlements, with Africa and Southeast Asia being affected first (see table on following page). Those extinctions were due to direct exploitation, hunting, and some habitat modification through the use of fire.

If environmental problems are to be minimized, many people have argued that population growth must slow severely. Among the most well-known proponents of this position are Stanford biologist Paul Ehrlich and University of California, Santa Barbara, biologist Garrett Hardin. Ehrlich founded the Zero Population Growth Organization to promote his goal. Clearly, the implication is that families with three or more children are a burden to the environment and should be discouraged. Is this reproductive persecution? The U.S. government has granted political asylum to Chinese people, who are subjected to a one-child-per-couple rule in their homeland. Furthermore, population control is contrary to many religious and political beliefs. How we should go about it in Western nations, short of giving economic incentives, is unclear. But it is in developing nations where population control is perhaps more needed. Many women would like fewer children than they have (Eckholm 1976). Here, the solution is easier—provide family-planning services. However, any slowing down of population growth rates to replacement levels (that is, two children per family) would preserve the existing religious and racial status quo in many countries. To minority groups with above-average birth rates, this looks like a policy to limit their proportion in the population. Even though that is clearly not the intended result, the suspicions raised create tensions.

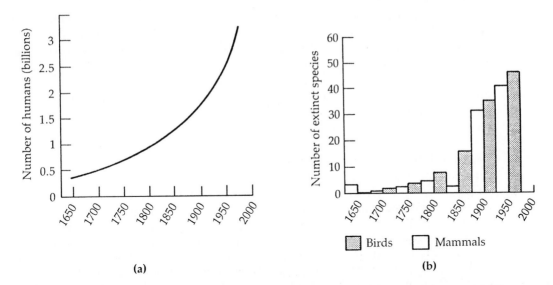

(a)                                                                    (b)

*Population growth and animal extinctions. (**a**) Geometric increase in the human population. (**b**) Increasing numbers of extinctions in birds and mammals. These figures suggest that, as human numbers increase, more and more living species are exterminated.* (Redrawn from Goudie, 1986.)

Hardin proposed the "lifeboat ethics" idea, that most developing nations are drowning in a sea of hopelessness and are beyond help. Developed countries, those in the lifeboat, should only share their resources with a few third-world countries that are not likely to sink the boat.

But not all biologists think this way. Barry Commoner, of Queens College, New York, has argued that most environmental problems are a result of the misuse of technology, not

**Dates of major episodes of generic extinction in the Late Pleistocene, probably due to human influence. (From data of Martin 1967.)**

| Area | Date (years before present) |
| --- | --- |
| North America | 11,000 |
| South America | 10,000 |
| West Indies | Mid–post Glacial |
| Australia | 13,000 |
| New Zealand | 900 |
| Madagascar | 800 |
| Northern Eurasia | 13,000–11,000 |
| Africa and southeast Asia (?) | 50,000–40,000 |

overpopulation. He advocates the use of cleaner, more environmentally benign technology and considers Ehrlich's population control strategy as inappropriate and Hardin's as barbaric. Ehrlich has countered that modern technology has arisen mostly because of the need to support vast increases in human populations. Commoner's solution might also limit industrial development in the third world and keep society there at a lower level of advancement than that of the West, even though that is not the intention. There is clearly no easy solution as to how to protect the environment. It is difficult, if not impossible, to separate the environmental effects of population growth versus new technology.

University of Maryland economist Julian Simon argues that the environment is actually becoming better for humans, and by many yardsticks, this is true (Simon 1990). While the environment may be improving for humans, it is certainly worsening for many other forms of life, many species of which are declining in population sizes. The present book has clearly focused on these other forms of life.

The global population level that can be supported without reliance on modern agricultural techniques such as the uses of fertilizers and pesticides and the conversion of land to agriculture is 200 million, a figure surpassed by the Middle Ages (though close to the 500 million suggested by Keyfitz [1975] as ideal for the planet in cultural terms). We clearly cannot reduce the human population to these levels any time soon. Our aims should be to reduce population growth so that organic agriculture does not rely so heavily on fossil fuels, to reduce pollution levels, to direct resources from weapon production to a more life-enhancing end, and to avoid all waste of energy until the large potential available from solar or other nonfossil fuel sources becomes omnipresent and cheap. There is now one soldier for every 43 people and one doctor for every 1,030; 40 percent of our research expenditure goes to defense, when there are already enough atomic weapons to kill everyone on Earth 67 times over. It takes the world's military only two and a half hours to spend the entire annual budget of the World Health Organization and two days to spend the cost of the UN plan to conquer desertification. To increase the number of contraceptive-using couples from 663 million in 1987 to 1 billion in the year 2000, we shall have to triple the annual family-planning budget from $3 billion to $9 billion (Myers 1990). But even $9 billion is only three days of military spending worldwide. There is clearly the money to protect the environment, only the political will is needed.

More money should be channeled into conservation of natural habitats. The World Bank's new wildlands policy may help; because of it, biological diversity in developing nations may stand a better chance of being preserved (Goodland 1987). In 1987, World Bank president Barber Conable announced sweeping reorganizations, including the addition of about 40 new environmental positions (Holden 1988). One of the most important additions to the staff was Herman Daly, a rare breed of economist who advocates the concept of "sustainability" rather than the pure mining of biological reserves. The World Bank's policy is now not to fund projects that would convert the most important wildlands. It should encourage development on already degraded lands. If conversion of ecologically valuable lands is justified, then equivalent lands within the same country must be protected. In reality, deforestation was still rampant in World Bank-financed projects in the Amazon after the policy change (Fearnside 1987). Highway development leads to deforestation and low-diversity cattle pastures—the cheapest way to secure land claims. Profits from land sales are increased by road development and by titling. This is a positive-feedback mechanism. Previous commitments to preserving natural habitats and tribal areas are frequently reneged on. Environmental measures are often merely symbolic actions serving to tranquilize public concern during the key period when the development is not yet a *fait accompli* (Fearnside 1987).

What would it cost to protect the environment? A 1990 estimate (Myers 1990) was: protecting topsoil $9 billion, restoring forests $3 billion, stopping desertification $4 billion, supplying clean water $30 billion, increasing energy efficiency $10 billion, developing renewable energy $5 billion, slowing population growth $18 billion, and retiring third-world debt $30 billion. This totals $109 billion. If this amount were doubled—to $200 billion, to allow for inflation by the year 2000, it would still amount to no more than a few weeks' military spending. To reiterate, the money is available to prevent environment degradation—only the political will is lacking.

It is important to note that society fails to price its environmental resources adequately. It is not widely appreciated that, even in hard economic terms, tropical forests may be worth more standing than logged. Peters, Gentry, and Mendelsohn (1989) presented data on inventory, production, and current market value of all the commercial tree species found in 1 ha of species-rich Amazonian forest near Iquitos, Peru. Of the total, 350 individuals (41.6 percent) of 72 tree species (26.2 percent) yielded products with a market value in Iquitos. Edible fruits were produced by 11 species, one species produced rubber, and 60 species produced commercial lumber. In addition, many medicinal plants were to be found, but these were not included in the sample. Based on current market prices in Iquitos, the total value of the fruit and rubber was U.S. $698 annually, $422 net. The value of the timber, after logging and transportation, was $1,000 total. Using a simple discounting procedure, Peters, Gentry, and Mendelsohn calculated the net current value of the forest, for sustainable yield of fruit and rubber year after year, to be $6,330 per ha, more than six times the value of the timber. The value of the forest in its unchanged state was also more than twice the worth of an equal area of fully stocked cattle pasture, figured at $2,960 annually by Peters, Gentry, and Mendelsohn, even before deduction of the costs of weeding, fencing, and animal care, which would lower the value of the pasture substantially. In a similar vein, Grimes et al. (1994) calculated that the value of a 3-ha plot of primary forest in Amazonia Equador was more for annual harvests of fruits, barks, and resins than for alternative land uses in the area. Unfortunately, tropical timber generates foreign exchange, whereas nonwood resources are collected and sold in local markets and are therefore hard to track in national accounting schemes. Even under the World Bank's new policy, worldwide aid to preserve forests will not be spent on protecting forests from timber companies; it will be spent on forestry (Pearce 1989). The World Bank believes, perhaps correctly, that only by taking charge of tropical forests and allowing companies to make money out of them can the world save the rain forests. With the exception of reserves set aside for unique biological features, the forests must cease to be a common resource. They must be fenced, owned, and managed for profit if they are to survive. Such a strategy has been argued to have saved the elephants in southern Africa, where some herds are increasing (see Chapter 18). A total ban on logging may lead to uncontrolled destruction of forests by farmers, ranchers, and loggers alike. In the Philippines, on islands where logging is barred, such as Mindanao and Marinduque, rates of deforestation have been higher (3–5 percent per year) than in areas that are managed for timber (0.5–2 percent) (Pearce 1989). However, Geist (1994) argues that private ownership of natural resources often equates to disaster. This is so because the best economic strategy is to harvest everything now and invest the profits in something else. Yet the opposite, public ownership, often leads to what Hardin (1968) termed the tragedy of the commons. Public land, being owned by no one, is abused by everyone.

Ridley and Low (1993) borrowed a mechanism from the study of animal behavior to explain how natural resources might best be managed. People generally are not willing pay for the long-term good of society or the planet. Biologist and economists agree that cooperation cannot be taken for granted. Richard Dawkins, of Oxford University, has stated the position of many biologists in his book, *The Selfish Gene*, (1989) which asserts that animals, including humans, act altruistically only when it brings some benefit to copies of their own genetic material. This happens under two circumstances: first, when the altruist and the beneficiary are close relatives, such as parents and offspring, and second, when the altruist is in a position to have the favor returned at a later date. For example, baboons remember past favors when coming to one another's aid in fights. In humans, money is often used to mediate reciprocal altruism, and economists have long argued that cooperation requires economic incentives. Ridley and Low argue that to protect the environment, we will have to find a way to reward individuals. This is at odds with much conventional theory. For example, Lester Brown has advocated a restructured global economy and major changes in technology, values, and lifestyles to permit sustainable use of the environment. This is an unlikely scenario. Ridley and Low are advocating just the opposite—one can protect the environment with the current system, if people have the incentive. They suggest it is wrong to assume that a fully informed public, when aware of the long-term collective consequences of their action on the environment, will impose restraint.

How can selfishness save the environment? The theory begins with a game known as the prisoner's dilemma (Axelrod and Hamilton 1981). Two guilty accomplices are held in separate cells. If they both confess (or "defect"), they will go to jail for three years. If they both stay silent (or "cooperate"), they will go to jail for a year on a lesser charge that the police can prove. But if one defects and the other does not, the defector will walk free and the cooperator, who stayed silent, will get a five-year sentence. The logical thing to do is to confess, for this reduces the maximum sentence from five to three years. If the accomplice stays silent, the defector will still reduce his or her sentence from one year to nothing. So what happens is that both confess and each gets three years. Strategies that seem best for the individual result in an outcome that was not actually the best (both accomplices staying silent would have been the best option). Biologists were interested in game theory because of its application to how and why animals might cooperate in nature rather than defect. They discovered that if the prisoner's dilemma game were played repeatedly, the most common strategy would not necessarily be to defect. The most common would be to cooperate on the first trial and then do whatever the other guy did last time. This became known as the tit-for-tat strategy. The threat of retaliation makes defection much less likely to payoff. Robert Axelrod, a political scientist, and William Hamilton, a biologist, both at the University of Michigan, convened a computer-based tournament to determine the best strategy in a repeated prisoners dilemma game. None beat tit-for-tat. Tit-for-two-tats, that is, cooperate unless other player defects twice, came the closest. Field biologists have been finding how commonly tit-for-tat occurs in nature ever since. Tit-for-tat also works to solve environmental issues. Local people can get together and solve environmental problems as long as the community is small, stable, and communicating and has a concern for the future. This is so because they use the tit-for-tat strategy. When local farmers extract water from a river or local people harvest a fish population, individuals can collectively set appropriate limits, and they can quickly identify cheaters. Cheaters risk ostracism and sanctions in the next round of the game, that is, the next growing season or the next harvest.

Self-interest strategy may even work with big business. DuPont seemingly willingly abandoned the profitable chlorofluorocarbons business. Was it for the good of the environment, or were they shamed into it? Perhaps loss of public good will would cost them more than the products were worth. More likely was that the CFC technology was mature and vulnerable (the patents were expiring). DuPont was in danger of losing market share to rivals. A ban on CFC production ten years hence would prevent competitors from entering the market and give DuPont ten years of income while it gained a dominant market share of chemicals to replace CFCs. Again, self-interest was most of the motive for environmental change. Shame was probably a smaller part.

Shame can sometimes be effective in stopping environmental crimes, as it has been in stopping people from wearing real fur coats or using ivory. But financial reward is probably more effective. Recycling rates in Michigan are less than 10 percent for nonrefundable glass, metal, and plastic but more than 90 percent for refundable objects. Charities have long known that people respond better to solicitations for money if they are rewarded with just a small token—a tag or a pin. Even at the international level, some countries may agree to treaties to limit the production of sulfur oxides or carbon dioxide but not adequately enforce them. Trade sanctions, blackmail, bribes, and even shame between governments can work to stop violations. Economic incentives should be used to promote good environmental policy as much as sanctions are used to stop destructive environmental policy. As long as obedience is cheaper than disobedience, tit-for-tat will work.

## Summary

1. The most serious effects on the distribution and abundance of plants and animals are from humans. Most often, these effects are detrimental to native species; only rarely are they beneficial.

2. To combat the ecological effects on humans, human population growth must slow severely. How this can be achieved is a highly contentious issue.

3. If human population levels do not change much from what they are at present, society has the capability to stop most environmental degradation and reduction in species richness and abundance. A substantial reduction in military spending would easily allow enough money to be available to stop deforestation, desertification, and other forms of habitat destruction.

4. Some ecologists have suggested that the best way to protect the environment and maintain the current distributions of plants and animals is to reward people who protect the environment. This contrasts with the viewpoint that a fully informed public, when aware of the potential ecological disasters that may rise if people do not change their habits, will impose restraint.

*Discussion Question:* Do you think that the desired number of children by most women is more or less than replacement levels? If it is more, how could the desired number be changed? Should it be changed? Should we put values on native species, and if so, how do we do it?

# *Glossary*

**Abiotic factors**   environmental influences produced other than by living organisms, for example temperature, humidity, pH, and other physical and chemical influences; contrast with "biotic factors."

**Acclimation**   changes by an organism, often biochemical, subjected to new environmental conditions that enable it to withstand those conditions.

**Acid rain**   rainfall acidified by contact with sulfur dioxide (a by-product of the burning of fossil fuels) in the atmosphere.

**Acquired character**   a character not inherited but acquired by an individual organism during its lifetime.

**Adaptation**   a change in an organism's structure or habits that produces better adjustment to the environment; a genetically determined characteristic that enhances the ability of an organism to cope with its environment.

**Adaptive radiation**   evolutionary diversification of species derived from a common ancestor into a variety of ecological roles.

**Aerobic**   pertaining to organisms or processes that require the presence of oxygen.

**Age class**   individuals in a population of a particular age

**Aggressive mimicry**   resemblance of predators or parasites to harmless species that causes potential prey or hosts to ignore them.

**Agricultural pollution**   contamination of the environment with liquid and solid wastes from all types of farming, including pesticides, fertilizers, runoff from feedlots, erosion and dust from plowing, animal manure and carcasses, and crop residues and debris.

**Air pollution**   the presence of contaminants in the air in concentrations that overcome the normal dispersive ability of the air and that interfere directly or indirectly with human health, safety, or comfort or with the full use and enjoyment of property.

**Algal bloom**   a proliferation of living algae in a lake, stream, or pond.

**Allele**   one of two or more alternative forms of a gene located at a single point (locus) on a chromosome.

**Allelopathy**   the negative chemical influence of plants, exclusive of microorganisms, upon one another.

**Allen's (1878) rule**   among homeotherms, the tendency for limbs and extremities to become shorter and more compact in colder climates than in warmer ones.

**Allochronic speciation**   separation of a population into two or more evolutionary units as a result of reproductive isolation arising from a difference in mating times.

**Allopatric speciation**   separation of a population into two or more evolutionary units as a result of reproductive isolation arising from geographic separation.

**Altruism**   in an evolutionary sense, enhancement of the fitness of an unrelated individual by acts that reduce the evolutionary fitness of the altruistic individual.

**Anadromous**   living in salt water but breeding in fresh water.

**Anaerobic**   pertaining to organisms or processes that occur in the absence of oxygen.

**Apatetic coloration**   coloration of an animal that causes it to resemble physical features of the habitat.

**Apomixis**   parthenogenetic reproduction in which offspring develop from unfertilized eggs or somatic cells.

**Aposematism**   conspicuous appearance of an organism warning that it is noxious or distasteful; warning coloration.

**Apostatic selection**   selective predation on the most abundant of two or more forms in a population, leading to balanced polymorphism (the stable occurrence of more than one form in a population).

**Apterous**   wingless.

**Aquaculture**   farming of aquatic or marine systems; rearing of organisms such as fish, algae, or shellfish under controlled conditions.

**Aquifer**   a layer of rock, sand, or gravel through which water can pass; an underground bed or stratum of earth, gravel, or porous stone that contains water; the place in the ground where groundwater is naturally stored.

**Assimilation efficiency**   the percentage of energy ingested in food that is assimilated into the protoplasm of an organism.

**Association**   a group of species occurring in the same place.

**Assortative mating**   nonrandom mating; the propensity to mate with others of like phenotypes.

**Autecology**   study of the individual in relation to environmental conditions; contrast "synecology."

**Autogamous**   able to produce offspring sexually by the fusion of gametes from the same individual, for example by fusion of pollen and ovules from the same plant.

**Autosome**   a chromosome other than a sex chromosome.

**Autotroph**   an organism that obtains energy from the sun and materials from inorganic sources; contrast with "heterotroph."

**Average GNP**   the gross national product of a country divided by its total population.

**Balanced polymorphism**   the stable occurrence of more than one form in a population.

**Barrier island**   a narrow, elongated, sandy island paralleling the coast, separated from the mainland by a bay or lagoon.

**Batesian mimicry**   resemblance of an edible (mimic) species to an unpalatable (model) species to deceive predators.

**Benthic**   pertaining to aquatic bottom or sediment habitats.

**Benthos**   bottom-dwelling aquatic organisms (for example, burrowing worms, molluscs, and sponges).

**Bergmann's (1847) rule**   among homeotherms, the tendency for organisms in colder climates to have larger body size (and thus smaller surface-to-volume ratio) than those in warm climates.

**Bioassay**   the use of living organisms to measure the biological effect of some substance, factor, or condition.

**Biochemical oxygen demand (BOD)**   the amount of oxygen that would be consumed if all the organic substances in a given volume of water were oxidized by bacteria and other organisms; reported in milligrams per liter.

**Biodegradable**   capable of being decomposed quickly by the action of microorganisms.

**Biogeochemical cycle**   the passage of a chemical element (such as nitrogen, carbon, or sulfur) from the environment into organic substances and back into the environment.

**Biogeography**   the branch of biology that deals with the geographic distribution of plants and animals.

**Biological control**   use of natural enemies (diseases, parasites, predators) to regulate populations of pest species.

**Biological magnification**   the concentration of a substance as it "moves up" the food chain from consumer to consumer.

**Biomass**   dry weight of living material in all or part of an organism, population, or community; commonly expressed as weight per unit area, biomass density.

**Biome**   a major terrestrial climax community; a major ecological zone or region corresponding to a climatic zone or region; a major community of plants and animals associated with a stable environmental life zone or region (for example, northern coniferous forest, Great Plains, tundra).

**Biosphere**   the whole Earth ecosystem.

**Biota**   all the living organisms occurring within a certain area or region.

**Biotic factors**   environmental influences caused by living organisms; contrast with "abiotic factors."

**Boreal**   occurring in the temperate and subtemperate zones of the Northern Hemisphere.

**Breeder reactor**   a nuclear reactor that produces more fissionable material than it consumes.

**Burner reactor**   a nuclear reactor that consumes more fissionable material than it produces.

**Calcareous**   in soil terminology, rich in calcium carbonate and having a basic reaction.

**Canonical distribution**   a lognormal distribution of the numbers of individuals or species according to the mathematical formulation of Preston (1962).

**Carbon-14**   a radioactive isotope of carbon (atomic weight 14) that can be used for dating organic materials.

**Carcinogen**   a chemical or physical agent capable of causing cancer.

**Cardinal index**   a cardinal number is used in counting or showing how many, e.g. two, five, ten. A cardinal index, therefore, treats all things as equal in imporance.

**Carnivore**   an animal (or plant) that eats other animals; contrast with "herbivore."

**Carrying capacity**   the amount of animal or plant life (or industry) that can be supported indefinitely on available resources; the number of individuals that the resources of a habitat can support.

**Catadromous**   living in fresh water but breeding in sea water.

**Character displacement**   divergence in the characteristics of two otherwise similar species where their ranges overlap, caused by selective effects of competition between the species in the area of overlap.

**Character divergence**   evolution of differences between similar species occurring in the same area.

**Chimera**   a piece of DNA incorporating genes from two different species; an organism whose cells are not all genetically alike.

**China syndrome**   a popular term for the consequence of core meltdown in a nuclear reactor in which a molten mass of intensely radioactive material plummets through vessel and containment and into the Earth beneath, in the direction (from the Western Hemisphere) of China.

**Clade**   the set of species descended from a particular ancestral species.

**Clearcutting**   the practice of cutting all trees in an area, regardless of species, size, quality, or age.

**Cleistogamy**   self-pollination within a flower that does not open.

**Climax community**   the community capable of indefinite self-perpetuation under given climatic and edaphic conditions.

**Cline**   a gradient of change in population characteristics over a geographic area, usually related to a corresponding environmental gradient.

**Clone**   a lineage of individuals reproduced asexually.

**Coadaptation**   evolution of characters of two or more species to their mutual advantage.

**Coevolution**   development of genetically determined traits in two species to facilitate some interaction, usually mutually beneficial.

**Coexistence**   occurrence of two or more species in the same habitat; usually applied to potentially competing species.

**Cohort**   those members of a population that are of the same age, usually in years or generations.

**Coliform index**   a measure of the purity of water based on a count of the coliform bacteria it contains.

**Commensalism**   an association between two organisms in which one benefits and the other is not affected.

**Community**   a group of populations of plants and animals in a given place; used in a broad sense to refer to ecological units of various sizes and degrees of integration.

**Competition**   the interaction that occurs when organisms of the same or different species use a common resource that is in short supply ("exploitation" competition) or when they harm one another in seeking a common resource ("interference" competition).

**Competitive exclusion principle**   the hypothesis, based on theoretical considerations and laboratory experiments, that two or more species cannot coexist and use a single resource that is scarce relative to demand for it.

**Conspecific**   belonging to the same species.

**Consumer**   an organism that obtains its energy from the organic materials of other organisms, living or dead; contrast with "producer."

**Continental drift**   the movement of the continents, by tectonic processes, from their original positions as parts of a common land mass to their present locations.

**Continental island**   an island that is near to and geologically part of a continent, for example the British Isles or Trinidad.

**Continental shelf**   the shallow part of the sea floor immediately adjacent to a continent.

**Convergent evolution**   the development of similar adaptations by genetically unrelated species, usually under the influence of similar environmental conditions.

**Coprolite**   fossil excrement.

**Core**   the heart of a nuclear reactor where energy is released; the region of a reactor containing fuel (and moderator, if any) and within which the fission reaction occurs.

**Courtship**   any behavioral interaction between individuals of opposite sexes that facilitates mating.

**Crop rotation**   the farming practice of planting the same field with a different crop each year to prevent nutrient depletion.

**Crypsis**   coloration or appearance that tends to prevent detection of an organism by others, especially predators.

**Cultural change**   any modification of characteristics specific to a population that is transmitted by learning rather than by genetic mechanisms.

**DDT**   1,1,1-trichloro-2,2-bis(p-chloriphenyl) ethane; the first of the modern chlorinated-hydrocarbon insecticides.

**Decomposers**   consumers, especially microbial consumers, that change their organic food into mineral nutrients.

**Deforestation**   removal of trees from an area without adequate replanting.

**Deme**   a local population, usually small and panmictic.

**Denitrification**   enzymatic reduction by bacteria of nitrates to nitrogen gas.

**Density**   the number of individuals per unit area.

**Density dependent**   having an influence on individuals that varies with the number of individuals per unit area in the population.

**Density independent**   having an influence on individuals that does not vary with the number of individuals per unit area in the population.

**Desalinization**   the process of removing salt from water.

**Desert**   a region receiving very small amounts of precipitation or where (for example, ice caps) the moisture present is unavailable to vegetation.

**Deterministic model**   a mathematical model in which all relationships are fixed and stochastic processes play no part; contrast "stochastic model."

**Diapause**   a period of suspended growth or development and reduced metabolism in the life cycle of many insects, during which the organism is more resistant to unfavorable environmental conditions than during other periods.

**Dimorphism**   the occurrence of two forms of individuals in a population.

**Dioecious**   characterized by individuals each of which has only male or only female reproductive organs.

**Diploid**   of cells or organisms, having two sets of chromosomes.

**Direct competition**   exclusion of individuals from resources, by other individuals, by aggressive behavior or use of toxins.

**Dispersal**   movement of organisms away from the place of birth or from centers of population density.

**Disruptive selection**   selection against individuals in a population that have intermediate values of a trait, leading to the divergence of subpopulations with extreme values of the trait.

**Distribution**   the area or areas (taken together) where a species lives and reproduces.

**Diversity**   the number of species in a community or region; alpha diversity is the diversity of a particular habitat; beta diversity is diversity of a region pooled across habitats.

**Dominance**   the influence or control exerted by one or more species in a community as a result of their greater number, coverage, or size.

**Ecologic**   pertaining to the living environment.

**Ecologic efficiency**   the percentage of energy in biomass produced by one trophic level that is incorporated into biomass by the next highest trophic level.

**Ecological impact**   the total effect of an environmental change, whether natural or man-made.

**Ecological release**   the expansion of habitat or food use by populations in regions of low species diversity, permitted by reduced interspecific competition.

**Ecology**   the branch of science dealing with the relationships of living things to one another and to their environment.

**Ecosystem**   a biotic community and its abiotic environment.

**Ecotone**   the transition zone between two diverse communities.

**Ecotype**   a subspecies or race that is specially adapted to a particular set of environmental conditions.

**Ecumene**   the portion of the Earth's surface occupied by permanent human settlement.

**Edaphic**   pertaining to soil.

**Emergent property**   a feature of a community or system not deducible from the features of single species or lower order processes.

**Emigration**   the movement of organisms out of a population.

**Endangered species**   a species with so few living members that it will soon become extinct unless measures are taken to slow its loss.

**Endemic**   an organism that is native to a particular region.

**Endogenous**   produced from within; originating from or due to internal causes; contrast "exogenous."

**Energy resource**   a natural supply of energy available for use, for example the Earth's internal heat, fossil fuels, hydropower, nuclear energy, solar energy, and wind.

**Enrichment**   the addition of nutrients to an ecosystem, for example the addition of nitrogen to waterways by agricultural runoff.

**Environment**   all the biotic and abiotic factors that affect an individual organism at any one point in its life cycle.

**Epidemiology**   the study of disease in populations or groups.

**Epilimnion**   the upper layer of water in a lake, usually warm and containing high levels of dissolved oxygen.

**Epiphyte**   a plant that lives on another plant but uses it only for support, drawing its water and nutrients from natural runoff and the air.

**Epistasis**   a synergistic effect whereby the effect of two or more gene loci is greater than the sum of their individual effects.

**Equilibrium**   a condition of balance, such as that between immigration and emigration or birth rates and death rates in a population of fixed size.

**Euphotic zone**   that part of the water column that receives sufficient sunlight to support photosynthesis; usually limited to the upper 60 m.

**Eutrophication**   the normally slow aging process by which a lake fills with organic matter, evolves into a bog or marsh, and ultimately disappears.

**Evapotranspiration**   the sum of the water lost from the land by evaporation and plant transpiration. Potential evapotranspiration is the evapotranspiration that would occur if water were unlimited.

**Evolution**   the gradual accumulation of genetic change that is thought to have given rise, beginning with common ancestors, to the diversity of life.

**Exponential growth**   the steepest phase in a growth curve, that in which the curve is described by an equation containing a mathematical exponent.

**Exponential rate of increase**   the rate at which a population is growing at a particular instant expressed as a proportional increase per unit time.

**Extant**   of a species, currently represented by living individuals.

**Extinct**   of a species, no longer represented by living individuals.

**Extrafloral nectaries**   nectar-secreting glands found on leaves and other vegetative parts of plants.

**Facilitation**   enhancement of a population of one species by another, often during succession, a type of one-way mutualism.

**Fecundity**   the potential of an organism to produce living offspring.

**Feral**   having reverted from domestication to the wild but remaining distinct from other wild species.

**Fission**   splitting or division; nuclear fission is the splitting of the nuclei of the atoms of certain elements into lighter nuclei and is accompanied by the release of relatively large amounts of energy.

**Fitness**   genetic contribution by an individual's offspring and heir offspring to future generations.

**Fixation**   attainment by an allele of a frequency of 1 (100 percent) in a population, which in effect becomes monomorphic for that allele.

**Food chain**   figure of speech describing the dependence for food of organisms upon others, in a series beginning with plants and ending with the largest carnivores.

**Forest**   a region that, because it receives sufficient average annual precipitation (usually 75 cm [30 inches] or more), supports trees and small vegetation.

**Fossil fuels**   coal, oil, and natural gas, so-called because they are derived from the fossil remains of ancient plant and animal life.

**Fossorial**   living in burrows.

**Founder effect**   the principle that a population started by a small number of colonists will contain only a fraction of the genetic variation of the parent population.

**Functional response**   a change in the rate of exploitation of prey by an individual predator resulting from a change in prey density (see also "numerical response").

**Fusion**   the combination of two atoms into a single atom as a result of a collision, usually accompanied by the release of energy.

**Gene**   a unit of genetic information.

**Gene flow**   the exchange of genetic traits between populations by movement of individuals, gametes, or spores.

**Generation time**   the time between the birth of a parent and the birth of its offspring.

**Genetic drift**   change in gene frequency caused solely by chance, usually unidirectional and more important in small populations.

**Genome**   the entire genetic complement of an individual.

**Genotype**   the genetic constitution of an organism or a species in contrast to its observable characteristics; contrast "phenotype."

**Genus**   the taxonomic category above the species and below the family; a group of species believed to have descended from a common direct ancestor.

**Geometric rate of increase**   the factor by which the size of a population changes over a specified period; contrast "exponential rate of increase."

**Geometric series**   a series in which each number is obtained by multiplying the previous one by one factor, e.g. 1, 3, 9, 27, etc.

**Glacial epoch**   the Pleistocene epoch, the earlier of the two epochs comprising the Quaternary period, characterized by the extensive glaciation of regions now free from ice.

**Global stability**   ability to withstand perturbations of a large magnitude without being affected; contrast "local stability."

**Grassland**   a region with sufficient average annual precipitation (25-75 cm [10-30 inches]) to support grass but not trees.

**Green belts**   areas from which buildings and houses are excluded, often serving as buffers between pollution sources and concentrations of population.

**Greenhouse effect**   the heating effect of the atmosphere upon the Earth, particularly as $CO_2$ concentration rises, caused by its ready admission of light waves but its slower release of the heat they generate on striking the ground.

**Gross production**   production before respiration losses are subtracted; photosynthetic production for plants and metabolizable production for animals.

**Group selection**   elimination of a group of individuals with a detrimental genetic trait, caused by competition with other groups lacking the trait.

**Guyot**   a flat-topped submarine volcano.

**Habitat**   the sum of the environmental conditions where an organism, population, or community lives; the place where an organism normally lives; the environment in which the life needs of an organism are supplied.

**Half-life**   the time it takes for one-half the atoms of a radioactive isotope to decay into another isotope; the time it takes certain materials, such as persistent pesticides, to lose half their strength.

**Haplodiploidy**   the presence of haploid males and diploid females in the same species, for example in the Hymenoptera.

**Haploid**   containing one set of chromosomes.

**Harem**   a group of females controlled by one male.

**Herbivore**   an organism that eats plants; contrast "carnivore."

**Heredity**   genetic transmission of traits from parents to offspring.

**Heterotroph**   an organism that obtains energy and materials from other organisms; contrast "autotroph."

**Hierarchy**   a rank order; the pecking order, leadership, or dominance patterns among the members of a population.

**Holotype**   the single specimen chosen by the original author of a species as the archetypical example of that species and which any revised description of the species must include;

contrast "lectotype" and "neotype," terms for such an archetypical specimen when it is chosen, not by the original author, but by a later author in the absence of a holotype.

**Home range**    the area in which an individual member of a population roams and carries on all of its activities.

**Horizon**    in a soil, a major stratification or zone, having particular structural and chemical characteristics.

**Host**    the organism that furnishes food, shelter, or other benefits to an organism of another species.

**Hybridization**    breeding (crossing) of individuals from genetically different strains, populations, or, sometimes, species.

**Hypolimnion**    the layer of cold, dense water at the bottom of a lake.

**Immigration**    the movement of individuals into a population.

**Inbreeding**    a mating system in which adults mate with relatives more often than would be expected by chance.

**Inclusive fitness**    The total genetic contribution of an individual by way of its sons, daughters, and all other relatives combined.

**Independent assortment**    the separate inheritances, without mutual influence, of genes occurring on different chromosomes.

**Indirect competition**    exploitation of a resource by one individual that reduces the availability of that resource to others.

**Innate capacity for increase $r$**    measure of the rate of increase of a population under controlled conditions (also referred to as intrinsic rate of increase).

**Interdemic selection**    group selection of populations within a species.

**Interspecific**    between species; between individuals of different species.

**Intraspecific**    within species; between individuals of the same species.

**Isolating mechanism**    any condition, for example a genetically determined difference or a mechanical or geographic separation, that prevents gene flow between two populations.

**Iterative evolution**    the repeated evolution of similar phenotypic characteristics at different times during the history of a clade.

**Kin selection**    a form of genic selection in which alleles differ in their rate of propagation because they influence the survival of kin who carry the same alleles.

**LDC**    less-developed country, typically with low GNP, high population growth, low literacy, and low industrialization.

**Landfill**    a waste-disposal site in which layers of solid waste are laid down in alternation with layers of soil.

**Leaching**    the process by which soluble materials in the soil, such as nutrients, pesticides, or contaminants, are washed into a lower layer of soil or are dissolved and carried away by water.

**Lek**    a communal courtship area on which several males hold courtship territories to attract and mate with females; sometimes called an arena.

**Lentic**    pertaining to standing freshwater habitats (ponds and lakes).

**Life table**    tabulation presenting complete data on the mortality schedule of a population.

**Ligase**    an enzyme that joins DNA together.

**Limiting resource**    the nutrient or substance that is in shortest supply in relation to organisms' demand for it.

**Linkage**    occurrence of two loci on the same chromosome; functional linkage occurs when two loci do not segregate independently at meiosis.

**Locus**   the site on a chromosome occupied by a specific gene.

**Logistic equation**   a model of population growth described by a symmetrical S-shaped curve with an upper asymptote.

**Lognormal distribution**   a frequency distribution of species abundance in which abundance, on the $x$ axis, is expressed on a logarithmic scale.

**Lotic**   pertaining to running freshwater habitats (streams and rivers).

**MDC**   more-developed country, typically with high GNP, low population growth, high literacy, and a strong economy.

**Malthusian theory of population**   the theory of English economist and religious leader Thomas Malthus that populations increase geometrically (2, 4, 8, 16) while food supply increases arithmetically (1, 2, 3, 4), leading to the conclusion that humans are doomed to overpopulation, misery, and poverty, and that population levels will be reduced by disease, famine, and war.

**Meiotic drive**   a preponderance (generally a frequency greater than 50 percent) of one allele among the gametes produced by a heterozygote.

**Melanism**   occurrence of black pigment, usually melanin.

**Meltdown**   of a reactor core, the consequence of overheating that allows part or all of the solid fuel to reach the temperature at which cladding and possibly fuel and support structure liquify and collapse.

**Modifier gene**   a gene that alters the phenotypic expression of genes at one or more other loci.

**Monoculture**   cultivation of a single crop to the exclusion of all other species on a piece of land.

**Monoecious**   having separate male and female reproductive organs on the same individual, used mainly of plants.

**Morph**   a specific form, shape, or structure.

**Mullerian mimicry**   mutual resemblance of two or more conspicuously marked, distasteful species to reinforce predator avoidance.

**Multivoltine**   having several generations during a single season; contrast "univoltine."

**Mutant**   an organism with a changed characteristic resulting from a genetic change.

**Mutation**   a change in the genetic makeup of an organism resulting from a chemical change in its DNA.

**Mutualism**   an interaction between two species in which both benefit from the association.

**Natural selection**   the natural process by which the organisms best adapted to their environment survive and those less well adapted are eliminated.

**Nektonic**   free-swimming in the upper zone of ocean water and strong enough to swim against the current.

**Neotenic**   exhibiting neoteny, the ability of species to reproduce sexually when still exhibiting juvenile characteristics.

**Neritic**   pertaining to the shallow, coastal marine zone.

**Net production**   production after respiration losses are subtracted.

**Net production efficiency**   percentage of assimilated energy that is incorporated into growth and reproduction.

**Net reproductive rate R**   the number of offspring a female can be expected to bear during her lifetime, for species with clearly defined discrete generations.

**Niche**   the place of an organism in an ecosystem; all the components of the environment with which an organism or population interacts.

**Nitrate**   a salt of nitric acid; a compound containing the radical $NO_3$; biologically, the final form of nitrogen from the oxidation of organic nitrogen compounds.

**Nitrogen cycle**   the biogeochemical processes that move nitrogen from the atmosphere into and through its various organic chemical forms and back to the atmosphere.

**Nitrogen fixing bacteria**   bacteria that can reduce atmospheric nitrogen to cell nitrogen.

**Nonrenewable resource**   a resource available in a fixed amount (such as minerals and oil), not replaceable after use.

**Norm of reaction**   the set of phenotypic expressions of a genotype under different environmental conditions.

**Numerical response**   change in the population size of a predatory species as a result of a change in the density of its prey.

**Oligotrophic**   low in nutrients and organisms; low in productivity.

**Omnivore**   an organism whose diet includes both plant and animal foods.

**Operational sex ratio**   ratio of sexually ready males to fertilizable females.

**Organic**   of biological origin; in chemistry, containing carbon.

**Palaeontology**   the science that deals with life of past geologic ages, treating fossil remains.

**Panmixis**   the condition in which mating in a population is entirely random.

**Parasite**   the organism that benefits in an interspecific interaction in which one organism benefits and the other is harmed.

**Parasitoid**   a specialized insect parasite that is usually fatal to its host and therefore might be considered a predator rather than a classical parasite.

**Particulate matter**   in air pollution, solid particles and liquid droplets, as opposed to material uniformly dispersed among the air molecules.

**Parthenogenesis**   reproduction without fertilization by male gametes, usually involving the formation of diploid eggs whose development is initiated spontaneously.

**PCBs**   polychlorinated biphenyls, a family of chemicals similar in structure to DDT. Pelagic: pertaining to the upper layers of the open ocean.

**Per-capita rate of population growth $r$**   rate of population growth per individual, used for species with overlapping, nondiscrete generations.

**Permafrost**   a permanently frozen layer of soil underlying the Arctic tundra biome.

**Persistence**   of pesticides, the length of time they remain in the soil or on crops after being applied.

**pH**   a measure of acidity or alkalinity.

**Phenology**   study of the periodic (seasonal) phenomena of animal and plant life (for example, flowering time in plants) and their relations to weather and climate.

**Phenotype**   the physical expression in an organism of the interaction between its genotype and the environment; the outward appearance of an organism.

**Phoresy**   the transport of one organism by another of a different species.

**Photic zone**   the surface zone of a body of water that is penetrated by sunlight.

**Photoperiodism**   seasonal response (for example, flowering, seed germination, reproduction, migration, or diapause) by organisms to change in the length of the daylight period.

**Photosynthesis**   synthesis, with the aid of chlorophyll and with light as the energy source, of carbohydrates from carbon dioxide and water, with oxygen as a by-product.

**Phyletic evolution**   genetic changes that occur within an evolutionary line.

**Phylogeny**   the line, or lines, of direct descent in a given group of organisms; also the study or the history of such relationships.

**Phylum**   one of the primary divisions of the animal and plant kingdoms; a group of closely related classes of animals or plants.

**Phytoplankton**   the plant community in marine and freshwater situations, containing many species of algae and diatoms, that floats free in the water.

**Plate tectonics**   the study of the global-scale movements of the Earth's crust that have resulted in continental drift and the creation of many geological formations.

**Pleiotropy**   the phenotypic effect of a gene on more than one characteristic.

**Pleistocene**   a geological epoch, characterized by alternating glacial and interglacial stages, that ended about 10,000 years ago, lasted one-to-two million years, and is subdivided into four glacial stages and three interglacials.

**Point source**   an individual, stationary source of large-volume pollution, usually industrial in origin.

**Pollutant**   any natural or artificial substance that enters the ecosystem in such quantities that it does harm to the ecosystem; any introduced substance that makes a resource unfit for a specific purpose.

**Polygamy**   a mating system in which a male pairs with more than one female at a time (polygyny) or a female pairs with more than one male (polyandry).

**Polymorphism**   occurrence in a population of more than two different forms, independent of sexual differences.

**Population**   a group of individuals of a single species.

**Primary production**   production by autotrophs, normally green plants.

**Production**   amount of energy (or material) formed by an individual, population, or community in a specific time period; see also "primary production," "secondary production," "gross production," "net production."

**Protandry**   the condition of an individual that, during the course of its development, changes from male to female.

**Proximate factors**   the mechanisms responsible for an evolutionary adaptation with reference to its physiological and behavioral operation; the mechanics of how an adaptation operates; contrast "ultimate factors."

**Punctuated equilibrium**   a model that depicts macroevolution as taking place in the form of short periods of rapid speciation alternating with long periods of relative stasis.

**$r$ and $K$ selection**   alternative expressions of selection on traits that determine fecundity and survivorship to favor rapid population growth at low population density $r$ or competitive ability at densities near the carrying capacity $K_r$.

**Recombinant DNA**   a single molecule combining DNA from two distinct sources.

**Recruitment**   addition, by reproduction, of new individuals to a population.

**Recycling**   the process by which waste materials are transformed into usable products.

**Red queen hypothesis**   the idea, named for the character in Lewis Carroll's *Through the Looking Glass*, that a species must continually evolve just to keep pace with environmental change and with other species, let alone to get ahead in the coevolutionary struggle.

**Resource**   a substance or object required by an organism for normal maintenance, growth, or reproduction.

**Respiration**   the complex series of chemical reactions in all organisms by which stored energy is made available for use and which produces carbon dioxide and water as by-products.

**Restriction enzyme**   an enzyme that recognizes a specific base sequence on DNA and cuts DNA. Some restriction enzymes cut the DNA at a particular point, others at random.

**Riparian**   related to, living in, or located on the bank of a natural watercourse, usually a river, sometimes a lake or tidewater.

**Runoff**   water entering rivers, lakes, reservoirs, or the ocean from land surfaces.

**Saprophyte**   a plant that obtains food from dead or decaying organic matter.

**Secondary production**   production by herbivores, carnivores, or detritus feeders; contrast "primary production."

**Search image**   a behavioral selection mechanism that enables predators to increase searching efficiency for prey that are abundant and worth capturing.

**Self-regulation**   a process of population regulation in which population increase is prevented by a deterioration in the quality of individuals that make up the population; population regulation by adjustments in behavior and physiology internal to the population rather than by external forces such as predators.

**Sere**   the series of successional communities leading from bare substrate to the climax community.

**Sessile**   of animals, attached to an object or fixed in place, as for example barnacles.

**Sewage**   the organic waste and waste water generated by residential and commercial establishments.

**Sewerage**   the entire system of sewage collection, treatment, and disposal; all effluent carried by sewers, whether sanitary sewage, industrial wastes, or storm-water runoff.

**Sex-linked gene**   a gene carried on one of the sex chromosomes (expressible phenotypically in either sex).

**Sexual selection**   selection by one sex for specific characteristics in individuals of the opposite sex, usually exercised through courtship behavior.

**Sibling species**   species that are difficult or impossible to distinguish by morphological characters.

**Sigmoid curve**   an S-shaped curve, for example the logistic curve.

**Slash-and-burn cultivation**   a primarily tropical practice in which forest vegetation is cut, left to dry, and burned to add nutrients to the soil before the land is planted with crops, then abandoned after two-to-five years as a result of falling yields.

**Smog**   a term coined from "smoke" and "fog" to describe photochemical air pollution.

**Sociobiology**   the study of the biological reasons behind animal behavior and sociology.

**Species (both singular and plural)**   organisms forming a natural population or group of populations that transmit specific characteristics from parent to offspring; a group of organisms reproductively isolated from similar organisms and usually producing infertile offspring when crossed with them.

**Stability**   absence of fluctuations in populations; ability to withstand perturbations without large changes in composition.

**Stochastic model**   a mathematical model incorporating factors determined by chance and providing not a single prediction but a range of predictions; contrast "deterministic model."

**Succession**   replacement of one kind of community by another; the progressive changes in vegetation and animal life that tend toward climax.

**Supergene**   a group of two or more loci between which recombination is so reduced that they are usually inherited together as a single entity.

**Symbiosis**   in a broad sense, the living together of two or more organisms of different species; in a narrow sense, synonymous with "mutualism."

**Sympatric**   occurring in the same place.

**Sympatric speciation**   formation of species without geographic isolation; reproductive isolation that arises between segments of a single population.

**Synecology**   the study (including population, community, and ecosystem ecology) of groups of organisms in relation to their environment; contrast "autecology."

**Synergism**   the situation in which the combined effect of two factors is greater than the sum of their separate effects.

**Taiga**   the northern boreal forest zone; a broad band of coniferous forest south of the Arctic tundra.

**Territory**   any area defined by one or more individuals and protected against intrusion by others of the same or different species.

**Thermocline**   the thin transitional zone in a lake that separates the epilimnion from the hypolimnion.

**Threatened species**   a species not yet endangered but whose population levels are low enough to cause concern.

**Timberline**   the uppermost altitudinal limit of forest vegetation.

**Time lag**   delay in response to a change.

**Topsoil**   the top few inches of soil, rich in organic matter and plant nutrients.

**Trophic level**   the functional classification of an organism in a community according to its feeding relationships.

**Tundra**   level or undulating treeless land, characteristic of Arctic regions and high altitudes, having permanently frozen subsoil.

**Turnover**   rate of replacement of resident species by new, immigrant species.

**Ultimate factors**   the evolutionary survival values of adaptations; the evolutionary reasons for an adaptation; contrast "proximate factors."

**Univoltine**   having only one generation per year; contrast "multivoltine."

**Upwelling**   the process whereby, as a result of wind patterns, nutrient-rich bottom waters rise to the surface of the ocean.

**Urban**   pertaining to city areas.

**Vector**   an organism (often an insect) that transmits a pathogen (for example, a virus, bacterium, protozoan, or fungus) acquired from one host to another.

**Vicarants**   disjunct species, closely related, assumed to have been created when the initial range of their common ancestor was split by some historical event.

**Vicariance biogeography**   the study of distribution patterns of organisms that attempts to reconstruct events through the study of shared characteristics (cladistics), often with little or no attention to dispersal capabilities or ecological properties.

**Watershed**   the land area that drains into a particular lake, river, or reservoir.

**Wilderness**   undisturbed area, as it was before human-made changes.

**Wild type**   the allele, genotype, or phenotype that is most prevalent in wild populations.

**Zoogeography**   the study of the distributions of animals.

**Zooplankton**   the animal community, predominantly single-celled animals, that floats free in marine and freshwater environments, moving passively with the currents.

# *Literature Cited*

Abrahamson, W. G. 1975. Reproductive strategies in dewberries. *Ecology* 56:721–726.

Achiron, M., and R. Wilkinson, 1986. Africa: the last safari? *Newsweek* 108(7):40–42.

Addicott, J. R. 1986. Variation in the costs and benefits of mutualism: The interaction between yuccas and yucca moths. *Oecologia* 70:486–494.

Addicott, J. F., and H. I. Freedman. 1984. On the structure and stability of materialistic systems: analysis of predator-prey and competition models as modified by the action of a slow growing mutualist. *Theoretical Population Biology,* 26:320–339.

Addicott, J. R., J. M. Aho, M. F. Antolin, D. K. Padilla, J. S. Richardson, and D. A. Soluk. 1987. Ecological neighborhoods: scaling environmental patterns. *Oikos* 49:340–346.

Ager, D. 1991. Ban the stone axe. *New Scientist 19,* January, 1991. pp. 61–62.

Alcock, J. 1979. *Animal Behavior: An Evolutionary Approach,* 2nd ed. Sinauer Associates, Sunderland, Massachusetts.

Alexander, R. D. 1974. The evolution of social behavior. *Annual Review of Ecology and Systematics* 5:325–383.

Alexander, R. D., and P. W. Sherman. 1977. Local mate competition and parental investment in social insects. *Science* 196:494–500.

Alford, R. A., and H. M. Wilbur. 1985. Priority effects in experimental pond communities: Competition between *Bufo* and *Rana. Ecology* 66:1097–1105.

Allee, W. C. 1931. *Animal Aggregations. A Study in General Sociology.* University of Chicago Press, Chicago.

Allee, W. C., A. E. Emerson, O. Park, and K. P. Schmidt. 1949. *Principles of Animal Ecology.* W. B. Saunders, Philadelphia.

Allen, R. 1980. *How to Save the World*. Barnes and Noble, Totowa, New Jersey.

Allendorf, F. W. 1994. Genetically effective sizes of grizzly bear populations. pp. 155–156 in G. K. Meffe and C. R. Carrol (eds.). *Principles of Conservation Biology*. Sinauer Associates, Sunderland, Massachusetts.

Allendorf, F. W., and R. F. Leary. 1986. Heterozygosity and fitness in natural populations of animals. pp. 57–76 in M. E. Soulé (ed.), *Conservation Biology: The Science of Scarcity and Diversity*. Sinauer Associates, Sunderland, Massachusetts.

Ames, B. H. 1983. Dietary carcinogens and anticarcinogens. *Science* 221:1256–1264.

Anderson, D. R., and K. P. Burnham. 1976. *Population Ecology of the Mallard: VI. The Effect of Exploitation on Survival*. U.S. Fish and Wildlife Service Publication 128.

Anderson, J. M. 1981. *Ecology for Environmental Sciences: Biosphere, Ecosystems and Man*. Wiley, New York.

Anderson, R. M. 1982. Epidemiology. pp. 204–251 in F. E. G. Cox (ed.) *Modern Parasitology*. Blackwell Scientific Publications, Oxford.

Andrewartha, H. G., and L. C. Birch. 1954. *The Distribution and Abundance of Animals*. University of Chicago Press, Chicago.

Anonymous. 1981. *List of Intercepted Plant Pests from October 1, 1978, through September 30, 1979*. U.S.D.A. APHIS Publication 82–7.

Antonovics, J., A. D. Bradshaw, and R. G. Turner. 1971. Heavy metal tolerance in plants. *Advances in Ecological Research* 7:1–85

Applegate, V. C. 1950. *Natural History of the Sea Lamprey*, Petromyzon marinus, *in Michigan*. U.S. Fish and Wildlife Service, Special Scientific Report, Fisheries, No. 55, Washington, D.C.

Armstrong, S. K. and F. Bridgland. 1989. Elephants and the ivory tower. *New Scientist* 1679:37–41.

Aron, W. I., and S. H. Smith. 1971. Ship canals and aquatic ecosystems. *Science* 174:13–20.

Askew, R. R. 1968. Considerations on speciation in Chalcidoidea (Hymenoptera). *Evolution* 22:642–645.

Askew, R. R. 1971. *Parasitic Insects*. Heineman, London.

Auerbach, M. J., and D. R. Strong. 1981. Nutritional ecology of *Heliconia* herbivores: Experiments with plant fertilization and alternative hosts. *Ecological Monographs* 51:63–83.

Auerbach, M. J., and D. R. Strong. 1981. Nutritional ecology of *Heliconia* herbivores: experiments with plant fertilization and alternative hosts. *Ecological Monographs* 51:63–83.

Axelford, R., and W. D. Hamilton. 1981. The evolution of cooperation. *Science* 211: 1390–1396.

Azim Abul-Atta, E. A. 1978. *Egypt and the Nile After the Construction of the High Aswan Dam*. Department of Irrigation and Land Reclamation, Egypt.

Bach, C. E. 1991. Direct and indirect interactions between ants, scales and plants. *Oecologia* 82: 233–239.

Bach C, E, 1994. Effects of herbivory and genotype on growth and survivorship in sand-dune willow (*Salix cordata*). *Ecological Entomology* 19:303–309.

Baker, C. S. and S. R. Palumbi. 1994. Which whales are hunted? A molecular genetic approach to monitoring whaling. *Science* 265: 1538-1539.

Baker, S. J., and C. N. Clarke. 1988. Cage trapping coypus (*Myocastor coypus*) on baited rafts. *Journal of Applied Ecology* 25:41–48.

Baldwin, N. S. 1964. Sea lamprey in the Great Lakes. *Canadian Audubon Magazine*, November-December: 142–147.

Ballard, R. D. 1977. Notes on a major oceanographic fluid. *Oceanus* 20:35–44.

Bambach, R. K. 1983. Ecospace utilization and guilds in marine communities through the Phanerozoic. pp. 719–746 in M. J. S. Tevesz and P. L. McCall (eds.). *Biotic Interactions in Recent and Fossil Benthic Communities*. Plenum, New York.

Barber, R. T., and F. P. Chavez. 1983. Biological consequences of El Niño. *Science* 222:1203–1210.

Barnola, J. M., D. Raynaud, Y. S. Korotkevich, and C. Lorius. 1987. Vostok ice core provides 160,000-year record of atmospheric $CO_2$. *Nature* 329:408–414.

Barrons, K. C. 1981. *Are Pesticides Really Necessary?* Regnery Gateway, Chicago.

Barrowclough, G. F. 1980. Gene flow, effective population sizes, and genetic variance components in birds. *Evolution* 34:789–798.

Bartholomew, G. A. 1986. The role of natural history in contemporary biology. *BioScience* 36:324–329.

Barton, A. M. 1986. Spatial variation in the effect of ants on an extrafloral nectary plant. *Ecology* 67:495–504.

Baskin, Y. Ecologists dare to ask: how much does diversity matter? *Science* 245:202–203.

Bates, H. W. 1862. Contributions to an insect fauna of the Amazon Valley. *Transactions of the Linnaean Society of London* 23:495–566.

Bateson, P. P. G., W. Lotwick, and D. K. Scott. 1980. Similarities between the faces of parents and offspring in Bewick's swans and the differences between mates. *Journal of the Zoological Society of London* 191:61–74.

Batzli, G. O. 1983. Responses of arctic rodent populations to nutritional factors. *Oikos* 40:396–406.

Batzli, G. O., R. G. White, S. F. Maclean, Jr., F. A. Pitelka, and B. D. Collier, 1980. The herbivore-based trophic system. pp. 335–340 in J. Brown, P. C. Miller, L. L. Tiezen, and F. L. Bunnell (eds.). *An Arctic Ecosystem*. Dowden, Hutchinson and Ross, Stroudsburg, Pennsylvania.

Baur, B. and H. Baur. 1993. Climatic warning due to thermal radiation from an urban area as possible cause for the local extinction of a land snail. *Journal of Applied Ecology* 30:333–340.

Beattie, A. J. 1985. *The Evolutionary Ecology of Ant-Plant Mutualisms*. Cambridge University Press, Cambridge.

Beattie, A. J., C. Turnbull, R. B. Knox, and E. G. Williams. 1984. Ant inhibition of pollen function: A possible reason why ant pollination is rare. *American Journal of Botany* 71:421–426.

Beaufait, W. R. 1960. Some effects of high temperatures on the cones and seeds of jack prine. *Forest Science* 6:194–199.

Beaver, R. A. 1985. Geographical variation in food web structure in *Nepenthes* pitcher plants. *Ecological Entomology* 10:241–248.

Beck, B. B., L. G. Rapaport, M. R. Stanley Price, and A. C. Wilson. 1994. Reintroduction of captive-born animals. pp. 265–286 in P. J. S. Olney, G. M. Mace, and A. T. C. Feistner (eds) *Creative Conservation: Interactive Management of Wild and Captive Animals*.

Beger, W. H., and Parker, F. L. 1970. Diversity of planktonic Farmonifera in deep sea sediments. *Science* 168:1345–1347.

Begon, M., J. L. Harper, and C. R. Townsend. 1990. *Ecology: Individuals, Populations and Communities*. Blackwell Scientific Publications, Oxford.

Belovsky, G. E. 1987. Extinction models and mammalian persistence. pp. 35–58 in M. E. Soulé (ed.). *Viable Populations for Conservation*. Cambridge University Press, Cambridge.

Belsky, A. J. 1986. Does herbivory benefit plants? A review of the evidence. *American Naturalist* 127:870–892.

Belsky, A. J. 1987. The effects of grazing: confounding of ecosystem, community, and organism scales. *American Naturalist* 129:777–783.

Bender, E. A., T. J. Case, and M. E. Gilpin. 1984. Perturbation experiments in community ecology: Theory and practice. *Ecology* 65:1–13.

Bennett, F. D., and I. W. Hughes. 1959. Biological control of insect pests in Bermuda. *Bulletin of Entomological Research* 50:423–426.

Berger, J. 1990. Persistence of different-sized populations: An empirical assessment of rapid extinctions in bighorn sheep. *Conservation Biology* 4:91–98.

Berger, W. H., and F. L. Parker. 1970. Diversity of planktonic foraminifera in deep sea sediments. *Science* 168:1345–1347.

Bergerud, A. T. 1980. A review of the population dynamics of caribou and wild reindeer in North America. pp. 556–581 in *Second International Reindeer/Caribou Symposium*. Roros, Norway.

Bernays, E. A. 1981. Plant tannins and insect herbivores: an appraisal. *Ecological Entomology* 6:353–360.

Bernays, E. A. 1983. Nitrogen in defence against insects. pp. 321–344 in J. A. Lee, S. McNeill, and I. H. Rorison (eds.). *Nitrogen as an Ecological Factor. 22nd Symposium of the British Ecological Society*. Blackwell Scientific Publications, Oxford.

Bernays, E. A., and E. Graham. 1988. On the evolution of host specificity in phytophagous arthropods. *Ecology* 69:886–892.

Berry, R. J. 1988. Natural history in the twenty-first century. *Archives of Natural History* 15:1–14.

Berry, R. J. 1990. Industrial melanism and peppered moths (*Biston betularia* [L.]). *Biological Journal of the Linnean Society* 39:301–322.

Bertness, M. D. 1987. Competitive and facilitative interactions in acorn barnacle populations in a sheltered habitat. *Ecology* 70:257–268.

Bertness, M. D., and S. D. Hacker. 1994. Physical stress and positive associations among marsh plants. *American Naturalist* 144:363–372.

Bertram, B. C. R. 1975. Social factors influencing reproduction in wild lions. *Journal of the Zoological Society of London* 177:463–482.

Bertram, B. C. R. 1976. Kin selection in lions and in evolution. pp. 281–301 in P. P. G. Bateson and R. A. Hinde (eds.). *Growing Points in Ethology*. Cambridge University Press, Cambridge.

Bertram, B. C. R. 1979. Serengeti predators and their social systems. pp. 221–248 in A. R. E. Sinclair and M. Morton-Griffiths (eds.). *Serengeti: Dynamics of an Ecosystem*. University of Chicago Press, Chicago.

Beverton, R. J. H. 1983. Science and decision-making in fisheries regulation. pp. 919–938 in G. D. Sharpe and J. Csirke (eds.). *Proceedings of the Expert Consultation to Examine Changes in Abundance and Species Composition of Neritic Fish Resources*. FAO Fishery Report 291. Food and Agriculture Organization, Rome.

Birch, L. C. 1953. Experimental background to the study of the distribution and abundance of insects. I. The influence of temperature, moisture and food on the innate capacity for increase of three grain beetles. *Ecology* 3:698–711.

Birks, H. J. B. 1980. British trees and insects: A test of the time hypothesis over the last 13,000 years. *American Naturalist* 115:600–605.

Bishop, J. A. 1981. A neo-Darwinian approach to resistance: examples from mammals. pp. 37–51 in J. A. Bishop and L. M. Cook (eds.). *Genetic Consequences of Man-made Change*. Academic Press, London.

Bishop, J. A., and L. M. Cook. 1980. Industrial melanism and the urban environment. *Advances in Ecological Research* 11:373-404.

Bishop, J. A., and L. M. Cook (eds.). 1981. *Genetic Consequences of Man-made Change*. Academic Press, London.

Black, F. L. 1975. Infectious diseases in primitive societies. *Science* 187:515–518.

Boecklen, W. J., and D. Simberloff. 1986. Area-based extinction models in conservation. pp. 247–276 in D. K. Elliott (ed.). *Dynamics of Extinction*. Wiley, New York.

Boersma, L. K., and J. A. Gulland. 1973. Stock assessment of the Peruvian anchovy (*Engraulis ringens*) and management of the fishery. *Journal of the Fisheries Research Board of Canada* 30:2226–2235.

Bonnell, M. L., and R. K. Selander. 1974. Elephant seals: genetic variation and near extinction. *Science* 184:908–909.

Bonner, J. 1994. Wildlife's roads to nowhere? *New Scientist*, 20 August, 1994, pp. 30–34.

Bonner, J. T. 1965. *Size and Cycle: An Essay on the Structure of Biology*. Princeton University Press, Princeton, New Jersey.

Booth, W. 1988a. Animals of invention. *Science* 240:718.

Borner, M., C. D. FitzGibbon, M. Borner, T. M. Caro, W. K. Lindsay, D. A. Collins, and M. E. Holt. 1987. The decline of Serengeti Thompson's gazelle population. *Oecologia* 73:32–40.

Boucher, D. H. 1985. *The Biology of Mutualism*. Croom Helm, London.

Bragg, J. R., R. C. Prince, E. J. Harner, and R. M. Atlas 1994. Effectiveness of bioremediation for the *Exxon Valdez* oil spill. *Nature* 368: 413–418.

Bragg, T. B., and L. C. Hurlbert. 1976. Woody plant invasion of unburned Kansas bluestem prairie. *Journal of Range Management* 29:19–29.

Brakefield, P. M. 1990. A decline of melanism in the peppered moth *Biston betularia* in the Netherlands. *Biological Journal of the Linnean Society* 39:327–334.

Brett, J. R. 1959. Thermal requirements of fish: Three decades of study, 1940–1960. (Transcript of the second seminar on Biological Problems in Water Pollution, April 1959.) U. S. Public Health Service, Taft Center, Cincinnati, Ohio.

Brewer, R. 1988. *The Science of Ecology*. Saunders, Philadelphia.

Briand, F. 1983. Environmental control of food web structure. *Ecology* 64:253–263.

Briand, F., and J. E. Cohen. 1984. Community food webs have scale-invariant structure. *Nature* 307:264–266.

Brodbeck, B. V., and D. R. Strong. 1987. Amino acid nutrition of herbivorous insects and stress to host plants. pp. 347–364 in P. Barbosa and J. C. Schultz (eds.). *Insect Outbreaks*. Academic Press, San Diego.

Brody, S. 1945. *Bioenergetics and Growth*. Van Nostrand Reinhold, New York.

Bronstein, J. L. 1991. Mutualism studies and the study of mutualism. *Bulletin of the Ecological Society of America* 72:6–7.

Brower, L. P. 1969. Ecological chemistry. *Scientific American* 220:22–29.

Brower, L. P. 1970. Plant poisons on a terrestrial food chain and implication for mimicry theory. pp. 69–82 in K. L. Chambers (ed.). Biochemical Coevolution. Proceedings of the 29th Annual Biological Colloquium. Oregon State University Press, Corvallis.

Brower, L. P., W. M. Ryerson, L. L. Coppinger, and S. C. Glazier. 1968. Ecological chemistry and the palatability spectrum. *Science* 161:1349–1351.

Brown, A. A., and K. P. Davis. 1973. *Forest Fire Control and Its Use*, 2nd ed. McGraw-Hill, New York.

Brown, C. R. 1988. Enhanced foraging efficiency through information transfer centers: a benefit of coloniality in cliff swallows. *Ecology* 59:602–613.

Brown, C. R., and M. Bomberger Brown. 1986. Ectoparasitism as a cost of coloniality in cliff swallows (*Hirundo pyrrhonota*). *Ecology* 67:1206–1218.

Brown, J. H. 1978. The theory of insular biogeography and the distribution of boreal birds and mammals. *Great Basin Naturalist Memoirs* 2:209–227.

Brown, J. H. 1989. Patterns, modes and extents of invasions by vertebrates. pp. 85–109 in J. A. Drake, H. A. Mooney, F. di Castri, R. H. Groves, F. J. Kruger, M. Rejmanek, and M. Williamson (eds.). *Biological Invasions: A Global Perspective*. Wiley, Chichester, England.

Brown, J. H., D. W. Davidson, J. C. Munger, and R. S. Inouye. 1986. Experimental community ecology: The desert granivore system. pp. 41–61 in J. Diamond and T. J. Case (eds.), *Community Ecology*. Harper and Row, New York.

Brown J. H., and A. Kodric-Brown. 1977. Turnover rates in insular biogeography: Effect of immigration and extinction. *Ecology* 58: 445–449.

Brown, J. H., and M. V. Lomolino. 1989. Independent discovery of the equilibrium theory of island biogeography. *Ecology* 70:1954–1957.

Brown, J. L. 1969. The buffer effect and productivity in tit populations. *American Naturalist* 103:374–354.

Brown, J. L., E. R. Brown, S. D. Brown, and D. D. Dow. 1982. Helpers: effects of experimental removal on reproductive success. *Science* 215:421–422.

Brown, J. L., D. D. Dow, E. R. Brown, and S. D. Brown. 1978. Effects of helpers on feeding and nestlings in the grey-crowned babbler, *Pomatostomus temporalis*. *Behavioral Ecology and Sociobiology* 4:43–60.

Brown, L. R. 1970. Human food production as a process in the biosphere. *Scientific American* 223(3):161–170.

Bruce, H. M. 1966. Smell as an exteroceptive factor. *Journal of Animal Science*, Supplement 25:83–89.

Bryson, R. A., and T. J. Murray, 1977. *Climates of Hunger*. University of Wisconsin Press, Madison, Wisconsin.

Buckingham, G. R. 1987. Florida's #1 weed: *Hydrilla* vs. biocontrol. pp. 22–25 in *Research 87*, IFAS Editorial Department, University of Florida, Gainesville.

Buckley, J. 1986. Environmental effects of DDT. pp. 358–374 in *Ecological Knowledge and Environmental Problem-Solving, Concepts and Case Studies*. National Academy Press, Washington, D.C.

Bunting, A. H. 1988. The humid tropics and the nature of development. *Journal of Biogeography* 15:5–10.

Burton, J. A. 1976. Illicit trade in rare animals. *New Scientist* 72:168.

Bush, G. L. 1975a. Modes of animal speciation. *Annual Review of Ecology and Systematics* 6:334–364.

Bush, G. L. 1975b. Sympatric speciation in phytophagous parasitic insects. pp. 187–206 in P. W. Price (ed.). *Evolutionary Strategies of Parasitic Insects and Mites*. Plenum, New York.

Bush, G. L. 1994. Sympatric speciation in animals: New wine in old bottles. *Trends in Ecology and Evolution* 9:285–288.

Bygott, J. D., B. C. R. Bertram, and J. P. Hanby. 1979. Male lions in large coalitions gain reproductive advantage. *Nature* 282:839–841.

Cain, A. J., and P. M. Sheppard. 1954a. Natural selection in *Cepaea*. *Genetics* 39:89–116.

Cain, A. J., and P. M. Sheppard. 1954b. The theory of adaptive polymorphism. *American Naturalist* 88:321–326.

Cain, M. L., W. P. Carson, and R. B. Root. 1991. Long-term suppression of insect herbivores increases the production and growth of *Solidago altissima* rhizomes. *Oecologia* 88:251–257.

Cairns, J. J. Overbaugh, and S. Miller. 1988. The origin of mutants. *Nature* 335:142–145.

Cameron, R. A. D. 1992. Change and stability on *Cepaea* populations over 25 years: A case of climatic selection. *Proceedings of the Royal Society* B248:181–187.

Cameron, R. D., K. R. Whitten, W. T. Smith, and D. D. Roby. 1979. Caribou distribution and group composition associated with construction of the trans-Alaska pipeline. *Canadian Field Naturalist* 93:155–162.

Caraco, T., and L. L. Wolf. 1975. Ecological determinants of group sizes of foraging lions. *American Naturalist* 109:343–352.

Carlton, J. T., and J. B. Geller. 1993. Ecological roulette: The global transport of nonindigenous marine organisms. *Science* 261:78–82.

Caro, T. M., and M. K. Laurenson. 1994. Ecological and genetic factors in conservation: a cautionary tale. *Science* 263:485–486.

Carothers, J. H. 1986. Homage to Huxley: On the conceptual origin of minimum size ratios among competing species. *The American Naturalist* 128:440–442.

Carson, R. 1962. *Silent Spring*. Houghton Mifflin, Boston.

Caswell, H. 1978. Predator mediated coexistence: A nonequilibrium model. *American Naturalist* 112:127–154.

Cates, R. G., and G. H. Orians. 1975. Successional status and the palatability of plants to generalist herbivores. *Ecology* 56:410–418.

Caughley, G. 1976. Wildlife management and the dynamics of ungulate populations. *Advances in Applied Biology* 1:183–246.

Caughley, G. 1977. *Analysis of Vertebrate Populations*. Wiley, Chichester, England.

Caughley, G., G. C. Grigg, J. Caughley, and G. J. E. Hill. 1980. Does dingo predation control the densities of kangaroos and emus? *Australian Wildlife Research* 7:1–12.

Chalk, P. M., and C. J. Smith. 1983. Chemodenitrification. In Gaseous loss of nitrogen from plant-soil systems. *Developments in Plant and Soil Sciences*. Vol. 9, ed. J. R. Freney and J. R. Simpson. pp. 65–89. The Hague. Martinus Nijhoff/Dr. W. Junk.

Chaloupka, M. J., and S. B. Domm. 1986. Role of anthropochory in the invasion of coral cays by alien flora. *Ecology* 67:1536–1547.

Chapin, F. S. III, L. R. Walker, C. L. Fastie, and L. C. Sharman. 1994. Mechanisms of primary succession following deglaciation at Glacier Bay, Alaska. *Ecological Monographs* 64:149–175.

Charlesworth, B. 1984. The cost of phenotypic evolution. *Paleobiology* 10:319–327.

Cherfas, J. 1986. What price whales? *New Scientist* 110(1511):36–40.

Cherif, A. H. 1990. Mutualism—the forgotten concept in teaching science. *American Biology Teacher* 52:206–208.

Chew, F. S., and S. P. Courtney. 1991. Plant apparency and evolutionary escape from insect herbivory. *The American Naturalist* 138:729–750.

Christensen, M. L. 1981. Fire regimes in southeastern ecosystems. pp. 117–136 in H. A. Mooney, T. M. Bonnicksen, M. L. Christensen, J. E. Lotan, and W. A. Reiners (eds.). *Fire Regimes and Ecosystem Properties*. U.S.D.A. Forest Service, General Technical Report WO-26, Washington, D.C.

Christie, W. J. 1974. Changes in the fish species composition of the Great Lakes. *Journal of the Fisheries Research Board of Canada* 31:827–854.

Clapham, W. B., Jr. 1981. *Human Ecosystems*. Macmillan, New York.

Claridge, M. F., and M. R. Wilson. 1978. British insects and trees: a study in island biogeography or insect/plant coevolution? *American Naturalist* 112:451–456.

Clark, B. C. 1962. Balanced polymorphism and the diversity of sympatric species. *Systematics Association Publication* 4:47–70.

Clarke, C. A., G. S. Mani, and G. Wynne. 1985. Evolution in reverse: Clean air and the peppered moth. *Biological Journal of the Linnean Society* 26:189–199.

Clarke, C. A., F. M. M. Clarke, and H. C. Dawkins. 1990. *Biston betularia* (The peppered moth) in West Kirby, Wirral, 1959–1989: Updating the decline in F. *carbonaria*. *Biological Journal of the Linnean Society* 39:323–326.

Clay, K. 1988. Fungal endophytes of grasses: A defense mutualism between plants and fungi. *Ecology* 69:10–16.

Clements, F. E. 1916. Plant succession: Analysis of the development of vegetation. *Carnegie Institute of Washington Publication* 242:1–512.

Clements, F. E. 1936. Nature and structure of the climax. *Journal of Ecology* 24:252–284.

Clements, F. E., J. Weaver, and H. Hansson. 1926. *Plant Competition: an Analysis of the Development of Vegetation*. Carnegie Institute, Washington, D. C.

Closs, G., G. A. Watterson, and P. J. Donnelly. 1993. Constant predatory-prey ratios: an arithmetical artifact? Ecology 74:238–243.

Cobb, N. S., and T. G. Whitham. 1993. Herbivore deme formation on individual trees: A test case. *Oecologia* 94:496–502.

Cock, M. J. W. 1985. *A Review of Biological Control of Pests in the Commonwealth Caribbean and Bermuda up to 1982*. Technical Communication of the Commonwealth Institute of Biological Control 9.

Cockburn, A. T. 1971. Infectious diseases in ancient populations. *Current Anthropology* 12:45–62.

Cohen, J. E. 1978. *Food Webs and Niche Space*. Princeton University Press, Princeton, New Jersey.

Cohen, J. E. and F. Briand. 1984. Trophic links of community foods webs. *Proceedings of the National Academy of Sciences of the United States of America* 81:4105–4109.

Cody, M. L. 1974. *Competition and the Structure of Bird Communities*. Princeton University Press, Princeton.

Cohen, J. E., F. Briand, and C. M. Newman. 1990. Community food webs: Data and theory. *Biomathematics*. Vol. 20. Springer, Berlin.

Coley, P. D. 1983. Herbivory and defensive characteristics of tree species in a lowland tropical forest. *Ecological Monographs* 53:209–233.

Coley, P. D. 1988. Effects of plant growth and leaf lifetime on the amount and type of anti-herbivore defense. *Oecologia* 74:531–536.

Collinson, A. S. 1977. *An Introduction to World Vegetation*. George Allen and Unwin, London.

Colwell, R. K. 1973. Competition and coexistence in a simple tropical community. *American Naturalist* 107:737–760.

Colwell, R. K. 1984. What's new? Community ecology discovers biology. pp. 387–396 in P. W. Price, C. N. Slobodchikoff, and W. S. Gaud (eds.). *A New Ecology: Novel Approaches to Interactive Systems*. Wiley, New York.

Compton, S. G., D. Newsome, and D. A. Jones. 1983. Selection for cyanogenesis in the leaves and petals of *Lotus corniculatus* L. at high latitudes. *Oecologia* 60:353–358.

Conant, R. 1975. *A field guide to the reptiles and amphibians of eastern and Central North America*. 2nd ed. Houghton Mifflin Co., Boston.

Connell, J. H. 1961. The influence of interspecific competition and other factors on the distribution of the barnacle *Chthamalus stellatus*. *Ecology* 42:710–723.

Connell, J. H. 1978. Diversity in tropical rain forests and coral reefs. *Science* 199:1302–1310.

Connell, J. H. 1980. Diversity and the coevolution of competitors, or the ghost of competition past. *Oikos* 35:131–138.

Connell, J. H. 1983. On the prevalence and relative importance of interspecific competition: evidence from field experiments. *American Naturalist* 122:661–696.

Connell, J. H. 1990. Personal communication.

Connell, J. H., I. R. Noble, and R. O. Slatyer. 1987. On the mechanisms of producing successional change. *Oikos* 50:136–137.

Connell, J. H., and R. O. Slatyer. 1977. Mechanisms of succession in natural communities and their role in community stability and organization. *American Naturalist* 111:1119–1144.

Connell, J. H., and W. P. Sousa. 1983. On the evidence needed to judge ecological stability or persistence. *American Naturalist* 121:789–824.

Connor, E., and E. D. McCoy. 1979. The statistics and biology of the species-area relationship. *American Naturalist* 113:791–833.

Connor, E. G., E. D. McCoy, and B. J. Cosby. 1983. Model discrimination and expected slope values in species-area studies. *The American Naturalist*. 122:789–96.

Connor, E., and D. Simberloff. 1979. The assembly of species communities: Chance or competition? *Ecology* 60:1132–1140.

Conover, W. J. 1980. *Practical Nonparametric Statistics*, 2nd ed. Wiley, New York.

Conway, G. R. 1976. Man versus pests. pp. 257–281 in R. M. May (ed.). *Theoretical Ecology: Principles and Applications*. Blackwell Scientific Publications, Oxford.

Cook, R. E. 1969. Variation in species diversity of North American birds. *Systematic Zoology* 18:63–84.

Cooke, E. 1975. Flow of energy through a technological society. pp. 30–62 in J. Lenihan and W. W. Fletcher (eds.). *Energy Resources and the Environment*. Blackie, Glasgow.

Cooper, J. P. (ed.). 1975. *Photosynthesis and Productivity in Different Environments*. Cambridge University Press, London.

Cooper, S. M., and N. Owen-Smith. 1985. Condensed tannins deter feeding by browsing ruminants on a South African savanna. *Oecologia* 67:142–146.

Cooper, S. M., and N. Owen-Smith. 1986. Effects of plant spinescence on large mammalian herbivores. *Oecologia* 68:446–455.

Cooper, W. S. 1923. The recent ecological history of Glacier Bay, Alaska. II. The present vegetation cycle. *Ecology* 4:223–246.

Cornell, H. 1974. Parasitism and distributional gaps between allopatric species. *American Naturalist* 108:880–883.

Cornell, H. V., and B. A. Hawkins. 1995. Survival patterns and mortality sources of herbivorous insects: some demographic trends. *American Naturalist* 145:563–593.

Cotter, D. J., and J. T. McGinnis, 1965. Recovery of hardwood strands 3–5 years following acute irradiation. *Health Physics* 11:1663–1673.

Cousins, S. 1987. The decline of the trophic level concept. *Trends in Ecology and Evolution* 2:312–215.

Cowles, H. C. 1899. The ecological relations of the vegetation on the sand dunes of Lake Michigan. *Botanical Gazette* 27:95–117, 167–202, 361–391.

Cox, C. B., I. N. Healey, and P. B. Moore. 1976. *Biogeography, an Ecological and Evolutionary Approach*, 2nd ed. Blackwell Scientific Publications, Oxford.

Cox, F. E. G. 1982. Immunology. pp. 173–203 in F. E. G. Cox (ed.). *Modern Parasitology*. Blackwell Scientific Publications, Oxford.

Coyne, J. A. 1976. Lack of genic similarity between two sibling species of *Drosophila* as revealed by varied techniques. *Genetics* 84:593–607.

Coyne, J. A., A. A. Felton, and R. C. Lewontin. 1978. Extent of genetic variation at a highly polymorphic esterase locus in *Drosophila pseudoobscura*. *Proceedings of the National Academy of Sciences of the United States of America* 75:5090–5093.

Cracraft, J. 1983. Species concepts and speciation analysis. pp. 159–187 in R. F. Johnson (ed.). *Current Ornithology* Vol. 1, Plenum Press, New York.

Crawley, M. J. 1983. *Herbivory, the Dynamics of Animal-Plant Interactions. Studies in Ecology*, Vol. 10. Blackwell Scientific Publications, Oxford.

Crawley, M. J. 1985. Reduction of oak fecundity by low-density herbivore populations. *Nature* 314:163–164.

Crawley, M. J. 1987. Benevolent herbivores? *Trends in Ecology and Evolution* 2:167–169.

Creel, S. R., and P. M. Waser. 1991. Failures of reproductive suppression in dwarf mongooses (*Helogale parvula*): Accident or adaptation? *Behavioral Ecology* 2:7–15.

Cristoffer, C., and J. Eisenberg. 1985. *On the Captive Breeding and Reintroduction of the Florida Panther in Suitable Habitats*. Task #1, Report #2, Florida Game and Fresh Water Fish Commission and Panther Technical Advisory Committee, Tallahassee.

Crocker, R. L., and J. Major. 1955. Soil development in relation to vegetation and surface age at Glacier Bay, Alaska. *Journal of Ecology* 43:427–448.

Crosby, A. W. 1986. *Ecological Imperialism, the Biological Expansion of Europe 900–1900.* Cambridge University Press, Cambridge.

Crow, J. F., and M. Kimura. 1970. *An Introduction to Population Genetics Theory.* Harper and Row, New York.

Currie, D. J. 1991. Energy and large-scale patterns of animal- and plant-species richness. *American Naturalist* 137:27–49.

Currie, D. J., and V. Paquin. 1987. Large-scale biogeographical patterns of species richness of trees. *Nature* 329:326–327.

Daday, H. 1954. Gene frequencies in wild populations of *Trifolium repens* L. I. Distribution by latitude. *Heredity* 8:61–78.

Dahlsten, D. L. 1986. Control of invaders. pp. 275–302 in H. A. Mooney and J. A. Drake (eds.). *Ecology of Biological Invasions of North America and Hawaii.* Springer-Verlag, New York.

Darlington, P. J., Jr. 1959. Area, climate and evolution. *Evolution* 13:488–510.

Darwin, C. 1859. *On the Origin of Species by Means of Natural Selection.* John Murray, London.

Dasmann, R. F. 1976. *Environmental Conservation,* 4th ed. Wiley, New York and Chichester, England.

Davidson, J., and H. G. Andrewartha 1948a. Annual trends in a natural population of *Thrips imaginis.* (Thysonoptera). *Journal of Animal Ecology* 17:193–199.

Davidson, J., and H. G. Andrewartha. 1948b. The influence of rainfall evaporation, and atmospheric temperature on fluctuations in the size of a natural population of *Thrips Imaginis.* *Journal of Animal Ecology* 17:200–222.

Davies, N. B. 1978. Territorial defense in the speckled wood butterfly (*Pararge aegeria*): The resident always wins. *Animal Behavior* 26:138–147.

Davies, N. B., B. J. Hatchwell, T. Robson, and T. Burke. 1992. Paternity and parental effort in dunnocks *Prunella modularis*: How good are male chick-feeding rules. *Animal Behavior* 43:729–45.

Davis, M. B. 1983. Holocene vegetational history of the eastern United States. pp. 166–181 in H. E. Wright, Jr. (ed.) *Late-Quaternary Environments of the United States,* Vol. II. The Holocene, University of Minnesota Press, Minneapolis.

Davis M. G., and C. Zabinski. 1992. Changes in geographical range resulting from greenhouse warming: effects on biodiversity in forests. pp. 297–308 in R. Peters and T. Lovejoy (eds.). *Global Warming and Biodiversity.* Yale University Press, New Haven, Connecticut.

Dawkins, R. 1986. *The Blind Watchmaker.* Norton, New York.

Dawkins, R. 1989. *The Selfish Gene.* 2nd ed. Oxford University Press, Oxford.

Dawkins, R., and J. R. Krebs. 1979. Arms races between and within species. *Proceedings of the Royal Society of London Series B* 205:489–511.

Dayan, T., and D. Simberloff. 1994. Character displacement, sexual dimorphism and morphological variation among British and Irish mustelids. *Ecology* 75:1063–1073.

Dayton, P. K., and M. J. Tegner. 1984. The importance of scale in community ecology: A kelp forest example with terrestrial analogs. pp. 457–481 in P. W. Price, C. N. Slobodchikoff, and W. S. Gaud (eds.). *A New Ecology: Novel Approaches to Interactive Systems.* Wiley, New York.

Debach, P. S., and H. S. Smith. 1941. The effect of host density on the rate of reproduction of entomophagous parasites. *Journal of Economic Entomology* 34:741–745.

Debach, P. S., and R. A. Sundby. 1963. Competitive displacement between ecological homologues. *Hilgardia* 43:105–166.

den Boer, P. J. 1968. Spreading of risk and stabilization of animal numbers. *Acta Biotheoretica* 18:165–194.

den Boer, P. J. 1981. On the survival of populations in a heterogeneous and variable environment. *Oecologia* 50:39–53.

Denniston, C. 1978. Small population size and genetic diversity: Implications for endangered species. pp. 281–289 in S. A. Temple (ed.). *Endangered Birds: Management Techniques for Preserving Threatened Species*. University of Wisconsin Press, Madison.

Denno, B.F., M. S. McClure, and J. R. Ott. 1995. Interspecific interactions in phytophagous insects: competition re-examined and resurrected. *Annual Review of Entomology* 40:297–331.

Desowitz, R. S. 1981. *New Guinea Tapeworms and Jewish Grandmothers: Tales of Parasites and People*. Norton, New York.

De Vos, A., R. H. Manville, and G. Van Gelder. 1956. Introduced mammals and their influence on native biota. *Zoologica* 41:163–194.

Diamond, J. 1986a. The environmentalist myth. *Nature*, 324:19–20.

Diamond, J. 1986b. Overview: laboratory experiments, field experiments, and natural experiments. pp. 3–22 in J. Diamond and T. J. Case (eds.). *Community Ecology*. Harper and Row, New York.

Dirzo, R., and J. L. Harper. 1982. Experimental studies on slug-plant interactions. IV. The performance of cyanogenic and acyanogenic morphs of *Trifolium repens* in the field. *Journal of Ecology* 70: 119–138.

Di Silvestro, R. L. 1988. U.S. demand for carved ivory hastens African elephant's end. *Audubon* 90(2):14.

Diver, C. 1929. Fossil records of Mendelian mutants. *Nature* 124:183.

Dobson, A., and M. Crawley. 1994. Pathogens and the structure of plant communities. *Trends in Ecology and Evolution* 9: 393–398.

Dobzhansky, T. 1936. Studies on hybrid sterility. II. Localization of sterility factors in *Drosophila pseudoobscura* hybrids. *Genetics* 21:113–135.

Dobzhansky, T. 1950. Evolution in the tropics. *American Scientist* 38:209–221.

Dobzhansky, T. 1970. *Genetics of the Evolutionary Process*. Columbia University Press, New York.

Donoghue, M. J. 1985. A critique of the biological species concept and recommendation for a phylogenetic alternative. *The Bryologist* 88: 172–181.

Drake, J. A. 1990. Communities as assembled structures: Do rules govern pattern? *Trends in Ecology and Evolution* 5: 159–164.

Drake, J. A. and 9 other authors. 1993. The construction and assembly of an ecological landscape. *Journal of Animal Ecology* 62:117–730.

Dunbar, M. J. 1980. The blunting of Occam's Razor, or to hell with parsimony. *Canadian Journal of Zoology* 58:123–128.

Duncan, P., and N. Vigne. 1979. The effect of group size in horses on the rate of attacks by blood-sucking flies. *Animal Behavior* 27:623–625.

Dyson, F. J. 1988. *Infinite in All Directions*. Harper and Row, New York.

Eadire, J. McA., L. Broekhoven, and P. Colgan. 1987. Size ratios and artifacts: Hutchinson's rule revisited. *American Naturalist* 129:1–17.

Ebenhard, T. 1988. Introduced birds and mammals and their ecological effects. *Swedish Wildlife Research* 13:1–107.

Eckholm, E. P. 1976. *Losing Ground*. Norton, New York.

Edmondson, W. T. 1979. Lake Washington and the predictability of limnological events. *Archiv für Hydrobiologie, Beiheft* 13:234–241.

Edmunds, M. 1974. *Defence in Animals*. Harlow, Essex, England.

Edwards, C. A., and G. W. Heath. 1963. The role of soil animals in breakdown of leaf material. pp. 76–84 in D. Doiksen and J. van der Pritt (eds.). *Soil Organisms*. North-Holland, Amsterdam.

Edwards, P. J., and S. P. Wratten. 1985. Induced plant defenses against insect grazing: fact or artifact? *Oikos* 44:70–74.

Egler, F. E. 1954. Vegetation science concepts: initial floristic composition—a factor in old-field development. *Vegetation* 4:412–417.

Ehler, L. E. 1979. Assessing competitive interactions in parasite guilds prior to introduction. *Environmental Entomology* 8:558–560.

Ehler, L. E., and R. W. Hall. 1982. Evidence for competitive exclusion of introduced natural enemies in biological control. *Environmental Entomology* 1:1–4.

Ehrlich, P. R. 1975. The population biology of coral reef fishes. *Annual Review of Ecology and Systematics* 6:213–247.

Ehrlich, P. R., and A. Ehrlich. 1970. *Population Resources Environment: Issues in Human Ecology*, 2nd ed. Freeman, San Francisco.

Ehrlich, P. R., A. Ehlrich, and J. P. Holden. 1977. *Ecoscience: Population, Resources, Environment*. Freeman, San Francisco.

Ehrlich, P. R., and P. H. Raven. 1964. Butterflies and plants: a study in coevolution. *Evolution* 18:586–608.

Ehrlich, P. R., and J. Roughgarden. 1987. *The Science* of Ecology. Macmillan, New York.

Einstein, A., and Infeld, L. 1938. *The Evolution of Physics: From Early Concepts to Relativity and Quanta*. Simon and Schuster, New York.

Eisner, T., and D. J. Aneshansley. 1982. Spray aiming in bombardier beetles: Jet deflection by the Coanda effect. *Science* 215:83–85.

Eisner, T., and J. Meinwald. 1966. Defensive secretions of arthropods. *Science* 153:1341–1350.

Elner, R. W., and R. N. Hughes. 1978. Energy maximization in the diet of the shore crab, *Carcinus maenas*. *Journal of Animal Ecology* 47:103–116.

Elton, C. 1927. *Animal Ecology*. Sidgwick and Jackson, London.

Elton, C. 1942. Voles, mice and lemmings. *Problems in Population Dynamics*. Clarendon Press, Oxford.

Elton, C. 1958. *The Ecology of Invasions by Animals and Plants*. Methuen, London.

Elton, C., and M. Nicholson. 1942. The ten-year cycle in numbers of the lynx in Canada. *Journal of Animal Ecology* 11:215–244.

Eltringham, S. K. 1984. *Wildlife Resources and Economic Development*. Wiley, Chichester, England.

Enright, J. T. 1976. Climate and population regulation. *Oecologia* 24:295–310.

Erickson, E., and S. L. Buchmann. 1983. Electrostatics and pollination. pp. 173–184 in C. E. Jones and R. J. Little (eds.). *Handbook of Experimental Pollination Biology*. Van Nostrand Reinhold, New York.

Erlinge, S., G. Göransson, G. Högstedt, G. Jansson, O. Liberg, J. Loman, I. N. Nilsson, T. von Schantz, and M. Sylvén. 1984. Can vertebrate predators regulate their prey? *American Naturalist* 123:125–133.

Erwin, P. H., J. W. Valentine, and J. J. Sepkoski. 1987. A comparative study of diversification events. The early Paloeozoic versus the Mesozoic. *Evolution* 41:1177–1186

Erwin, T. L. 1982. Tropical forests: their richness in Coleoptera and other arthropod species. *Coleopterists Bulletin* 36:74–75.

Erwin, T. L. 1983. Beetles and other insects of tropical forest canopies at Manaus, Brazil, sampled by insecticidal fogging. pp. 59–75 in S. L. Sutton, T. C. Whitmore, and A. C. Chadwick (eds.). *Tropical Rain Forest: Ecology and Management.* Blackwell Scientific Publications, Oxford.

Estes, J. A. 1991. Catastrophes and conservation lessons from sea otters and the *Exxon Valdez. Science* 254:1596.

Estes, J. A., R. J. Jameson, and E. B. Rhode. 1982. Activity and prey selection in the sea otter: Influence of population status on community structure. *American Naturalist* 120:242–258.

Facelli, J. M., and E. Facelli 1993. Interactions after death: plant litter controls priority effects in a successional plant community. *Oecologia* 95:277–282.

Faeth, S. H. 1985. Quantitative defense theory and patterns of feeding by oak insects. *Oecologia* 68:34–40.

Faeth, S. H. 1988. Plant-mediated interactions between seasonal herbivores: Enough for evolution or coevolution? pp. 391–414 in K. C. Spencer (ed.). *Chemical Mediation of Coevolution.* Academic Press, New York.

Fagerström, T. 1987. On theory, data and mathematics in ecology. *Oikos* 50:258–261.

Fahim, H. M. 1981. *Dams, People and Development, the Aswan High Dam Case.* Pergamon, New York.

Fay, P. A. and D. C. Hartnett. 1991. Constraints on growth and allocation patterns of *Silphium integrifolium* (Asteraceae) caused by a cynipid gall wasp. *Oecologia* 88: 243-250.

Fearnside, P. M. 1987. Deforestation and international economic development projects in Brazilian Amazonia. *Conservation Biology* 1:214–221.

Feeny, P. 1970. Seasonal changes in the oak leaf tannins and nutrients as a cause of spring feeding by winter moth caterpillars. *Ecology* 51:565–581.

Feeny, P. 1976. Plant apparency and chemical defense. *Recent Advances in Phytochemistry* 10:1–40.

Fenchel, T. 1974. Intrinsic rate of natural increase: The relationship with body size. *Oecologia* 14:317–326.

Fenner, F., and F. Ratcliffe. 1965. *Myxamatosis.* Cambridge University Press, Cambridge.

Ferguson, K. I., and P. Stiling. 1995. Non-additive effects of multiple natural enemies on aphid populations. *Oecologia* (in review).

Firey, W. J. 1960. *Man, Mind and Land: Theory of Resource Use.* Free Press, Glencoe, Illinois. Greenwood Press, London.

Fischer, J., N. Simon, and J. Vincent. 1969. *The Red Book—Wildlife in Danger.* Collins, London.

Fisher, R. A. 1930. *The Genetical Theory of Natural Selection.* Clarendon Press, Oxford.

Fitzgibbon, C. D. 1989. A cost to individuals with reduced vigilance in groups of Thompson's gazelles hunted by cheetahs. *Animal Behavior* 37:508–510.

Fleischer, S., G. Anersson, Y. Brodin, W. Dickson, J. Hermann, and I. Muniz. 1993. Acid water research in Sweden—knowledge for tomorrow? *Ambio* 22:258–263.

Fleiss, J. L. 1981. *Statistical Methods for Rates and Proportions*. Wiley, New York.

Flessa, K. W., and D. Jablonski. 1985. Declining Phanerozoic background extinction rates: Effects of taxonomic structure? *Nature* 313:216–218.

Food and Agriculture Organization. 1986. *Production Yearbook*. FAO, Rome.

Ford, R. G., and F. A. Pitelka. 1984. Resource limitation in the California vole. *Ecology* 65:122–136.

Foster, G. M., and B. G. Anderson. 1979. *Medical Anthropology*. Wiley, New York.

Foster, M. S. 1991. Rammed by the *Exxon Valdez*: A reply to Paine. *Oikos* 62:93–96.

Foulds, W., and J. P. Grime. 1972a. The influence of soil moisture on the frequency of cyanogenic plants in populations of *Trifolium repens* and *Lotus corniculatus*. *Heredity* 28:143–146.

Foulds, W., and J. P. Grime. 1972b. The response of cyanogenic and acyanogenic phenotypes of *Trifolium repens* to soil moisture supply. *Heredity* 28:181–187.

Fowler, S. V., and J. H. Lawton. 1985. Rapidly induced defenses and talking trees: The devil's advocate position. *American Naturalist* 126:181–195.

Fowler, S. V., and M. MacGarvin. 1985. The impact of hairy wood ants, *Formica lugubris*, on the guild structure of herbivorous insects on birch, *Beteula pubescens*. *Journal of Animal Ecology* 54:847–855.

Fox, B. J., and J. H. Brown. 1993. Assembly rules for functional groups in North American desert rodent communities. *Oikos* 67:358–370.

Fox, L. R., and P. A. Morrow. 1992. Eucalypt responses to fertilization and reduced herbivory. *Oecologia* 89:214–222.

Fraley, L., and F. W. Whicker. 1971. Response of a native shortgrass plant stand to ionizing radiation. pp. 999–1006 in D. J. Nelson (ed.). *Radionuclides in Ecosystems. Proceedings of the Third National Symposium on Radioecology*. U.S. Atomic Energy Commission, Washington, D.C.

Frankel, O. H., and M. E. Soulé. 1981. *Conservation and Evolution*. Cambridge University Press, Cambridge.

Franklin, I. R. 1980. Evolutionary change in small populations. pp. 135–139 in M. E. Soulé and B. A. Wilcox (eds.). *Conservation Biology: An Evolutionary-Ecological Perspective*. Sinauer Associates, Sunderland, Massachusetts.

Frenkel, R. E. 1970. *Ruderal Vegetation Along Some California Roadsides*. University of California Publication in Geography 20, Berkeley.

Freudenberger, D. 1982. Southern Africa. pp. 3–192 in D. Yerex (ed.). *The Farming of Deer*. Agricultural Associates, Wellington, New Zealand.

Fryer, G., and T. D. Iles. 1972. *The Chiclid Fishes of the Great Lakes of Africa*. T. F. H. Publications, Neptune City, New Jersey.

Futuyma, D. J. 1983. Evolutionary interactions among herbivorous and plants. pp. 207–231 in D. J. Futuyma and M. Slatkin (eds.). *Coevolution*. Sinauer Associates, Sunderland, Massachusetts.

Futuyma, D. J. 1986. *Evolutionary Biology*, 2nd ed. Sinauer Associates, Sunderland, Massachusetts.

Futuyma, D. J., and S. C. Peterson. 1985. Genetic variation in the use of resources by insects. *Annual Review of Entomology* 30:217–238.

Gause, G. F. 1932. Experimental studies on the struggle for existence. I. Mixed population of two species of yeast. *Journal of Experimental Biology* 9:389–402.

Gause, G. F. 1934. *The Struggle for Existence*. Macmillan (Hafner Press), New York (reprinted 1964).

Geist, V. 1994. Wildlife conservation as wealth. *Nature* 368:491–492.

Gentle, W., F. R. Humphreys, and M. J. Lambert, 1965. An examination of *Pinus radiata* phosphate fertilizer trial fifteen years after treatment. *Forest Science* 11:315–324.

Gentry, A. H. 1988. Tree species of upper Amazonian forests. *Proceedings of the National Academy of Science* of the United States of America 85:156.

George, C. J. 1972. The role of the Aswan High Dam in changing the fisheries of the southeastern Mediterranean. pp. 159–178 in M. Taghi Farvar and J. P. Milton (eds.). *The Careless Technology: Ecology and International Development*. Natural History Press, New York.

Gerrish, G., D. Mueller-Dombois, and K. W. Bridges. 1988. Nutrient limitation and *Metrosideros* forest dieback in Hawaii. *Ecology* 69:723–727.

Gessel, S. P. 1962. Progress and problems in mineral nutrition of forest trees. pp. 221-235 in T. T. Kozlowski (ed.), *Tree Growth*. Ronald Press, New York.

Gibbens, R. P., K. M. Havstad, D. D. Billheimer, and C. H. Herbel. 1993. Creosotebush vegetation after 50 years of lagomorph exclusion. *Oecologia* 94:210–217.

Gilbert, F. S. 1980. The equilibrium theory of island biogeography: Fact or fiction. *Journal of Biogeography* 7:209–235.

Gill, D. E. 1974. Intrinsic rate of increase, saturation density, and competitive ability. II. The evolution of competitive ability. *American Naturalist* 108:103–116.

Gilman, A. P., D. P. Peakall, D. J. Hallett, G. A. Fox, and R. J. Norstrom. 1979. *Animals as Monitors of Environmental Pollutants*. National Academy Press, Washington, D.C.

Givnish, T. J. 1982. On the adaptive significance of leaf height in forest herbs. *American Naturalist* 120:353–381.

Gleason, H. A. 1926. The individualistic concept of the plant association. *Torrey Botanical Club Bulletin* 53:7–26.

Godfrey, L. R. 1983. *Scientists Confront Creationism*. Norton, New York.

Goldberg, E. D., and K. K. Bertine. 1975. Marine pollution. pp. 273–295 in W. W. Murdoch (ed.), *Environment*, 2nd ed. Sinauer Associates, Sunderland, Massachusetts.

Goldsmith, F. B. 1983. Evaluating nature. pp. 233–246 in A. Warren and F. B. Goldsmith (eds.). *Conservation in Perspective*. Wiley, Chichester, England.

Goodland, R. J. 1975. The tropical origin of ecology: Eugen Warming's jubilee. *Oikos* 26:240–245.

Goodland, R. J. A. 1987. The World Bank's wildlands policy: A major new means of financing conservation. *Conservation Biology* 1:210–213.

Goodman, D. 1975. The theory of diversity-stability relationships in ecology. *Quarterly Review of Biology* 50:237–266.

Gordon, H. S. 1954. The economic theory of a common property resource: The fishery. *Journal of Political Economics* 62:124–142.

Gotelli, N. 1995. *A Primer of Ecology*. Sinauer Associates, Sunderland, Massachusetts.

Götmark, F., M. Åhlund, and M. O. G. Eriksson. 1986. Are indices reliable for assessing conservation value of natural areas? An avian case study. *Biological Conservation* 38:55–73.

Goudie, A. 1986. *The Human Impact on the Natural Environment*, 2nd ed. MIT Press, Cambridge, Mass.

Gould, S. J., and R. C. Lewontin. 1979. The spandrels of San Marco and the Panglossian paradigm: A critique of the adaptationist programme. *Proceedings of the Royal Society of London B* 205:581–598.

Goulding, M. 1980. *The Fishes and the Forest: Explorations in Amazonian Natural History*. University of California Press, Berkeley and Los Angeles.

Grant, B., and R. J. Howlett. 1988. Background selection by the peppered moth (*Biston betularia* Linn): Individual differences. *Biological Journal of the Linnean Society* 33:217–232.

Grant, K. A., and V. Grant. 1964. Mechanical isolation of *Salvia apiana* and *Salvia mellifera* (Labiatae). *Evolution* 18:196–212.

Grant, P. R., and B. R. Grant. 1987. The extraordinary El Niño event of 1982–1983: Effects on Darwin's finches on Isla Genovesa, Galápagos. *Oikos* 49:55–66.

Grant, V. 1977. *Organismic Evolution*. Freeman, San Francisco.

Grant, V. 1981. *Plant Speciation*, 2nd ed. Columbia University Press, New York.

Grant, V. 1985. *The Evolutionary Process, a Critical Review of Evolutionary Theory*. Columbia University Press, New York.

Graveland, J., R. van der Wal, J. H. van Balen, and A. J. van Noordwijk. 1994. Poor reproduction in forest passerines from decline of snail abundance on acidified soils. *Nature* 368:446–448.

Gray, J. S. 1981. *The Ecology of Marine Sediments*. Cambridge University Press, Cambridge.

Gray, J. S. 1987. Species-abundance patterns. pp. 53–68 in J. H. R. Gee and P. S. Giller (eds.). *Organization of Communities: Past and Present*. Blackwell Scientific Publications, Oxford.

Greene, J. C. 1959. *The Death of Adam: Evolution and Its Impact on Western Thought*. Iowa State University Press, Ames, Iowa.

Greenslade, P. J. M. 1983. Adversity selection and the habitat templet. *American Naturalist* 122:352–365.

Greig, J. C. 1979. Principles of genetic conservation in relation to wildlife management in Southern Africa. *S. Afr. Tydskr. Naturnau.* 9: 57–78.

Griffith, B., J. M. Scott, J. W. Carpenter, and C. Reed. 1989. Translocation as a species conservation tool: Status and strategy. *Science* 245: 477–480.

Grime, J. P. 1977. Evidence for the existence of three primary strategies in plants and its relevance to ecological and evolutionary theory. *American Naturalist* 111:1169–1194.

Grime, J. P. 1979. *Plant Strategies and Vegetation Process*. Wiley, New York.

Grime, J. P. 1993. Ecology sans frontiers. *Oikos* 68:385–392.

Grimes, A., and 11 other authors. 1994. Valuing the rain forest: The economic value of nontimber forest products in Ecuador. *Ambio* 23:405–410.

Gulland, J. 1988. The end of whaling? *New Scientist* 120(1636):42–47.

Gunn, A. S. 1990. Preserving rare species. pp. 289–308 in T. Regan (ed.). *Earthbound, Introductory Essays in Environment Ethics*. Waveland Press, Inc. Prospect Heights, Illinois.

Gupta, A. P., and R. C. Lewontin. 1982. A study of reaction norms in natural populations of *Drosophila pseudoobscura*. *Evolution* 36:934–948.

Hadley, J. L., and W. K. Smith. 1986. Wind effects on needles of timberline conifers: Seasonal influence on mortality. *Ecology* 67:12–19.

Hairston, J. G., Sr. 1989. *Ecological Experiments: Purpose, Design, and Execution*. Cambridge University Press, Cambridge.

Hairston, N. G. 1991. The literature glut: Causes and consequences: Reflections of a dinosaur. *Bulletin of the Ecological Society of America* 72:171–174.

Hairston, N. G., J. D. Allen, R. K. Colwell, D. J. Futuyma, J. Howell, M. D. Lubin, J. Mathias, and J. H. Vandermeer. 1968. The relationship between species diversity and stability: An experimental approach with protozoa and bacteria. *Ecology* 49:1091–1101.

Hairston, N. G., F. E. Smith, and L. B. Slobodkin. 1960. Community structure, population control, and competition. *American Naturalist* 44:421–425.

Haldane, J. B. S. 1953. Animal populations and their regulation. *Penguin Modern Biology* 15:9–24.

Haldane, J. B. S. 1963. The acceptance of a scientific idea. Reprinted in R. L. Weber (ed.). 1982, *More Random Walks in Science*. Institute of Physics, Bristol.

Hall, B. G. 1991. Adaptive evolution that requires multiple spontaneous mutations: Mutations involving base substitutions. Proceedings of the National Academy of the United States of America 88:5882–5886.

Hall, R. W., L. E. Ehler, and B. Bisabri-Ershadi. 1980. Rates of success in classical biological control of arthropods. *Bulletin of the Entomological Society of America* 26:111–114.

Hamburg, S. P., and C. V. Cogbill. 1988. Historical decline of red spruce populations and climatic warming. *Nature* 331:428–431.

Hamilton, W. D. 1964. The genetical evolution of social behaviour. I, II. *Journal of Theoretical Biology* 7:1–52.

Hamilton, W. D. 1967. Extraordinary sex ratios. *Science* 156:477–488.

Hamilton W. D. 1971. Geometry for the selfish herd. *Journal of Theoretical Biology* 31:295–311.

Hamrick, J. L., and M. J. W. Godt. 1990. Allozyme diversity in plant species. pp. 43–63 in A. D. Brown, M. T. Clegg, A. L. Kehles, and B. S. Weir (eds.). *Plant Population Genetics, Breeding, and Genetic Resources*. Sinauer Associates, Sunderland, Mass.

Hannah, L., D. Lotise, C. Hutchinson, J. L. Carr, and H. Lankerani. 1994. A preliminary inventory of human disturbance of world ecosystems. *Ambio* 23:246–250.

Harcourt, A. H., P. H. Harvey, S. G. Larson, and R. V. Short. 1981. Testis weight, body weight and breeding system in primates. *Nature* 293:55–57.

Hardin, G. 1957. The threat of clarity. *American Journal of Psychiatry* 114:392–396.

Hardin, G. 1968. The tragedy of the commons. *Science* 162:1243-1248.

Harley, J. L., and S. E. Smith. 1983. *Mycorrhizal Symbiosis*. Academic Press, London.

Harris, H. 1966. Enzyme polymorphisms in man. *Proceedings of the Royal Society of London Series B* 164:298–310.

Harrison, S. 1991. Local extinction in a metapopulation context: an empirical evaluation. *Biological Journal of the Linnean Society* 42:73–88.

Harvey, P. H., J. J. Bull, and R. J. Paxton. 1983. Looks pretty nasty. *New Scientist* 97:26–27.

Harvey, P. H., M. Kavanagh, and T. H. Clutton-Brock. 1978. Sexual dimorphism in primate teeth. *Journal of Zoology* 186:475–486.

Hasler, A. D. 1947. Eutrophication of lakes by domestic drainage. *Ecology* 28:383–395.

Hassell, M. F., and G. C. Varley. 1969. New inductive population model for insect parasites and its bearing on biological control. *Nature* 223:1133–1137.

Hassell, M. P., J. Latto, and R. M. May. 1989. Seeing the wood for the trees: Detecting density dependence from existing life-table studies. *Journal of Animal Ecology* 58:883–892.

Hastings, J. R., and R. M. Turner. 1965. *The Changing Mile*. University of Arizona Press, Tucson.

Haukioja, E. 1980. On the role of plant defenses in the fluctuation of herbivore populations. *Oikos* 35:202–213.

Haury, L. R., J. A. McGowan, and P. H. Wiebe. 1978. Patterns and processes in the time-space scales of plankton distributions. pp. 277–327 in J. H. Steele (ed.). *Spatial Patterns in Plankton Communities*. Plenum, New York.

Hawkins, B. A. 1992. Parasitoid-host food webs and donor control. *Oikos* 65:159–162.

Hay, M. 1981. The functional morphology of turf-forming seaweeds: Persistence in stressful marine habitats. *Ecology* 62:739–750.

Heal, O. W., and S. F. Maclean. 1975. Comparative productivity in ecosystems—secondary productivity. pp. 89–108 in W. H. van Dobben and R. H. Lowwe-McConnell (eds.). *Unifying Concepts in Ecology*. Dr. W. Junk The Hague.

Heatwole, H., and R. Levins. 1972. Trophic structure stability and faunal change during recolonization. *Ecology* 53:531–534.

Hedrick, P. W., and P. S. Miller. 1992. Conservation genetics: techniques and fundamentals. *Ecological Applications* 2: 30–46.

Heinrich, B. 1976. Flowering phenologies: Bog, woodland, and disturbed habitats. *Ecology* 57:890–899.

Heinrich, B. 1979. *Bumblebee Economics*. Harvard University Press, Cambridge, Massachusetts.

Heron, A. C. 1972. Population ecology of a colonizing species: The pelagic tunicate *Thalia democratica*. *Oecologia* 10:269–293, 294–312.

Hershkovitz, P. 1977. *Living New World Monkeys (Platyrhini), Vol. 1*. University of Chicago Press, Chicago.

Heske, E. J. , J. H. Brown, and S. Mistry. 1994. Long-term experimental study of chihuahuan desert rodent community: 13 years of competition. *Ecology* 75:438–445.

Hessler, R., P. Lonsdale, and J. Hawkins. 1988. Patterns on the ocean floor. *New Scientist* 117(1605):47–51.

Heywood, V. H. 1989. Patterns, extents and modes of invasions by terrestrial plants. pp. 31–60 in J. A. Drake, H. A. Mooney, F. di Castri, R. H. Groves, F. J. Kruger, M. Rejmanek, and M. Williamson (eds.), *Biological Invasion: A Global Perspective*. Wiley, Chichester, England.

Hibler, C. P., R. E. Lange, and C. Metzger. 1972. Transplacental transmission of *Protostrongylus* spp. in big-horn sheep. *Journal of Wildlife Diseases* 8:389.

Hilborn, R., and C. J. Walters. 1992. *Quantitative Fisheries Stock Assessment*. Chapman and Hall, New York.

Hokkanen, H., and D. Pimentel. 1984. New approach for selecting biological control agents. *Canadian Entomologist* 116:1109–1121.

Holden, C. 1988. The greening of the World Bank. *Science* 240:1610–1611.

Holdgate, M. W. 1960. The fauna of the mid-Atlantic islands. *Proceedings of the Royal Society of London Series B* 152:550–567.

Holling, C. S. (ed.). 1978. *Adaptive Environmental Assessment and Management*. John Wiley and Sons, Chichester, England.

Holm, L. G., L. W. Weldon, and R. D. Blackburn. 1969. Aquatic weeds. *Science* 166:699–709.

Holt, R. D. 1977. Predation, apparent competition and the structure of prey communities. *Theoretical Population Biology* 12:197–229.

Holyoak, K., and P. H. Crowley. 1993. Avoiding erroneously high levels of detection in combinations of semi-independent tests. *Oecologia* 95:103–114.

Hopcraft, D. 1982. Wildlife ranching in perspective. *Tigerpaper* 9:17–20.

Horn, H. S. 1971. *Adaptive Geometry of Trees*. Princeton University Press, Princeton, New Jersey.

Horn, H. S. 1975. Markovian processes of forest succession. pp. 196–213 in M. L. Cody and J. Diamond (eds.). *Ecology and Evolution of Communities*. Harvard University Press, Cambridge, Massachusetts.

Horn, H. S. 1976. Succession. pp. 187–204 in R. M. May (ed.). *Theoretical Ecology: Principles and Applications*. Saunders, Philadelphia.

Horn, H. S., and R. M. May. 1977. Limits to similarity among coexisting competitors. *Nature* 270:660–661.

Houston, A. I., and N. B. Davies. 1985. The evolution of cooperation and life history in the dunnock, *Prunella modularis*. pp. 471–487 in R. M. Sibly and R. H. Smith (eds.). *Behavioural Ecology, Ecological Consequences of Adaptive Behaviour*. Blackwell Scientific Publications, Oxford.

Howard, R. D. 1978. The evolution of mating strategies in bullfrogs, *Rana catesbeiana*. *Evolution* 32:850–871.

Howard, W. E. 1949. Dispersal, amount of inbreeding, and longevity in a local population of prairie deer mice on the George Reverse, southern Michigan. *Contributions from the Laboratory of Vertebrate Biology of the University of Michigan* 43:1–50.

Howarth, F. G. 1983. Classical biological control: Panacea or Pandora's box? *Proceedings of the Hawaii Entomological Society* 24:239–244.

Hudson, P. J., A. P. Dobson, and D. Newborn. 1992. Do parasites make prey vulnerable to predation? Red grouse and parasites. *Journal of Animal Ecology* 61:681–692.

Huffaker, C. B., and C. E. Kennett. 1959. A ten year study of vegetational changes associated with biological control of Klamath weed. *Journal of Range Management* 12:69–82.

Huffaker, C. B., and C. E. Kennett. 1969. Some aspects of assessing efficiency of natural enemies. *Canadian Entomologist* 101:425-440.

Hungate, R. E. 1975. The rumen microbial system. *Annual Review of Ecology and Systematics* 6:39–66.

Hunter, M. D., and P. W. Price 1992. Playing chutes and ladders: heterogeneity and the relative roles of bottom-up and top-down forces in natural communites. *Ecology* 73:724–732.

Hurlbert, S. H. 1971. The nonconcept of species diversity: A critique and alternative parameters. *Ecology* 52:577–586.

Hutchinson, G. E. 1959. Homage to Santa Rosalia, or why are there so many kinds of animals? *American Naturalist* 93:145–159.

Hutchinson, J. B. 1965. Crop-plant evolution: A general discussion. pp. 166–181 in J. B. Hutchinson (ed.). *Essays on Crop Plant Evolution*. Cambridge University Press, New York.

Iason, G. R., C. D. Duck, and T. H. Clutton-Brock. 1986. Grazing and reproductive success of red deer: the effect of local enrichment by gull colonies. *Journal of Animal Ecology* 55:507–515.

Inouye, R. S., N. J. Huntly, D. Tilman, J. R. Tester, M. Stillwell, and K. C. Zinnel. 1987. Old-field succession on a Minnesota sand plain. *Ecology* 68:12–26.

International Union for Conservation of Nature and Natural Resources. 1980. *World Conservation Strategy*. International Union for Conservation of Nature and Natural Resources, United National Environmental Program, World Wildlife Fund. Gland, Switzerland.

International Union for Conservation of Nature and Natural Resources. 1985. *The United Nations List of National Parks and Protected Areas*. Gland, Switzerland.

Jablonski, D. 1986. Evolutionary consequences of mass extinctions. pp. 313–330 in D. M. Raup and D. Jablonski (eds.), *Patterns and Processes in the History of Life*. Springer-Verlag, Berlin.

Jablonski, D. 1993. The tropics as a source of evolutionary novelty through geological time. *Nature* 364:142-144.

Jackson, M. V., and T. E. Williams. 1979. Response of grass swards to fertilizer N under cutting or grazing. *Journal of Agricultural Science*, Cambridge, 92:549–562.

Jackson, W., and J. Piper. 1989. The necessary marriage between ecology and agriculture. *Ecology* 70:1591–1593.

Jaksic, F. M. 1981. Abuse and misuse of the term "guild" in ecological studies. *Oikos* 37:397–400.

Jaksic, F. M., and R. G. Medel. 1990. Objective recognition of guilds: testing for statistically significant species clusters. *Oecologia* 82:87–92.

Janzen, D. H. 1966. Coevolution of mutualism between ants and acacias in Central America. *Evolution* 20:249–275.

Janzen D. H. 1968. Host plants as islands in evolutionary and contemporary time. *American Naturalist* 102:592–595.

Janzen, D. H. 1970. Herbivores and the number of tree species in tropical forests. *American Naturalist* 104:501–528.

Janzen, D. H. 1973. Host plants as islands II. Competition in evolutionary and contemporary time. *American Naturalist* 107:786–790.

Janzen, D. H. 1979. How to big a fig. *Annual Review of Ecology and Systematics* 10:13–51.

Janzen, D. H. 1979b. New horizons in the biology of plant defenses. pp. 331–350 in G. A. Rosenthal and D. H. Janzen (eds.). *Herbivores: Their Interaction with Secondary Plant Metabolites*. Academic Press, New York.

Janzen, D. H. 1979c. Why fruit rots. *Natural History Magazine* 88(6):60–64.

Janzen, D. H., and P. S. Martin. 1982. Neotropical anachronisms: the fruits the gomphotheres ate. *Science* 215:19–27.

Jarvis, J. V. M. 1981. Eusociality in a mammal: co-operative breeding in naked mole rat colonies. *Science* 212:241–250.

Jarvis, J. U. M., M. J. O'Riain, N. C. Bennett, and P. W. Sherman 1994. Mammalian eusociality: A family affair. *Trends in Ecology and Evolution* 9:47–51.

Jarvis, J. V. M., and J. B. Sale. 1971. Burrowing and burrow patterns of East African mole rats— *Tachyoryctes, Heliophobius*, and *Heterocephalus*. *Journal of Zoology* 163:451–479.

Jeffreys, A. J., V. Wilson, and S. L. Thein. 1985. Individuals-specific "fingerprints" of human DNA. *Nature* 316:76–79.

Jenny, J. 1980. *The Soil Resource: Origin and Behavior*. Springer-Verlag, New York.

Johnson, C. G. 1969. *Migration and Dispersal of Insects by Flight*. Methuen, London.

Johnson, D. M., and P. Stiling. 1995. Host specificity of *Cactoblastis cactorum* Berg, on exotic *Opuntia*—feeding moth, in Florida. *Annals of the Entomological Society of America* (in review).

Johnson, E. 1990. Treating the dirt: Environmental ethics and moral theory. pp. 336–358 in T. Reagan (ed.). *Earthbound, Introductory Essays in Environmental Ethics*. Waveland Press, Inc. Prospect Heights, Illinois.

Johnson, M. S. 1978. The botanical significance of derelict industrial sites in Britain. *Environmental Conservation* 5:223–238.

Johnson, M. S., P. D. Putwain, and R. J. Holliday. 1978. Wildlife conservation values of derelict metalliferous mine workings in Wales. *Biological Conservation* 14:131–148.

Jones, D. A. 1973. Co-evolution and cyanogenesis. pp. 213–242 in V. H. Heywood (ed.), *Taxonomy and Ecology*. Academic Press, New York.

Jones, D. F. 1924. The attainment of homozygosity in inbred strains of maize. *Genetics* 9:405–418.

Jones, J. S., B. H. Leith, and P. Rawllings. 1977. Polymorphism in *Cepaea*: A problem with too many solutions. *Annual Review of Ecology and Systematics* 8:109–143.

Jordan, C. F., and J. R. Kline. 1972. Mineral cycling: Some basic concepts and their application in a tropical rain forest. *Annual Review of Ecology and Systematics* 3:33–49.

Jouzel, D. Raymond, J. M. Barnola, J. Chappellaz, J. Delmas, and C. Lourius. 1993. The ice-core record of greenhouse gases. *Science* 259:926–929.

Jukes, T. H. 1983. Evolution of the amino acid code. pp. 191–207 in M. Nei and R. K. Koehn (eds.). *Evolution of Genes and Proteins*. Sinauer Associates, Sunderland, Massachusetts.

Karban, R. 1980. Periodical cicada nymphs impose periodical oak tree wood accumulation. *Nature* 287:326–327.

Karban, R. 1982. Experimental removal of 17-year cicada nymphs and growth of host apple trees. *Journal of the New York Entomological Society* 90:74–81.

Karban, R. 1985. Addition of periodical cicada nymphs to an oak forest: Effects on cicada density, acorn production and rootlet density. *Journal of the Kansas Entomological Society* 58:269–276.

Karban, R. 1987. Effects of clonal variation of the host plant, interspecific competition, and climate on the population size of a folivorous thrips. *Oecologia* 74:298–303.

Karban, R. 1989a. Fine-scale adaptation of herbivorous thrips to individual host plants. *Nature* 340:60–61.

Karban, R. 1989b. Community organization of *Erigeron glaucus* folivores: effects of competition, predation, and host plant. *Ecology* 70:1028–1039.

Karban, R. 1992. Plant variation: Its effects on populations of herbivorous insects. pp. 195–215 in R. S. Fritz and E. L. Simms (eds.). *Plant resistance to herbivores and pathogens: Ecology, Evolution and Genetics*. University of Chicago Press, Chicago.

Karban, R., and J. R. Carey. 1984. Induced resistance of cotton seedlings to mites. *Science* 225:53–54.

Karban, R., and R. E. Ricklefs. 1984. Leaf traits and species richness and abundance of lepidopteran larvae on deciduous trees in southern Ontario. *Oikos* 43:165–170.

Karr, J. R. 1991. Avian survival rates and the extinction process on Barro Colorado Island, Panama. *Conservation Biology* 4:391–397.

Kataev, G., J. Suomela, and P. Palokangas. 1994. Densities of microtine rodents along a pollution gradient from a copper-nickel smelter. *Oecologia* 97:491–498.

Keddy, P. 1990. Is mutualism really irrelevant to ecology? *Bulletin of the Ecological Society of America* 71:101–102.

Keith, L. B. 1983. Role of food in hare population cycles. *Oikos* 40:385–395.

Keller, M. A. 1984. Reassessing evidence for competitive exclusion of introduced natural enemies. *Environmental Entomology* 13:192–195.

Kennedy, G. G., and J. D. Barbour. 1992. Resistance variation in natural and managed systems. pp. 13–41 in R. S. Fritz and E. L. Simms (eds.). *Plant resistance to herbivores and pathogens: Ecology, Evolution and Genetics*. The University of Chicago Press, Chicago.

Kenward, R. E. 1978. Hawks and doves: Factors affecting success and selection in goshawk attacks on wood-pigeons. *Journal of Animal Ecology* 47:449–460.

Kerr, R. A. 1988. Whom to blame for the Great Storm? *Science* 239:1238–1239.

Kethley, J. B., and D. E. Johnston. 1975. Resource tracking patterns in bird and mammal ectoparasites. *Miscellaneous Publications of the Entomological Society of America* 9:231–236.

Kettlewell, H. B. D. 1955. Selection experiments on industrial melanism in the Lepidoptera. *Heredity* 10:287–301.

Kettlewell, H. B. D. 1973. *The Evolution of Melanism*. Clarendon Press, Oxford.

Keyfitz, M. 1975. Population growth: causes and consequences. pp. 39–64 in W. W. Murdoch (ed.). *Environment*, 2nd ed. Sinauer Associates, Sunderland, Massachusetts.

Kilburn, P. D. 1966. Analysis of the species-area relation. *Ecology* 47:831–843.

Kimura, M. 1983a. *The Neutral Theory of Molecular Evolution*. Cambridge University Press, Cambridge.

Kimura, M. 1983b. The neutral theory of molecular evolution. pp. 208–233 in M. Nei and R. K. Koehn (eds.). *Evolution of Genes and Proteins*. Sinauer Associates, Sunderland, Massachusetts.

King, C. E. 1964. Relative abundance of species and MacArthur's Model. *Ecology* 45:716–727.

King, D. A. 1986. Tree form, height growth, and susceptibility to wind damage in *Acer saccharum*. *Ecology* 67:980–990.

Kishk, M. A. 1986. Land degradation in the Nile Valley. *Ambio* 15:226–230.

Klein, D. R. 1968. The introduction, increase, and crash of reindeer on St. Matthew Island. *Journal of Wildlife Management* 32:350–367.

Klomp, H. 1966. The dynamics of a field population of the pine looper, *Bupalus piniarius* (Lepidoptera: Geom.). *Advances in Ecological Research* 3:207–305.

Koblentz-Mishke, I. J., V. V. Volkounsky, R. J. B. Kabanova. 1970. Plankton primary production of the world ocean. pp. 183–193 in W. S. Wooster (ed.). *Scientific Exploration of the South Pacific*. National Academy of Sciences, Washington, D.C.

Koehn, R. , W. J. Diehl, and T. M. Scott. 1988. The differential contribution by individual enzymes of glycolysis and protein catabolism to the relationship between heterozygosity and growth rate in the coot clam *Mulinia lateralis*. *Genetics* 118:121–130.

Koehn, R. K., A. J. Zera, and J. G. Hall. 1983. Enzyme polymorphism and natural selection. pp. 115–136 in M. Nei and R. K. Koehn (eds.). *Evolution of Genes and Proteins*. Sinauer Associates, Sunderland, Massachusetts.

Kozhov, M. 1963. Lake Baikal and its life. *Monographs in Biology* 11:1–344.

Krebs, C. J. 1985a. *Ecology, the Experimental Analysis of Distribution and Abundance*, 3rd ed. Harper and Row, New York.

Krebs, C. J. 1985b. Do changes in spacing behaviour drive population cycles in small mammals? pp. 295–312 in R. M. Sibly and R. H. Smith (eds.). *Behavioural Ecology: Ecological Consequences of Adaptive Behaviour*. Blackwell Scientific Publications, Oxford.

Krebs, C. J. 1988. The experimental approach to rodent population dynamics. *Oikos* 52:143–149.

Krebs, J. R., and N. B. Davies. 1993. *An Introduction to Behavioural Ecology*. 3rd ed. Blackwell Scientific Publications, Oxford.

Krementz, D. G., M. J. Conroy, J. E. Hines, and H. F. Percival. 1988. The effects of hunting on survival rates of American blank ducks. *Journal of Wildlife Management* 52:214–226.

Kruuk, H. 1964. Predators and anti-predator behaviour of the black headed gull, *Larus ridibundus*. *Behaviour Supplement* 11:1–129.

Kruuk, H., and M. Turner. 1967. Comparative notes on predation by lion, leopard, cheetah and wild dog in the Serengeti area, East Africa. *Mammalia* 31:1–27.

Kucera, C. L. 1981. Grasslands and fire. pp. 9–111 in H. A. Mooney, T. M. Bonnicksen, M. L. Christensen, J. E. Lotan, and W. A. Reiners (eds.) *Fire Regimes and Ecosystem Properties*. U.S.D.A. Forest Service General Technical Report WO–26, Washington, D.C.

Kuris, A. M., A. R. Blaustein, and J. J. Alio. 1980. Hosts as islands. *American Naturalist* 116:570–586.

Kurtén, B. 1963. Return of a lost structure in the evolution of the field dentition. *Societas Scientiarum Fennica Arsbok-Vuosikirja* 26:3–11.

Lack, D. 1968. *Ecological Adaptations for Breeding in Birds*. Methuen, London.

Lacy, R. C. 1987. Loss of genetic diversity from a managed populations: Interacting effects of drift, mutation, immigration, selection and population subdivision. *Conservation Biology* 1:143–158.

Lamarck, J. B. P. de. 1809. *Philosophie Zoologique*. Paris.

Lande, R. 1976. The maintenance of genetic variability by mutation in a polygenic character with linked loci. *Genetic Research* 26:221–235.

Larkin, P. A. 1984. A commentary on environmental impact assessment for large projects affecting lakes and streams. *Canadian Journal of Fisheries and Aquatic Science* 41:1121–1127.

Law, R. 1979. Optional life histories under age-specific predation. *American Naturalist* 114:399–417.

Lawrence, W. F. 1991. Ecological correlates of extinction proneness in Australian tropical rain forest mammals. *Conservation Biology* 5:79–89.

Laws, R. M., I. S. C. Parker, and R. C. B. Johnstone. 1975. *Elephants and Their Habitat*. Clarendon Press, Oxford.

Lawton, J. H. 1984. Non-competitive populations, non-convergent communities, and vacant niches: The herbivores of bracken. pp. 67–100 in D. R. Strong, D. Simberloff, L. G. Abele, and A. B. Thistle (eds.). *Ecological Communities: Conceptual Issues and the Evidence*. Princeton University Press, Princeton.

Lawton, J. H., and S. McNeill. 1979. Between the devil and the deep blue sea: On the problem of being a herbivore. pp. 223–244 in R. M. Anderson, B. D. Taylor, and L. R. Taylor (eds.). *Population Dynamics*. Blackwell Scientific Publications, Oxford.

Lawton, J. H., and D. R. Strong. 1981. Community patterns and competition in folivorous insects. *American Naturalist* 118:317–338. Leader-Williams, N., and S. D. Albon. 1988. Allocation of resources for conservation. *Nature* 336:533–535.

Leader-William, N., S. Dalban, and P. S. M. Berry. 1990. Illegal exploitation of black rhinoceros and elephant populations: Patterns of decline, law enforcement and patrol effort in Luangwa Valley, Sambia. *Journal of Applied Ecology* 27:1055–1087.

Le Boeuf, B. J. 1974. Male-male competition and reproductive success in elephant seals. *American Zoologist* 14:163–176.

Le Boeuf, B. J., and S. Kaza (eds.). 1981. *The Natural History of Año Nuevo*. Boxwood Press, Pacific Grove, California.

Lee, J. A., and B. Greenwood. 1976. The colonization by plants of calcareous wastes from the salt and alkali industry in Cheshire. *Biological Conservation* 1:131–149.

Lees, D. R. 1981. Industrial melanism: Genetic adaptation of animals to air pollution. pp. 129–176 in J. A. Bishop and L. M. Cook (eds.). *Genetic Consequences of Man-made Change*. Academic Press, London.

Lees, D. R. and Creed, E. R. 1975. Industrial melanism in *Biston betularia*: The role of selective predation. *Journal of Animal Ecology* 44:67–83.

Lehman, J. T. 1986. Control of eutrophication in Lake Washington. pp. 301–316 in *Ecological Knowledge and Environmental Problem-Solving, Concepts and Case Studies*. National Academy Press, Washington, D.C.

Leibold, M. A. 1989. Resource edibility and the effects of predators and productivity on the outcome of trophic interactions. *American Naturalist* 134:922–949.

Lenski, R. E., and J. E. Mittler. 1993. The directed mutation controversy and Neo-Darwinism. *Science* 259:188–194.

Lerner, I. M. 1954. *Genetic Homeostasis*. Oliver and Boyd, Edinburgh.

Levin, D. A. 1979. The nature of plant species. *Science* 204:381–384.

Levin, D. A. 1981. Dispersal versus gene flow in plants. *Annals of the Missouri Botanical Garden* 68:233–253.

Levin, D. A. 1983. Polyploidy and novelty in flowering plants. *American Naturalist* 122:1–25.

Levin, S. A., 1992. The problem of pattern and scale in ecology. *Ecology* 73:1943–1967.

Levins, R. 1968. *Evolution in Changing Environments*. Princeton University Press, Princeton, New Jersey.

Lewin, R. 1983. Santa Rosalia was a goat. *Science* 221:636–639.

Lewin, R. 1986. In ecology, change brings stability. *Science* 234:1071–3.

Lewis, J. G. 1977. Game domestication for animal production in Kenya: Activity patterns of eland, oryx, buffalo and zebu cattle. *Journal of Agricultural Science*, Cambridge, 89:551–563.

Lewis, J. G. 1978. Game domestication for animal production in Kenya: Shade behavior and factors affecting the herding of eland, oryx, buffalo and zebu cattle. *Journal of Agricultural Science*, Cambridge, 90:587–595.

Lewontin, R. C. 1974. The analysis of variance and the analysis of causes. *American Journal of Human Genetics* 26:400–411.

Lewontin, R. C., and J. L. Hubby. 1966. A molecular approach to the study of genic heterozygosity in natural populations. II. Amount of variation and degree of heterozygosity in natural populations of *Drosophila pseudoobscura*. *Genetics* 54:595–609.

Lieberei, R., B. Biehl, A. Giesemann, and N. T. V. Junqueira. 1989. Cyanogenesis inhibits active defense reactions in plants. *Plant Physiology* 90:33–36.

Liebert, T. G., and P. M. Brakefield. 1987. Behavioral studies on the peppered moth *Biston betularia* and a discussion of the role of pollution and epiphytes in industrial melanism. *Biological Journal of the Linnean Society* 31:129–150.

Lieth, H. 1975. Primary productivity in ecosystems: Comparative analysis of global patterns. pp. 67–88 in W. H. van Dobben and R. H. Lowe-McConnell (eds.). *Unifying Concepts in Ecology*. Dr. W. Junk, The Hague.

Lindeman, R. L. 1942. The trophic-dynamic aspect of ecology. *Ecology* 23:399–418.

Lindow, S. E. 1985. Ecology of *Pseudomonas syringae* relevant to the field use of ice-deletion mutants constructed in vitro for plant frost control. pp. 23–35 in H. O. Halvorson, D. Pramer, and M. Rogul (eds.). *Engineered Organisms in the Environment: Scientific Issues*. American Society for Microbiology, Washington, D.C.

Lloyd, J. E. 1975. Aggressive mimicry in *Photuris* fireflies: Signal repertoires by femme fatales. *Science* 187:452–453.

Loaker, C., and H. Damman. 1991. Nitrogen content of food plants and vulnerability of *Pieris rapae* to natural enemies. *Ecology* 72:1586–1590.

Loman, J., and T. von Schantz. 1991. Birds in a farmland—more species in small than in large habitat island. *Conservation Biology* 5:176–188.

Losos, J. B., S. Naeem, and R. K. Colwell. 1989. Hutchinsonian ratios and statistical power. *Evolution* 43:1820–1826.

Lotka, A. J. 1925. *Elements of Physical Biology*. Reprinted in 1956 by Dover Publications, New York.

Lovelock, J. E. 1979. *Gaia: A New Look at Life on Earth*. Oxford University Press, Oxford.

Lucas, G., and H. Synge. 1978. *Red Data Book*. International Union for the Conservation of Nature and Natural Resources, Marges, Switzerland.

Luck, R. F., and H. Podoler. 1985. Competitive exclusions of *Aphytis lignanensis* by *A. melinus:* potential role of host size. *Ecology* 66:904–913.

Luria, S. E., and M. Delbruck. 1943. Of bacteria from virus sensitivity to virus resistance. *Genetics* 28:491–511.

Lyell, C. 1969. *Principles of Geology*. 3 volumes. Reprint of 1830-1833 ed. Introduction by M. J. S. Rudwick. Johnson Reprint Collection, New York.

Lynch, J. M. 1990. *The Rhizosphere*. Wiley, New York.

Mabey, R. 1980. *The Common Ground: A Place for Nature in Britain's Future?* Hutchinson, London.

MacArthur, R. H. 1955. Fluctuations of animal populations, and a measure of community stability. *Ecology* 36:538–536.

MacArthur, R. H., and J. W. MacArthur. 1961. On bird species diversity. *Ecology* 42:594–598.

MacArthur, R. H. 1957. On the relative abundance of bird species. *Proceedings of the National Academy of Science, USA* 43:293–5.

MacArthur, R. H. 1960. On the relative abundance of species. *American Naturalist* 94:25–36.

MacArthur, R. H., and J. W. MacArthur. 1961. On bird species diversity. *Ecology* 42:594–598.

MacArthur, R. H., and E. O. Wilson. 1963. An equilibrium theory of insular biogeography. *Evolution* 17:373–387.

MacArthur, R. H., and E. O. Wilson. 1967. *The Theory of Island Biogeography*. Princeton University Press, Princeton, New Jersey.

MacDonald, I. A., L. L. Loope, M. B. Usher, and O. Hamann. 1989. Wildlife conservation and the invasion of nature reserves by introduced species: A global perspective. pp. 215–255 in J. A. Drake, H. A. Mooney, F. di Castri, R. H. Groves, F. J. Kruger, M. Rejmanek, and M. Williamson (eds.). *Biological Invasions: A Global Perspective*. Wiley, Chichester, England.

MacEwen, A., and M. MacEwen. 1983. National parks: A cosmetic conservation system. pp. 391–409 in A. Warren and F. B. Goldsmith (eds.). *Conservation in Perspective*. Wiley, Chichester, England.

Maddux, G. D., and N. Cappuccino. 1986. Genetic determination of plant susceptibility to an herbivorous insect depends on environmental context. *Evolution* 40:863–866.

Magurran, A. E. 1988. *Ecological Diversity and Its Management*. Princeton University Press, Princeton.

Mahoney, M. J. 1977. Publication prejudices: An experimental study of confirmatory bias in the peer review system. *Cognitive Therapy and Research* 1:161–175.

Maiorana, V. C. 1978. An explanation of ecological and developmental constants. *Nature* 273:375–377.

Mallet, J., J. T. Longino, D. Murawski, A. Murawski, and A. Simpson de Gamboa. 1987. Handling effects in *Heliconius*: where do all the butterflies go? *Journal of Animal Ecology* 56:377–386.

Malthus, T. R. 1798. *An Essay on the Principle of Population, as It Affects the Future Improvement of Society, with Remarks on the Speculations of Mr. Godwin, M. Condorcet and Other Writers*. J. Johnson, London.

Margalef, R. 1969. Diversity and Stability: A practical proposal and a model of interdependence. Brookhaven Symposium of Biology 22:25–37.

Marquis, R. J., and C. J. Whelan. 1994. Insectivorous birds increase growth of white oak through consumption of leaf-chewing insects. *Ecology* 75:2007–2014.

Martin, F. W., R. S. Pospahala, and J. D. Nichols. 1979. Assessment and population management of North American migratory birds. pp. 187–239 in J. Cairns, Jr., G. P. Patil, and W. E. Waters (eds.). *Environmental Biomonitoring, Assessment, Prediction and Management—Certain Case Studies and Related Quantitative Issues*. International Co-operative Publishing House, Fairland, Md.

Martin, P. S. 1967. Prehistoric overkill. pp. 75–120 in P. S. Martin and H. E. Wright (eds.). *Pleistocene Extinctions*. Yale University Press, New Haven, Connecticut.

Martin, P. S. 1982. The pattern and meaning of Holarctic mammoth extinction. pp. 399–408 in D. M. Hopkins, J. V. Matthews, C. S. Schweger, and S. B. Young (eds.), *Paleoecology of Beringia*. Academic Press, New York.

Martin, P. S., and R. G. Klein. 1984. *Pleistocene Extinctions*. University of Arizona Press, Tucson.

Martinat, P. J. 1987. The role of climatic variation and weather in forest insect outbreaks. pp. 241–268 in P. Barbosa and J. C. Schultz (eds.). *Insect Outbreaks*. Academic Press, San Diego.

Martinez, N. D. 1992. Constant connectance in community food webs. *The American Naturalist* 139:1208–1218.

Massey, A. B. 1925. Antagonism of the walnuts (*Juglans nigra* L. and *J. cinerea* L.) in certain plant associations. *Phytopathology* 15:773–784.

Mattson, W. J. 1980. Herbivory in relation to plant nitrogen content. *Annual Review of Ecology and Systematics* 11:119–161.

Mattson, W. J., and N. D. Addy. 1975. Phytophagous insects as regulators of forest primary production. *Science* 190:515–522.

Mauffette, Y., and W. C. Oechel. 1989. Seasonal variation in leaf chemistry of the coast live oak *Quercus agrifolia* and implications for the California oak moth *Phrygaridia californica*. *Oecologia* 79:439–445.

May, R. M. 1973. *Stability and Complexity in Model Ecosystems*. Princeton University Press, Princeton, New Jersey.

May, R. M. 1975. Patterns of species abundance and diversity. pp. 81–120 in M. L. Cody and J. Diamond (ed.). *Ecology and Evolution of Communities*. Harvard University Press, Cambridge, Massachusetts.

May, R. M. 1976a. Models for single populations. pp. 4–25 in R. M. May (ed.). *Theoretical Ecology: Principles and Applications*. Saunders, Philadelphia.

May, R. M. 1976b. Models for two interacting populations. pp. 49–70 in R. M. May (ed.). *Theoretical Ecology: Principles and Applications*. Saunders, Philadelphia.

May, R. M. 1977. Food lost to pests. *Nature* 267:669–670.

May, R. M. 1978. The dynamics and diversity of insect faunas. pp. 188–204 in L. A. Mound and N. Waloff (eds.). *Diversity of Insect Faunas*. Blackwell Scientific Publications, Oxford.

May, R. M. 1979. Fluctuations in abundance of tropical insects. *Nature* 278:505–507.

May, R. M. 1980. Mathematical models in whaling and fisheries management. American Mathematics Society Lectures. *Mathematics in the Life Sciences* 13:1–64.

May, R. M. 1981. Population biology of parasitic infections. pp. 208–235 in K. S. Warren and E. F. Purcell (eds.). *The Current Status and Future of Parasitology*. Josiah Macy, Jr., Foundation, New York.

May, R. M. 1983. Parasitic infections as regulators of animal populations. *American Scientist* 71:36–45.

May, R. M. 1986. Experimental control of malaria in west Africa. pp. 190–204 in *Ecological Knowledge and Environmental Problem-Solving, Concepts and Case Studies*. National Academy Press, Washington, D.C.

May, R. M. 1990. Taxonomy as destiny. *Nature* 347:129–130.

May, R. M., and R. M. Anderson. 1979. Population biology of infectious diseases. *Nature* 280:455–461.

May, R. M., and R. H. MacArthur. 1972. Niche overlap as a function of environmental viability. *Proceedings of the National Academy of Sciences of the United States of America* 69:1109–1113.

Maynard Smith, J. 1968. *Mathematical Ideas in Biology*. Cambridge University Press, New York.

Maynard Smith, J. 1976. Group selection. *Quarterly Review of Biology* 51:277–283.

Maynard Smith, J. 1978. The ecology of sex. pp. 159–179 in J. R. Krebs and N. B. Davies (eds.). *Behavioural Ecology, an Evolutionary Approach*. Sinauer Associates, Sunderland, Massachusetts.

Mayr, E. 1942. *Systematics and the Origin of Species*. Columbia University Press, New York.

Mayr, E. 1963. *Animal Species and Evolution*. Harvard University Press, Cambridge, Massachusetts.

McCoy, E. D., and J. R. Rey. 1987. Terrestrial arthropods on northwest Florida saltmarshes: Hymenoptera (Insecta). *Florida Entomologist* ,70:90–97.

McIntosh, R. P. 1967. The continuum concept of vegetation. *Botanical Review* 33: 130–187.

McIntosh, R. P. 1987. Pluralism in ecology. *Annual Review of Ecology and Systematics* 18:321–341.

McKaye, K. R., S. M. Louda, and J. R. Stauffer. 1990. Bower size and male reproductive success in a chichlid fish lek. *American Naturalist* 135:597–613.

McKey, D. 1979. The distribution of secondary compounds within plants. pp. 55–133 in G. A. Rosenthal and D. H. Janzen (eds.). *Herbivores: Their Interaction with Secondary Plant Metabolites.* Academic Press, New York.

McKitrick, M. C., and R. M. Zink. 1988. Species concepts in ornithology. *Condor* 90:1–14.

McLaren, B. E., and R. C. Peterson. 1994. Wolves, moose and tree rings on Isle Royale. *Science* 266:1555–1558.

McNaughton, S. J. 1986. On plants and herbivores. *American Naturalist* 128:765–770.

McNaughton, S. J. 1988. Mineral nutrition and spatial concentrations of African ungulates. *Nature* 334:343–345.

McNaughton, S. J., M. Oesterheld, D. A. Frank, and K. J. Williams. 1989. Ecosystem-level patterns of primary productivity and herbivory in terrestrial habitats. *Nature* 341:142–144.

McNaughton, S. J., M. Oesterheld, D. A. Frank, and K. J. Williams. 1991. Primary and secondary production in terrestrial ecosystems. pp. 120–139 in J. Cole, G. Lovett, and S. Findlay (eds.). *Comparative Analyses of Ecosystems: Patterns, Mechanisms, and Theories.* Springer-Verlag, New York.

McQueen, D. J., M. R. S. Johannes, J. R. Post, T. J. Stewart, and D. R. S. Lean. 1989. Bottom-up and top-down impacts on freshwater pelagic community structure. *Ecological Monographs* 59:289–309.

Medvedev, Z. A. 1994. Chernobyl: Eight years after. *Trends in Ecology and Evolution* 10:369–371.

Meffe, G. K., and C. R. Carroll. 1994. *Principles of Conservation Biology.* Sinauer Associates, Inc., Sunderland, Massachusetts.

Mellanby, K. 1967. *Pesticides and Pollution.* Fontana, London.

Menger B. A., and T. M. Farrell. 1989. Community structure and interaction webs in shallow marine hard-bottom communities: Test of an environmental stress model. *Advances in Ecological Research* 19:189–262.

Menge, B. A. 1995. Indirect effects in marine rocky intertidal interaction webs: patterns and importance. *Ecological Monographs* 65:21–74.

Menge, B. A., and J. P. Sutherland. 1976. Species diversity gradients: Synthesis of the roles of predation, competition, and temporal heterogeneity. *American Naturalist* 110:351–369.

Menges, E. 1990. Population viability analysis for an endangered plant. *Conservation Biology* 4:52–62.

Menkinick, E. F. 1964. A comparison of some species-individuals diversity indices applied to samples of field insects. *Ecology* 45:859–861.

Menzel, D. W., and J. H. Ryther. 1961. Nutrients limiting the production of phytoplankton in the Sargasso Sea, with special reference to iron. *Deep-Sea Research* 7:?276–281.

Meyer, G. A. 1993. A comparison of the impacts of leaf- and sap-feeding insects on growth and allocation of goldenrod. *Ecology* 74:1101–1116.

Miller, D. 1970. Biological control of weeds in New Zealand 1927-48. *New Zealand Department of Scientific and Industrial Research Information Series* 74:1–104.

Milne, A. 1961. Definition of competition among animals. *Society for Experimental Biology (Symposia)* 15:40–61.

Milner-Gulland, E. J., and N. Leader-Williams. 1992. A model of incentives for the illegal exploitation of black rhinos and elephants: Poaching pays in Luangwa Valley, Zambia. *Journal of Applied Ecology* 29:388–401.

Mitchell, G. C. 1986. Vampire bat control in Latin America. pp. 151–164 in *Ecological Knowledge and Environmental Problem-Solving, Concepts and Case Studies.* National Academy Press, Washington, D.C.

Mitchell, J. G. 1994. Uncle Sam's undeclared war against wildlife. *Wildlife Conservation* 97:20–31.

Mittlebach, G. G., C. W. Osenberg, and M. A. Leibold. 1988. Trophic relations and ontogenic niche shifts in aquatic ecosystems. pp. 219–235 in B. Ebenman and L. Person (eds.). Size Structured Populations. Springer-Verlag, Berlin.

Mittlebach, G. W. 1988. Competition among refuging sunfishes and effects of fish density on littoral zone invertebrates. *Ecology* 61:614–623.

Mitter, C., and D. R. Brooks. 1983. Phylogenetic aspects of coevolution. pp. 65–98 in D. J. Futuyma and M. Slatkin (eds.). *Coevolution.* Sinauer Associates, Sunderland, Massachusetts.

Mlot, C. 1993. Predators, prey and natural disasters attract ecologists. *Science* 261:1115.

Moav, R., T. Brody, and G. Hulata. 1978. Genetic improvement of wild fish populations. *Science* 204:1090-1094.

Moffat, A. S. 1994. Theoretical ecology: Winning its spurs in the real world. *Science* 263:1090–1092.

Molineaux, L., and G. Gramiccia (eds.). 1980. *The Garki Project.* World Health Organization, Geneva.

Moon, R. D. 1980. Biological Control through interspecific competition. *Environmental Entomology* 9:723–728.

Moore, D. M. 1983. Human impact on island vegetation. pp. 237–248 in W. Holzner, M. J. A. Werger, and I. Ikusima (eds.). *Man's Impact on Vegetation.* Dr. W. Junk, The Hague.

Moore, J. A. 1961. A cellular basis for genetic isolation. pp. 62–68 in W. F. Blair (ed.). *Vertebrate Speciation.* University of Texas Press, Austin.

Moore, J. A. 1985. Science as a way of knowing—human ecology. *American Zoologist* 25:483–637.

Mopper, S., M. Beck, D. Simberloff and P. Stiling. 1995. Local adaptation and agents of selection in a mobile insect. *Evolution* (in press).

Morrell, V. 1993. Australian pest control by virus causes concern. *Science* 261:683–684.

Morris, R. F. 1959. Single-factor analysis in population dynamics. *Ecology* 40:580–588.

Morris, R. F. 1963. The dynamics of epidemic spruce budworm populations. *Memoirs of the Entomological Society of Canada* 31:1–332.

Morris, R. F., and C. A. Miller. 1954. The development of life tables for the spruce budworm. *Canadian Journal of Zoology* 32:283–301.

Morris, W. F. 1992. The effects of natural enemies, competition, and host plant water availability on an aphid population. *Oecologia* 90:359–365.

Morse, D. R., N. E. Stork, and J. H. Lawton. 1988. Species number, species abundance and body length relationships of arboreal beetles in Bornean lowland rain forest trees. *Ecological Entomology* 13:25–37.

Morton, N. E., J. F. Crow, and H. J. Muller. 1956. An estimate of the mutational damage in man from data on consanguineous marriages. *Proceedings of the National Academy of Sciences of the United States of America* 42:855–863.

Mosby, H. S. 1969. The influence of hunting on the population dynamics of a woodlot gray squirrel population. *Journal of Wildlife Management* 33:59–73.

Muller, C. H. 1966. The role of chemical inhibition (allelopathy) in vegetational composition. *Bulletin of the Torrey Botanical Club* 93:332–351.

Muller, C. H. 1970. Phytotoxins as plant habitat variables. *Recent Advances in Phytochemistry* 3:105–121.

Muller, F. 1879. *Ituna* and *Thyridis*, a remarkable case of mimicry in butterflies (translated from the German by R. Meldola). *Proceedings of the Entomological Society of London* 27:20–29.

Mumme, R. L. 1992. Do helpers increase reproductive success? An experimental analysis in the Florida scrub jay. *Behavioral Ecology and Sociobiology* 31:319–328.

Munroe, E. G. 1948. The geographical distribution of butterflies in the West Indies. Dissertation, Cornell University, Ithaca, New York.

Murdoch, W. W. 1975. Diversity, complexity, stability and pest control. *Journal of Applied Ecology* 12:795–807.

Murie, A. 1944. *Wolves of Mount McKinley*. Fauna of National Parks, U.S. Fauna Series Number 5, Washington, D.C.

Murphy, P. G., R. R. Sharitz, and A. J. Murphy. 1977. Response of a forest ecotone to ionizing radiation. pp. 43–48 in J. Zavitkovski (ed.). *The Enterprise, Wisconsin, Radiation Forest*. USERDATID–26113–p2. U.S. Energy Research and Development Administration, Washington, D.C.

Murray, B. G., Jr. 1986. The structure of theory, and the role of competition in community dynamics. *Oikos* 46:145–158.

Murray, B. G., Jr. 1992. Research methods in physics and biology. *Oikos* 64:594–596.

Murton, R. K., N. J. Westwood, and A. J. Isaacson. 1974. A study of wood-pigeon shooting: The exploitation of a natural animal population. *Journal of Applied Ecology* 11:61–81.

Myers, J. H., and K. S. Williams. 1984. Does tent caterpillar attack reduce the food quality of red alder foliage? *Oecologia* 62:74–79.

Myers, J. H., and K. S. Williams. 1987. Lack of short or long term inducible defenses in the red alder-western test caterpillar system. *Oikos* 48:73–78.

Myers, J. P., P. G. Connors, and F. A. Pitelka. 1981. Optimal territory size and the sanderling: Compromises in a variable environment. pp. 135–158 in A. C. Kamil and T. D. Sargent (eds.). *Foraging Behavior, Ecological, Ethological and Psychological Approaches*. Garland STPM Press, New York.

Myers, N. 1983. *A Wealth of Wild Species: Storehouse for Human Welfare*. Westview Press, Boulder, Colorado.

Myers, N. 1990. *The Gaia Atlas of Future Worlds: Challenge and Opportunity in an Age of Change*. Doubleday, New York.

Naeem, S., L. J. Tompson, S. P. Lawler, J. H. Lawton, and R. M. Woodfin. 1994. Declining biodiversity can alter the performance of ecosystems. *Nature* 368:734–737.

National Academy of Sciences. 1984. *Science and Creationism. A View from the National Academy of Sciences*. National Academy Press, Washington, D.C.

National Research Council. 1986. *Ecological Knowledge and Environmental Problem-Solving*. National Academy Press, Washington, D.C.

Neel, J. V. 1983. Frequency of spontaneous and induced "point" mutations in higher eukaryotes. *Journal of Heredity* 74:2–15.

Nei, M. 1975. *Molecular Population Genetics and Evolution*. North-Holland, Amsterdam.

Nei, M. 1983. Genetic polymorphism and the role of mutation in evolution. pp. 165–190 in M. Nei and R. K. Koehn (eds.). *Evolution of Genes and Proteins*. Sinauer Associates, Sunderland, Massachusetts.

Neil, M. 1983. Genetic polymorphism and the role of mutation in evolution. pp. 165–190 in M. Nei and R. K. Koehn (eds.), *Evolution of Genes and Proteins*. Sinauer Associates, Sunderland, Massachusetts.

Neill, S. R. St. J., and J. M. Cullen. 1974. Experiments on whether schooling by their prey affects the hunting behaviour of cephalopods and fish predators. *Journal of the Zoological Society of London* 172:549–569.

Newsome, A. 1990. The control of vertebrate pests by vertebrate predators. *Trends in Ecology and Evolution* 5:187–191.

Newsome, A. E., I Parer, and P. C. Catling. 1989. Prolonged prey suppression by carnivores-predator-removal experiments. *Oecologia* 78:458–467.

Newton, I. and I. Wyllie. 1992. Recovery of a sparrowhawk population in relation to declining pesticide contamination. *Journal of Applied Ecology* 29:476–484.

Ngugi, A. W. 1988. Cultural aspects of fuelwood shortage in the Kenyan highlands. *Journal of Biogeography* 15:165–170.

Nichols, J. P., M. J. Conroy, D. R. Anderson, and K. P. Burnham. 1984. Compensatory mortality in waterfowl populations: A review of the evidence and implications for research and management. *Transactions of the North American Wildlife and Natural Resources Conference* 49:535–554.

Nicholson, A. J. 1933. The balance of animal populations. *Journal of Animal Ecology* 2:132–178.

Nicholson, A. J., and V. A. Bailey. 1935. The balance of animal populations. Part I. *Proceedings of the Zoological Society of London* 1935:551–598.

Niemelä, P., J. Tuomi, R. Mannila, and P. Ojala. 1984. The effect of previous damage on the quality of Scots pine foliage as food for dipronid sawflies. *Zeitschift angewandte Entomologie* 98:33–43.

Niklas, K. J. 1986. Large-scale changes in animal and plant terrestrial communities. pp. 383–405 in D. M. Raup and D. Jablonski (eds.). *Patterns and Processes in the History of Life*. Springer-Verlag, Berlin.

Niklas, K. J., B. H. Tiffney, and A. H. Knoll. 1980. Apparent changes in the diversity of fossil plants. *Evolutionary Biology* 12:1–89.

Nilsson, S. G., and U. Wästljung. 1987. Seed predation and cross-pollination in mast-seeding beech (*Fagus sylvatica*) patches. *Ecology* 68:260–265.

Noble, I. R. 1981. Predicting successional change. pp. 278–300 in H. A. Mooney (ed.). *Fire Regimes and Ecosystem Properties*. U.S. Dept. Agriculture Forest Service, General Technical Report WO–26.

Numbers, R. L. 1982. Creationism in 20th-century America. *Science* 218:538–544.

O'Brien, S. J. and nine others. 1985. Genetic basis for species vulnerability in the cheetah. *Science* 227:1428–1434.

O'Brien, S. J., D. E., Wildt, D. Goldman, C. R. Merril and M. Bush. 1983. The cheetah is depauperate in genetic variation. *Science* 221:459–462.

Odum, E. P. 1957. The ecosystem approach in the teaching of ecology illustrated with simple class data. *Ecology* 38:531–535.

Odum, E. P. 1959. *Fundamentals of Ecology,* 2nd ed. Saunders, Philadelphia.

Odum, E. P. 1969. The strategy of ecosystem development. *Science* 164:262–270.

Odum, E. P. 1971. *Fundamentals of Ecology,* 3rd edition. Saunders, Philadelphia.

Odum, H. T., J. E. Cantlon, and L. S. Kornicher. 1960. An organization hierarchy postulate for the interpretation of species-individual distributions, species entropy, ecosystem evolution, and the meaning of a species-variety index. *Ecology* 41:395–399.

Odum, H. T., P. Murphy, G. Drewry, F. McCormick, C. Schinan, E. Morales, and J. A. McIntyre. 1970. Effects of gamma radiation on the forest at El Verde. pp. D–3–D–75 in H. T. Odum and R. F. Pigeon (eds.). *A Tropical Rain Forest.* U.S. Atomic Energy Commission, Washington, D.C.

Oksanen, L., S. D. Fretwell, J. Arruda, and P. Niemelä. 1981. Exploitation ecosystems in gradients of primary productivity. *American Naturalist* 118:240–261.

Olson, J. S. 1958. Rates of succession and soil changes on southern Lake Michigan sand dunes. *Botanical Gazette* 119:125–170.

Onuf, C. P. 1978. Nutritive value as a factor in plant-insect interactions with an emphasis on field studies. pp. 85–96 in G. G. Montgomery (ed.). *The Ecology of Arboreal Folivores.* Smithsonian Press, Washington, D.C.

Onuf, C. P., J. M. Teal, and I. Valiela. 1977. Interactions of nutrients, plant growth and herbivory in a mangrove ecosystem. *Ecology* 58:514–526.

Opler, P. A. 1974. Oaks as evolutionary islands for leaf-mining insects. *American Scientist* 62:67–73.

Orians, G. H. 1969. The number of bird species in some tropical forests. *Ecology* 50:783–797.

Orians, G., 1975. Diversity, stability and maturity in natural ecosystems. pp. 139–158 in W. H. van Dobben and R. H. Lowe-McConnell. *Unifying Concepts in Ecology.* Dr. W. Jank, The Hague.

Osmond, C. H., and J. Monro. 1981. Prickly pear. pp. 194–222 in D. J. Carr and S. G. M. Carr (eds.). *Plants and Man in Australia.* Academic Press, New York.

Oster, G. F., and E. O. Wilson. 1978. *Caste and Ecology in the Social Insects. Monographs in Population Biology No. 12.* Princeton University Press, Princeton, New Jersey.

Owen, D. F. 1966. Polymorphism in Pleistocene land snails. *Science* 152:71–72.

Owen, D. F. 1980. *Camouflage and Mimicry.* Oxford University Press, Oxford.

Owen, D. F. 1961. Industrial melanism in North American moths. *American Naturalist* 95:227–233.

Owen, D. F. 1962. The evolution of melanism in six species of North American geometrid moths. *Annals of the Entomological Society of America* 55:699–703.

Owen, D. F. 1980. *Camouflage and Mimicry.* Oxford University Press. Oxford.

Owen, D. F., and D. L. Whiteley. 1986. Reflexive selection: Moment's hypothesis resurrected. *Oikos* 47:117–120.

Owen, D. F., and R. G. Wiegert. 1987. Leaf eating as mutualism. pp. 81–95 in P. Barbosa and J. C. Schultz (eds.). *Insect Outbreaks.* Academic Press, San Diego, California.

Owen, O. S. 1975. *Natural Resource Conservation, an Ecological Approach,* 2nd ed. Macmillan, New York.

Pacala, S. W., M. P. Hassell, and R. M. May. 1990. Host-parasitoid associations in patchy environments. *Nature* 344:150–153.

Packer, C. 1977. Reciprocal altruism in *Papio anubis*. *Nature* 265:441–443.

Packer, C., D. A. Gilbert, A. E. Pusey, and S. J. O'Brien. 1991. A molecular genetic analysis of kinship and cooperation in African lions. *Nature* 351:562–565.

Paine, R. T. 1966. Food web complexity and species diversity. *American Naturalist* 100:65-75.

Paine, R. T. 1988. Food webs: road maps of interactions of grist for theoretical development? *Ecology* 69:1648–1654.

Paine, R. T. 1991. Between sylla and charybdis: Do some kinds of criticism merit a response? *Oikos* 62:90–92.

Paine, R. T. 1992. Food-web analysis through field measurement of per capita interaction strength. *Nature* 355:73–75.

Paine, R. T., and S. A. Levin. 1981. Intertidal landscapes: Disturbance and the dynamics of pattern. *Ecological Monographs* 51:145–178.

Painter, R. H. 1951. *Insect Resistance in Crop Plants*. MacMillan, New York.

Park, T. 1948. Experimental studies of interspecies competition. I. Competition between populations of the flour beetles *Tribolium confusum* Duval and *Tribolium castaneum* Herbst. *Ecological Monographs* 18:265–307.

Park, T. 1954. Experimental studies of interspecies competition. II. Temperature, humidity, and competition in two species of *Tribolium*. *Physiological Zoology* 27:177–238.

Park, T., P. H. Leslie, and D. B. Metz. 1964. Genetic strains and competition in populations of *Tribolium*. *Physiological Zoology* 37:97–162.

Parry, G. D. 1981. The meanings of *r*- and *K*-selection. *Oecologia* 48:260–264.

Parsons, K. A., and A. A. de la Cruz. 1980. Energy flow and grazing behavior of canocephaline grasshoppers in a *Juncus roemerianus* marsh. *Ecology* 61:1045–1050.

Payne, I. 1987. A lake perched on piscine peril. *New Scientist* 115(1575):50–54.

Payne, R. 1973. Decline of whales. pp. 143 in W. Jackson (ed.). *Man and the Environment*, 2nd ed. William C. Brown Company, Dubuque, Iowa.

Peabody, R. R. 1931. *The Common Sense of Drinking*. Little, Brown, Boston.

Peakall, R., A. J. Beattie, and S. J. James. 1987. Pseudocopulation of an orchid by male ants: A test of two hypotheses accounting for the rarity of ant pollination. *Oecologia* 78:522–524.

Pearce, F. 1989. Kill or cure? Remedies for the rainforest. *New Scientist* 1682:40–43.

Pearl, R. 1928. *The Rate of Living*. Knopf, New York.

Pearl, R., and L. J. Reed. 1920. On the rate of growth of the population of the United States since 1790 and its mathematical representation. *Proceedings of the National Academy of Sciences of the United States of America* 6:275–288.

Pearsall, W. H. 1954. Biology and land use in East Africa. *New Biology* 17:9–26.

Peet, R. K. 1974. The measurement of species diversity. *Annual Review of Ecology and Systematics* 5:285–307.

Persson, L. 1985. Asymmetrical competition: Are larger animals competitively superior? *American Naturalist* 126:261–266.

Peters, C. M., A. H. Gentry, and R. O. Mendelsohn. 1989. Valuation of an Amazonian rainforest. *Nature* 339:655–656.

Petts, G. E. 1985. *Impounded Rivers: Perspectives for Ecological Management.* Wiley, Chichester, England.

Pianka, E. R. 1970. On *r-* and *K*-selection. *American Naturalist* 104:592–597.

Pianka, E. R. 1976. Competition and niche theory. pp. 114–141 in R. M. May (ed.). *Theoretical Ecology: Principles and Applications.* Blackwell Scientific Publications, Oxford.

Pilgram, T., and D. Western. 1986. Inferring hunting patterns on African elephants from tusks in the international ivory trade. *Journal of Applied Ecology* 23:503–514.

Pimentel, D. 1986. Biological invasions of plants and animals in agriculture and forestry. pp. 149–162 in H. A. Mooney and J. A. Drake (eds.). *Ecology of Biological Invasions of North America and Hawaii.* Springer-Verlag, New York.

Pimentel, D. 1988. Herbivore population feeding pressure on plant hosts: Feedback evolution and host conservation. *Oikos* 53:289–302.

Pimentel, D., D. Andow, R. Dyson-Hudson, D. Gallahan, S. Jacobson, M. Irish, G. Kroop, A. Moss, I. Schreiner, M. Shepard, T. Thompson, and B. Vinzant. 1980. Environmental and social costs of pesticides: A preliminary assessment. *Oikos* 34:126–140.

Pimm, S. L. 1979. The structure of food webs. *Theoretical Population Ecology* 16:144–158.

Pimm, S. L. 1980. Food web design and the effect of species deletion. *Oikos* 35:139–147.

Pimm, S. L. 1982. *Food Webs.* Chapman and Hall, London.

Pimm, S. L. 1984. Food chains and return times. pp. 397–412 in D. R. Strong, D. Simberloff, L. G. Abele, and A. B. Thistle (eds.). *Ecological Communities: Conceptual Issues and the Evidence.* Princeton University Press, Princeton, New Jersey.

Pimm, S. L., and R. L. Kitching. 1987. The determinants of food chain lengths. *Oikos* 50:302–307.

Pimm, S. L., J. H. Lawton, and J. E. Cohen. 1991. Food web patterns and their consequences. *Nature* 350:669–674.

Pimm, S. L., and A. Redfearn. 1988. The variability of population densities. *Nature* 334:613–614.

Pirie, N. W. 1969. *Food Resources, Conventional and Novel.* Penguin, Harmondsworth.

Pitelka, F. A. 1964. The nutrient-recovery hypothesis for Arctic microtine cycles. I. Introduction. pp. 55–56 in D. J. Crisp (ed.). *Grazing in Terrestrial and Marine Environments.* Blackwell Scientific Publications, Oxford.

Platt, A. W. 1941. The influence of some environmental factors on the expressions of the solid stem character in certain wheat varieties. *Scientific Agriculture* 22:216–223.

Podoler, H., and D. Rogers. 1975. A new method for the identification of key factors from life-table data. *Journal of Animal Ecology* 44:85–115.

Polis, G. A. 1991. Complex trophic interactions in deserts: an empirical critique of food-web theory. *The American Naturalist* 138:123–155.

Pollard, A. J. 1992. The importance of deterrence: Responses of grazing animals to plant variation. pp. 216–239 in R. S. Fritz and E. L. Simms (eds.). Plant resistance to herbivores and pathogens: *Ecology, Evolution, and Genetics.* The University of Chicago Press, Chicago.

Popper, K. R. 1972a. *The Logic of Scientific Discovery*, 3rd ed. Hutchinson, London.

Popper, K. R. 1972b. *Objective Knowledge: An Evolutionary Approach.* Clarendon Press, Oxford.

Portwood, D. 1978. *Commonsense Suicide: The Final Right.* Dodd, Mead, New York.

Potter, D. A., and T. W. Kimmerer. 1988. Do holly leaf spines really deter herbivory? *Oecologia* 75:216–221.

Powell, G. V. M., A. H. Powell, and N. K. Paul. 1988. Brother, can you spare a fish? *Natural History* 97(2):34–39.

Power, M. E. 1984. Depth distributions of armored catfish: predator-induced resource avoidance? *Ecology* 65:523–528.

Power, M. E. 1992. Top-down and bottom-up forces in food webs: Do plants have primacy? *Ecology* 73:733–746.

Prescott-Allen, C., and R. Prescott-Allen. 1986. *The First Resource: Wild Species in the North American Economy.* Yale University Press, New Haven, Connecticut.

Prestidge, R. A., and S. McNeill. 1983. The role of nitrogen in the ecology of grassland Auchenorrhyncha. pp. 257–283 in J. A. Lee, S. McNeil, and I. H. Rorison (eds.). *Nitrogen as an Ecological Factor. 22nd Symposium of the British Ecological Society.* Blackwell Scientific Publications, Oxford.

Preston, F. W. 1948. The commonness and rarity of species. *Ecology* 29:254–283.

Preston, F. W. 1960. Time and space and the variation of species. *Ecology* 41:611–627.

Preston, F. W. 1962. The canonical distribution of commonness and rarity. *Ecology* 43:185–215, 410–432.

Price, P. W. 1970. Characteristics permitting coexistence among parasitoids of a sawfly in Quebec. *Ecology* 51:445–454.

Price, P. W. 1980. *Evolutionary Biology of Parasites.* Princeton University Press, Princeton, New Jersey.

Prins, H. H. T., and I. Douglas-Hamilton. 1990. Stability in a multi-species assemblage of large herbivores in East Africa. *Oecologia* 83:392–400.

Prins, H. H. T., and F. J. Weyerhaeuser. 1987. Epidemics in populations of wild ruminants: Anthrax and impala, rinderpest and buffalo in Lake Manyara National Park, Tanzania. *Oikos* 49:28–38.

Proctor, J., and S. Proctor. 1978. *Color in Plants and Flowers.* Everest, New York.

Pullin, A. S., and J. E. Gilbert. 1989. The stinging nettle, *Urtica dioica*, increases trichome density after herbivore and mechanical damage. *Oikos* 54:275–280.

Pusey, A. E. 1987. Sex-biased dispersal and inbreeding avoidance in birds and mammals. *Trends in Ecology and Evolution* 2:295–300.

Quinn, J. F. 1983. Mass extinctions in the fossil record. *Science* 219:1239–1240.

Quinn, J. F., and A. E. Dunham. 1983. On hypothesis testing in ecology and evolution. *American Naturalist* 122:602–617.

Quinn, J. F., and S. P. Harrison. 1988. Effects of habitat fragmentation and isolation on species richness: Evidence from biogeographic patterns. *Oecologia* 75:132–140.

Rabinowitz, D. 1981. Seven forms of rarity. pp. 205–217 in H. Synge (ed.). *The Biological Aspects of Rare Plant Conservation.* Wiley, London.

Ralls, K., and J. Ballou. 1983. Extinction: lessons from zoos. pp. 164–184 in C. M. Schonewald-Cox, S. M. Chambers, B. MacBryde, and L. Thomas (eds.). *Genetics and Conservation: A Reference for Managing Wild Animal and Plant Populations.* Benjamin/Cummings, Menlo Park, California.

Ratcliffe, D. A. 1980. *The Peregrine Falcon.* Buteo Books, Vermillion, South Dakota.

Raup, D. M. 1962. Computer as aid in describing form in gastropod shells. *Science* 138:150–152.

Raup, D. M. 1966. Geometric analysis of shell coiling: General problems. *Journal of Paleontology* 40:1178–1190.

Raup, D. M. 1979. Biases in the fossil record of species and genera. *Bulletin of the Carnegie Museum of Natural History* 13:85–91.

Raup, D. M. 1984. Evolutionary radiations and extinctions. pp. 5–14 in H. D. Holl and A. F. Trendall (eds.). *Patterns of Change in Earth Evolution.* Springer-Verlag, Berlin.

Raup, D. M., S. J. Gould, T. J. M. Schopf, and D. Simberloff. 1973. Stochastic models of phylogeny and the evolution of diversity. *Journal of Geology* 81:525–542.

Raup, D. M., and J. J. Sepkoski, Jr. 1984. Periodicities of extinctions in the geologic past. *Proceedings of the National Academy of Sciences of the United States of America* 81:801–805.

Rausher, M. D., K. Iwao, E. L. Simms, N. Ohsaki, and D. Hall. 1993. Induced resistance in *Ipomoea purpurea*. *Ecology* 74:20–29.

Reed, J. M., P. D. Doerr, and J. R. Walters. 1988. Minimum viable population size of the red-cockaded woodpecker. *Journal of Wildlife Management* 52:385–391.

Reice, S. R. 1994. Nonequilibrium determinants of biological community structure. *American Scientist* 82:424–435.

Reiners, W. A. 1986. Complementary models for ecosystems. *The American Naturalist* 127:59–73.

Rennie, P. J. 1957. The uptake of nutrients by timber forest and its importance to timber production in Britain. *Quarterly Journal of Forestry* 51:101–115.

Revelle, P., and C. Revelle. 1984. *The Environment, Issues and Choices for Society,* 2nd ed. Willard Grant Press, Boston.

Rey, J. R. 1981. Ecological biogeography of arthropods on *Spartina* islands in northwest Florida. *Ecological Monographs* 51:237–265.

Rey, J. R., E. D. McCoy, and D. R. Strong, Jr. 1981. Herbivore pests, habitat islands and the species-area relation. *American Naturalist* 117:611–622.

Reynolds, S. G. 1988. Some factors of importance in the integration of pastures and cattle with coconuts (*Cocos nucifera*). *Journal of Biogeography* 15:31–39.

Rhoades, D. F. 1979. Evolution of plant chemical defense against herbivores. pp. 3–54 in G. A. Rosenthal and D. H. Janzen (eds.), *Herbivores: Their Interaction with Secondary Plant Metabolites.* Academic Press, New York.

Rhoades, D. F., and R. G. Cates. 1976. Toward a general theory of plant antiherbivore chemistry. *Recent Advances in Phytochemistry* 10:168–213.

Richer, W. E. 1981. Changes in the average size and average age of Pacific salmon. *Canadian Journal of Fisheries and Aquatic Science* 38:1636–1656.

Ricklefs, R. 1990. *Ecology,* 3rd ed. W. H. Freeman, San Francisco.

Ridley, M., and B. Low. 1993. Can selfishness save the environment? *Atlantic Monthly* 272:76–79.

Ringwood, A. E., S. E. Kesson, N. G. Ware, W. Hibberson, and A. Major. 1979. Immobilisation of high level nuclear reactor wastes in SYNROC. *Nature* 278:219–223.

Ritland, D. B. 1994. Variation in palatability of queen butterflies (*Danaus gilippus*) and implications regarding mimicry. *Ecology* 75:732–746.

Ritland, D. B., and L. P. Brower. 1991. The viceroy butterfly is not a batesian mimic. *Nature* 350:497–498.

Roberts, L. 1988. Is their life after climate change? *Science* 242:1010–1012.

Rohde, K. 1982. *Ecology of Marine Parasites.* University of Queensland Press, St. Lucia, Australia.

Rohde, K. 1992. Latitudinal gradients in species diversity: the search for the primary cause. *Oikos* 65:514–527.

Rojas, M. 1992. The species problem and conservation: What are we protecting? *Conservation Biology* 6:170–178.

Roland, J. 1988. Decline of winter moth populations in North America: Direct versus indirect effect of introduced parasites. *Journal of Animal Ecology* 57:523–531.

Rood, J. P. 1978. Dwarf mongoose helpers at the den. *Zeitschrift für Tierpsychologie* 48:277–287.

Rood, J. P. 1990. Group size, survival, reproduction and routes to breeding in dwarf mongooses. *Animal Behavior* 39:566–572.

Room, P. M., K. L. S. Harley, I. W. Forno, and P. P. D. Sands. 1981. Successful biological control of the floating weed *Salvinia. Nature* 294:78–80.

Root, R. 1967. The niche exploitation pattern of the blue-gray gnatcatcher. *Ecological Monographs* 37:317–350.

Root, R. 1973. Organization of a plant-arthropod association in simple and diverse habitats: The fauna of collards (*Brassica oleracea*). *Ecological Monographs* 43:95–124.

Root, T. L. 1993. Effects of global climate change on north American birds and their communities. pp. 280–292 in P. M. Kareiva, J. G. Kingsolver, and R. B. Huey (eds.). *Biotic Interactions and Global Change.* Sinauer Associates, Sunderland, Massachusetts.

Rosenzweig, M. L. 1968. Net primary productivity of terrestrial communities: Prediction from climatological data. *American Naturalist* 102:67–74.

Rosenzweig, M. L. 1971. Paradox of enrichment: Destabilization of exploitation ecosystems in ecological time. *Science* 171:385–387.

Rosenzweig, M. L., and R. H. MacArthur. 1963. Graphical representation and stability conditions of predator-prey interactions. *American Naturalist* 97:209–223.

Rotheray, G. E. 1981. Host searching and oviposition behavior of some parasitoids of aphidophagous Syrphidae. *Ecological Entomology* 6:79–87.

Rotheray, G. E. 1986. Colour, shape and defense in aphidophagous syrphid larvae (Diptera). *Zoological Journal of the Linnaean Society* 88:201–216.

Rotheray, G. E., and P. Barbosa. 1984. Host related factors affecting oviposition behavior in *Brachymeria intermedia. Entomologia Experimentalis et Applicata* 35:141–145.

Roughgarden, J. 1983. Competition and theory in community ecology. *American Naturalist* 122:583–601.

Royal Society Study Group. 1983. *The Nitrogen Cycle of the United Kingdom.* The Royal Society, London.

Ryan, F. J. 1955. Spontaneous mutation in non-dividing bacteria. *Genetics* 40:726–738.

Ryther, J. H. 1963. Geographic variation in productivity. pp. 347–380 in M. N. Hill (ed.). *The Sea*, Vol. 2, 2nd ed. Wiley-Interscience, New York.

Saffo, M. B. 1992. Coming to terms with a field: words and concepts in symbiosis. *Symbiosis* 14:17–31.

Sailer, R. I. 1983. History of insect introductions. pp. 15–38 in C. Graham and C. Wilson (eds.). *Exotic Plant Pests and North American Agriculture*. Academic Press, New York.

Sala, O. S., W. J. Parton, L. A. Joyce, and W. K. Lauenroth. 1988. Primary production of the central grassland region of the United States. *Ecology* 69:40–45.

Salt, G. 1970. *The Cellular Defense Reactions of Insects*. Cambridge University Press, New York.

Sanders, H. L. 1968. Marine benthic diversity: a comparative study. *American Naturalist* 102:243–282.

Sanderson, I. T. 1945. *Living Treasure*. Viking, New York.

Sargent, T. D. 1968. Cryptic moths: effects on background selection of painting the circumocular scales. *Science* 159:100–101.

Schaffer, M. L., and F. BH. Samson. 1985. Population size and extinction: A note on determining critical population sizes. *American Naturalist* 125:144–152.

Scheffer, V. B. 1951. The rise and fall of a reindeer herd. *Scientific Monthly* 73:356–362.

Schemske, D. W. 1980. The evolutionary significance of extrafloral nectar production by *Costus woodsonii* (Zingiberaceae): An experimental analysis of ant protection. *Journal of Ecology* 68:959–967.

Schemske, D. W. 1983. Limits to specialization and coevolution in plant-animal mutualisms. pp. 67–109 in M. H. Nitecki (ed.). *Coevolution*. University of Chicago Press, Chicago.

Schindler, D. W. Eutrophication and recovery in experimental lakes: implications for lake management. *Science* 184:397–399.

Schindler, D. W. 1977. Evolution of phosphorus limitation in lakes. *Science* 195:260–262.

Schmitz, O. J., and T. D. Nudds. 1994. Parasite-mediated competition in deer and moose: How strong is the effect of meningeal worm on moose? *Ecological Applications* 4:91–103.

Schoener, A. 1974. Experimental zoogeography: Colonization of marine mini-islands. *American Naturalist* 108:715–737.

Schoener, T. W. 1974. Resource partitioning in ecological communities. *Science* 185:27–39.

Schoener, T. W. 1976. The species-area relation within archipelagos: models and evidence from island land birds. *Proceeding of the 16th International Ornithology Conference* (Canberra, 1974), pp. 629–642.

Schoener, T. W. 1983. Field experiments on interspecific competition. *American Naturalist* 122:240–285.

Schoener, T. W. 1985. Some comments on Connell's and my reviews of field experiments in interspecific competition. *American Naturalist* 125:730–740.

Schoener, T. W. 1986. Overview: Kinds of ecological communities—ecology becomes pluralistic. pp. 467–479 in J. Diamond and T. J. Case (eds.). *Community Ecology*. Harper and Row, New York.

Schoener, T. W. 1987. The geographical distribution of rarity *Oecologia* 74:161–173.

Schoener, T. W. 1989. Food webs from the small to the large *Ecology* 70:1559–1589.

Schoener, A., E. R. Long, and J. R. DePalma. 1978. Geographic variation in artificial island colonization curves. *Ecology* 59:367–382.

Schoener, T. W., and D. A. Spiller. 1987. Effect of lizards on spider populations: Manipulative reconstruction of a natural experiment. *Science* 236:949–952.

Schrader-Frechette, K. S., and E. D. McCoy. 1993. *Method in Ecology: Strategies for Conservation*. Cambridge University Press, Cambridge.

Schultz, A. M. 1964. The nutrient-recovery hypothesis for Arctic microtine cycles. II. Ecosystem variables in relation to Arctic microtine cycles. pp. 57–68 in D. J. Crisp (ed.). *Grazing in Terrestrial and Marine Environments*. Blackwell Scientific Publications, Oxford.

Schultz, J. C., and I. T. Baldwin. 1982. Oak leaf quality declines in response to defoliation by gypsy moth larvae. *Science* 217:149–151.

Schupp, E. W. 1986. *Azteca* protection of *Cecropia*: Ant occupation benefits juvenile trees. *Oecologia* 70:379–385.

Schutt, D. A. 1976. The effect of plant oestrogens on animal reproduction. *Endeavour* 35:110–113.

Scott, G. R. 1970. Rinderpest. pp. 20–35 in J. W. Davis, L. H. Karstad, and D. O. Trainer (eds.). *Infectious Diseases of Wild Mammals*. Iowa State University Press, Ames, Iowa.

Scriber, J. M., and F. Slansky. 1981. The nutritional ecology of immature insects. *Annual Review of Entomology* 26:183–211.

Seghers, B. H. 1974. Schooling behaviour in the guppy *Poecilia reticulata*: An evolutionary response to predation. *Evolution* 28:486–489.

Selander, R. K. 1976. Genic variation in natural populations. pp. 21–45 in F. J. Ayala (ed.). *Molecular Evolution*. Sinauer Associates, Sunderland, Massachusetts.

Sepkoski, J. J., Jr. 1978. A kinetic model of Phanerozoic taxonomic diversity, I. Analysis of marine orders. *Paleobiology* 4:223–251.

Sepkoski, J. J., Jr. 1979. A kinetic model of Phanerozoic taxonomic diversity. II. Early Phanerozoic families and multiple equilibria. *Paleobiology* 5:222–251.

Sepkoski, J. J., Jr. 1984. A kinetic model of Phanerozoic taxonomic diversity. III. Post-Paleozoic families and mass extinctions. *Paleobiology* 10:246–267.

Shaffer, M. L. 1990. Population viability analysis. *Conservation Biology* 4:39–40.

Shannon, C. E., and W. Weaver. 1949. *The Mathematical Theory of Communication*. University of Illinois Press, Urbana.

Sherlock, R. L. 1922. *Man as a Geological Agent*. Witherby, London.

Sherman, P. W. 1977. Nepotism and the evolution of alarm cells. *Science* 197:1246–1253.

Sibly, R. M. 1983. Optimal group size is unstable. *Animal Behavior* 31:947–948.

Sih, A. 1982. Foraging strategies and the avoidance of predation by an aquatic insect, *Notoneeta hoffmani*. *Ecology* 68:786–796.

Sih, A. 1991. Reflections on the power of a grand paradigm. *Bulletin of the Ecological Society of America* 72:174–178.

Sih, A., P. Crawley, M. McPeek, J. Petranka, and K. Strohmeir. 1985. Predation, competition, and prey communities: A review of field experiments. *Annual Review of Ecology and Systematics*. 16:269–311.

Silvertown, J. W. 1980. The evolutionary ecology of mast seeding in trees. *Biological Journal of the Linnaean Society* 14:235–250.

Simberloff, D. 1972. Properties of rarefaction diversity measures. *American Naturalist* 106:414–415.

Simberloff, D. 1976a. Experimental zoogeography of islands: effects of island size. *Ecology* 57:629–648.

Simberloff, D. 1976b. Species turnover and equilibrium island biogeography. *Science* 194:572–578.

Simberloff, D. 1978. Colonization of islands by insects: Immigration, extinction, and diversity. pp. 139–153 in L. A. Mound and N. Waloff (eds.). *Diversity of Insect Faunas.* Blackwell Scientific Publications, Oxford.

Simberloff, D. 1983. Competition theory, hypothesis-testing, and other community ecological buzz-words. *American Naturalist* 122:626–635.

Simberloff, D. 1986a. Design of natural reserves. pp. 316–337 in M. B. Usher (ed.). *Wildlife Conservation Evaluation.* Chapman and Hall, London.

Simberloff, D. 1986b. The proximate causes of extinction. pp. 259–276 in D. M. Raup and D. Jablonski (eds.). *Patterns and Processes in the History of Life.* Springer-Verlag, Berlin.

Simberloff, D. 1988. The contribution of population and community biology to conservation science. *Annual Review of Ecology and Systematics* 19:473–512.

Simberloff, D. S. 1974. Equilibrium theory of island biogeography and ecology. *Annual Review of Ecology and Systematics* 5:161–179.

Simberloff, D. S. 1978. Using island biographic distributions to determine if colonization is stochastic. *American Naturalist* 112:713–726.

Simberloff, D. 1995. Why do introduced species appear to devastate islands more than mainland areas? *Pacific Science* 49:87–97.

Simberloff, D., and W. J. Boecklen. 1981. Santa Rosalia reconsidered: Size ratios and competition. *Evolution* 35:1206–1228.

Simberloff, D., B. J. Brown, and S. Lowrie. 1978. Isopod and insect root borers may benefit Florida mangroves. *Science* 201:630–632.

Simberloff, D., and T. Dyan. 1991. The guild concept and the structure of ecological communities. *Annual Review of Ecology and Systematics* 22:115–143.

Simberloff, D., J. A. Farr, J. Cox, and D. W. Mehlman. 1993. Movement corridors: Conservation bargains or poor investments? *Conservation Biology* 6:493–504.

Simberloff, D., and E. O. Wilson. 1970. Experimental zoogeography of islands: A two year record of re-colonization. *Ecology* 51:934–937.

Simberloff, D., and E. O. Wilson. 1969. Experimental zoogeography of islands: The colonization of empty islands. *Ecology* 50:278–296.

Simmons, I. G. 1981. *The Ecology of Natural Resources,* 2nd ed. Edward Arnold, London.

Simon, J. L. 1986. Disappearing species, deforestation and data. *New Scientist* 110(1508):60–63.

Simon, J. L. 1990. Population matters, people, resources, environment and immigration. *Transaction Publishers*, New Brunswick, New Jersey.

Simpson, G. G. 1964. Species density of North American Recent mammals. *Systematic Zoology* 13:57–73.

Simpson, G. H. 1949. Measurement of diversity. *Nature* 163:688.

Sinclair, A. R. E. 1986. Testing multi-factor causes of population limitation: An illustration using snowshoe hares. *Oikos* 47:360–364.

Sinclair, A. R. E. , J. M. Gosline, G. Holdsworth, C. J. Krebs, S. Boutin, J. N. M. Smith, R. Boonstra, and M. Dale. 1993. Can the solar cycle and climate synchronize the snowshoe hare cycle in Canada? Evidence from tree rings and ice cones. *American Naturalist* 141:173–198.

Sinclair, A. R. E., and M. Norton-Griffiths. 1979. *Serengeti. Dynamics of an Ecosystem.* University of Chicago Press, Chicago.

Skinner, G. J., and J. B. Whittaker. 1981. An experimental investigation of interrelationships between the wood-ant (*Formica rufa*) and some tree-canopy herbivores. *Journal of Animal Ecology* 50:313–326.

Sláma, K. 1969. Plants as a source of materials with insect hormone activity. *Entomologia Experimentalis et Applicata* 12:721–728.

Slobodkin, L. B. 1986*a*. The role of minimalization in art and science. *American Naturalist* 127:257–265.

Slobodkin. L. B. 1986*b*. Natural philosophy rampant. *Paleobiology* 12:111–118.

Slobodkin, L. B. 1988. Intellectual problems of applied ecology. *BioScience* 38:337-342.

Slobodkin, L. C., F. E. Smith, and N. G. Hairston. 1967. Regulation in terrestrial ecosystems in gradients of primary productivity. *American Naturalist* 118:240–261.

Smallwood, P. D., and W. D. Peters. 1986. Grey squirrel food preferences: the effects of tannin and fat concentration. *Ecology* 67:168–174.

Smiley, J. 1986. Ant constancy at *Passiflora* extrafloral nectaries: Effects on caterpillar survival. *Ecology* 67:516–521.

Smith, J. N. M. 1962. Detoxification mechanisms. *Annual Review of Entomology* 7:465–480.

Smith, J. N. M., C. J. Krebs, A. R. E. Sinclair, and R. Boonstra. 1988. Population biology of snowshoe hares. II. Interactions with winter food plants. *Journal of Animal Ecology* 57:269–286.

Smith, K. L., Jr. 1985. Deep-sea hydrothermal vent mussels: Nutritional state and distribution at the Galápagos rift. *Ecology* 66:1067–1080.

Smith, R. L. 1974. *Ecology and Field Biology*, 2nd ed. Harper and Row, New York.

Sober, E., and R. C. Lewontin. 1982. Artifact, cause and genic selection. *Philosophy of Science* 49:157–180.

Solow, A. R., and J. H. Steele. 1990. On sample size, statistical power, and the detection of density dependence. *Journal of Animal Ecology* 59:1073–1076.

Soulé, M. E. 1980. Thresholds for survival: maintaining fitness and evolutionary potential. pp. 151–170 in M. E. Soulé and B. A. Wilcox (eds.). *Conservation Biology: An Evolutionary-Ecological Perspective.* Sinauer Associates, Sunderland, Massachusetts.

Soulé, M. E., and D. Simberloff. 1986. What do genetics and ecology tell us about the design of nature reserves? *Biological Conservation* 35:19-40.

Soulé, M. E., and B. A. Wilcox (eds.). 1980. *Conservation Biology.* Sinauer Associates, Sunderland, Massachusetts.

Sousa, W. P. 1979. Disturbance in marine intertidal boulder fields: The nonequilibrium maintenance of species diversity. *Ecology* 60:1225–1239.

Southwick, C. H. (ed.). 1985. *Global Ecology.* Sinauer Associates, Sunderland, Massachusetts.

Southwood, T. R. E. 1961. The number of species of insect associated with various tree. *Journal of Animal Ecology* 30:1–8.

Southwood, T. R. E. 1976. Bionomic strategies and population parameters. pp. 26–48 in *Theoretical Ecology, Principles and Applications.* Saunders, Philadelphia.

Southwood, T. R. E. 1977. The relevance of population dynamic theory to pest status. pp. 35–54 in J. M. Cherrett and G. R. Sagor (eds.). *Origins of Pest, Parasite, Disease and Weed Problems*. Blackwell Scientific Publications, Oxford.

Southwood, T. R. E. 1978. *Ecological Methods with Particular Reference to the Study of Insect Populations*, 2nd ed. Methuen, London.

Sparks, J. 1982. *Discovering Animal Behaviour*. BBC Publications, London.

Spear, L. 1988. A naturalist at large, the halloween mask episode. *Natural History* 97(6):4–8.

Spencer, W. P. 1957. Genetic studies on *Drosophila mulleri*. I. Genetic analysis of a population. *Texas University Publication* 5721:186–205.

Stangel, P. W., M. R. Lennartz, and M. H. Smith. 1992. Genetic variation and the population structure of red-cockaded woodpeckers. *Conservation Biology* 6:283–292.

Stanley, S. M. 1975. A theory of evolution above the species level. *Proceedings of the National Academy of Sciences of the United States of America* 72:646–650.

Stanley, S. M. 1979. *Macroevolution: Pattern and Process*. Freeman, San Francisco.

Stark, R. W. 1964. Recent trends in forest entomology. *Annual Review of Entomology* 10:303–324.

Stebbins, G. L. 1950. *Variation and Evolution in Plants*. Columbia University Press, New York.

Stebbins, G. L. 1974. *Flowering Plants: Evolution Above the Species Level*. Harvard University Press, Cambridge, Massachusetts.

Steele, J. H. 1974. *The Structure of Marine Ecosystems*. Blackwell Scientific Publications, Oxford.

Steele, J. H., and E. W. Henderson. 1984. Modeling long-term fluctuations in fish stocks. *Science* 224:985–987.

Stenseth, N. C. 1983. Causes and consequences of dispersal in small mammals. pp. 63–101 in I. R. Swingland and P. J. Greenwood (eds.). *The Ecology of Animal Movement*. Oxford University Press, Oxford.

Stephens, D. W., and J. R. Krebs. 1986. *Foraging Theory*. Princeton University Press, Princeton.

Stern, C. 1973. *Principles of Human Genetics*, 3rd ed. Freeman, San Francisco.

Sternberg, J. G., G. P. Waldbauer, and M. R. Jeffords. 1977. Batesian mimicry: Selective advantage of color pattern. *Science* 195:681–683.

Stevenson-Hamilton, J. 1957. Tsetse fly and the rinderpest epidemic of 1896. *South African Journal of Science* 58:216.

Stewart, A. J. A. 1986a. Nymphal colour/pattern polymorphism in the leafhoppers *Eupteryx urticae* (F.) and *E. cyclops* Matsumara (Hemiptera: Auchenorrhyncha): Spatial and temporal variation in morph frequencies. *Biological Journal of the Linnaean Society* 27:79–101.

Stewart, A. J. A. 1986b. The inheritance of nymphal colour/pattern polymorphism in the leafhoppers *Eupteryx urticae* (F.) and *E. cyclops* Matsumara (Hemiptera: Auchenorrhyncha). *Biological Journal of the Linnaean Society* 27:57–77.

Stiling, P. D. 1980. Colour polymorphism in some nymphs of the genus *Eupteryx*. *Ecological Entomology* 5:175–178.

Stiling, P. D. 1987. The frequency of density dependence in insect-host-parasitoid systems. *Ecology* 68:844–856.

Stiling, P. D. 1988a. Eating a thin line. *Natural History* 97(2):62–67.

Stiling, P. D. 1988b. Key factors and density-dependent processes in insect populations. *Journal of Animal Ecology* 57:581–593.

Stiling, P. D. 1990. Calculating the establishment rates of parasitoids in classical biological control. *American Entomologist* 36:225–230.

Stiling, P. 1993. Why do natural enemies fail in classical biological control programs? *American Entomologist* 39: 31-37.

Stiling, P. D. 1994. What do ecologists do? *Bulletin of the Ecological Society of America.* 75:116–121.

Stiling, P. D., B. V. Brodbeck, and D. R. Strong. 1982. Foliar nitrogen and larval parasitism as determinants of leafminer distribution patterns on *Spartina alterniflora. Ecological Entomology* 7:447–452.

Stiling, P. D., B. V. Brodbeck, and D. Simberloff. 1991. Rates of leaf abscission between plants may affect the distribution patterns of sessile insects. *Oecologia* 88:367–370.

Stiling, P., and A. M. Rossi. 1995a. Coastal insect herbivore communities are affected more by local environmental conditions than by plant genotype. *Ecological Entomology* 20:184–190.

Stiling, P., and A. M. Rossi. Complex effects of genotype and environment on insect herbivores and their natural enemies on coastal plants. *Ecology* (in press).

Stiling, P., and A. M. Rossi. 1995. Local adaptation in a dispersive gall-forming midge in *Genetic Structure of Specialist Insect Populations: Effects of Host Plant and Life History.* S. Mopper and S. Strauss (eds.). Chapman and Hall (in press).

Stiling, P. D., and D. Simberloff. 1989. Leaf abscission: Induced defense against pests or response to damage? *Oikos* 55:43–49.

Stiling, P. D., and D. R. Strong. 1983. Weak competition among *Spartina* stem borers by means of murder. *Ecology* 64:770–778.

Stiling, P. D., A. Throckmorton, J. Silvanima, and D. Strong. 1991. Does scale affect the incidence of density dependence? A field test with insect parasitoids. *Ecology* 72:2143–2154

Strassman, J. E. 1989. Altruism and relatedness at colony foundation in social insects. *Trends in Ecology and Evolution* 4:371–374.

Strathmann, R. R. 1978. Progressive vacating of adaptive types during the Phanerozoic. *Evolution* 32:907–914.

Strauss, S.Y. 1986. Personal communication.

Strauss, S. Y. 1987. Direct and indirect effects of host-plant fertilization on an insect community. *Ecology* 68:1670–1678.

Strauss, S. Y. 1988. Determining the effects of herbivory using damaged plants. *Ecology* 69:1628–1630.

Strauss, S. Y. 1991. Indirect effects in community ecology: Their definition, study and importance. *Trends in Ecology and Evolution* 6:206–210.

Strickberger, M. W. 1986. *Genetics,* 3rd ed. Macmillan, New York.

Strong, D. R. 1974a. Nonasymptotic species richness models and the insects of British trees. *Proceedings of the National Academy of Sciences of the United States of America* 71:2766–2769.

Strong, D. R. 1974b. The insects of British trees: Community equilibrium in ecological time. *Annals of the Missouri Botanical Garden* 61:692–701.

Strong, D. R. 1979. Biogeographic dynamics of insect-host plant communities. *Annual Review of Entomology* 24:89–119.

Strong, D. R. 1980. Null hypotheses in ecology. *Synthése* 43:271–285.

Strong, D. R. 1984. Density-vague ecology and liberal population regulation in insects. pp. 313–327 in *A New Ecology*, P. W. Price, C. N. Slobodchikoff, and W. S. Gaud (eds.). Wiley, New York.

Strong, D. R. 1986. Density-vague population change. *Trends in Ecology and Evolution* 1:39–42.

Strong, D. R. 1988. Insect host range (Special Feature). *Ecology* 69:885.

Strong, D. R. 1992. Are trophic cascades all wet? Differentiation and donor control in speciose ecosystems. *Ecology* 73:747–754.

Strong, D. R. Jr. 1983. Natural variability and the manifold mechanisms of ecological communities. *American Naturalist* 122:636–660.

Strong, D. R., J. H. Lawton, and T. R. E. Southwood. 1984. *Insects on Plants, Community Patterns and Mechanisms*. Blackwell Scientific Publications, Oxford.

Strong, D. R., E. D. McCoy, and J. R. Rey. 1977. Time and the number of herbivore species: The pests of sugarcane. *Ecology* 58:167–175.

Strong, D. R., L. Szyska, and D. Simberloff. 1979. Tests of community-wide character displacement against null hypotheses. *Evolution* 33:897–913.

Struhsaker, T. T. 1967. Social structure among vervet monkeys (*Cercopithecus aethiops*). *Behaviour* 29:83–121.

Stubbs, M. 1977. Density dependence in the life-cycles of animals and its importance in $K$- and $r$-strategies. *Journal of Animal Ecology* 46:677–688.

Sugden, A. M. 1994. 100 issues of TREE. *Trends in Ecology and Evolution* 9:353–354.

Sugihara, G. 1981. $S = cA^z$, $z = \frac{1}{4}$: a reply to Connor and McCoy. *The American Naturalist* 117:790–793.

Sunquist, F. 1988. Zeroing in on keystone species. *International Wildlife* 18(5):18–23.

Suppe, F. 1977. *The Structure of Scientific Theories*, 2nd ed. University of Illinois Press, Urbana.

Swift, M. J., O. W. Heal, and J. M. Anderson. 1979. *Decomposition in Terrestrial Ecosystems*. Blackwell Scientific Publications, Oxford.

Tansley, A. G. 1935. The use and abuse of vegetational concepts and terms. *Ecology* 16:284–307.

Teal, J. M. 1962. Energy flow in the salt marsh ecosystem of Georgia. *Ecology* 43:614–624.

Temple, S. A. 1977. Plant-animal mutualism: Co-evolution with dodo leads to near extinction of plant. *Science* 197:885–886.

Templeton, A. R., and B. Read. 1983. The elimination of inbreeding depression in a captive herd of Speke's gazelle. pp. 241–261 in C. M. Schonewald-Cox, S. M. Chambers, B. MacBryde, and L. Thomas (eds.). *Genetics and Conservation*. Benjamin/Cummings, Menlo Park, California.

Terborgh, J. 1973. On the notion of favorableness in plant ecology. *American Naturalist* 107:481–501.

Thomas, C. D., D. Ng, C. M. Singer, J. L. B. Mallet, C. Parmesan, and H. L. Billington. 1987. Incorporation of a European weed into the diet of a North American herbivore. *Evolution* 41:892–901.

Thompson, D. W. 1942. *On Growth and Form*, 2nd ed. Cambridge University Press, Cambridge.

Thompson, R. 1988. What's going wrong with the weather? *New Scientist* 117(1605):65.

Tilman, D. 1982. *Resource Competition and Community Structure*. Princeton University Press, Princeton, New Jersey.

Tilman, D. 1987. The importance of mechanisms of interspecific competition. *American Naturalist* 129:769–774.

Tinbergen, L. 1960. The natural control of insects in pine woods. I. Nérlandaise de Zoologie 13:265–343.

Toksöz, M. N. 1975. The subduction of the lithosphere. *Scientific American* 233(5):88–98.

Toone, W. D., and M. P. Wallace. 1994. The extinction in the wild and reintroduction of the California condor (*Gymnogyps californianus* ). pp. 411–419 in P. J. S. Olney, G. M. Mace, and A. T. C. Keistner (eds.). *Creative Conservation: Interactive Management of Wild and Captive Animals*. Chapman & Hall, London.

Tranquillini, W. 1979. *Physiological Ecology of the Alpine Timberline*. Springer-Verlag, Berlin.

Travis, J. 1989. Results of the survey of the membership of the Ecological Society of America: 1987–1988. *Bulletin of the Ecological Survey of America* 70:78–88.

Trewartha, G. T. 1969. *A Geography of Population: World Patterns*. Wiley, New York.

Trivers, R. L. 1971. The evolution of reciprocal altruism. *Quarterly Review of Biology* 46:35–37.

Trivers, R. L., and H. Hare. 1976. Haploidiploidy and the evolution of social insects. *Science* 191:249–263.

Trostel, K., A. R. E. Sinclair, C. J. Walters, and C. J. Krebs. 1987. Can predation cause the 10-year hare cycle? *Oecologia* 74:185–192.

Tscharntke, T. 1989. Changes in shoot growth of *Phragmites australis* caused by the gall maker *Giraudiella inclusa* (Diptera: Cecidomyiidae). *Oikos* 54:370–377.

Tscharntke, T. 1992. Coexistence, tritrophic interactions and density dependence in a species-rich parasitoid community. *Journal of Animal Ecology* 61:59–67.

Tschinkel, W. R. 1990. Personal communication.

Turchin, P. 1990. Rarity of density dependence or population regulation with lays? *Nature* 344:660–663.

Turner, J. 1962. The *Tilia* decline: An anthropogenic interpretation. *New Phytologist* 61:328–341.

Turner, J. R. G., C. M. Gratehouse, and C. A. Carey. 1987. Does solar energy control organic diversity? Butterflies, moths and the British climate. *Oikos* 48:195–205.

Turner, J. R. G., J. J. Lennon, and J. A. Lawrenson. 1988. British bird species distributions and the energy theory. *Nature* 335:539–541.

Uhazy, L. S., J. C. Holmes, and J. G. Stelfox. 1973. Lungworms in the rocky mountain bighorn sheep of western Canada. *Canadian Journal of Zoology* 51:817–824.

U.S. Congress Office of Technology Assessment. 1987. *Technologies to Maintain Biological Diversity*, OTA–F–330. U.S. Government Printing Office, Washington, D.C.

U.S. Department of the Interior and U.S. Department of Commerce, U.S. Fish and Wildlife Service, and U.S. Bureau of the Census. 1982. *1980 National Survey of Fishing, Hunting, and Wildlife-Associated Recreation*. Washington, D.C.

Uyeda, S. 1978. *The New View of the Earth: Moving Continents and Moving Oceans*. Freeman, San Francisco.

van der Schalie, H. 1972. WHO project Egypt 10: A case history of a schistosomiasis control project. pp. 116–136 in M. Taghi Farvar and J. P. Milton (eds.). *The Careless Technology: Ecology and International Development*. Natural History Press, New York.

van Hylckama, T. E. A. 1975. Water resources. pp. 147–165 in W. W. Murdoch (ed.). *Environment*, 2nd ed. Sinauer Associates, Sunderland, Massachusetts.

van Lenteren, J. C. 1980. Evaluation of control capabilities of natural enemies: Does art have to become science? *Netherlands Journal of Zoology* 30:369–381. van Riper, C., III, S. G.

van Riper, M. L. Goff, and M. Laird. 1986. The epizootiology and ecological significance of malaria in Hawaiian land birds. *Ecological Monographs* 56:327–344.

Van Scoy, K., and K. Coale. 1994. Dumping iron in the Pacific. *New Scientist* December 4, 1994. pp. 32–35.

Van Valen, L. 1975. Time to ecological equilibrium. *Nature* 253:684.0

Van Valen, L. M. 1973. A new evolutionary law. *Evolutionary Theory* 1:1–30.

Vane-Wright, R. I. , C. J. Humphries, and P. H. Williams. 1991. What to protect?-Systematics and the agony of choice. *Biological Conservation* 55:235–254.

Varley, G. C. 1970. The concept of energy flow applied to a woodland community. pp. 389–405 in A. Watson (ed.). *Animal Populations in Relation to Their Food Resources*. Blackwell Scientific Publications, Oxford.

Varley, G. C. 1971. The effects of natural predators and parasites on winter moth populations in England. pp. 103–116 in *Proceedings, Tall Timbers Conference on Ecological Animal Control by Habitat Management No. 2*. Tall Timbers Research Station, Tallahassee, Florida.

Varley, G. C., G. R. Gradwell, and M. P. Hassell. 1973. *Insect Population Ecology, an Analytical Approach*. Blackwell Scientific Publications, Oxford.

Verhulst, P. F. 1838. Notice sur la loi que la population suit dans son accroissement. *Correspondence in Mathematics and Physics* 10:113–121.

Vickery, W. L., and T. D. Nudds. 1991. Testing for density-dependent effects in sequential censuses. *Oecologia* 85:419–423.

Vince, S. W., I. Valiela, and J. M. Teal. 1981. An experimental study of the structure of herbivorous insect communities in a salt marsh. *Ecology* 62:1662–1678.

Volterra, V. 1926. Fluctuations in the abundance of a species considered mathematically. *Nature* 118:558–560.

Vuilleumier, F. 1970. Insular biogeography in continental regions. I. The northern Andes of South America. *American Naturalist* 104:373–388.

Wade, D., J. Ewel, and R. Hotstetler. 1980. *Fire in South Florida Ecosystems*. USDA Forest Service General Technical Report SE–17, Asheville, North Carolina.

Waldbauer, G. P. 1988. Aposematism and Batesian mimicry. *Evolutionary Biology* 22:261–286. Walde, S. J., and W. W. Murdoch. 1988. Spatial density dependence in parasitoids. *Annual Review of Entomology* 33:441–466.

Walde, S. J., and W. W. Murdoch. 1988. Spatial density dependence in parasitoids. *Annual Review of Entomology* 33:441–466.

Walker, B. A. 1976. An assessment of the ecological basis of game ranching in southern African savannas. *Proceedings of the Grassland Society of South Africa* 11:125–130.

Walker, L. R., and F. S. Chapin III. 1987. Interactions among processes controlling successional change. *Oikos* 50:131–135.

Walker, T. J. 1982. Sound traps for sampling mole crickets (Orthoptera: Gryllotalpidae: *Scapteriscus*). *Florida Entomologist* 65:13–25.

Wall, G., and C. Wright. 1977. *The Environmental Impact of Outdoor Recreation*. Waterloo, Ontario, University. Department of Geography Publication Series No. 11.

Wallace, J. B., and J. O'Hop. 1985. Life on a fast pad: waterlily leaf beetle impact on water lilies. *Ecology* 66:1534–1544.

Waloff, N., and O. W. Richards. 1977. The effect of insect fauna on growth, mortality, and natality of broom, *Sorothamnus scoparius*. *Journal of Applied Ecology* 14:787–798.

Walsh, J. 1973. The wake of the *Torrey Canyon*. pp. 60–62 in W. Jackson (ed.). *Man and the Environment*, 2nd ed. William C. Brown Company, Dubuque, Iowa.

Walton, S. 1981. Aswan revisited: U.S.-Egypt Nile project studies high dam's effects. *BioScience* 31:9–13.

Ward, P., and A. Zahavi. 1973. The importance of certain assemblages of birds as "information-centres" for food finding. *Ibis* 115:517–534.

Warming, J. E. B. 1895. Plantesamfund. *Grundtraek af den okologiske Plantegegraft*. Philipsen, Kjobenhavn, Denmark.

Warren, A., and F. B. Goldsmith. 1983. An introduction to nature conservation. pp. 1–15 in A. Warren and F. B. Goldsmith (eds.). *Conservation in Perspective*. Wiley, Chichester, England.

Warren, S. D., H. L. Black, D. A. Eastmond, and W. H. Whaley. 1988. Structure function of buttresses of *Tachigalia versicolor*. *Ecology* 69:532–536.

Wasserman, A. O. 1957. Factors affecting interbreeding in sympatric species of spadefoots (genus *Scaphiopus*). *Evolution* 11:320–338.

Watson, A., R. Moss, and R. Parr. 1984. Effects of food enrichment on numbers and spacing behavior of red grouse. *Journal of Animal Ecology* 53:663–678.

Weck, J., and C. Wiebecke. 1961. *Weltwirtschaft and Deutschlands Forst-und Holzwirtschaft*. Bayerischer Landwirtschaftsverlag, Munich.

Weismann, A. 1893. *The Germ-Plasm: A Theory of Heredity* (translation). Walter Scott, London.

West Eberhard, M. J. 1975. The evolution of social behaviour by kin selection. *Quarterly Review of Biology* 50:1–33.

Whicker, F. W., and V. Schultz. 1982. *Radioecology: Nuclear Energy and the Environment, Vols. I and II*. CRC Press, Boca Raton, Florida.

White, G. G. 1981. Current status of prickly pear control by *Cactoblastis cactorum* in Queensland. pp. 609–616 in *Proceedings of the Fifth International Symposium for the Biological Control of Weeds, Brisbane, 1980*.

White, M. J. D. 1978. *Modes of Speciation*. Freeman, San Francisco.

White, T. C. R. 1978. The importance of relative shortage of food in animal ecology. *Oecologia* 33:71–86.

White, T. C. R. 1984. The abundance of invertebrate herbivores in relation to the availability of nitrogen in stressed food plants. *Oecologia* 63:90–105.

Whitehead, D. R., and C. E. Jones. 1969. Small islands and the equilibrium theory of insular biogeography. *Evolution* 23:171–179.

Whittaker, R. H. 1961. Experiments with radiophosphorus tracer in aquarium microcosms. *Ecological Monographs* 31:157–188.

Whittaker, R. H. 1975. *Communities and Ecosystems*, 2nd ed. Macmillan, New York.

Whittaker, R. H., and G. E. Likens. 1975. The biosphere and man. pp. 305–328 in H. Lieth and R. H. Whittaker (eds.). *Primary Productivity of the Biosphere.* Springer-Verlag Ecological Studies Vol. 14, Berlin.

Whittaker, R. J., M. B. Bush, and F. K. Richards, 1989. Plant recolonization and vegetation succession on the Krakatau Islands, Indonesia. *Ecological Monographs* 59:59–123.

Wickler, W. 1968. *Mimicry in Plants and Animals.* McGraw-Hill, New York.

Wiens, J. A. 1977. On competition and variable environments. *American Scientist* 65:590–597.

Wiens, J. A., J. F. Addicott, T. J. Case, and J. Diamond. 1986. Overview: the importance of spatial and temporal scale in ecological investigations. pp. 145–153 in J. Diamond and T. J. Case (eds.). *Community Ecology.* Harper and Row, New York.

Wigglesworth, V. B. 1955. The contribution of pure science to applied biology. *Annals of Applied Biology* 42:34–44.

Wigley, T. M. L., P. D. Jones, and P. M. Kelly. 1980. Scenarios for a warm high-$CO_2$ world. *Nature* 238:17–21.

Wigley, T. M. L., and S. C. B. Raper. 1991. Detection of the enhanced greenhouse effect on climate. pp. 231–242 in J. Jager and H. L. Gerguson (eds.). *Climate Change: Science, Impacts and Policy.* Cambridge University Press, Cambridge.

Wilcove, D., S. McMillan, and K. C. Winston. 1993. What exactly is an endangered species? An analysis of the U.S. endangered species list: 1985–1991. *Conservation Biology* 7:87–93.

Wilcox, B. A. 1986. Extinction models and conservation. *Trends in Ecology and Evolution* 1:46–48.

Williams, A. G., and T. G. Whitham. 1986. Premature leaf abscission: An induced plant defense against gall aphids. *Ecology* 67:1619–1627.

Williams, C. B. 1964. *Patterns in the Balance of Nature and Related Problems in Quantitative Ecology.* Academic Press, New York.

Williams, E. D. 1978. *Botanical Composition of the Park Grass Plots at Rothamstead 1856–1976.* Rothamstead Experimental Station, Harpenden, U.K.

Williams, G. C. 1966. *Adaptation and Natural Selection.* Princeton University Press, Princeton, New Jersey.

Williams, N. H. 1983. Floral fragrances as cues in animal behavior. pp. 50–72 in C. E. Jones and R. J. Little (eds.). *Handbook of Experimental Pollination Biology.* Van Nostrand Reinhold, New York.

Williamson M. 1987. Are communities ever stable? pp. 353–371 in A. J. Gray, M. J. Crawley and P. J. Edwards (eds.). Colonization, Succession and Stabiity. Blackwell Scientific Publications, Oxford.

Williamson, M. 1989. The MacArthur and Wilson theory today: True but trivial. *Journal of Biogeography* 16:3–4.

Williamson, M. H., and K. C. Brown. 1986. the analysis and modelling of British invasions. *Proceedings and Transactions of the Royal Society of London B* 314:502-522.

Willis, E. O. 1972. Do birds flock in Hawaii, a land without predators? *California Birds* 3:1–8.

Wilson, A. C., and M. R. Stanley Price. 1994. Reintroduction as a reason for captive breeding. pp. 243–264 in P. J. S. Olney, G. M. Mace and A. T. C. Feistner (eds.). Creative Conservation: Interactive Management of Wild and Captive Animals. Chapman and Hall, London.

Wilson, D. B. (ed.). 1983. *Did the Devil Make Darwin Do It?* Iowa State University Press, Ames, Iowa.

Wilson, E. O. 1971. *The Insect Societies.* Belknap Press, Cambridge, Massachusetts.

Wilson, E. O. 1980. Caste and division of labor in leaf-cutting ants (Hymenoptera: Formicidae: *Atta*). I. The overall pattern in *A. sexdens. Behavioral Ecology and Sociobiology* 7:143–156.

Wilson, E. O. 1985. The biological diversity crisis. *BioScience* 35:700–706.

Wilson, E. O. 1988. The current state of biological diversity. pp. 3–18 in E. O. Wilson and F. M. Peter (eds.). *Biodiversity*. National Academy Press, Washington, D.C.

Wilson, E. O., and W. L. Brown. 1953. The subspecies concept and its taxonomic applications. *Systematic Zoology* 2:97–111.

Wilson, E. O., and E. O. Willis. 1975. Applied biogeography. pp. 522–534 in M. L. Cody and J. Diamond (eds.). *Ecology and Evolution of Communities*. Harvard University Press, Cambridge, Massachusetts.

Wint, G. R. W. 1983. The effect of foliar nutrients upon the growth and feeding of a lepidopteran larva. pp. 301–320 in J. A. Lee, S. McNeill, and I. H. Rorison (eds.). *Nitrogen as an Ecological Factor. 22nd Symposium of the British Ecological Society*. Blackwell Scientific Publications, Oxford.

Wirens, J. A. 1977. On competition and variable environments. *American Scientist* 65:590–597.

Wise, D. H. 1981. A removal experiment with darkling beetles: lack of evidence for interspecific competition. *Ecology* 62:727–738.

Witmer, M. C. 1991. The dodo and the tambaldcoque tree: An obligate mutualism reconsidered. *Oikos* 61:133–137.

Witz, B. W. 1989. Antipredator mechanisms in arthropods: A twenty year literative survey. *Florida Entomologist* 73:71–99.

Wolda, H. 1983. "Long-term" stability of tropical insect populations. *Researches on Population Ecology, Tokyo*, Supplement No. 3:112–126.

Wolda, H. 1986. Spatial and temporal variation in abundance in tropical animals. pp. 93–105 in S. L. Sutton, T. C. Whitmore, and A. C. Chadwick (eds.). *Tropical Rain Forest: Ecology and Management*. Blackwell Scientific Publications, Oxford.

Wolda, H., and E. Broadhead. 1985. The seasonality of Psocoptera in two tropical forests in Panama. *Journal of Animal Ecology* 54:519–530.

Wolda, H., and B. Dennis. 1993. Density dependence tests, are they? *Oecologia* 95:581–591.

Wood, D. M., and R. del Moral. 1987. Mechanisms of early primary succession in sabal pine habitats on Mount St. Helens. *Ecology* 8:780–790.

Wood, T. K., and S. I. Guttman. 1983. *Enchenopa binotata* complex: sympatric speciation? *Science* 220:310–312.

Woodley, J. D., E. A. Chornesky, P. A. Clifford, J. B. C. Jackson, L. S. Kaufman, N. Knowlton, J. C. Lang, M. P. Pearson, J. W. Porter, M. C. Rooney, K. W. Rylaarsdam, V. J. Tunnicliffe, C. M. Wahle, J. L. Wulft, A. S. G. Curtis, M. D. Dallmeyer, B. P. Jupp, M. A. R. Koehl, J. Neigel, and E. M. Sides. 1981. Hurricane Allen's impact on Jamaican coral reefs. *Science* 214:749–755.

Woodwell, G. M. 1967. Radiation and the pattern of nature. *Science* 156:461–470.

Woodwell, G. M. 1970. Effects of pollution of the structure and physiology of ecosystems. *Science* 168:429–433.

Woodwell, G. M., and A. L. Rebuck. 1967. Effects of chronic gamma radiation on the structure and diversity of an oak-pine forest. *Ecological Monographs* 37:53–64.

Woodwell, G. M., C. F. Warster, and P. A. Isaacson. 1967. DDT residues in an east coast estuary. *Science* 156:821–824.

Woolfenden, G. E. 1975. Florida scrub jay helpers at the nest. *Auk* 92:1–15.

Woolfenden, G. E., and J. W. Fitzpatrick. 1984. *The Florida Scrub Jay: Demography of a Cooperative Breeding Bird.* Princeton University Press, Princeton, New Jersey.

Woolfenden, G. E., and J. W. Fitzpatrick. 1984. *The Florida Scrub Jay: Demography of a Cooperative Breeding Bird.* Princeton University Press, Princeton, New Jersey.

World Conservation Monitoring Centre. 1992. *Global Biodiversity: Status of the Earth's Living Resources.* Chapman and Hall, London.

Worthen, W. B., S. Mayrose, and R. G. Wilson. 1994. Complex interactions between biotic and abiotic factors: effects on mycophagous fly communities. *Oikos* 69:277–286.

Wright, D. A. 1983. Species-energy theory: an extension of species-area theory. *Oikos* 41:496–506.

Wright, M. 1988. Mixed blessings of the flooding in Sudan. *New Scientist* 119(1631)44–47.

Wynne-Edwards, V. C. 1962. *Animal Dispersion in Relation to Social Behaviour.* Oliver and Boyd, Edinburgh.

Wynne-Edwards, V. C. 1977. Intrinsic population control and introduction. pp. 1–22 in F. J. Ebling and D. M. Stoddart, (eds.). *Population Control by Social Behavior.* Institute of Biology, London.

Young, A. 1998. Agroforestry and its potential to contribute to land development in the tropics. *Journal of Biogeography* 15:19–30.

Young, T. P. 1987. Increased thorn length in *Acacia depranolobium*—an induced response to browsing. *Oecologia* 71:436–438.

Yule, G. U. 1949. Measurement of diversity. *Nature* 163:688.

Zangerl, A. R., and F. A. Bazzaz. 1992. Theory and pattern in plant defense allocation. pp. 383–391 in R. S. Fritz and E. L. Simms (eds.). *Plant Resistance to Herbivores and Pathogens: Ecology, Evolution, and Genetics.* University of Chicago Press, Chicago.

Zar, J. H. 1984. *Biostatistical Analysis*, 2nd ed. Prentice-Hall, Englewood Cliffs, New Jersey.

Zaret, T. M., and R. T. Paine. 1973. Species introduction in a tropical lake. *Science* 182:449–455.

Zimmerman, B. L., and R. O. Bierregaard. 1986. Relevance of the equilibrium theory of island biogeography and species-area relations to conservation with a case from Amazonia. *Journal of Biogeography* 13:133–143.

Zucker, W. V. 1983. Tannins: does structure determine function? An ecological perspective. *American Naturalist* 121:335–365.

Academic Press Ltd., London: Figure 1 from S. Harrison, 1991, "Local Extinction in a Metapopulation Context," *Biological Journal of the Linnean Society* 42:73; Figures 1 and 2 from C.A. Clarke et al., 1990, "*Biston betularia* (Peppered Moth) in West Kirby, Wirral, 1959-1989," *Biological Journal of the Linnean Society* 39:323; R.M. Sibly, 1983, "Optimal Group Size Is Unstable," *Animal Behavior* 31:947; Figures from C.B. Williams, 1964, *Patterns in the Balance of Nature and Related Problems in Quantitative Ecology.* Reprinted by permission.

Annual Reviews Inc.: Figure 1 from R. K. Peet, 1974, "The Measurement of Species Diversity," *Annual Review of Ecology and Systematics* 5:285. Reproduced, with permission, from the Annual Review of Ecology and Systematics, Volume 5, © 1974, by Annual Reviews Inc.

Benjamin/Cummings Publishing Co. (Addison-Wesley): Figure from K. Ralls and J. Ballou, 1983, "Extinction Lessons from Zoos," in C.M. Schonewald-Cox et al. (eds.), *Genetics and Conservation: A Reference for Managing Wild Animal and Plant Populations.* Reprinted by permission of the publishers.

Blackwell Science, Inc.: Figures from C. B. Cox et al., 1976, *Biogeography: An Ecological and Evolutionary Approach*; Figure 1 from R.A. Beaver, 1985, "Geographical Variation in Food Web Structure in Pitcher Plants," *Ecological Entomology* 10:241; K. Williamson, 1987, "Are Communities Ever Stable?" in *Colonization, Succession and Stability,* A.J. Gray et al. (eds.); R.C. Lacey, 1987, "Loss of Genetic Diversity from a Managed Population," *Conservation Biology* 1:143-158; J. Berger, 1990, "Persistence of Different-Sized Populations," *Conservation Biology* 4:91-98; Figure from E.R. Pianka, 1976, "Competition and Niche Theory," in *Theoretical Ecology,* R.M. May (ed.); Figures from M. Begon et al., 1986, *Ecology, Individuals, Populations , and Communities*; Figures from N.J. Crawley, 1983, "Herbivory, the Dynamics of Animal-Plant Interactions," *Studies in Ecology,* vol. 10; Figures from G. C. Varley et al., 1973, *Insect Population Ecology.* Reprinted by permission of the Society for Conservation Biology and Blackwell Scientific Publications, Inc.

Buteo Books: Figure from D.A. Ratcliff, 1980, *The Peregrine Falcon,* Buteo Books, Vermillion, South Dakota. Reprinted by permission.

Cambridge University Press: Figure from T.M.L. Wigley and S.C.B. Raper, "Observed Global Mean Temperature Changes 1861-1869," in *Climate Change,* Jill Jager and H.L. Ferguson (eds.). Reprinted with the permission of Cambridge University Press.

Columbia University Press: From V. Grant, 1985, *The Origin of Adaptations.* Reprinted by permission.

Company of Biologists Ltd.: Figure 2.3 from G.F. Gause, 1932, "Experimental Studies on the Struggle for Existence," *Journal of Experimental Biology* 9:389-402. Reprinted by permission of the Company of Biologists Ltd.

Ecological Society of America: Table 1 from M.E. Power, 1992, "Top-down and Bottom-up Forces in Food Webs," *Ecology* 73:733; Figure from J.H. Connell, 1961, "The Influence of Interspecific Competition on Distribution of the Barnacle *Chthamalus stellatus,*" *Ecology* 42:710; Figure from P.W. Price, 1970, "Characteristics Permitting Coexistence Among Parasitoids of a Sawfly," *Ecology* 51:445. Copyright © by the Ecological Society of America. Reprinted by permission.

Elsevier Trends Journals: Figure 2 from J.A. Drake, 1990, "Communities as Assembled Structures," *Trends in Ecology and Evolution* 5:159; Figure 2 from A. Newsome, 1990, "The Control of Vertebrate Pests by Vertebrate Predators," *Trends in Ecology and Evolution* 5:187-191. Reprinted by permission.

Entomological Society of Canada: Figure from R.F. Morris, 1963, "The Dynamics of Epidemic Spruce Budworm Populations," *Memoirs of the Entomological Society of Canada* 31:19.

Evolution: Figure from P. H. Erwin et al., 1987, "A Comparative Study of Diversification Events," *Evolution* 41:1177-1186; Figure from B.H. Seghers, 1974, "Schooling Behaviour in the Guppy," *Evolution* 28:488. Reprinted by permission.

HarperCollins Publishers: Figures 8.6, 12.20, 18.21, and 26.12 from *Ecology: The Experimental Analysis of Distribution and Abundance,* 3rd ed., by Charles J. Krebs. Copyright © 1985 by Harper & Row, Publishers, Inc.; Table 1.1 from *Comunity Ecology* by Jared Diamond and Ted J. Case, Copyright © 1986 by Harper & Row, Publishers, Inc.; Figure from James F. Crow and Motoo Kimura, 1970, *An Introduction to Population Genetics Theory,* Copyright © 1970 by James F. Crow and Motoo Kimura. Reprinted by permission of HarperCollins Publishers, Inc.

Harvard University Press: Figure from E.O. Wilson, 1971, *The Insect Societies,* Belknap Press; Table from "Markovian Properties of Forest Succession" by H.S. Horn, from *Ecology and Evolution of Communities,* edited by M.C. Cody and J.M. Diamond, Cambridge, Mass.: Harvard University Press, Copyright © 1975 by the President and Fellows of Harvard College. Reprinted by permission of the publishers.

Heredity: From H. Daday, 1954,"Gene Frequencies in Wild Populations of *Trifolium repens,*" *Heredity* 8:65. Reprinted by permission.

Houghton Mifflin Co.: Map from A Field Guide to Reptiles and Amphibians of Eastern and Central North America, Copyright © 1958, 1975 by Roger Conant. Reprinted by permission of Houghton Mifflin Co. All rights reserved.

John Wiley & Sons: Figure 3 from D.R. Strong, 1984, "Density-Vague Ecology and Liberal Population Regulation in Insects," in *A New Ecology,* P.W. Price et al. (eds.), Wiley, New York; D. Jablonski, 1986, "Causes and Consequences of Mass Extinctions," in *Dynamics of Extinctions,* D. K. Elliott (ed.). Reprinted by permission of John Wiley & Sons, Inc.

Journal of Zoology: Figure from J.V.M. Jarvis and J.B. Sale, 1971, "Burrowing and Burrow Patterns of East African Mole Rats," *Journal of Zoology,* 163:468. Reprinted by permission.

Macmillan Magazines Ltd.: Figure from R. M. May, 1990, "Taxonomy as Destiny," *Nature* 347:129, Copyright 1990 Macmillan Magazines Ltd.; Figure 1 from S. L. Pimm and A. Redfearn, 1988, "The Variability of Population Densities," *Nature* 334:613, Copyright 1988 Macmillan Magazines Ltd.; Figure 4 from N. Leader-Williams and S.D. Albon, 1988, "Allocation of Resources for Conservation," *Nature* 336:533, Copyright 1988 Macmillan Magazines Ltd.; Figure 2 from J.M. Barnola et al., 1987, "Vostok Ice Core," *Nature*

329:408, Copyright 1988 Macmillan Magazines Ltd; Map from D.J. Currie and V. Paquin, 1987, "Large-Scale Biogeographical Patterns of Species Richness of Trees," *Nature* 329:326, Copyright 1987 Macmillan Magazines Ltd. Reprinted by permission.

National Academy Press: Table from *Ecological Knowledge and Environmental Problem Solving*, Copyright 1986 by the National Academy of Sciences. Courtesy of the National Academy Press, Washington, D.C.

Plenum Publishing Corp.: Figure from L.R. A. Haury, J.A. McGowan, and P. J. Wiebe, "Patterns and Process in the Time-Space of Plankton Distributions," in *Spatial Patterns in Plankton Communities*, J.H. Steele (ed.), 1978. Reprinted by permission.

Prentice Hall (Macmillan): Figures from P.R. Ehrlich and J. Roughgarden, 1987, *The Science of Ecology*, Macmillan, New York; Figure 3.8 and Table 4.2 from R.H. Whittaker, 1975, *Communities and Ecosystems*, 2nd ed., Macmillan, New York; Copyright © 1975 by Robert H. Whitaker; Figures from Monroe W. Strickberger, 1985, *Genetics*, 3rd ed., Macmillan, New York, Copyright © 1985 by Monroe W. Strickberger. Reprinted by permission.

Princeton University Press: Figure 6.3 in Lawton, J.H., 1984, "Non-competitive Population, Non-convergent Communities, and Vacant Niches," in D.R. Strong et al. (eds.), *Ecological Communities*, Copyright © 1984 by Princeton University Press; Figure 1 from John Tyler Bonner, 1965, *Size and Cycle*, Copyright © 1965 by Princeton University Press; Figure 2 from R.H. MacArthur and E.O. Wilson, 1967, *The Theory of Island Biogeography*, Copyright © 1967 by Princeton University Press; Table 4.5 and Figure 2.4 from A.E. Magurran, 1988, *Ecological Diversity and Its Management*, Copyright © 1988 by Princeton University Press; Figures 19A, 24, and 25 from D. Tilman, 1982, *Resource Competition and Community Structure*, Copyright © 1982 by Princeton University Press. Reprinted by permission.

Routledge: Figure 30 from W. Morton Wheeler, 1928, *The Social Insects*, Routledge and Kegan Paul. Reprinted by permission.

Saunders: Figures from E.P. Odum, 1971, *Fundamentals of Ecology*, 3rd ed, Saunders, Philadelphia. Reprinted by permission of the publisher.

Science: Figure 1 and Table 1 from B. Griffith et al., 1989, "Translocation as a Species Conservation Tool," *Science* 245:477, Copyright 1989 by the AAAS; Figure from J.H. Connell, 1978, "Diversity in Tropical Rain Forests and Coral Reefs, *Science* 199:1302, Copyright 1978 by the AAAS; Figure from C. Mlot, 1993, "Predators, Prey and Natural Disasters Attract Ecologists," *Science* 261:1115, Copyright 1992 by the AAAS; Figure 1 from D. Simberloff et al., 1978, "Isopod and Insect Root Borers May Benefit Florida Mangroves," *Science* 201:630, Copyright 1978 by the AAAS; Table 1 from E.P. Odum, 1969, "The Strategy of Ecosystem Development," *Science* 164:262, Copyright 1969 by the AAAS; Figure 1 from V.B. Scheffer, 1951, "The Rise and Fall of a Reindeer Herd," *Scientific Monthly* 73:356. Copyright 1951 by the AAAS. Reprinted with permission.

Sinauer Associates, Inc.: Figure 4 from M.E. Soule and B.A. Wilcox, 1980, *Conservation Biology*; Figure 2.5 from N. J. Gotelli, 1995, *A Primer of Ecology*, Sunderland, Mass., Sinauer Associates. Reprinted by permission.

Springer-Verlag: A.E. Newsome et al., 1989, "Prolonged Prey Suppression by Carnivores-Predator-Removal Experiments," *Oecologia* 78:458-467, Copyright 1989 by Springer-Verlag GmbH & Co. KG; Figure from T. Fenchel, 1974, "Intrinsic Rate of Natural Increase," *Oecologia* 14:319, Copyright 1974 by Springer-Verlag GmbH & Co. KG; Figure from A.C. Heron, 1972, "Population Ecology of a Colonizing Species," *Oecologia* 10:305, Copyright 1972 by Springer-Verlag GmbH & Co. KG. Reprinted by permission.

Svenska Jagareforbundet: Tables 11, 12, and 18, and Figure 1 from T. Ebenhard, 1988, "Introduced Birds and Mammals and Their Ecological Effects," *Swedish Wildlife Research* 131:1-107. Reprinted by permission.

Times Mirror Higher Education Group: Table 17.1 from James H. Brown and Arthur C. Gibson, *Biogeography*, Copyright © 1983 by the C.V. Mosby Company. Reprinted by permission of Times Mirror Higher Education Group, Inc., Dubuque, Iowa. All Rights Reserved.

University of Arizona Press: From J.R. Hastings and R.M. Turner, 1965, *The Changing Mile*, Copyright 1965, University of Arizona Press, Tucson. Reprinted by permission.

University of Chicago Press: Figure from J. Terborgh, 1973, "On the Notion of Favorableness in Plant Ecology," *American Naturalist* 107:491; Table 2 from T.W. Schoener, 1983, "Field Experiments on Interspecific Competition," *American Naturalist* 122:240; Table 3 from G.A. Polis, 1991," Complex Trophic Interactions in Deserts," *American Naturalist* 138:123; Table 1 from G.E. Hutchinson, 1959, "Homage to Santa Rosalia," *American Naturalist* 93:145; Figure 1 from E.R. Pianka, 1970, "On r- and K- Selection," *American Naturalist* 104:591; Figure 2B from J.H. Connell, 1983, "On the Prevalence and Relative Importance of Interspecific Competition," *American Naturalist* 122:661; Table 18 from T. Park, 1954, "Experimental Studies of Interspecies Competition," *Physiological Zoology* 27:177; Figure 10.1 from R.S. Fritz and E.L. Simms, *Plant Resistance to Herbivore and Pathogens*. Reprinted by permission of the University of Chicago Press.

Unwin Hyman Ltd.: Figure from A.S. Collinson, 1977, *An Introduction to World Vegetation*. Reprinted by permission.

W.H. Freeman: Figure 30.18 from R. Ricklefs, 1990, *Ecology*, 3rd ed. Copyright © 1990 by W. H. Freeman and Company. Used with permission.

WMO/UNEP Intergovernmental Panel on Climate Change: J.T. Houghton, G. J. Jenkins and J. J. Ephraums, 1990. *Climate Change: The IPCC Scientific Assessment*.

World Conservation Monitoring Centre, England: Tables 17.4 and 29.4, and Figures 17.1, 17.3, 17.4, and 17.6 from *Global Biodiversity: Status of the Earth's Living Resources*, 1992. Reprinted by permission of the World Conservation Monitoring Centre, Cambridge, U.K.

Yale University Press: Figure 22 from M.G. Davis and C. Zabinski, 1992, "Changes in Geographical Range Resulting from Greenhouse Warming," in *Global Warming and Biodiversity*, ed. R. Peters and T. Lovejoy. Reprinted by permission of Yale University Press.

# Index

518        Index